# Food Science Text Series

The Food Science Text Series provides faculty with the leading teaching tools. The Editorial Board has outlined the most appropriate and complete content for each food science course in a typical food science program and has identified textbooks of the highest quality, written by the leading food science educators.

More information about this series at http://www.springer.com/series/5999

Romeo T. Toledo · Rakesh K. Singh
Fanbin Kong

# Fundamentals of Food Process Engineering

## Fourth Edition

 Springer

Romeo T. Toledo
Department of Food Science &
Technology
University of Georgia
Athens, GA, USA

Rakesh K. Singh
Department of Food Science &
Technology
University of Georgia
Athens, GA, USA

Fanbin Kong
Department of Food Science &
Technology
University of Georgia
Athens, GA, USA

ISSN 1572-0330                    ISSN 2214-7799   (electronic)
Food Science Text Series
ISBN 978-3-030-07933-8           ISBN 978-3-319-90098-8   (eBook)
https://doi.org/10.1007/978-3-319-90098-8

This Springer imprint is published by the registered company Springer International Publishing AG part of Springer Nature.
The registered company address is: Gewerbestrasse 11, 6330 Cham, Switzerland

# Preface

Since the publication of the first edition in 1981, the second edition in 2001, and the third edition in 2007, this textbook has been widely adopted for Food Engineering courses worldwide. The authors express their gratitude to colleagues who have adopted this textbook and to those who have made constructive criticisms on the materials in the previous editions. This new edition not only incorporates changes suggested by colleagues, but additional materials have been added to include facilitated problem-solving using a computer and new food processing and food product technologies, such aseptic processing and emerging food processing technologies. New sections have been added in most of the chapters reflecting the current state of the technology. The expanded coverage may result in not enough time available in a school term to cover all areas; therefore, instructors are advised to carefully peruse the book and select the most appropriate sections to cover in a school term. The advantage of the expanded coverage is the elimination of the need for a supplementary textbook.

The success of this textbook has been attributed to the expansive coverage of subject areas specified in the Institute of Food Technologists model curriculum for food science majors in the USA and the use of examples utilizing conditions encountered in actual food processing operations. This theme continues in the fourth edition. In addition to the emphasis on problem-solving, technological principles that form the basis for a process are presented so that the process can be better understood and selection of processing parameters to maximize product quality and safety can be made more effective. The fourth edition incorporates most of what was in the third edition with most of the material updated to include the use of computers in problem-solving. Use of the spreadsheet and macros such as the determinant for solving simultaneous linear equations, the solver function, and programming in Visual BASIC are used throughout the book. The manual problem-solving approach has not been abandoned in favor of the computer approach. Thus, users can still apply the concepts to better understand a process rather than just mechanically entering inputs into a preprogrammed algorithm.

Entirely new sections include enthalpy change calculations in freezing based on the freezing point depression, evaporative cooling, interpretation of pump performance curves, determination of shape factors in heat exchange by radiation, unsteady-state heat transfer, kinetic data for thermal degradation of foods during thermal processing, pasteurization parameters for shelf-stable high-acid foods and long-life refrigerated low-acid foods, high-pressure processing of fluid and packaged foods, concentration of juices, environmentally friendly refrigerants, modified atmosphere packaging of produce, sorption equations for water activity of solid foods, the osmotic pressure and water activity relationships, vacuum dehydration, new membranes commercially available for food processing and waste treatment, and supercritical fluid extraction.

This edition contains much new hard-to-find data needed to conduct food process engineering calculations and will be very useful as a sourcebook of data and calculation techniques for practicing food engineers.

Athens, GA, USA                                                              Romeo T. Toledo
                                                                                     Rakesh Singh
                                                                                     Fanbin Kong

# Contents

# Units and Dimensions

Units used to designate magnitude of a dimension have evolved based on common usage and instruments available for measurement. Two major systems for measurement have been used: the English system, which was used primarily in industry, and the metric system, which was used in the sciences. The confusion that results from the use of various terms to represent the same dimension has led to the development of a common system of units that is proposed for use in both science and industry. The *Système International d'Unites* (International System of Units) and the official international designation SI was adopted in 1960 by the General Conference on Weights and Measures. This body consists of delegates from member countries of the Meter Convention, and it meets at least once every 6 years. There are at least 44 countries represented in this convention, one of which is the United States.

The use of SI is now widespread in the scientific community. However, it is still often necessary to convert data from one system to another, as tables in handbooks may be in a different unit from what is needed in the calculations, or experimental data may be obtained using instruments calibrated in a different unit from what is desired in reporting the results.

In this chapter, the various units in SI are discussed, and techniques for conversion of units using the dimensional equation are presented. Also emphasized in this chapter is the concept of dimensional consistency of mathematical equations involving physical quantities and how units of variables in an equation are determined to ensure dimensional consistency.

## 1.1 Definition of Terms

Dimension: used to designate a physical quantity under consideration (e.g., time, distance, weight).

Unit: used to designate the magnitude or size of the dimension under consideration (e.g., m for length, kg for weight).

Base unit: base units are dimensionally independent. They are used to designate only one dimension (e.g., units of length, mass, and time).

Derived units: a combination of various dimensions. An example of a derived unit is the unit of force, which includes the dimensions of mass, length, and time.

Precision: synonymous with reproducibility, the degree of deviation of the measurements from the mean. This is often expressed as a $\pm$ term

© Springer International Publishing AG, part of Springer Nature 2018
R. T. Toledo et al., *Fundamentals of Food Process Engineering*,
https://doi.org/10.1007/978-3-319-90098-8_1

or as the smallest value of unit that can be consistently read in all determinations.

Accuracy: refers to how a measured quantity relates to a known standard. To test for accuracy of a measurement, the mean of a number of determinations is compared against a known standard. Accuracy depends on proper calibration of an instrument.

## 1.2    Systems of Measurement

The various systems in use are shown in Table 1.1. These systems vary in the base units used. Under the English system, variations exist in expressing the unit of force. The chemical and food industries in the United States use the American Engineering System, although SI is the preferred system in scientific articles and textbooks.

The metric system uses prefixes on the base unit to indicate magnitude. Industry adopted the "mks" system, whereas the sciences adopted the "cgs" system. SI is designed to meet the needs of both science and industry.

## 1.3    The SI System

The following discussion of the SI system and the convention followed in rounding after conversion is based on the American National Standard, Metric Practice, adopted by the American National Standards Institute, the American Society for Testing and Materials, and the Institute of Electrical and Electronics Engineers.

### 1.3.1    Units in SI and Their Symbols

SI uses base units and prefixes to indicate magnitude. All dimensions can be expressed in either a base unit or combinations of base units. The latter are called derived units and some have specific names. The base units and the derived units with assigned names are shown in Table 1.2.

### 1.3.2    Prefixes Recommended for Use in SI

Prefixes are placed before the base multiples of 10. Prefixes recommended for general use are shown in Table 1.3.

A dimension expressed as a numerical quantity and a unit must be such that the numerical quantity is between 0.1 and 1000. Prefixes should be used only on base units or named derived units. Double prefixes should not be used.

Examples:
1. 10,000 cm should be 100 m, not 10 kcm.
2. m should be 1 μm.

**Table 1.1** Systems of measurement

| Dimension | | | | | | | |
|---|---|---|---|---|---|---|---|
| System | Use | Length | Mass | Time | Temperature | Force | Energy |
| English | | | | | | | |
| English absolute | Scientific | Foot | Pound mass | Second | °F | Poundal | BTU ft (poundal) |
| British engineering | Industrial | Foot | Slug | Second | °F | Pound force | BTU ft (pound force) |
| American engineering | US industrial | Foot | Pound mass | Second | °F | Pound force | BTU ft (pound force) |
| Metric | | | | | | | |
| Cgs | Scientific | Centimeter | Gram | Second | °C | Dyne | Calorie, erg |
| Mks | Industrial | Meter | Kilogram | Second | °C | Kilogram force | Kilocalorie joule |
| SI | Universal | Meter | Kilogram | Second | °K | Newton | Joule |

**Table 1.2**   Base units of SI and derived units with assigned names and symbols

| Quantity | Unit | Symbol[a] | Formula |
|---|---|---|---|
| Length | Meter | m | – |
| Mass | Kilogram | kg | – |
| Electric current | Ampere | A | – |
| Temperature | Kelvin | K | – |
| Amount of substance | Mole | mol | – |
| Luminous intensity | Candela | cd | – |
| Time | Second | s | – |
| Frequency (of a periodic phenomenon) | Hertz | Hz | $1/s$ |
| Force | Newton | N | $kg\ m/s^2$ |
| Pressure, stress | Pascal | Pa | $N/m^2$ |
| Energy, work, quantity of heat | Joule | J | $N \cdot m$ |
| Power, radiant flux | Watt | W | $J/s$ |
| Quantity of electricity, electric charge | Coulomb | C | $A \cdot s$ |
| Electric potential, potential difference, electromotive force | Volt | V | $W/A$ |
| Capacitance | Farad | F | $C/V$ |
| Electric resistance | Ohm | | $V \cdot A$ |
| Conductance | Siemens | S | $A/V$ |
| Magnetic flux | Weber | Wb | $V \cdot s$ |
| Magnetic flux density | Tesla | T | $Wb/m^2$ |
| Inductance | Henry | H | $Wb/A$ |
| Luminous flux | Lumen | lm | $cd \cdot sr^b$ |
| Illuminance | Lux | lx | $lm/m^2$ |
| Activity (of radionuclides) | Becquerel | Bq | $1/s$ |
| Absorbed dose | Gray | Gy | $J/kg$ |

*Source*: American National Standard (1976)

[a]Symbols are written in lowercase letters unless they are from the name of a person

[b]sr stands for steradian, a supplementary unit used to represent solid angles

**Table 1.3**   Prefixes recommended for use in SI

| Prefix | Multiple | Symbol[a] |
|---|---|---|
| Tera | $10^{12}$ | T |
| Giga | $10^9$ | G |
| Mega | $10^6$ | M |
| Kilo | 1000 | k |
| Milli | $10^{-3}$ | m |
| Micro | $10^{-6}$ | μ |
| Nano | $10^{-9}$ | n |
| Pico | $10^{-12}$ | p |
| Femto | $10^{-15}$ | f |

[a]Symbols for the prefixes are written in capital letters when the multiplying factor is $10^6$ and larger. Prefixes designating multiplying factors less than $10^6$ are written in lowercase letters

3. 3000 $m^3$ should *not* be written as 3 $km^3$.
4. 10,000 $N/m^2$ can be written as 10 kPa but not 10 $kN/m^2$.

## 1.4   Conversion of Units

### 1.4.1   Precision, Rounding-Off Rule, and Significant Digits

Conversion from one system of units to another should be done without gain or loss of precision. Results of measurements must be reported such that a reader can determine the precision. The easiest way to convey precision is in the number of significant figures.

Significant figures include all nonzero digits and nonterminal zeroes in a number. Terminal zeroes are significant in decimals, and they may be significant in whole numbers when specified. Zeroes preceding nonzero digits in decimal fractions are not significant.

Examples:

123 has three significant figures.

103 has three significant figures.

103.03 has five significant figures.

10.030 has five significant figures.

0.00230 has three significant figures.

1500 has two significant figures unless the two terminal zeroes are specified as significant.

When the precision of numbers used in mathematical operations is known, the answer should be rounded off following these rules recommended by the American National Standards Institute. All conversion factors are assumed to be exact.

1. In addition or subtraction, the answer shall not contain a significant digit to the right of the least precise of the numbers, for example, 1.030 + 1.3 + 1.4564 = 3.8. The least precise of the numbers is 1.3. Any digit to the right of the tenth digit is not significant.
2. In multiplication or division, the number of significant digits in the answer should not exceed that of the least precise among the original numbers. For example, $123 \times 120 = 15{,}000$ if the terminal zero in 120 is not significant. The least precise number has two significant figures, and the answer should also have two significant figures.
3. When rounding off, raise the terminal significant figure retained by 1 if the discarded digits start with 5 or larger; otherwise, the terminal significant figure retained is unchanged. If the start of the digits discarded is 5, followed by zeroes, make the terminal significant digit retained even.

If the precision of numbers containing terminal zeroes is not known, assume that the number is exact. When performing a string of mathematical operations, round off only after the last operation is performed.

Examples:

1253 rounded off to two significant figures = 1300

1230 rounded off to two significant figures = 1200 (2 is even)

1350 rounded off to two significant figures = 1400 (3 is odd)

1253 rounded off to three significant figures = 1250

1256 rounded off to three significant figures = 1260

## 1.5   The Dimensional Equation

The magnitude of a numerical quantity is uncertain unless the unit is written along with the number. To eliminate this ambiguity, make a habit of writing both a number and its unit.

An equation that contains both numerals and their units is called a dimensional equation. The units in a dimensional equation are treated just like algebraic terms. All mathematical operations done on the numerals must also be done on their corresponding units. The numeral may be considered as a coefficient of an algebraic symbol represented by the unit. Thus, $(4 \text{ m})^2 = (4)^2 (\text{m})^2 = 16 \text{ m}^2$ and

$$\left[5_{\text{J kg K}}\right](10 \text{ kg})(5 \text{ K}) = 5(10)(5)\left[\frac{\text{J kg K}}{\text{kg K}}\right] = 250 \text{ J}$$

Addition and subtraction of numerals and their units also follow the rules of algebra; that is, only like terms can be added or subtracted. Thus, $5 \text{ m} - 3 \text{ m} = (5{-}3) \text{ m}$, but $5 \text{ m} - 3 \text{ cm}$ cannot be simplified unless their units are expressed in like units.

## 1.6   Conversion of Units Using the Dimensional Equation

Determining the appropriate conversion factors to use in conversion of units is facilitated by a dimensional equation. The following procedure may be used to set up the dimensional equation for conversion:

1. Place the units of the final answer on the left side of the equation.
2. The number being converted and its unit are the first entries on the right-hand side of the equation.
3. Set up the conversion factors as a ratio using Appendix 1.
4. Sequentially multiply the conversion factors such that the original units are systematically eliminated by cancelation and replacement with the desired units.

**Example 1.1** Convert BTU/(lb °F) to J/(g K):

$$\frac{J}{g \cdot K} = \frac{BTU}{lb.^\circ F} \times \text{appropriate conversion factors}$$

The numerator J on the left corresponds to BTU on the right-hand side of the equation. The conversion factor is 1054.8 J/BTU. Because the desired unit has J in the numerator, the conversion factor must have J in the numerator. The factor $9.48 \times 10^4$ BTU/J may be obtained from the table, but it should be entered as J/9.48 $\times 10^4$ BTU in the dimensional equation. The other factors needed are $2.2046 \times 10^{-3}$ lb/g or lb/453.6 g and 1.8°E/K.

The dimensional equation is:

$$\frac{J}{g \cdot K} = \frac{BTU}{lb.^\circ F} \cdot \frac{1054.8\,J}{BTU} \cdot \frac{2.2046 \times 10^{-3} lb}{g} \cdot \frac{1.8^\circ F}{K}$$

Another form for the dimensional equation is:

$$\frac{J}{g \cdot K} = \frac{BTU}{lb.^\circ F} \cdot \frac{J}{9.48 \times 10^{-4} BTU} \cdot \frac{lb}{453.6\,g} \cdot \frac{1.8^\circ F}{K}$$

Canceling out units and carrying out the arithmetic operations:

$$\frac{J}{g \cdot K} = \frac{BTU}{lb.^\circ F} \times 4.185$$

**Example 1.2** The heat loss through the walls of an electric oven is 6500 BTU/h. If the oven is operated for 2 h, how many kilowatt-hours of electricity will be used just to maintain the oven temperature (heat input = heat loss)?

To solve this problem, rephrase the question. In order to supply 6500 BTU/h for 2 h, how many kilowatt-hours are needed? Note that power is energy/time; therefore the product of power and time is the amount of energy. Energy in BTU is to be converted to J.

The dimensional equation is:

$$J = \frac{6500\,BTU \cdot 2h}{h} \cdot \frac{1054.8\,J}{BTU}$$

$$W = \frac{J}{s}$$

Therefore, $W \cdot s = J$

Convert $W \cdot s$ to kWh;

$$kWh = W \cdot s \frac{1kW}{1000W} \cdot \frac{1h}{3600s.}$$

Since $W \cdot s = J$, substituting the expression for J results in the following expression for kWh which is the desired answer to the problem.

Thus:

$$kWh = \frac{6500BTU}{h} \cdot 2h \cdot \frac{1054.8J}{BTU} \cdot \frac{1kW}{1000w} \cdot \frac{1h}{3600s}$$

$$kWh = 3.809$$

An alternative dimensional equation is:

$$kW \cdot h = \frac{6500\,BTU \cdot 2h}{h} \cdot \frac{h}{60\,min} \cdot \frac{1.757 \times 10^{-2} kW}{BTU/min} = 3.809$$

The conversion factors in Appendix 1 express the factors as a ratio and are most convenient to use in a dimensional equation. Although column I is labeled "denominator" and columns II and III are labeled "numerator," the reciprocal may be used. Just make sure that the numerals in column II go with the units in column III.

## 1.7 The Dimensional Constant ($G_c$)

In the American Engineering System of measurement, the units of force and the units of mass are both expressed in pounds. It is necessary to differentiate between the two units because they have actually different physical significance. To eliminate confusion between the two units, the pound force is usually written as $lb_f$, and the pound mass is written as $lb_m$.

Because the force of gravity on the surface of the earth is the weight of a given mass, in the American Engineering System, the weight of 1 $lb_m$ is exactly 1 $lb_f$. Thus, the system makes it easy to conceptualize the magnitude of a 1 $lb_f$.

The problem in the use of $lb_f$ and $lb_m$ units is that an additional factor is introduced into an equation to make the units dimensionally consistent, if both units of force and mass are in the same equation. This factor is the dimensional constant $g_c$.

The dimensional constant $g_c$ is derived from the basic definition of force: force = mass × acceleration.

If $lb_f$ is the unit of force and $lb_m$ is the unit of mass, substituting ft/s$^2$ for acceleration results in a dimensional equation for force that is not dimensionally consistent:

$$lb_f = lb_m \cdot \frac{ft}{s^2}$$

The American Engineering System of measurement is based on the principle that the weight, or the force of gravity on a pound mass on the surface of the earth, is a pound force. Introducing a dimensional constant, the equation for force becomes:

$$lb_f = lb_m \cdot \frac{ft}{s^2} \cdot \frac{1}{g_c}$$

To make $lb_f$ numerically equal to $lb_m$ when acceleration due to gravity is 32.174 ft/s$^2$, the dimensional constant $g_c$ should be a denominator in the force equation and have a numerical value of 32.174. The units of $g_c$ is:

$$g_c = \frac{ft \cdot lb_m}{lb_f \cdot s^2}$$

The dimensional constant, $g_c$, should not be confused with the acceleration due to gravity, g. The two quantities have different units. In SI, the unit of force can be expressed in the base units following the relationship between force, mass, and acceleration; therefore, $g_c$ is not needed. However, if the unit $kg_f$ is to be converted to SI, the equivalent $g_c$ of 9.807 kg · m/$kg_f$ · s$^2$ should

be used. Use of conversion factors in Appendix 1 eliminates the need for $g_c$ in conversion from the American Engineering System of units to SI.

## 1.8    Determination of Appropriate SI Units

A key feature of SI is expression of any dimension in terms of the base units of meter, kilogram, and second. Some physical quantities having assigned names as shown in Table 1.2 can also be expressed in terms of the base units when used in a dimensional equation. When properly used, the coherent nature of SI ensures dimensional consistency when all quantities used for substitution into an equation are in SI units. The following examples illustrate selection of appropriate SI units for a given quantity.

**Example 1.3** A table for viscosity of water at different temperatures lists viscosity in units of $lb_m$/(ft· h). Determine the appropriate SI unit and calculate a conversion factor.

The original units have units of mass ($lb_m$), distance (ft), and time (h). The corresponding SI base units should be kg/(m · s). The dimensional equation for the conversion is:

$$
\begin{aligned}
\frac{kg}{m \cdot s} &= \frac{lb_m}{ft \cdot h} \cdot conversion factor \\
&= \frac{lb_m}{ft \cdot h} \cdot \frac{1 kg}{2.2046 \, lb_m} \frac{3.281 \, ft}{1 \, m} \cdot \frac{1 h}{3600 s} \frac{kg}{m \cdot s} \\
&= \frac{lb_m}{ft \cdot h} \left( 3.84027 \times 10^{-5} \right)
\end{aligned}
$$

Viscosity in SI is also expressed in Pa · s. Show that this has the same base units as in the preceding example:

$$Pa \cdot s = \frac{N}{m^2} \cdot s = \frac{kg \cdot m}{s^2} \frac{s}{m^2} = \frac{kg}{m \cdot s}$$

**Example 1.4** Calculate the power available in a fluid that flows down the raceway of a reservoir at a rate of 525 $lb_m$/min from a height of 12.3 ft. The potential energy (PE) is:

$$PE = m\,g\,h$$

where m is the mass, g is the acceleration due to gravity which is 32.2 ft./s$^2$, and h is the height.

From Table 1.2, the unit of power in SI is the watt and the formula is J/s. Expressing in base units:

$$P = W = \frac{N \cdot m}{s} = \frac{kg \cdot m}{s^2} \cdot \frac{m}{s} = \frac{kg \cdot m^2}{s^3}$$

$$\frac{kg \cdot m^2}{s^3} = \frac{525\,lb_m}{min} \cdot \frac{32.2\,ft}{s^2} \cdot (12.3\,ft) \cdot \frac{kg}{2.2046\,lb_m}$$

$$\cdot \frac{1\,min}{60s} \cdot \frac{m^2}{(3.281)^2 ft^2} = 146W$$

## 1.9   Dimensional Consistency of Equations

All equations must have the same units on both sides of the equation. Equations should be tested for dimensional consistency before substitution of values of variables. A dimensional equation would easily verify dimensional consistency. Consistent use of SI for units of variables substituted into an equation ensures dimensional consistency.

**Example 1.5** The heat transfer equation expresses the rate of heat transfer (q = energy/time) in terms of the heat transfer coefficient h, the area A, and the temperature difference $\Delta$T.

Test the equation for dimensional consistency and determine the units of h:

$$q = hA\,\Delta T$$

The dimensional equation will be set up using SI units of the variables:

$$\frac{J}{s} = W = (-)(m^2)(K)$$

The equation will be dimensionally consistent if the units of h = W/m$^2$·K. Then m5 and K will cancel out on the right-hand side of the equation, leaving W on both sides, signifying dimensional consistency.

**Example 1.6** Van der Waal's equation of state is as follows:

$$\left(P + \frac{n^2 a}{V^2}\right)(V - nb) = nRT$$

where P is pressure, n is moles, V is volume, R is gas constant, and T is temperature. Determine units of the constants a, b, and R and test for dimensional consistency.

Because the two terms in the left-hand side of the equation involve a subtraction, both terms in the first parenthesis will have units of pressure and in the second parenthesis units of volume. Product of pressure and volume will have the same units as terms on the right side of the equation.

Using SI units for pressure of N/m$^2$ and volume, m$^3$, the dimensional equation is:

$$\frac{N}{m^2}(m^3) = (kg\,mole)(- - -)(^\circ K)$$

The equation will be dimensionally consistent if the units of R = N · m/(kg mol · K) because that will give the same units, Nm, on both sides of the equation. The units of a and b are determined as follows. The first term, which contains a, has units of pressure:

$$\frac{N}{m^2} = \frac{(kg\,mole)^2(- - -)}{(m^3)^2}$$

To make the right-hand side have the same units as the left, a will have units of N(m)$^4$/(kg mole)$^2$. The second term, which contains b, has units of volume:

$$m^3 = (kg\,mole)(- - -)$$

Thus, b will have units of m$^3$/kg mole.

## 1.10   Conversion of Dimensional Equations

Equations may be dimensionless (i.e., dimensionless groups are used in the equation). Examples of dimensionless groups are Reynolds number (Re), Nusselt number (Nu), and Prandtl number (Pr),

Fourier number (Fo) and Biot number (Bi) are defined below, and the dimensional equations for their units show that each group is dimensionless. The variables are as follows: V is velocity, D is diameter, $\rho$ is density, $\mu$ is viscosity, k is thermal conductivity, h is heat transfer coefficient, $\alpha$ is thermal diffusivity, $C_p$ is specific heat, L is thickness, and t is time:

$$\text{Re} = \frac{DV\rho}{\mu} = m\frac{m\,kg}{s\,m^3}\frac{1}{kg/(m\cdot s)}$$

$$\text{Nu} = \frac{hD}{k} = \frac{W}{m^2\cdot K}\,m\,\frac{1}{W/(m\cdot K)}$$

$$\text{Pr} = \frac{C_p\mu}{k} = \frac{J}{kg\cdot K}\cdot\frac{kg}{m\cdot s}\cdot\frac{1}{J/(s\cdot m\cdot K)}$$

$$\text{Fo} = \frac{\propto t}{L^2} = \frac{m^2}{s}\cdot s\cdot\frac{1}{m^2}$$

$$\text{Bi} = \frac{hL}{k} = \frac{W}{m^2\cdot K}\cdot m\cdot\frac{1}{W/(m\cdot K)}$$

An example of a dimensionless equation is the Dittus-Boelter equation for heat transfer coefficients in fluids flowing inside tubes:

$$\text{Nu} = 0.023(\text{Re})^{0.8}(\text{Pr})^{0.3}$$

Because all terms are dimensionless, there is no need to change the equation regardless of what system of units is used in determining the values of the dimensionless groups.

Dimensional equations result from empirical correlations (e.g., statistical analysis of experimental data). With dimensional equations, units must correspond to those used on the original data, and substitution of a different system of units will require a transformation of the equation.

When converting dimensional equations, the following rules will be useful:

1. When variables appear in the exponent, the whole exponent must be dimensionless; otherwise, it will not be possible to achieve dimensional consistency for the equation.
2. When variables are in arguments of logarithmic functions, the whole argument must be dimensionless.

3. Constants in an equation may not be dimensionless. The units of these constants provide for dimensional consistency.

**Example 1.7** The number of surviving microorganisms in a sterilization experiment is linear in a semi-logarithmic graph; therefore, the equation assumes the form:

$$\log N = -at + b$$

where N is number survivors, t is time, and a and b are constants. The equation will be dimensionally consistent if b is expressed as log N at time 0, designated log $N_o$, and a will have units of reciprocal time. Thus, the correct form of the equation that is dimensionally consistent is:

$$\log\frac{N}{N_o} = -at$$

In this equation, both sides are dimensionless.

**Example 1.8** An equation for heat transfer coefficient between air flowing through a bed of solids and the solids is:

$$h = 0.0128\,G^{0.8}$$

where G is mass flux of air in lb/(ft$^2$ • h) and h is heat transfer coefficient in BTU/(h • ft$^2$ • F). Derive an equivalent equation in SI.

The dimensional equation is:

$$\frac{BTU}{h\cdot ft^2\cdot F} = (---)\left(\frac{lb}{ft^2\cdot h}\right)^{0.8}$$

An equation must be dimensionally consistent; therefore, the constant 0.0128 in the above equation must have units of:

$$\frac{BTU}{h\cdot ft^2\cdot F} = \left(\frac{ft^{1.6}h^{0.8}}{lb^{0.8}}\right) = \frac{BTU}{h^{0.2}ft^{0.4}lb^{0.8}F}$$

Converting the equation involves conversion of the coefficient. The equivalent SI unit for h = W/(m$^2$ K) and G = kg/(m$^2$ s). Thus, the coefficient will have units of J/(s$^{0.2}$m$^{0.4}$kg$^{0.8}$K). The dimensional equation for the conversion is:

$$\frac{J}{s^{0.2}m^{0.4}kg^{0.8}K} = \frac{0.0128\,BTU}{h^{0.2}ft^{0.4}lb^{0.8}F} \cdot \frac{1054.8\,J}{BTU}$$

$$\cdot \frac{ft^{0.4}}{(0.3048)^{0.4}m^{0.4}(3600)^{0.2}(s^{0.2})} \cdot \frac{h^{0.2}}{}$$

$$\cdot \frac{(2.2048)^{0.8}lb^{0.8}}{kg^{0.8}} \cdot \frac{1.8F}{K}$$

$$= 14.305$$

The converted equation is:

$$h = 14.305 G^{0.8}$$

where both h and G are in SI units. To check if $G = 100$ lb/(ft$^2$ h) and $h = 0.0128(100)^{0.8}$, $h = 0.5096$ BTU/(h ft$^2$ °F).

The equivalent SI value is:

$$0.5096\,BTU h \cdot ft^2 \cdot F \cdot \frac{5.678263 W/(m^2 \cdot K)}{BTU/(h \cdot ft^2 \cdot °F)} = 2.893 \frac{W}{(m^2 \cdot K)}$$

In SI, $G = [100 lb/(ft^2 \cdot h)][ft^2/0.3048)^2\,m^2]$ $(h/3600s)(kg/2.2046lb) = 0.1356$. Using converted equation,

$$h = 14.305(0.1356)0^8 = 2.893\ W/(m^2K).$$

---

## Problems

1.1. Set up dimensional equation and determine the appropriate conversion factor to use in each of the following:

$$\frac{lb}{ft^3} = \frac{lb}{gal} \times \text{conversion factor}$$

$$\frac{lb}{in^2} = \frac{lb}{ft^2} \times \text{conversion factor}$$

$$W = \frac{cal}{s} \times \text{conversion factor}$$

1.2. The amount of heat required to change the temperature of a material from T1 to T2 is given by:

$$q = mC_p(T_2 - T_1)$$

where q is BTU, m is mass of material in lb, $C_p$ is specific heat of material in BTU/lb·°F,

and $T_1$ and $T^2$ are initial and final temperatures in °F.

(a) How many BTUs of heat are required to cook a roast weighing 10 lb from 40 °F to 130 °F? $Cp = 0.8$ BTU/(lb · °F)

(b) Convert the number of BTUs of heat in (a) into watt-hours.

(c) If this roast is heated in a microwave oven having an output of 200 watts, how long will it take to cook the roast?

1.3. How many kilowatt-hours of electricity will be required to heat 100 gallons of water (8.33 lb/gal) from 60 °F to 100 °F? $C_p$ of water is 1 BTU/(lb · °F).

1.4. Calculate the power requirements for an electric heater necessary to heat 10 gallons of water from 70 °F to 212 °F in 10 min. Express this in Joules/min and in watts. Use the following conversion factors in your calculations:
Specific heat of water = 1 BTU/(lb · °F)
3.414 BTU/(W · h)
60 min/h
3600 s/h
8.33 lb water/gal
$1.054 \times 10^3$ J/BTU

1.5. One ton of refrigeration is defined as the rate of heat withdrawal from a system necessary to freeze 1 ton (2000 lb) of water at 32 °F in 24 h. Express this in watts and in BTU/h. Heat of fusion of water = 80 cal/g.

1.6. (a) In the equation $\tau = \mu(\gamma)$, what would be the units of $\tau$ in the equation if $\mu$ is expressed in dyne · s/cm$^2$ and $\gamma$ is in s$^{-1}$?
(b) If $\mu$ is to be expressed in lb$_m$/(ft · s), $\tau$ is expressed in lb$_f$/ft$^2$, and $\gamma$ is in s$^{-1}$, what is needed in the equation to make it dimensionally consistent?

1.7. In the equation

$$\bar{V} = \frac{1000(\rho_1 - \rho_2)}{m\rho_1\rho_2} + \frac{M}{\rho_2}$$

what units should be used for the density, $\rho$, such that $\bar{V}$ would have the units ml/mole?

(a) m = moles/1000 g

(b) M = g/mole

1.8. Express the following in SI units. Follow the rounding-off rule on your answer.

   (a)   The pressure at the base of a column of fluid 8.325 in. high when the acceleration due to gravity is 32.2 ft/s$^2$, and the fluid density is 1.013 g/cm$^3$:

   P = density × height
          × acceleration due to gravity

   (b)   The compressive stress (same units as pressure) on a specimen having a diameter of 0.525 in. when the applied force is 5.62 pound force:

   Stress = Force/area

   (c)   The force needed to restrain a piston having a diameter of 2.532 in. when a pressure of 1500 (exact) lb$_f$/in$^2$ is in the cylinder behind the piston:

   Force = Pressure × area

1.9. An empirical equation for heat transfer coefficient in a heat exchanger is:

$$h = a(V)^{0.8}(1 + 0.011T)$$

where h = BTU/(h · ft$^2$ · $\Delta°$F), V = ft/s, and T = °F. In one experimental system, a had a value of 150. What would be the form of the equation and the value of a if h, V, and T are in SI units?

1.10. A correlation equation for the density of a liquid as a function of temperature and pressure is as follows:

$$d = (1.096 + 0.0086\ T)(e)^{0.000953P}$$

where d is density in g/cm$^3$, T is temperature in Kelvin, P is pressure in atm, and e is the base of natural logarithms. A normal atmosphere is 101.3 kPa. Determine the form of the equation if all variables are to be expressed in SI.

1.11. The Arrhenius equation for the temperature dependence of diffusivity (D) is given by D = D$_o$[e]$^{-E/RT}$. R is a constant with a value of 1.987 cal/mole K. (T is temperature in degrees Kelvin). If D is in cm$^2$/s, determine the units of D$_o$ and E.

1.12. The heat of respiration of fresh produce as a function of temperature is q = a e$^{bT}$. If q has units of BTU/(ton · 24 h), and T is in °F, determine the units of a and b. The values of a and b for cabbage are 377 and 0.041, respectively. Calculate the corresponding values if q is expressed in mW/kg and T is in °C.

## Suggested Reading

American National Standards Institute. (1976). *American National Standard, metric practice. Am. Natl. Stand. Inst. IIEE Std. 268–1976*. New York: IEEE.

American Society for Agricultural Engineers. (1978). *Use of customary and SI (metric) units. American Society for Agricultural Engineers Yearbook—1978*. St. Joseph: ASAE.

Benson, S. W. (1971). *Chemical calculations* (3rd ed.). New York: Wiley.

Feldner, R. M., & Rousseau, R. W. (1999). *Elementary principles of chemical processes* (2nd ed.). New York: Wiley.

Himmelblau, D. M. (1967). *Basic principles and calculations in chemical engineering* (2nd ed.). Englewood Cliffs: Prentice-Hall.

Kelly, F. H. C. (1963). *Practical mathematics for chemists*. London: Butterworths.

McCabe, W. L., Smith, J. C., & Harriott, P. (1985). *Unit operations of chemical engineering* (4th ed.). New York: McGraw-Hill Book.

Obert, E., & Young, R. L. (1962). *Elements of thermodynamics and heat transfer*. New York: McGraw-Hill Book.

Watson, E. L., & Harper, J. C. (1988). *Elements of food engineering* (2nd ed.). New York: Van Nostrand Reinhold.

# Material Balances

Material balance calculations are employed in tracing the inflow and outflow of material in a process and thus establish quantities of components or the whole process stream. The procedures are useful in formulating products to specified compositions from available raw materials, evaluating final compositions after blending, evaluating processing yields, and evaluating separation efficiencies in mechanical separation systems.

## 2.1 Basic Principles

### 2.1.1 Law of Conservation of Mass

Material balances are based on the principle that matter is neither created nor destroyed. Thus, in any process, a mass balance can be made as follows:

$$\text{Inflow} = \text{Outflow} + \text{Accumulation}$$

Inflow may include formation of material by chemical reaction or microbial growth processes, and outflow may include material depletion by chemical or biological reactions.

If accumulation is 0, inflow = outflow and the process is at steady state. If the accumulation term is not 0, then the quantity and concentration of components in the system could change with timer and the process is an unsteady state.

### 2.1.2 Process Flow Diagrams

Before formulating a material balance equation, visualize the process and determine the boundary of the system for which the material balance is to be made. It is essential that everything about the process that affects the distribution of components is known. The problem statement should be adequate to enable the reader to draw a flow diagram. However, in some cases, basic physical principles associated with a process may affect the distribution of components in the system but may not be stated in the problem. It is essential that a student remembers the physical principles applied in the processes used as examples. Knowing these principles not only allows the student to solve similar material balance problems but also provides information that may be used later as a basis for the design of a new process or for evaluation of parameters affecting efficiency of a process.

For example, consider a problem in crystallization. The problem may be simply stated: determine the amount of sugar (water-free basis) that can be produced from 100 kg of sugar solution that contains 20% by weight of sugar and 1% of a water-soluble uncrystallizable impurity. The solution is concentrated to 75% sugar, cooled to 20 °C, centrifuged, and the crystals dried.

The problem statement is indeed adequate to draw a process flow diagram. This is shown in Fig. 2.1.

© Springer International Publishing AG, part of Springer Nature 2018
R. T. Toledo et al., *Fundamentals of Food Process Engineering*,
https://doi.org/10.1007/978-3-319-90098-8_2

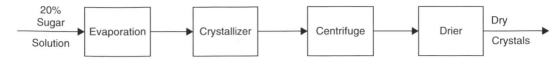

**Fig. 2.1**  Process flow diagram for a crystallization problem

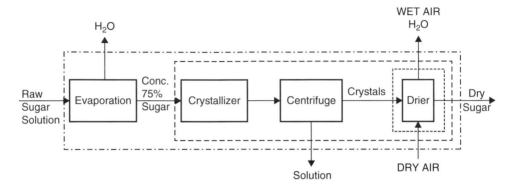

**Fig. 2.2**  Process flow diagram for a crystallization problem showing input and exit streams and boundaries enclosing subsystems for analyzing sections of the process

However, this flow diagram does not give a complete picture of where various streams separate and leave the system.

Figure 2.2 is a flow diagram of the same process, but after taking into consideration how components partition in various steps in the process, additional streams leaving the system are drawn in the diagram. To concentrate a 20% solution to 75% requires the removal of water. Thus, water leaves the system at the evaporator. The process of cooling does not alter the mass; therefore, the same process stream enters and leaves the crystallizer. Centrifugation separates most of the liquid phase from the solid phase, and the crystals, the solid phase containing a small amount of retained solution, enter the drier. A liquid phase leaves the system at the centrifuge. Water leaves the system at the drier.

Three physical principles involved in this problem are not stated:

(1) Crystals will crystallize out of a solution when solute concentration exceeds the saturation concentration. The solute concentration in the liquid phase is forced toward the saturation concentration as crystals are formed. Given enough time to reach

equilibrium, the liquid phase that leaves the system at the centrifuge is a saturated sugar solution.

(2) The crystals consist of pure solute, and the only impurities are those adhering to the crystals from the solution.

(3) It is not possible to completely eliminate the liquid from the solid phase by centrifugation. The amount of impurities that will be retained with the sugar crystals depends on how efficiently the centrifuge separates the solid from the liquid phase. This principle of solids purity being dependent on the degree of separation of the solid from the liquid phase applies not only in crystallization but also in solvent extraction.

In order to solve this problem, the saturation concentration of sugar in water at 20 °C, and the water content of the crystals fraction after centrifugation must be known.

### 2.1.3   System Boundaries

Figure 2.2 shows how the boundaries of the system can be moved to facilitate solving the problem. If the boundary completely encloses the whole process, there will be one stream entering and four streams leaving the system. The boundary can also be set just around the evaporator in which case there is one stream entering and two leaving. The boundary can also be set around the centrifuge or around the drier. A material balance can be carried out around any of these subsystems or around the whole system. The material balance equation may be a total mass balance or a component balance.

### 2.1.4   Total Mass Balance

The equation in section "Law of Conservation of Mass," when used on the total weight of each stream entering or leaving a system, represents a total mass balance. The following examples illustrate how total mass balance equations are formulated for systems and subsystems.

**Example 2.1** In an evaporator, dilute material enters and concentrated material leaves the system. Water is evaporated during the process. If I is the weight of the dilute material entering the system, W is the weight of water vaporized, and C is the weight of the concentrate, write an equation that represents the total mass balance for the system. Assume that a steady state exists.

**Solution**
The problem statement describes a system depicted in Fig. 2.3.
   The total mass balance is:

$$\text{Inflow} = \text{Outflow} + \text{Accumulation}$$
$$\text{Accumulation} = 0 \text{ in a steady state system}$$
$$I = W + C$$

**Fig. 2.3** Input and exit streams in an evaporation process

**Example 2.2** Construct a diagram and set up a total mass balance for a dehydrator. Air enters at the rate of A lb/min, and wet material enters at W lb/min. Dry material leaves the system at D lb/min. Assume steady state.

**Solution**
The problem statement describes a system (dehydrator) where air and wet material enters and dry material leaves. Obviously, air must leave the system also, and water must leave the system. A characteristic of a dehydrator not written into the problem statement is that water removed from the solids is transferred to air and leaves the system with the air stream. Figure 2.4 shows the dehydrator system and its boundaries. Also shown are two separate subsystems—one for the solids and the other for air—with their corresponding boundaries. Considering the whole dehydrator system, the total mass balance is:

$$W + A = \text{wet air} + D$$

Considering the air subsystem:

$$A + \text{water} = \text{wet air}$$

The mass balance for the solids subsystem is:

$$W = \text{water} + D$$

**Example 2.3** Orange juice concentrate is made by concentrating single-strength juice to 65% solids followed by dilution of the concentrate to 45% solids using single-strength juice. Draw a diagram for the system and set up mass balances for the whole system and for as many subsystems as possible.

**Solution**
The problem statement describes a process depicted in Fig. 2.5. Consider a hypothetical proportionator that separates the original juice (S) to that which is fed to the evaporator (F) and that (A) which is used to dilute the 65% concentrate. Also, introduce a blender to indicate that part of the process where the 65% concentrate ($C_{65}$) and the single-strength juice are mixed to

**Fig. 2.4** Diagram of
material flow in a
dehydration process

**Fig. 2.5** Diagram of an orange juice concentrate process involving evaporation and blending of concentrate with freshly squeezed juice

produce the 45% concentrate ($C_{45}$). The material balance equations for the whole system and the various subsystems are:

$$\text{Overall}: S = W + C_{45}$$
$$\text{Proportionator}: S = F + A$$
$$\text{Evaporator}: F = W + C_{65}$$
$$\text{Blender}: C_{65} + A = C_{45}$$

### 2.1.5   Component Mass Balance

The same principles apply as in the total mass balance except that components are considered individually. If there are n components, n independent equations can be formulated, one equation for total mass balance and n−1 component balance equations.

Because the object of a material balance problem is to identify the weights and composition of various streams entering and leaving a system, it is often necessary to establish several equations and simultaneously solve these equations to evaluate the unknowns. It is helpful to include the known quantities of process streams and concentrations of components in the process diagram in order that all streams where a component may be present can be easily accounted for. In a material balance, use mass units and concentration in mass fraction or mass percentage. If the quantities are expressed in volume units, convert to mass units using density.

A form of a component balance equation that is particularly useful in problems involving concentration or dilution is the expression for the mass fraction or weight percentage.

Mass fraction A =

$$= \frac{\text{mass of component A}}{\text{mass fraction of A}}$$

Rearrange the equation:

Total mass of mixture containing A

$$= \frac{\text{mass of component A}}{\text{mass fraction of A}}$$

Thus, if the weight of component A in a mixture is known, and its mass fraction in that mixture is known, the mass of the mixture can be easily calculated.

**Example 2.4** Draw a diagram and set up a total mass and component balance equation for a crystallizer where 100 kg of a concentrated sugar solution containing 85% sucrose and 1% inert, water-soluble impurities (balance, water) enters. Upon cooling, the sugar crystallizes from solution. A centrifuge then separates the crystals from a liquid fraction, called the mother liquor. The crystal slurry fraction has, for 20% of its weight, a liquid having the same composition as the mother liquor. The mother liquor contains 60% sucrose by weight.

**Solution**
The diagram for the process is shown in Fig. 2.6. Based on a system boundary enclosing the whole process of crystallization and centrifugation, the material balance equations are as follows:
    Total mass balance:

$$S = C + M$$

Component balance on sucrose:

$$S(0.85) = M(0.6) + C(0.2)(0.6) + C(0.8)$$

The term on the left is sucrose in the inlet stream. The first term on the right is sucrose n the mother liquor. The second term on the right is sucrose in mother liquor carried by crystals. The last term on the right is sucrose in the crystals.
    Component balance on water:
    Let $x$ = mass fraction of impurity in the mother liquor

$$S(0.14) = M(0.4 - x) + C(0.2)(0.4 - x)$$

The first term on the left is water in the inlet stream. The first term on the right is water in the mother liquor. The last term on the right is water in the mother liquor adhering to the crystals.
    Component balance on impurity:

$$S(0.01) = M(x) + C(0.2)(x)$$

Note that a total of four equations can be formulated but there are only three unknown quantities (C, M, and x). One of the equations is redundant.

**Example 2.5** Draw a diagram and set up equations representing total mass balance and component mass balance for a system involving the mixing of pork (15% protein, 20% fat, and 63% water) and back fat (15% water, 80% fat, and 3% protein) to make 100 kg of a mixture containing 25% fat.

**Solution**
The diagram is shown in Fig. 2.7.
    Total mass balance:

$$P + B = 100$$

Fat balance:

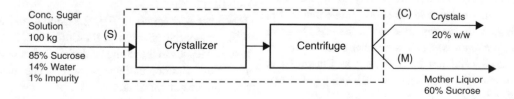

**Fig. 2.6** Diagram showing composition and material flow in a crystallization process

**Fig. 2.7** Composition and material flow in a blending process

**Fig. 2.8** Composition and material flow for a dilution process

$$0.2\,P + 0.8\,B = 0.25(100)$$

These two equations are solved simultaneously by substituting P = 100 − B into the second equation.

$$0.2(100 - B) + 0.8\,B = 25$$

$$B = \frac{25 - 20}{0.8 - 0.2} = 8.33\,\text{kg}$$

$$P = 100 - 8.33 = 91.67\,\text{kg}$$

### 2.1.6  Basis and "Tie Material"

A "tie material" is a component used to relate the quantity of one process stream to another. It is usually the component that does not change during a process. Examples of tie material are solids in dehydration or evaporation processes and nitrogen in combustion processes. Although it is not essential that these tie materials are identified, the calculations are often simplified if it is identified and included in one of the component balance equations. This is illustrated in Example 2.6 of the section "A Steady State" where the problem is solved rather readily using a component mass balance in the solid (the tie material in this system) compared to when the mass balance was made on water. In a number of cases, the tie material need not be identified as illustrated in examples in the section "Blending of Food Ingredients."

A "basis" is useful in problems where no initial quantities are given and the answer required is a ratio or a percentage. It is also useful in continuous flow systems. Material balance in a continuous flow system is done by assuming as a basis a fixed time of operation. A material balance problem can be solved on any assumed basis. After all the quantities of process streams are identified, the specific quantity asked in the problem can be solved using ratio and proportion. It is possible to change basis when considering each subsystem within a defined boundary inside the total system.

## 2.2  Material Balance Problems Involved in Dilution, Concentration, and Dehydration

### 2.2.1  Steady State

These problems can be solved by formulating total mass and component balance equations and solving the equations simultaneously.

**Example 2.6**  How many kilograms of a solution containing 10% NaCl can be obtained by diluting 15 kg of a 20% solution with water?

**Solution**
The process diagram in Fig. 2.8 shows that all NaCl enters the mixer with the 20% NaCl solution and leaves in the diluted solution. Let x = kg 10% NaCl solution and y = kg water. The material balance equations are:

$$\text{Total mass}: 15 = X - Y$$
$$\text{Component}: 15(0.20) = X(0.10)$$

The total mass balance equation is redundant because the component balance equation alone can be used to solve the problem.

$$x = \frac{3}{0.1} = 30\,kg$$

The mass fraction equation can also be used in this problem. Fifteen kilograms of a 20% NaCl solution contains 3 kg NaCl. Dilution would not change the quantity of NaCl so that 3 kg of NaCl is in the diluted mixture. The diluted mixture contains 10% NaCl; therefore:

$$x = \frac{3\,kg\,NaCl}{mass fraction\,NaCl} = \frac{3}{0.1} = 30\,kg$$

**Example 2.7** How much weight reduction would result when a material is dried from 80% moisture to 50% moisture?

**Solution**
The process diagram is shown in Fig. 2.9. Dehydration involves removal of water, and the mass of solids remains constant. There are two components, solids and water, and a decrease in the concentration of water, indicating a loss, will increase the solids concentration. Let $W$ = mass of 80% moisture material, $D$ = mass of 50% moisture material, and $X$ = mass water lost, which is also the reduction in mass.

No weights are specified; therefore, express reduction in weight as a ratio of final to initial weight.

The material balance equations are:

$$Total\,mass : W = X + D$$
$$Water : 0.8W = 0.5D + X$$

Solving simultaneously:

**Fig. 2.9** Composition and material flow in a dehydration process

$$W = X + D; X = W - D$$
$$0.8W = 0.5D + (W - D)$$
$$0.5D = 0.2W;$$
$$\frac{D}{W} = \frac{0.2}{0.5} = 0.4$$
$$\%wt.reduction = \frac{W - D}{W}(100)$$
$$= \left(1 - \frac{D}{W}\right)100$$
$$= (1 - 0.4)(100) = 60\%$$

The problem can also be solved by using as a basis 100 kg of 80% moisture material. Let $W = 100$.

$$Total\,mass\,balance : 100 = X + D$$
$$Water\,balance : 0.8(100) = X + 0.5D$$

Solving simultaneously, by subtracting one equation from the other:

$$100 - 80 = D(1 - 0.5); D = 20/0.5 = 40\,kg.$$
$$\%wt.reduction = \frac{100 - 40}{100}(100) = 60\%$$

## 2.2.2   Volume Changes on Mixing

When two liquids are mixed, the volumes are not always additive. This is true with most solutions and miscible liquids. Sodium chloride solution, sugar solutions, and ethanol solution all exhibit volume changes on mixing. Because of volume changes, material balances must be done on mass rather than volume of components. Concentrations on a volume basis must be converted to a mass basis before the material balance equations are formulated.

**Example 2.8** Alcohol content in beverages is reported as percent by volume. A "proof" is twice the volume percent of alcohol. The density of absolute ethanol is 0.7893 g/cm$^3$. The density of a solution containing 60% by weight of ethanol is 0.8911 g/cm$^3$. Calculate the volume of absolute ethanol that must be diluted with water to produce

1 liter of 60% by weight, ethanol solution. Calculate the "proof" of a 60% ethanol solution.

**Solution**

Use as a basis: 1 liter of 60% w/w ethanol.

Let X = volume of absolute ethanol in liters. The component balance equation on ethanol is:

$$X(1000)(0.7983) = 1(1000)(0.8911)(0.6)$$

$$X = 0.677\,L$$

$$g\,water = 1000(0.8911) - 0.677(0.7983)$$

$$= 356.7\,g\,or\,356.7\,mL$$

$$Total\ volume\ components = 0.677 + 0.3567$$

$$= 1.033\,L$$

This is a dilution problem, and a component balance on ethanol was adequate to solve the problem. Note that a mass balance was made using the densities given. There is a volume loss on mixing as more than 1 liter of absolute ethanol and water produced 1 liter of 60% w/w ethanol solution.

To calculate the "proof" of 60% w/w ethanol, use as a basis 100 g of solution.

Volume of solution : $100/0.8911 = 112.22\,cm^3$

Volume of ethanol : $100(0.6)/0.7893 = 76.016\,cm^3$

Volume percent : $(76.016/112.22) \times (100) = 67.74\%$

Proof : $2\,(volume\ percent) = 135.5\,proof$

### 2.2.3  Continuous Versus Batch

Material balance calculations are the same regardless whether a batch or continuous process is being evaluated. In a batch system, the total mass considered includes what entered or left the system at one time. In a continuous system, a basis of a unit time of operation may be used, and the material balance will be made on what entered or left the system during that period of time. The previous examples were batch operations. If the process is continuous, the quantities given will all be mass/time (e.g., kg/h). If the basis used is 1 hour of operation, the problem is reduced to the same form as a batch process.

**Example 2.9**  An evaporator has a rated evaporation capacity of 500 kg water/h. Calculate the rate of production of juice concentrate containing 45% total solids from raw juice containing 12% solids.

**Solution**

The diagram of the process is shown in Fig. 2.10. Use as a basis 1 hour of operation. Five hundred kg of water leaves the system. A component balance on solids and a total mass balance will be needed to solve the problem. Let F = the feed, 12% solids juice, and C = concentrate containing 45% solids. The material balance equations are:

Total mass:

$$F = C + 500$$

Solids:

$$0.12\,F = 0.45\,C;\,F = \frac{0.45C}{0.12} = 3.75C$$

$$C = \frac{500}{3.75 - 1} = 181.8\,kg$$

Substituting and solving for C: 3.75 C = C + 500.

Because the basis is 1 hour of operation, the answer will be: rate of production of concentrate = 181.8 kg/h.

### 2.2.4  Recycle

Recycle is evaluated similarly as in the previous examples, but the boundaries of subsystems analyzed are moved around to isolate the process streams being evaluated. A system is defined that

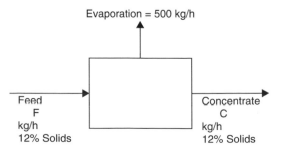

**Fig. 2.10**  Flow diagram of an evaporation process

has a boundary surrounding the recycle stream. If this total system is analyzed, the problem may be reduced to a simple material balance problem without recycle.

**Example 2.10** A pilot plant model of a falling film evaporator has an evaporation capacity of 10 kg water/h. The system consists of a heater through which the fluid flows down in a thin film and the heated fluid discharges into a collecting vessel maintained under a vacuum where flash evaporation reduces the temperature of the heated fluid to the boiling point. In continuous operation, a recirculating pump draws part of the concentrate from the reservoir, mixes this concentrate with feed, and pumps the mixture through the heater. The recirculating pump moves 20 kg of fluid/h. The fluid in the collecting vessel should be at the desired concentration for withdrawal from the evaporator at any time. If feed enters at 5.5% solids and a 25% concentrate is desired, calculate: (a) the feed rate and concentrate production rate, (b) the amount of concentrate recycled, and (c) concentration of mixture of feed and recycled concentrate.

**Solution**
A diagram of the system is shown in Fig. 2.11. The basis is 1 hour of operation.

A mass and solids balance over the whole system will establish the quantity of feed and concentrate produced per hour.

Total mass:

$$F = C + V; F = C + 10$$

Solids:

$$F(0.055) = C(0.25); F = C\left(\frac{0.25}{0.055}\right) = 4.545\,C$$

Substituting F:

$$4.545\,C = C + 10; C = \frac{10}{4.545 - 1} = 2.82\,kg$$

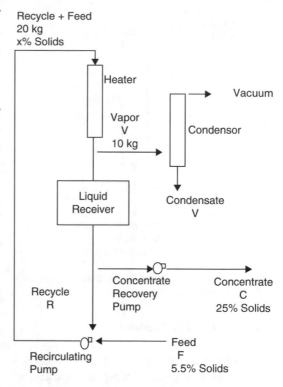

**Fig. 2.11** Diagram of material flow in a falling film evaporator with product recycle

(a)   Solving for F: $F = 4.545(2.82) = 12.82\ kg/h$. The concentrate production rate is 2.82 kg/h.

(b)   Material balance around the recirculating pump:

$$R + F = 20; R = 20 - 12.82 = 7.18\,kg$$
$$\text{Recycle rate} = 7.18\ kg/h$$

(c)   A material balance can be made around the part of the system where the vapor separates from the heated fluid as shown in Fig. 2.12.

Solids balance : $20(x) = 10(0.25); x = 2.5/20 = 0.125$.

The fluid entering the heater contains 12.5% solids.

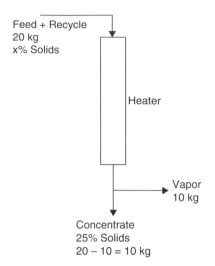

Feed + Recycle
20 kg
x% Solids

Heater

Vapor
10 kg

Concentrate
25% Solids
20 − 10 = 10 kg

**Fig. 2.12** Diagram of material balance around the heater of a falling film evaporator

### 2.2.5   Unsteady State

Unsteady-state material balance equations involve an accumulation term in the equation. Accumulation is expressed as a differential term of the rate of change of a variable with respect to time. A mass balance is made in the same manner as steady-state problems. Because a differential term is involved, the differential equation must be integrated to obtain an equation for the value of the dependent variable as a function of time.

**Example 2.11** A stirred tank with a volume of 10 liters contains a salt solution at a concentration of 100 g/L. If salt-free water is continuously fed into this tank at the rate of 12 L/h, and the volume is maintained constant by continuously overflowing the excess fluid, what will be the concentration of salt after 90 min.

**Solution**
This is similar to a dilution problem, except that the continuous overflow removes salt from the tank, thus reducing not only the concentration but also the quantity of salt present. A mass balance will be made on the mass of salt in the tank and in the streams entering and leaving the system.

Let x = the concentration of salt in the vessel at any time. Representing time by the symbol t, the accumulation term will be dx/dt, which will have units of g salt/(L · min). Convert all time units to minutes in order to be consistent with the units. Multiplying the differential term by the volume will result in units of g salt/min. The feed contains no salt; therefore, the input term is zero. The overflow is at the same rate as the feed, and the concentration in the overflow is the same as inside the vessel; therefore, the output term is the feed rate multiplied by the concentration inside the vessel. The mass balance equation is:

$$0 = Fx + V\left(\frac{dx}{dt}\right); \frac{dx}{dt} = -\frac{F}{V}x$$

The negative sign indicates that x will be decreasing with time. Separating variables:

Integrating:

$$\frac{dx}{x} = -\frac{F}{V}dt$$

Integrating:

$$\ln x = -\left(\frac{F}{V}\right)t + C$$

The constant of integration is obtained by substituting x = 100 at t = 0; C = ln(100); F = 12 L/h = 0.2 L/min; V = 10 L.

$$\ln\left(\frac{x}{100}\right) = -\frac{0.2t}{10}$$

At 90 min: ln(0.01 x) = −1.800.

$$x = 100\left(e^{-1.80}\right) = 16.53\,g/L$$

**Example 2.12** The generation time is the time required for cell mass to double. The generation time of yeast in a culture broth has been determined from turbidimetric measurements to be 1.5 h. (a) If this yeast is used in a continuous fermentor, which is a well-stirred vessel having a volume of 1.5 L, and the inoculum is 10,000 cells/mL, at what rate can cell-free substrate be fed into this fermentor in order that the yeast cell concentration will remain constant? The fermentor volume is maintained constant by continuous overflow. (b) If the feed rate is 80% of what is needed for a steady-state cell mass, calculate the cell mass after 10 hours of operation.

**Solution**

This problem is a combination of dilution with continuous cell removal but with the added factor of cell generation by growth inside the vessel. The diagram shown in Fig. 2.13 represents the material balance on cell mass around a fermentor. The balance on cell mass or cell numbers is as follows:

$$\text{Input(from feed} + \text{cell growth)} = \text{output} + \text{accumulation}$$

(a)  The substrate entering the fermentor is cell-free; therefore, this term in the material balance equation is zero. Let $R$ = substrate feed rate = overflow rate, because fluid volume in fermentor is maintained constant. Let $x$ = cells/mL. The material balance on cell numbers with the appropriate time derivatives for rate of increase in cell numbers, $(dx/dt)_{gen}$, and accumulation $(dx/dt)_{acc}$ is:

$$V\left[\frac{dx}{dt}\right]_{gen} = Rx + V\left[\frac{dx}{dt}\right]_{acc}$$

If cell mass is constant, then the last term on the right is zero. Cell growth is often expressed in terms of a generation time, $g$, which is the average time for doubling of cell numbers. Let $t$ = time in hours. Then:

$$x = x_o(2)^{(t/g)}$$

Differentiating to obtain the generation term in the material balance equation:

$$\left[\frac{dx}{dt}\right]_{gen} = \frac{d}{dt}\left[x_0(2)^{(t/g)}\right]$$
$$= x_0(g^{-1})(2)^{(t/g)}\ln 2 = xg^{-1}\ln 2$$

Substituting in the material balance equation and dropping out zero terms:

$V \cdot x \cdot g^{-1}\ln 2 = Rx; x$ cancels out on both sides.
$R = V(\ln 2)g^{-1}$

Substituting known quantities:

$$R = \frac{1.5\ln 2}{1.5} = \frac{0.693\,\text{L}}{\text{h}}$$

In continuous fermentation, the ratio $R/V$ is the dilution rate, and when the fermentor is at a steady state in terms of cell mass, the dilution rate must equal the specific growth rate. The specific growth rate is the quotient: rate of growth/cell mass $= (1/x)(dx/dt)_{gen} = \ln 2/g$. Thus, the dilution rate to achieve steady state is

**Fig. 2.13** Material balance around a fermentor

X = 0

Overflow
R
X Cells/ml

X
Cells/ml

Fermentor
V 1.5 L

Feed
F

Pump

Cell-free
Substrate

Internal
Cell
Generation

strictly a function of the rate of growth of the organism and is independent of the cell mass present in the fermentor.

(b) When the feed rate is reduced, the cell number will be in an unsteady state. The material balance will include the accumulation term. Substituting the expression for cell generation into the original material balance equation with accumulation:

$$V\frac{x \ln 2}{g} = Rx + V\frac{dx}{dt}$$

$$x\left[\frac{V \ln 2}{g} - R\right] = v\frac{dx}{dt}$$

Separating variables and integrating:

$$\frac{dx}{x} = \left[\frac{\ln 2}{g} - \frac{R}{V}\right]dt$$

$$\ln x = \left[\frac{\ln 2}{g} - \frac{R}{V}\right]t + C$$

At t = 0, x = 10,000 cells/mL; C = ln(10,000):

$$\ln\left(\frac{x}{10,000}\right) = \left[\frac{\ln 2}{g} - \frac{R}{V}\right]t$$

Substituting: V = 1.5 L; R = 0.8(0.693 L/h) = 0.554 L/h; g = 1.5 h; t = 10 h:

$$\ln\left(\frac{x}{10,000}\right) = 0.927; x = 10,000(2.5269)$$
$$x = 25,269 \text{ cells/ml}$$

## 2.3    Blending of Food Ingredients

### 2.3.1    Total Mass and Component Balances

These problems involve setting up total mass and component balances and involve simultaneously solving several equations.

**Example 2.13** Determine the amount of a juice concentrate containing 65% solids and single-

**Fig. 2.14** Material flow and composition in a process for blending of juice concentrates

strength juice containing 15% solids that must be mixed to produce 100 kg of a concentrate containing 45% solids.

**Solution**
The diagram for the process is shown in Fig. 2.14.

Total mass balance : X + Y = 100;

X = 100 − Y

Solid balance

0.65X + 0.15Y = 100(0.45) = 45

Substituting (100 − Y) for X :

0.65(100 − Y) + 0.15Y = 45

65 − 0.65Y + 0.15Y = 45

65 − 45 = 0.65Y − 0.15Y

20 = 0.5Y

Y = 40 kg single strength juice

X = 60 kg 65% concentrate

**Example 2.14** Determine the amounts of lean beef, pork fat, and water that must be used to make 100 kg of a frankfurter formulation. The compositions of the raw materials and the formulations are:

Lean beef : 14%fat, 67%water, 19%protein

Pork fat : 89%fat, 8%water, 3%protein

Frankfurter : 20%fat, 15%protein, 65%water

**Solution**
The diagram representing the various mixtures being blended is shown in Fig. 2.15.

**Fig. 2.15** Composition and material flow in blending of meats for a frankfurter formulation

Total mass balance : $Z + X + Y = 100$

Fat balance : $0.14X + 0.89Y = 20$

Protein balance : $0.19X + 0.03Y = 15$

These equations will be solved by determinants. Note that the fat and protein balance equations involve only X and Y; therefore, solving these two equations first will give values for X and Y. Z will be solved using the total mass balance equation. Please see an Algebra textbook or Chap. 1 in the 3rd Edition of Fundamentals of Food Process engineering for help in setting-up determinants.

The matrices for the fat and protein balance equations are:

$$\begin{bmatrix} 0.14 & 0.89 \\ 0.19 & 0.03 \end{bmatrix} \begin{bmatrix} X \\ Y \end{bmatrix} = \begin{bmatrix} 20 \\ 15 \end{bmatrix}$$

$$X = \begin{vmatrix} 20 & 0.89 & 0.14 & 0.89 \\ 15 & 0.03 & 0.19 & 0.03 \end{vmatrix}$$

$$= \frac{(20)(0.03) - (15)(0.89)}{(0.14)(0.03) - (0.19)(0.89)}$$

$$X = \frac{-12.75}{-.1649} = 77.31$$

$$Y = \begin{vmatrix} 0.14 & 20 & 0.14 & 0.89 \\ 0.19 & 14 & 0.19 & 0.03 \end{vmatrix}$$

$$= \frac{(0.14)(14) - (0.19)(0.20)}{(0.14)(0.03) - (0.19)(0.89)}$$

$$Y = \frac{-1.7}{-0.1649} = 10.3 \, \text{kg}$$

Total mass balance : $Z = 100 - 77.3 - 10.3$
$$= 12.4 \, \text{kg}.$$

The solution using the Solver macro in Excel is shown in Fig. 2.16.

**Example 2.15** A food mix is to be made that would balance the amount of methionine (MET), a limiting amino acid in terms of food protein nutritional value, by blending several types of plant proteins. Corn, which contains 15% protein, has 1.2 g MET/100 g protein; soy flour with 55% protein has 1.7 g MET/100 g protein; and nonfat dry milk with 36% protein has 3.2 g MET/100 g protein. How much of these ingredients must be used to produce 100 kg of formula that contains 30% protein and 2.2 g MET/100 g protein?

**Solution**
This will be solved by setting up a component balance on protein and methionine. Let C = kg corn, S = kg soy flour, and M = kg nonfat dry milk. The material balance equations are:

Total mass : $C + S + M = 100$

Protein : $0.15C + 0.55S + 0.36 \, M = 30$

MET : $\dfrac{(1.2)(0.15)}{100} C + \dfrac{(1.7)(0.55)}{100} S$

$$+ \frac{(3.2)(0.36)}{100} M = \frac{2.2}{100}(30)$$

$$0.18C + 0.935 \, S + 1.152 \, M = 66$$

The matrices representing the three simultaneous equations are:

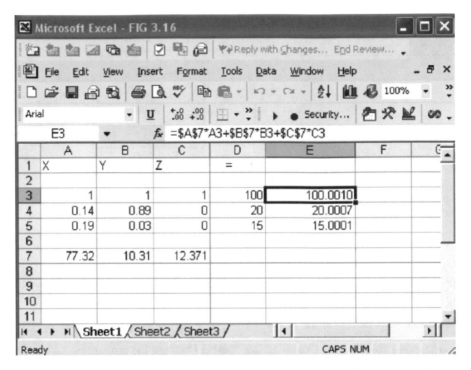

**Fig. 2.16** Solution to meats blending problem for frankfurter formulation using the Solver macro in Excel

$$\begin{bmatrix} 1 & 1 & 1 \\ 0.15 & 0.55 & 0.36 \\ 0.18 & 0.935 & 1.152 \end{bmatrix} \begin{bmatrix} C \\ S \\ M \end{bmatrix} = \begin{bmatrix} 100 \\ 30 \\ 66 \end{bmatrix}$$

Thus:

$$C = \frac{\begin{bmatrix} 100 & 1 & 1 \\ 30 & 0.55 & 036 \\ 66 & 0.935 & 1.152 \end{bmatrix}}{\begin{bmatrix} 1 & 1 & 1 \\ 0.15 & 0.55 & 0.36 \\ 0.18 & 0.935 & 1.152 \end{bmatrix}}$$

The denominator matrix is resolved to:

$$1\begin{bmatrix} 0.5 & 0.36 \\ 0.935 & 1.152 \end{bmatrix} - 0.15\begin{bmatrix} 1 & 1 \\ 0.935 & 1.152 \end{bmatrix}$$

$$+0.18\begin{bmatrix} 1 & 1 \\ 0.55 & 0.36 \end{bmatrix}$$

$$= 1[0.55(1.152) - 0.935(0.36)]$$
$$-0.15[1(1.152) - 1(0.935)]$$
$$+0.18[(1)(0.36) - (0.55)(1)]$$
$$= -0.03255 - 0.0342 = 0.23025$$

The numerator matrix resolves as follows:

$$100\begin{bmatrix} 0.55 & 0.36 \\ 0.935 & 1.152 \end{bmatrix} - 30\begin{bmatrix} 1 & 1 \\ 0.935 & 1.152 \end{bmatrix}$$

$$+66\begin{bmatrix} 1 & 1 \\ 0.55 & 0.36 \end{bmatrix} = 100[0.55(1.152) - 0.935(0.36)]$$

$$-30[1(1.152) - 0.935(1)] + 66[1(0.36) - 0.55(1)]$$
$$= 100(0.297) - 30(0.217) + 66(-0.19) = 10.65$$

$$C = \frac{10.65}{0.23025} = 46.25 \text{ kg}$$

$$S = \frac{\begin{bmatrix} 1 & 100 & 1 \\ 0.15 & 30 & 0.36 \\ 0.18 & 66 & 1.152 \end{bmatrix}}{0.23025}$$

The numerator matrix resolves as follows:

$$1\begin{bmatrix} 30 & 0.36 \\ 66 & 1.152 \end{bmatrix} - 0.15\begin{bmatrix} 100 & 1 \\ 66 & 1.152 \end{bmatrix}$$

$$+0.18\begin{bmatrix} 100 & 1 \\ 30 & 0.36 \end{bmatrix} = 1[30(1.152) - 66(0.36)]$$

$$-0.15[100(1.152) - 66(1) + 0.18[100(.36) - 30(1)]$$
$$= 1(10.9) - 0.15(49.2) + 0.18(6) = 4.5S$$

$$= \frac{4.5}{0.23025} = 19.54 \text{ kg}$$

**Fig. 2.17** Solution to food protein blending problem using the Solver macro in Excel

The total mass balance may be used to solve for M, but as a means of checking the calculations, the matrix for M will be formulated and resolved:

$$M = \frac{\begin{bmatrix} 1 & 1 & 100 \\ 0.15 & 0.55 & 30 \\ 0.18 & 0.935 & 66 \end{bmatrix}}{0.23025}$$

The numerator matrix is resolved as follows:

$$1\begin{bmatrix} 0.55 & 30 \\ 0.935 & 66 \end{bmatrix} - 0.15\begin{bmatrix} 1 & 100 \\ 0.935 & 66 \end{bmatrix}$$

$$+ 0.18\begin{bmatrix} 1 & 100 \\ 0.55 & 30 \end{bmatrix} = 1[0.55(66) - 0.935(30)]$$

$$- 0.15[1(66) - 0.935(100)]$$

$$+ 0.18[1(30) - 0.55(100)]$$

$$M = \frac{7.875}{0.23025} = 34.2\,kg = 1(8.25)$$

$$- 0.15(-27.5) + 0.18(-25) = 7.875$$

Check : $C + S + M = 46.25 + 19.54 + 34.2$
$= 99.99.$

The solution using the Solver macro in Excel is shown in Fig. 2.17.

### 2.3.2   Use of Specified Constraints in Equations

When blending ingredients, constraints may be imposed either by the functional properties of the ingredients or its components or by regulations. Examples of constraints are limits on the amount of nonmeat proteins that can be used in meat products like frankfurters, moisture content in meat products, or functional properties such as fat or water binding properties. Another constraint may be cost. It is necessary that these constraints specified be included as one of the equations to be resolved.

In some cases, a number of ingredients may be available for making a food formulation. To effectively solve the quantity of each ingredient using the procedures discussed in the previous sections, the same number of independent equations must be formulated as there are unknown quantities to be solved. If there are three components that can be used to formulate component balance equations, only three independent equations can be formulated. Thus, a unique solution for components of a blend, using the solution to simultaneous equations, will be possible only if there are three ingredients to be used.

When there are more ingredients possible for use in a formulation than there are independent equations that can be formulated using the total mass and component balances, constraints will have to be used to allow consideration of all possible ingredients. The primary objective is minimizing cost and maximizing quality. Both cost and quality factors will form the basis for specifying constraints.

The constraints may not be specific (i.e., they may only define a boundary rather than a specific value). Thus, a constraint cannot be used as a basis for an equation that must be solved simultaneously with the others.

One example of a system of constraints to include as many ingredients as possible into a formulation is the "least-cost formulation" concept used in the meat industry. Several software companies market "least-cost formulation" strategies, and the actual algorithm within each of these computer programs may vary from one company to the other. However, the basis for these calculations is basically the same:

(a) The composition must be met, usually 30% fat, and water and protein must satisfy the USDA requirement of water not to exceed four times the protein content plus 10% (4P + 10), or some other specified moisture range allowable for certain category of products.

(b) The bind values for fat must satisfy the needs for emulsifying the fat contained

within the formulation. Software companies who market least-cost sausage formulation strategies vary in how they use these fat holding capacities.

(c) Water holding capacity must be maximized; that is, the water and protein content relationship should stay within the 4P + 10 allowed in the finished product. This requirement is not usually followed because of the absence of reliable water holding capacity data, and processors simply add an excess of water during formulation to compensate for the water loss that occurs during cooking.

To illustrate these concepts, a simple least-cost formulation will be set up for frankfurters utilizing a choice of five different meat types. Note that, as in the previous examples, a total mass balance and only two component balance equations can be formulated (water can be determined from the total mass balance). Thus, only three independent equations can be formulated. Product textural properties may be used as a basis for a constraint. If previous experience has shown that a minimum of the protein present in the final blend must come from a type of meat (e.g., pork) to achieve a characteristic flavor and texture, then this constraint will be one factor to be included in the analysis. Other constraints are that no negative values of the components are acceptable.

The solution will be obtained using the Solver macro in Microsoft Excel (see Chapter 1, "Simultaneous Equations").

**Example 2.16** Derive a least-cost formulation involving four meat types from the five types shown. The composition and pertinent functional data, as bind values, for each of the ingredients are shown in Table 2.1. The fat content in the blend is 30% and protein is 15%.

The constraints are (1) for product color and textural considerations, at least 20% of the total protein must come from pork trim and pork cheeks; (2) the amount of beef cheek meat should not exceed 15% of the total mixture; and (3) the bind value (sum of the product of mass of each

**Table 2.1**  Data for the least-cost formulation problems

| Composition | | | | | |
|---|---|---|---|---|---|
| Meat type | Fat | Protein | Water | Bind constant kg fat/100 kg | Cost ($/kg) |
| Bull meat | 11.8 | 19.1 | 67.9 | 30.01 | 1.80 |
| Pork trm. | 25.0 | 15.9 | 57.4 | 19.25 | 1.41 |
| Beef chk. | 14.2 | 17.3 | 68.0 | 13.96 | 1.17 |
| Pork chk. | 14.1 | 17.3 | 67.0 | 8.61 | 1.54 |
| Back fat | 89.8 | 1.9 | 8.3 | 1.13 | 0.22 |

ingredient and its bind constant) must be greater than 15. These constraints are established by experience as the factors needed to impart the desired texture and flavor in the product.

**Solution**

Enter data from Table 2.1 in the spreadsheet in block A5:G10 as shown in Fig. 2.18. Designate column block H6:H10 to hold values of the weights of the selected meat components. Designate row block B12:H12 to hold the constraints. Enter formulas for the total mass balance in H12 and component mass balances on fat and protein, respectively, in B12 and C12. Designate F12 for the resulting bind value and G12 for the resulting least cost. Enter the following formulas:

1. Total weight: Enter in H12

$$= SUM(H6 : H10)$$

and $H$12 = 100.

2. Constraint 1: 15% protein in 100 kg mix will require 15 kg of protein. Enter 15 in C12. Constraint 1 specifies 20% of the 15 kg protein or 3 kg protein must come from pork trim and pork cheeks. This constraint is represented by the following equation entered in C14:

$$= \$H\$7^*0.159 + \$H\$9^*0.173$$
$$\text{and } \$C\$14 = 3$$

3. Because Constraint 2 refers to a meat weight (pork cheeks), the equation representing this constraint is entered in H8 as follows:
$H$8 < or = 15

4. Constraint 3, the cumulative bind value of the meat components is entered in F12 as:

$$= \$H\$6^*0.3001 + \$H\$7^*0.1925$$
$$+ \$H\$8^*0.1396 + \$H\$9^*0.0861$$
$$+ \$H\$10^*0.0113$$

and E12 > 15.

5. Cumulative fat from the meat components is entered in B12 as the fat balance formula:

$$= \$H\$6^*0.118 + \$H\$7^*0.25$$
$$+ \$H\$8^*0.142 + \$H\$9^*0.141$$
$$+ \$H\$10^*0.898$$

and B12 = 30.

6. Cumulative protein from the meat components is entered in C12:

$$= \$H\$3^*0.191 + \$H\$4^*0.159$$
$$+\$H\$5^*0.173 + \$H\$6^*0.173$$
$$+\$H\$7^*0.019$$

and: C12 = 15.

7. Other constraints are that no acceptable solution will have any meat component with values less than zero; these will be entered in the constraints box as follows:

H6 >= 0; H7 >= 0;   H9 >= 0; H10 >= 0

8. Formula for total cost is entered in G12:

$$= \$H\$6^*0.018 + \$H\$7^*0.0141$$
$$+ \$H\$8^*0.0117 + \$H\$9^*0.0154$$
$$+ \$H\$10^*0.0022$$

The value of G12 must be minimized.

Access the Solver by clicking on Tools and selecting Solver. The Solver Parameters box will

**Fig. 2.18** Solution to a least-cost meat formulation problem using the Solver macro in Excel

appear. In the Set Target Cell box, enter $G$12, and in the row Equal To, choose Min. In the By Changing Cells box, enter $H$6:$H$10. Add all the constraints into the "Subject to the Constraints" box. Click Solve. Results are displayed in the spreadsheet in Fig. 2.18.

The least = cost formulation has the following meats: 58.54 kg bull meat, 18.87 kg pork trimmings, 2.5 kg beef cheeks, 0 kg pork cheeks, and 20.06 kg back fat. The total mass of meats is 100 kg, and the cost of the formulation is 139.35 $/100 kg.

The least-cost formulation will change as the cost of the ingredients changes; therefore, processors need to update cost information regularly and determine the current least cost.

## 2.4    Multistage Processes

Problems of this type require drawing a process flow diagram and moving the system boundaries for the material balance formulations around parts of the process as needed to the unknown quantities. It is always good to define the basis used in each stage of the calculations. It is possi-

ble to change basis as the calculations proceed from an analysis of one subsystem to the next. A "tie material" will also be helpful in relating one part of the process to another.

**Example 2.17** The standard of identity for jams and preserves specify that the ratio of fruit to added sugar in the formulation is 45 parts fruit to 55 parts sugar. A jam also must have a soluble content of at least 65% to produce a satisfactory gel. The standard of identity requires soluble solids of at least 65% for fruit preserves from apricot, peach, pear, cranberry, guava, nectarine, plum, gooseberry, figs, quince, and currants. The process of making fruit preserves involves mixing the fruit and sugar in the required ratio, adding pectin, and concentrating the mixture by boiling in a vacuum, steam-jacketed kettle until the soluble solids content is at least 65%. The amount of pectin added is determined by the amount of sugar used in the formulation and by the grade of the pectin (a 100 grade pectin is one that will form a satisfactory gel at a ratio 1 kg pectin to 100 kg sugar).

If the fruit contains 10% soluble solids and 100 grade pectin is used, calculate the weight of

**Fig. 2.19** Process and material flow for manufacturing fruit preserves

fruit, sugar, and pectin necessary to produce 100 kg of fruit preserve. For quality control purposes, soluble solids are those that change the refractive index and can be measured on a refractometer. Thus, only the fruit soluble solids and sugar are considered soluble solids in this context, and not pectin.

**Solution**

The process flow diagram is shown in Fig. 2.19. The boundary used for the system on which the material balance is made encloses the whole process.

This problem uses data given in the problem, 100 kg fruit preserve, as a basis. The tie material is the soluble solids because pectin is unidentifiable in the finished product. The designation "fruit" is meaningless in the finished product because the water that evaporates comes from the fruit, and the soluble solids in the fruit mixes with the rest of the system.

Soluble solids balance:

$$0.1(X) + Y = 100(0.65)$$

Because the ratio 45 parts fruit to 55 parts sugar is a requirement, the other equation would be:

$$\frac{X}{Y} = \frac{45}{55}; \quad X = \frac{45}{55}Y$$

Substituting the expression for X in the equation representing the soluble solids balance and solving for Y:

$$0.1\left(\frac{45}{55}Y\right) + Y = 65$$

$$0.1(45)(Y) + 55Y = 55(65)$$

$$4.5Y + 55Y = 55(65)$$

$$59.5Y = 55(65)$$

$$Y = \frac{55(65)}{59.5} = 60 \text{kg sugar}$$

Because $X = 45Y/55$, $X = 45(60)/55 = 49$ kg fruit:

$$Z = \frac{\text{kg sugar}}{\text{grade of pectin}} = \frac{60}{100} = 0.6 \text{ kg pectin}$$

The amount of pectin, Z, can be calculated from the weight of sugar used. The problem can also be solved by selecting a different variable as a basis: 100 kg fruit. Equation based on the required ratio of fruit to sugar:

$$\text{Sugar} = 100 \text{ kg fruit} \times \frac{55 \text{ kg sugar}}{45 \text{ kg fruit}}$$
$$= 122 \text{ kg}$$

Let X = kg of jam produced.

Soluble solids balance: 100(0.1) + 122 = X (0.65)

$$X = \frac{132}{0.65} = 203 \text{ kg}$$

Because 100 kg fruit will produce 203 kg jam, the quantity of fruit required to produce 100 kg jam can be determined by ratio and proportion:

**Fig. 2.20** Process and material flow for a solvent extraction process involving simultaneous removal of fat and moisture

$$100 : 203 = X : 100$$

$$X = \frac{100(100)}{203} = 49 \text{ kg fruit}$$

$$\text{Sugar} = 49 \times \frac{55}{45} = 60 \text{ kg}$$

$$\text{Pectin} = \frac{1}{100}(60) = 0.6 \text{ kg}$$

**Example 2.18** In solvent extractions, the material to be extracted is thoroughly mixed with a solvent. An ideal system is one where the component to be extracted dissolves in the solvent, and the ratio of solute to solvent in the liquid phase equals the ratio of solute to solvent in the liquid absorbed in the solid phase. This condition occurs with thorough mixing until equilibrium is reached and if sufficient solvent is present such that the solubility of solute in the solvent is not exceeded.

Meat (15% protein, 20% fat, 64% water, and 1% inert insoluble solids) is extracted with five times its weight of a fat solvent that is miscible in all proportions with water. At equilibrium, the solvent mixes with water, and fat dissolves in this mixture. Assume there is sufficient solvent such that all the fat dissolves.

After thorough mixing, the solid is separated from the liquid phase by filtration, and the solid phase is dried until all the volatile material is removed. The weight of the dry cake is only 50% of the weight of the cake leaving the filter.

Assume none of the fat, protein, and inerts are removed from the filter cake by drying, and no nonfat solids are in the liquid phase leaving the filter. Calculate the fat content in the dried solids.

**Solution**

The diagram of the process is shown in Fig. 2.20. Using 100 kg of meat as a basis, the ratio of solvent to meat of five would require 500 kg of solvent. Consider as a system one that has a boundary that encloses the filter and the drier. Also draw a subsystem with a boundary that encloses only the drier. Write the known weights of process stream and the weights of components in each stream in the diagram. Entering the system at the filter will be 600 kg of material containing 15 kg protein, 1 kg inerts, 64 kg water, 20 kg of fat, and 500 kg of solvent. Because all nonfat solids leave at the drier, the weight of the dried solids D includes 16 kg of the protein and inert material. From the condition given in the problem that the weight of dried solids is 50% of the weight of the filter cake, the material entering the drier should be D/0.5.

All fat dissolves in the solvent-water mixture. Fat, however, will also appear in the solids fraction because some of the fat-water-solvent solution will be retained by the solids after filtration. Although the filtrate does not constitute all of the fat-water-solvent solution, the mass fraction of fat in the filtrate will be the same as in the whole solution, and this condition can be used to calculate the mass fraction of fat in the filtrate.

Mass fraction fat in filtrate

$$= \frac{\text{wt fat}}{\text{wt fat} + \text{wt solv} + \text{wtH}_2\text{O}}$$

$$= \frac{20}{20 + 64 + 500}$$

$$= 0.034246$$

Consider the system of the filter and drier. Let $x$ = mass fraction of fat in D. The following component balance can be made:

Fat balance:

$$F(0.034246) + D x = 20 \qquad (2.1)$$

Protein and inerts balance: Sixteen kilograms of proteins and inerts enter the system, and all of these leave the system with the dried solids D. Because D consists only of fat + protein + inert, and x is the mass fraction of fat, then $1 - x$ would be the mass fraction of protein and inert.

$$D(1 - x) = 16; \quad D = \frac{16}{1 - x} \qquad (2.2)$$

Solvent and water balance: Five hundred kilograms of solvent and 64 kg of water enter the system. These components leave the system with the volatile material V at the drier and with the filtrate F at the filter. The mass fraction of fat in the filtrate is 0.034246. The mass fraction of water and solvent $= 1 - 0.034246 = 0.965754$.

$$F(0.965754) + V = 564 \qquad (2.3)$$

There are four unknown quantities, F, D, V, and x, and only three equations have been formulated. The fourth equation can be formulated by considering the subsystem of the drier. The condition given in the problem that the weight of solids leaving the drier is 50% of the weight entering would give the following total mass balance equation around the drier:

$$\frac{D}{0.5} = D + V; \quad D = V \qquad (2.4)$$

The above four equations can be solved simultaneously. Substituting D for V in Eq. (2.3):

$$F(0.965754) + D = 564 \qquad (2.5)$$

Substituting Eq. (2.2) in Eq. (2.5):

$$F(0.965754) + \frac{16}{1 - x}(x) = 564 \qquad (2.6)$$

Substituting Eq. (2.2) in Eq. (2.1):

$$F(0.034246) + \frac{16}{1x}(x) = 20 \qquad (2.7)$$

Solving for F in Eqs. (2.6) and (2.7) and equating:

$$\frac{20 - 36\,x}{(0.034246)(1 - x)} = \frac{548 - 564x}{(0.965754)(1 - x)}$$

Simplifying and solving for x:

$$0.965754(20 - 36x) = 0.034246(548 - 564x)$$

$$19.31508 - 34.767144x = 18.7668 - 19.314744x$$

$$x(34.76144 - 19.314744) = 19.31508 - 18.7668$$

$$x = \frac{0.548272}{15.446696} = 0.03549$$

The percentage fat in the dried solids $= 3.549\%$.

The solution can be shortened and considerably simplified if it is realized that the ratio of fat/(solvent + water) is the same in the filtrate as it is in the liquid that adheres to the filter cake entering the drier. Because the volatile material leaving the drier is only solvent + water, it is possible to calculate the amount of fat carried with it from the fat/(solvent + water) ratio.

$$\frac{\text{fat}}{\text{solvent} + \text{water}} = \frac{20}{500 + 64} = 0.03546$$

The amount of fat entering the drier $= 0.03546$ V. Because $D = 16 + \text{fat}$; $D = 16 + 0.03546$ V. Total mass balance around the drier gives: $D = V$.

Substituting D for V gives:

$$D = 16 + 0.03546\,D$$

$$= \frac{16}{1 - 0.03546} = 16.5882$$

wt.fat in $D = 16.5882 - 16 = 0.5882$

%fat in $D = (0.5882/16.5882)(100) = 3.546\%$

A principle that was presented in an example at the beginning of this chapter was that in solvent extraction or crystallization, the purity of the product depends strongly on the efficiency of separation of the liquid from the solid phase. This is demonstrated here in that the amount of

**Fig. 2.21** Composition
and material flow in a
crystallization process

Mother Liquor
66% Sucrose
34% Water

fat entering the drier is a direct function of the
amount of volatile matter present in the wet material. Thus, the more efficient the solid-liquid separation process before drying the solids, the less
fat will be carried over into the finished product.

**Example 2.19** Crystallization. Determine the
quantity of sucrose crystals that will crystallize
out of 100 kg of a 75% sucrose solution after
cooling to 15 °C. The mother liquor contains
66% sucrose.

**Solution**
The diagram for crystallization is shown in
Fig. 2.21. A 75% sucrose solution enters the
crystallizer, and the material separates into a solution containing 66% sucrose and crystals of 100%
sucrose.

Figure 2.21 shows sucrose appearing in the
crystals and saturated solution fractions and
water appears only in the saturated solution fraction. Water can be used as a tie material in this
process.

The mother liquor contains 66% sucrose. The
balance (34%) is water. Using the mass fraction
principle, 25 kg of water enters the system, and all
this leaves with the mother liquor where the mass
fraction of water is 0.34. Thus, the weight of the
mother liquor would be 25/0.34 = 73.52 kg.

Total mass balance:

$$\text{wt crystals} = 100 - 73.52 = 26.48 \, \text{kg}$$

**Example 2.20** For the problem statement and
the process represented by the diagram in
Fig. 2.21, calculate the yield and the purity of
the sugar crystals. The mother liquor contains

67% sucrose w/w. The crystals fraction from the
centrifuge loses 15% of its weight in the drier and
emerges moisture-free.

**Solution**
The problem can be separated into two sections.
First, a material balance around the evaporator
will be used to determine the weight and composition of the material entering the crystallizer.
The next part, which involves the crystallizer,
centrifuge, and drier, can be solved utilizing the
principles demonstrated in the two preceding
example problems.

Material balance around the evaporator:
Twenty kilograms of sucrose enters, and this
represents 75% of the weight of the concentrate.

$$\text{wt concentrate} = 20/0.75 = 26.66 \, \text{kg}$$

The concentrate entering the crystallizer consists
of 20 kg sucrose, 1 kg inert, and 5.66 kg water.

Consider the centrifuge as a subsystem where
the cooled concentrate is separated into a pure
crystals fraction and a mother liquor fraction.
The weights of these two fractions and the composition of the saturated solution fraction can be
determined using a similar procedure as in the
preceding example. In Fig. 2.22, the known
quantities are indicated on the diagram.

The composition of the mother liquor fraction
is determined by making a material balance
around the crystallizer. The mass and composition of streams entering and leaving the crystallizer are shown in Fig. 2.23.

All the water and inert material entering the
centrifuge go into the mother liquor fraction. The
weight of water and inert material in the mother
liquor are 5.66 and 1 kg, respectively.

**Fig. 2.22**  Material flow and system boundaries used for analysis of crystal purity from a crystallizer

**Fig. 2.23**  Material balance around the crystallizer to determine the quantity of mother liquor and pure crystals that separate on crystallization

The material balance around the crystallizer is:

$$\text{Total mass} : M + S = 26.66$$

$$\text{Sucrose balance} : M(0.67) + S = 20$$

Eliminating S by subtracting the two equations:

$$M(1 - 0.67) = 26.66 - 20; M = 6.66/0.33$$
$$= 20.18 \text{ kg}$$

From the known weights of components, the composition of mother liquor, in mass fraction, is as follows:

Sucrose = 0.67, as defined
Inert: 1.00 kg; mass fraction = 1/20.18 = 0.04955
Water: 5.66 kg; mass fraction = 5.66/20.18 = 0.28048

The solids fraction C leaving the centrifuge is a mixture of the mother liquor adhering to the crystals, A, and pure sucrose crystals S. Because 26.66 kg of material entered the crystallizer and 20.18 kg is in the mother liquor, the pure sucrose crystals, S = 26.66 − 20.18 = 6.48 kg.

$$C = 6.48 + A$$

Water balance around the drier:

$$A(\text{mass fraction water in A}) = 0.15 \text{ C}$$

$$A(0.28048) = 0.15 \text{ C}$$

Substituting C:

$$A(0.28048) = 0.15(6.48 + A)$$
$$A(0.28048 - 0.15) = 0.15(6.48)$$
$$A = 7.4494 \text{ kg}$$
$$C = 6.48 + 7.4494 = 13.9294 \text{ kg}$$

Wt dry sugar, D = 0.85(6.48 + 7.4494) = 11.84 kg
Wt inert material = A (mass fraction inert in A) = 7.4494(0.04955) = 0.3691 kg
% inert in dry sugar = (0.3691/11.84)(100) = 3.117%

The dry sugar is only 96.88% sucrose.

**Example 2.21**  An ultrafiltration system has a membrane area of 0.75 m² and has a water permeability of 180 kg water/h(m²) under conditions used in concentrating whey from 7% to 25% total solids at a pressure of 1.033 MPa. The system is fed by a pump that delivers 230 kg/h, and the appropriate concentration of solids in the product is obtained by recycling some of the product through the membrane. The concentrate contained 11% lactose and the unprocessed when contained 5.3% lactose. There is no protein in the permeate.

Calculate (1) the production rate of 25% concentrate through the system, (2) the amount of product recycled/h, (3) the amount of lactose removed in the permeate/h, (4) the concentration of lactose in the mixture of fresh and recycled whey entering the membrane unit, and (5) the average rejection factor by the membrane for lactose based on the average lactose concentration entering and leaving the unit.

The rejection factor ($F_r$) of solute through a membrane is defined expressed by:

$$F_r = \frac{X_f - X_p}{X_f}$$

where $X_f$ is solute concentration on the feed side of the membrane, which may be considered as the mean of the solute concentration in the fluid entering and leaving the membrane unit, and $X_p$ is the solute concentration in the permeate.

**Solution**

The process flow diagram is shown in Fig. 2.24.

Let L, Pr, and W represent lactose, protein, and water, respectively. Let F = feed, P = permeate or fluid passing through the membrane, and R = recycled stream, which has the same composition as the concentrate, C. The fluid stream leaving the membrane unit which is the sum of the concentrate and recycle stream, is also referred to as the retentate.

Whole system material balance: basis—1 hour of operation.

$$\text{Water in permeate} = \frac{180 \text{ kg}}{\text{m}^2\text{h}}(0.75\text{m}^2)(1 \text{ h})$$

$$= 180(0.75) = 135 \text{ kg}$$

Water balance:

$$F(0.93) = 135 + C(0.75); 0.93F - 0.75C = 135$$

**Fig. 2.24** Diagram of a spiral wound membrane module in an ultrafiltration system for cheese whey

Protein balance:

$$F(0.017) = C(0.14); 0.017F - 0.14C = 0$$

Solving for F and C by determinants:

$$\begin{bmatrix} 0.93 & -0.75 \\ 0.017 & -0.14 \end{bmatrix} \begin{bmatrix} F \\ C \end{bmatrix} = \begin{bmatrix} 135 \\ 0 \end{bmatrix} F$$

$$= \frac{135(-0.14) - 0}{0.93(-0.14) - 0.017(-0.75)}$$

$$= \frac{-18.9}{-0.1302 + 0.01275} C$$

$$= \frac{0 - 0.017(135)}{-0.1302 + 0.01275} = \frac{-2.295}{-0.11745}$$

$$= 19.54 \text{ kg/h}$$

$$F = 160.9 \text{ kg/h}$$

1. Amount of product = 19.54 kg/h.
2. Amount recycled is determined by a material balance around the pump. The diagram is shown in Fig. 2.25.

$$R = 230 - 160.9 = 69.1 \text{ kg/h}$$

3. Lactose in permeate—lactose balance around whole system:

**Fig. 2.25** Diagram of material balance during mixing of feed and recycle streams in an ultrafiltration system

$L_p$ = lactose in permeate = $F(0.053)\,C(0.11)$

    = $160.9(0.053) - 19.54(0.11) = 6.379$

$X_p$ = mass fraction lactose in permeate.

    = $\dfrac{6.379}{135 + 6.379} = 0.04511$

4. Lactose balance around pump:

$F(0.053) + R(0.11) = 230(X_{f1})$

$X_{fl}$ = mass fraction lactose in fluid

        leaving the membrane unit.

$X_{fl} = \dfrac{160.9(0.053) + 69.1(0.11)}{230} = 0.0702$

$X_f$ = mean mass fraction lactose in the

        retentate side of the membrane unit.

$X_f = 0.5(0.0702 + 0.11) = 0.09.$

5. Rejection factor for lactose:

$F_r = (0.09 - 0.045)/0.09 = 0.498$

## Problems

2.1. A frankfurter formulation is to be made from the following ingredients:

Lean beef : 14%fat, 67%water, 19%protein

Pork fat : 89%fat, 8%water, 3%protein

Soy protein isolate : 90%protein, 8%water

Water needs to be added (usually in the form of ice) to achieve the desired moisture content. The protein isolate added is 3% of the total weight of the mixture. How much lean beef, pork fat, water, and soy isolate must be used to obtain 100 kg of a formulation having the following composition: protein, 15%; moisture, 65%; and fat, 20%.

2.2. If 100 kg of raw sugar containing 95% sucrose, 3% water, and 2% soluble uncrystallizable inert solids is dissolved in 30 kg of hot water and cooled to 20°C, calculate:

  (a) Kilograms of sucrose that remains in solution.

  (b) Crystalline sucrose.

  (c) The purity of the sucrose (in % sucrose) obtained after centrifugation and dehydration to 0% moisture. The solid phase contained 20% water after separation from the liquid phase in the centrifuge.

A saturated solution of sucrose at 20 °C contains 67% sucrose (w/w).

2.3. Tomato juice flowing through a pipe at the rate of 100 kg/min is salted by adding saturated salt solution (26% salt) into the pipeline at a constant rate. At what rate would the saturated salt solution be added to have 2% salt in the product?

2.4. If fresh apple juice contains 10% solids, what would be the solids content of a concentrate that would yield single-strength juice after diluting one part of the concentrate with three parts of water? Assume densities are constant and are equal to the density of water.

2.5. In a dehydration process, the product, which is at 80% moisture, initially has lost half its weight during the process. What is the final moisture content?

2.6. Calculate the quantity of dry air that must be introduced into an air drier that dries 100 kg/h of food from 80% moisture to 5% moisture. Air enters with a moisture content of 0.002 kg water per kg of dry air and leaves with a moisture content of 0.2 kg $H_2O$ per kg of dry air.

2.7. How much water is required to raise the moisture content of 100 kg of a material from 30% to 75%?

2.8. In the section "Multistage Process," Example 2.18, solve the problem if the meat to solvent ratio is 1:1. The solubility of fat in the water-solvent mixture is such that the maximum fat content in the solution is 10%.

2.9. How many kilograms of peaches would be required to produce 100 kg of peach preserves? The standard formula of 45 parts fruit to 55 parts sugar is used, the soluble solids content of the finished product is 65%, and the peaches have 12% initial soluble solids content. Calculate the weight of 100 grade pectin required and the amount of water removed by evaporation.

2.10. The peaches in Problem 9 come in a frozen form to which sugar has been added in the ratio of three parts fruit to one part sugar. How much peach preserves can be produced from 100 kg of this frozen raw material?

2.11. Yeast has a proximate analysis of 47% C, 6.5% H, 31% O, 7.5% N, and 8% ash on a dry weight basis. Based on a factor of 6.25 for converting protein nitrogen to protein, the protein content of yeast on a dry basis is 46.9%. In a typical yeast culture process, the growth medium is aerated to convert substrate primarily to cell mass. The dry cell mass yield is 50% of a sugar substrate. Nitrogen is supplied as ammonium phosphate.

The cowpea is a high-protein, low-fat legume that is a valuable protein source in the diet of many Third World nations. The proximate analysis of the legume is 30% protein, 50% starch, 6% oligosaccharides, 6% fat, 2% fiber, 5% water, and 1% ash. It is desired to produce a protein concentrate by fermenting the legume with yeast. Inorganic ammonium phosphate is added to provide the nitrogen source. The starch in cowpea is first hydrolyzed with amylase and yeast is grown on the hydrolyzate.

(a)  Calculate the amount of added inorganic nitrogen as ammonium phosphate to provide the stoichiometric amount of nitrogen to convert all the starch present to yeast mass. Assume none of the cowpea protein is utilized by the yeast.

(b)  If the starch is 80% converted to cell mass, calculate the proximate analysis of the fermented cowpea on a dry basis.

2.12. This whey is spray dried to a final moisture content of 3%, and the dry whey is used in an experimental batch of summer sausage. In summer sausage, the chopped meat is inoculated with a bacterial culture that converts sugars to lactic acid as the meat is allowed to ferment prior to cooking in a smokehouse. The level of acid produced is controlled by the amount of sugar in the formulation. The lactic acid level in the summer sausage is 0.5 g/100 g dry solids. Four moles of lactic acid ($CH_3$ $CHOHCOOH$) is produced from 1 mole of lactose ($C_1{}^2H^{22}O_{11}$). The following formula is used for the summer sausage:

3.18 kg lean beef
   (16%fat, 16%protein, 67.1%water, 0.9%ash)
1.36 kg pork
   (25%fat, 12%protein, 62.4%water, 0.6%ash)
0.91 kg ice
0.18 kg soy protein isolate
   (5%water, 1%ash, 94%protein)

Calculate the amount of dried whey protein that can be added into this formulation in order that, when the lactose is 80% converted to lactic acid, the desired acidity will be obtained.

2.13. Osmotic dehydration of blueberries was accomplished by contacting the berries with a corn syrup solution containing 60% soluble solids for 6 hours and draining the syrup from the solids. The solid fraction left on the screen after draining the syrup is 90% of the original weight of the berries. The berries originally contained 12% soluble solids, 86.5% water, and 1.5% insoluble solids. The sugar in the syrup penetrated the berries such that the berries themselves remaining on the screen, when washed free of the adhering solution, showed a soluble solids gain of 1.5% based on the original dry solids content. Calculate:

(a) The moisture content of the berries and adhering solution remaining on the screen after draining the syrup.

(b) The soluble solids content of the berries after drying to a final moisture content of 10%.

(c) The percentage of soluble solids in the syrup drained from the mixture. Assume none of the insoluble solids are lost in the syrup.

2.14. The process for producing dried mashed potato flakes involves mixing wet mashed potatoes with dried flakes in a 95:5 weight ratio, and the mixture is passed through a granulator before drying on a drum dryer. The cooked potatoes after mashing contained 82% water and the dried flakes contained 3% water. Calculate:

(a) The amount of water that must be removed by the dryer for every 100 kg of dried flakes produced.

(b) The moisture content of the granulated paste fed to the dryer.

(c) The amount of raw potatoes needed to produce 100 kg of dried flakes; 8.5% of the raw potato weight is lost on peeling.

(d) Potatoes should be purchased on a dry matter basis. If the base moisture content is 82% and potatoes at this moisture content cost $200/ton, what would be the purchase price for potatoes containing 85% moisture.

2.15. Diafiltration is a process used to reduce the lactose content of whey recovered using an ultrafiltration membrane. The whey is passed through the membrane first and concentrated to twice the initial solids content, rediluted, and passed through the membrane a second time. Two membrane modules in series, each with a membrane surface area of 0.5 $m^2$, are to be used for concentrating and removing lactose from

acid whey that contains 7.01% total solids, 5.32% lactose, and 1.69% protein. The first module accomplishes the initial concentration, and the retentate is diluted with water and reconcentrated in the second module to a solids content of 14.02%. Under the conditions used in the process, the membrane has an average water permeation rate of 254 kg/(h $\cong$ $m^2$). The rejection factor for lactose by the membrane based on the arithmetic mean of the feed and retentate lactose concentration and the mean permeate lactose concentration is 0.2. Protein rejection factor is 1. The rejection factor is defined as: $F_r = (C_r C_p)/C_r$ where $C_r$ is the concentration on the retentate side and $C_p$ is the concentration on the permeate side of the membrane. Use Visual BASIC to determine:

(a) The amount of 14.02% solids delactosed whey concentrate produced from the second module per hour.

(b) The lactose content of the delactosed whey concentrate.

2.16. An orange juice blend containing 42% soluble solids is to be produced by blending stored orange juice concentrate with the current crop of freshly squeezed juice. The following are the constraints:

The soluble solids to acid ratio must equal 18; and the currently produced juice may be concentrated first before blending, if necessary. The currently produced juice contains 14.5% soluble solids, 15.3% total solids, and 0.72% acid. The stored concentrate contains 60% soluble solids, 62% total solids, and 4.3% acid. Calculate:

(a) The amount of water that must be removed or added to adjust the concentration of the soluble solids in order to meet the specified constraints.

(b) The amounts of currently processed juice and stored concentrate needed to produce 100 kg of the blend containing 42% soluble solids.

2.17. The process for extraction of sorghum juice from sweet sorghum for production of sorghum molasses, which is still practiced in some areas in the rural south of the United States, involves passing the cane through a three-roll mill to squeeze the juice out. Under the best conditions, the squeezed cane (bagasse) still contains 50% water.

  (a) If the cane originally contained 13.4% sugar, 65.6% water, and 32% fiber, calculate the amount of juice squeezed from the cane per 100 kg of raw cane, the concentration of sugar in the juice, and the percentage of sugar originally in the cane that is left unrecovered in the bagasse.

  (b) If the cane is not immediately processed after cutting, moisture and sugar loss occurs. Loss of sugar has been estimated to be as much as 1.5% within a 24-hour holding period, and total weight loss for the cane during this period is 5.5%. Assume sugar is lost by conversion to $CO^2$; therefore, the weight loss is attributable to water and sugar loss. Calculate the juice yield based on the freshly harvested cane weight of 100 kg, the sugar content in the juice, and the amount of sugar remaining in the bagasse.

2.18. In a continuous fermentation process for ethanol from a sugar substrate, the sugar is converted to ethanol and part of it is converted to yeast cell mass. Consider a 1000 L continuous fermentor operating at a steady state. Cell-free substrate containing 12% glucose enters the fermentor. The yeast has a generation time of 1.5 hours, and the concentration of yeast cells within the fermentor is $1 \times 10^7$ cells/mL. Under these conditions, a dilution rate (F/V, where F is the rate of feeding of cell-free substrate and V is the volume of the fermentor) that causes the cell mass to stabilize at a steady state results in a residual sugar content in the overflow of 1.2%. The stoichiometric ratio for sugar to dry cell mass is 1:0.5 on a weight basis, and that for sugar to ethanol is based on 2 moles of ethanol produced per mole of glucose. A dry cell mass of 4.5 of the g/L fermentor is equivalent in g to ethanol/(La cell count $\cong$ h) of $1.6 \times 10^7$ cells/mL. Calculate the ethanol productivity of the fermentor in g ethanol/(L $\cong$ h).

2.19. A protein solution is to be demineralized by dialysis. The solution is placed inside a dialysis tube immersed in continuously flowing water. For all practical purposes, the concentration of salt in the water is zero, and the dialysis rate is proportional to the concentration of salt in the solution inside the tube. The contents of the tube may be considered to be well mixed. A solution that initially contained 500 µg/mL of salt and 15 mg protein/mL contained 400 µg/mL salt and 13 mg protein/mL at the end of 2 hours. Assume no permeation of protein through the membrane, and density of the solution is constant at 0.998 g/mL. The rate of permeation of water into the membrane and the rate of permeation of salt out of the membrane are both directly proportional to the concentration of salt in the solution inside the membrane. Calculate the time of dialysis needed to drop the salt concentration inside the membrane to 10 µg/mL. What will be the protein concentration inside the membrane at this time?

2.20. A fruit juice-based natural sweetener for beverages is to be formulated. The sweetener is to have a soluble solids to acid ratio of 80. Pear concentrate having a soluble solids to acid ratio of 52 and osmotically concentrated/deacidified apple juice with a soluble solids to acid ratio of 90 are available. Both concentrates contain 70% soluble solids.

  (a) Calculate the ratio of deacidified apple and pear concentrates that must be mixed to obtain the desired sol. solids/acid ratio in the product.

  (b) If the mixture is to be diluted to 45% soluble solids, calculate the amounts of deacidified apple and pear juice concentrates needed to make 100 kg of the desired product.

2.21. A spray drier used to dry egg whites produces 1000 kg/h of dried product containing 3.5% moisture from a raw material that contains 86% moisture.

(a) If air used for drying enters with a humidity of 0.0005 kg water per kg dry air and leaves the drier with a humidity of 0.04 kg water per kg dry air, calculate the amount of drying air needed to carry out the process.

(b) It is proposed to install a reverse osmosis system to remove some moisture from the egg whites prior to dehydration to increase the drying capacity. If the reverse osmosis system changes the moisture content of the egg whites prior to drying to 80% moisture, and the same inlet and outlet air humidities are used on the drier, calculate the new production rate for the dried egg whites.

2.22. An ultrafiltration system for concentrating milk has a membrane area of 0.5 m$^2$, and the permeation rate for water and low molecular weight solutes through the membrane is 3000 g/(m$^2 \cong$ min). The solids content of the permeate is 0.5%. Milk flow through the membrane system must be maintained at 10 kg/min to prevent fouling. If milk containing initially 91% water enters the system, and a concentrate containing 81% solids is desired, calculate:

(a) The amount of concentrate produced by the unit/h.

(b) The fraction of total product leaving the membrane that must be recycled to achieve the desired solids concentration in the product.

(c) The solids concentration of feed entering the unit after the fresh and recycled milk is mixed.

2.23. A recent development in the drying of blueberries involves an osmotic dewatering prior to final dehydration. In a typical process, grape juice concentrate with 45% soluble solids (1.2% insoluble solids; 53.8% water) in a ratio 2 kg juice/kg berries is used to osmoblanch the berries, followed by draining the berries and final drying in a tunnel drier to a moisture content of 12%.

During osmoblanching (the juice concentrate is heated to 80 °C, the berries added, temperature allowed to go back to 80 °C, 5-minute hold, and the juice is drained), the solids content of blueberries increases. When analyzed after rinsing the adhering solution, the total solids content of the berries was 15%. The original berries contained 10% total solids and 9.6% soluble solids. Assume leaching of blueberry solids into the grape juice is negligible.

The drained berries carry about 12% of their weight of the solution that was drained.

Calculate:

(a) The proximate composition of the juice drained from the blueberries.

(b) The yield of dried blueberries from 100 kg raw berries.

(c) The drained juice is concentrated and recycled. Calculate the amount of 45% solids concentrate that must be added to the recycled concentrate to make enough for the next batch of 100 kg blueberries.

2.24. A dietetic jelly is to be produced. In order that a similar fruit flavor may be obtained in the dietetic jelly as in the traditional jelly, the amount of jelly that can be made from a given amount of fruit juice should be the same as in the standard pectin jelly.

If 100 kg of fruit juice is available with a soluble solids content of 14%, calculate the amount of standard pectin jelly with a soluble solids content of 65% that can be produced.

Only the sugar in jelly contributes the caloric content.

(a) What will be the caloric equivalent of 20 g (1 tsp) of standard pectin jelly?

(b) The dietetic jelly should have a caloric content 20% that of the standard pectin jelly.

Fructose may be added to provide sweetness. The low methoxy pectin can be used in the same amount for the same quantity of fruit as in the standard jelly. Calculate the soluble solids content in the dietetic jelly to give the caloric reduction.

(c)  In part (b), calculate the amount of additional fructose that must be used.

## Suggested Reading

Charm, S. E. (1971). *Fundamentals of food engineering* (2nd ed.). Westport: AVI Publishing Co..

Felder, R. M., & Rousseau, R. W. (1999). *Elementary principles of chemical processes* (2nd ed.). New York: Wiley.

Himmelblau, D. M. (1967). *Basic principles and calculations in chemical engineering* (2nd ed.). Englewood Cliffs: Prentice-Hall.

Hougen, O. A., & Watson, K. M. (1946). *Chemical process principles. Part I. Material and energy balances*. New York: Wiley.

Sinnott, R. K. (1996). *Chemical engineering* (Vol. 6, 2nd ed.). Oxford: Butterworth-Heinemann.

Toledo, R. T. (2007). *Fundamentals of food process engineering* (3rd ed.). Springer, New York: NY.

# Gases and Vapors

Gases and vapors are naturally associated with foods and food processing systems. The equilibrium between food and water vapor determines temperatures achieved during processing. Dissolved gases in foods such as oxygen affect shelf life. Gases are used to flush packages to eliminate oxygen and prolong shelf life. Modified atmospheres in packages have been used to prolong shelf life of packaged foods. Air is used for dehydration. Gases are used as propellants in aerosol cans and as refrigerants. The distinction between gases and vapors is very loose because theoretically all vapors are gases. The term "vapor" is generally used for the gaseous phase of a substance that exists as a liquid or a solid at ambient conditions.

## 3.1 Equations of State for Ideal and Real Gases

Equations of state are expressions for the relationship between pressure, volume, temperature, and quantity of gases within a given system. The simplest equation of state, the ideal gas equation, closely approximates the actual behavior of gases at near ambient temperature and pressure where the effect of molecular interactions is minimal. At high pressures and temperatures, however, most gases deviate from ideal behavior, and several equations of state have been proposed to fit experimental data. In this section, two equations of state will be discussed: the ideal gas equation and one of the most used equations of state for real gases, van der Waals equation.

### 3.1.1 The Kinetic Theory of Gases

The fundamental theory governing the behavior of gases, the kinetic theory, was first proposed by Bernoulli in 1738 and was tested and extended by Clausius, Maxwell, Boltzmann, van der Waals, and Jeans. The postulates of the kinetic theory are as follows:

1. Gases are composed of discreet particles called *molecules*, which are in constant random motion, colliding with each other and with the walls of the surrounding vessel.
2. The force resulting from the collision between the molecules and the walls of the surrounding vessel is responsible for the *pressure* of the gas.
3. The lower the pressure, the farther apart the molecules; thus, attractive forces between molecules have reduced influence on the overall properties of the gas.

R. T. Toledo et al., *Fundamentals of Food Process Engineering*,
https://doi.org/10.1007/978-3-319-90098-8_3

4. The average kinetic energy of the molecules is directly proportional to the absolute temperature.

### 3.1.2    Absolute Temperature and Pressure

The pressure (P) of a gas is the force of collisions of gas molecules against a surface in contact with the gas. Because pressure is force per unit area, pressure is proportional to the number of gas molecules and their velocity. This pressure is the absolute pressure.

Pressure is often expressed as *gauge pressure* when the measured quantity is greater than atmospheric pressure and as *vacuum* when below atmospheric. Measurement of gauge and atmospheric pressure is shown in Fig. 3.1.

Two types of pressure measuring devices are depicted: a manometer and a Bourdon tube pressure gauge. On the left of the diagram, the pressure of the gas is counteracted by atmospheric pressure, such that if pressure is atmospheric, the gauges will read zero. The reading given by these types of gages is the *gauge pressure*. In the American Engineering System of measurement, gauge pressure is expressed as "psig" or pound force per square inch gauge. In the SI system, gauge pressure is expressed as "kPa above atmospheric."

The diagram on the right in Fig. 3.1 shows the measuring element completely isolated from the atmosphere. The pressure reading from these gauges represents the actual pressure or force of collision of gas molecules, the *absolute pressure*. In the American Engineering System, absolute pressure is expressed as "psia" or pound force per square inch absolute. In SI, it is expressed as "kPa absolute."

Conversion from gauge to absolute pressure is done using the following equation:

$$\mathbf{P}_{gauge} = \mathbf{P}_{absolute} - \mathbf{P}_{atmospheric} \tag{3.1}$$

The term "vacuum" after a unit of pressure indicates how much the pressure is below atmospheric pressure. Thus, vacuum may be interpreted as a negative gauge pressure and is related to the atmospheric pressure and absolute pressure as follows:

$$\mathbf{P}_{absolute} = \mathbf{P}_{atmospheric} - \mathbf{P}_{vacuum} \tag{3.2}$$

Pressure is sometimes expressed in *atmospheres* (atm) instead of the traditional force/area units. This has led to some confusion between a pressure specified in atmospheres and atmospheric pressure. Unless otherwise specified, the pressure term "atm" refers to a *standard atmosphere*, the mean atmospheric pressure at sea level, equivalent to 760 mm Hg, 29.921 in. Hg, 101.325 kPa, or 14.696 $lb_f$/in.$^2$. A *technical atmosphere* is a pressure of 1 $kg_f$/cm$^2$. Some technical articles express pressure in *Bar*, which is equivalent to 100 kPa. *Atmospheric pressure* is actual pressure exerted by the atmosphere in a particular location and varies with time and location. Atmospheric pressure must be specified if it is different from a standard atmosphere.

**Example 3.1:** Calculate the absolute pressure inside an evaporator operating under 20 in. Hg vacuum. Atmospheric pressure is 30 in. Hg. Express this pressure in SI and in the American Engineering System of units.

From the table of conversion factors, Appendix Table A.1, the following conversion factors are obtained:

$$\frac{0.4912 \frac{lb_f}{in^2}}{in.Hg} \qquad \frac{3.38638 \times 10^3 Pa}{in.Hg}$$

$$\begin{aligned} P_{absolute} &= P_{atmospheric} - P_{vacuum} \\ &= (30 - 20) \text{ in.Hg} \\ &= 10 \text{ in.Hg} \end{aligned}$$

$$P_{absolute} = 10 \text{ in.Hg} \cdot \frac{0.4912 \frac{lb_f}{in^2}}{in.Hg} = 4.912 \text{ psia}$$

$$\begin{aligned} P_{absolute} &= 10 \text{ in.Hg} \cdot \frac{3.38638 \times 10^3}{in.Hg} \\ &= 33.863 \text{ kPa absolute} \end{aligned}$$

Temperature (T) is a thermodynamic quantity related to the velocity of motion of molecules. The temperature scales are based on the boiling (liquid-gas equilibrium) and freezing points

**Fig. 3.1** Diagram showing differences in measurement of gauge and absolute pressure

(solid, liquid, gas equilibrium) of pure substances such as water at 1 standard atmosphere (101.325 kPa) pressure. The Kelvin (K) is defined as the fraction, 1/273.16, of the thermodynamic temperature of the triple point of water. The absolute temperature scales are based on a value on the scale that is zero at the temperature when the internal energy of molecules is zero. These are the Kelvin, in SI, and the Rankine (R) in the American Engineering System. Conversion from the commonly used Celsius and Fahrenheit temperature scales to the absolute or thermodynamic scales is as follows:

$$K = {}^{\circ}C + 273.16$$
$${}^{\circ}R = {}^{\circ}F + 460$$

In common usage, the conversion to Kelvin is rounded off to:

$$K = {}^{\circ}C + 273$$

The absolute temperature is used in equations of state for gases.

### 3.1.3 Quantity of Gases

The quantity of gases is often expressed as volume at a specific temperature and pressure. Because molecules of a gas have very weak attraction toward each other, they will disperse and occupy all space within a confining vessel. Thus, the volume (V) of a gas is the volume of the confining vessel and is not definitive of the quantity of the gas unless the pressure and temperature are also specified.

The most definitive quantification of gases is by mass, or as the number of moles, n, which is the quotient of mass and the molecular weight. The unit of mass is prefixed to mole (e.g., kgmole) to indicate the ratio of mass in kg and the molecular weight. The number of molecules in 1 gmole is the *Avogadro number*, $6.023 \times 10^{23}$ molecules/gmole.

At 273 K and 760 mm Hg pressure (101.325 kPa), 1 gmole, kgmole, and lbmole occupies 22.4 L, 22.4 m³, and 359 ft³, respectively.

## 3.1.4 The Ideal Gas Equation

The ideal gas equation is the simplest equation of state and was originally derived from the experimental work of Boyle, Charles, and Gay-Lussac. The equation will be derived based on the kinetic theory of gases.

Pressure, the force of collision between gas molecules and a surface, is directly proportional to temperature and the number of molecules per unit volume. Expressing the proportionality as an equation and using a proportionality constant, R:

$$P \propto \frac{n}{V}T; P = R\frac{n}{V}T$$

Rearranging:

$$PV = nRT \tag{3.3}$$

Equation (3.3) is the *ideal gas equation*. R is the gas constant and has values of 0.08206 L(atm)/ (gmole . K); or 8315 N(m)/(kgmole . K) or 1545 ft(lb$_f$)/(lbmole . °R).

### 3.1.4.1 P-V-T Relationships for Ideal Gases

When a fixed quantity of a gas that follows the ideal gas equation (Eq. 3.3) undergoes a process where the volume, temperature, or pressure is allowed to change, the product of the number of moles n and the gas constant R is a constant and:

$$\frac{PV}{T} = \text{constant}$$

If initial temperature, pressure, and volume are known and designated by subscript 1, these properties at another point in the process will be expressed by:

$$\frac{PV}{T} = \frac{P_1V_1}{T_1} \tag{3.4}$$

Ideality of gases predicted by Eqs. (3.3) and (3.4) occur only under conditions close to ambient or under processes that span a relatively narrow range of temperature and pressure. Thus, these equations are useful in problems usually encountered when using gases in food processing or packaging.

**Example 3.2:** Calculate the quantity of oxygen entering a package in 24 h if the packaging material has a surface area of 3000 cm$^2$ and an oxygen permeability 100 cm$^3$/(m$^2$)(24 h) STP (standard temperature and pressure = 0 °C and 1 standard atmosphere of 101.325 kPa).

**Solution**

Solving for the volume of oxygen permeating through the package in 24 h:

$$V = \frac{100\ \text{cm}^3}{\text{m}^2(24\text{h})}(24\text{h})\frac{1\,\text{m}^2}{(100)^2\text{cm}^2}(3000\,\text{cm}^2)$$

$$= 30\,\text{cm}^3$$

Using Eq. (3.3): n = PV/RT
Use R = 0.08206 L(atm)/(gmole . K)

moles oxygen =

$$n = \frac{1\,\text{atm}(30\ \text{cm}^3)}{[0.08206\,\text{L(atm)}/(\text{gmole}\cdot\text{K})](273\,\text{K})}\frac{1\,\text{L}}{1000\,\text{cm}^3}$$

$$= 0.001339\,\text{gmoles}$$

**Example 3.3:** Calculate the volume of $CO_2$ in ft$^3$ at 70 °F and 1 atm, which would be produced by vaporization of 1 lb of dry ice.

**Solution**

Dry ice is solid $CO_2$ (mol. wt. = M = 44 lb/ lbmole). T = 70 °F + 460 = 530 °R.

$$n = \frac{W}{M} = \frac{1\,\text{lb}}{44\text{lb/lbmole}} = 0.02273\,\text{lbmole}$$

$$P = 1\,\text{atm} = \frac{14.7\text{lb}_f}{\text{in.}^2}\frac{144\,\text{in}^2}{\text{ft}^2} = 2116.8\,\text{lb}_f/\text{ft}^2$$

Substituting in the ideal gas equation:
V = nRT/P

$$V = 0.02273\ \text{lbmole}\frac{1545\,\text{ft}\,\text{lb}_f}{\text{lbmole}(°R)}\frac{530°\text{R}}{2116\,\text{lb}_f/\text{ft}^2}$$

$$= 8.791\,\text{ft}^3$$

**Example 3.4:** Calculate the density of air (M = 29) at 70 ° F and 1 atm in (a) American Engineering and (b) SI units.

## Solution

Density is mass/volume = W/V. Using the ideal gas equation:

$$\text{Density} = \frac{W}{V} = \frac{PM}{RT}$$

(a) In the American Engineering System of units: $P = 2116.8 \text{ lb}_f/\text{ft}^2$; $M = 29 \text{ lb/lbmole}$; $R = 1545 \text{ ft lb}_f/\text{lbmole (°R)}$; and $T = 70 + 460 = 530 \text{ °R}$.

$$\text{Density} = \frac{2116.8 \, \text{lb}_f/\text{ft}^2}{1545 \text{ ft lb}_f/(\text{lbmole} \cdot \text{R})} \frac{29 \, \text{lb/lbmole}}{530° \text{R}}$$

$$= 0.07498 \ \text{lb/ft}^3$$

(b) in SI, $P = 101,325 \text{ N/m}^2$; $T = (70 - 32)/1.8 = 21.1 \text{ °C} = 21.1 + 273 = 294.1 \text{ K}$; $V = 1 \text{ m}^3$; $M = 29 \text{ kg/kgmole}$; and $R = 8315 \text{ Nm/(kgmole . K)}$.

$$\text{Density} = \frac{101,325 \, \text{N/m}^2}{8315 \ \text{Nm/(kgmole} \cdot \text{K)}} \frac{29 \, \text{kg/kgmole}}{294.1 K}$$

$$= 1.202 \, \text{kg/m}^3$$

**Example 3.5**  A process requires $10 \text{ m}^3/\text{s}$ at 2 atm absolute pressure and 20 °C. Determine the rating of a compressor in $\text{m}^3/\text{s}$ at STP (0 °C and 101,325 Pa) that must be used to supply air for this process.

## Solution

Unless otherwise specified, the pressure term "atm" means a standard atmosphere, or 101.325 kPa. Use the P-V-T equation (Eq. 3.4). $V_1 = 10 \text{ m}^3$; $T_1 = 293 \text{ K}$; $P_1 = 202,650 \text{ N/m}^2$; $P = 101,325 \text{ N/m}^2$; and $T = 273 \text{ K}$. Solving for V:

$$V = \frac{P_1 T_1}{T_1} \frac{T}{P}$$

$$= \frac{(202,650 \ \text{N/m}^2)(2\text{m}^3)}{293 \text{ K}} \frac{273 \text{ K}}{101,325 \text{N/m}^2}$$

$$= 18.635 \text{m}^3/\text{s}$$

**Example 3.6:**  An empty can was sealed in a room at 80 °C and 1 atm pressure. Assuming that only air is inside the sealed can, what will be the vacuum after the can and contents cool to 20 °C?

## Solution

The quantity of gas remains constant; therefore, the P-V-T equation can be used. The volume does not change; therefore, it will cancel out of Eq. (3.4). $P_1 = 101,325 \text{ N/m}^2$; $T_1 = 353 \text{ K}$; and $T = 293 \text{ K}$. The pressures to be used in Eq. (3.4) will be absolute.

$$P = \frac{101,325 \, \text{N/m}^2}{353 \text{ K}} (293 \, \text{K})$$

$$= 84,103 \, \text{Pa absolute}.$$

$$\text{Vacuum} = 17,222 \, \text{Pa} \left( \frac{1 \text{cm Hg}}{1333.33 \text{ Pa}} \right)$$

$$= 12.91 \, \text{cm Hg vacuum}$$

The vacuum in Pa will be

$$101,325 - 84,103$$

$$= 17,222 \, \text{Pa}.$$

### 3.1.5   Van Der Waals Equation of State

The ideal gas equation is based on unhindered movement of gas molecules within the confined space; therefore, at constant temperature when molecular energy is constant, the product of pressure and volume is constant. However, as pressure is increased, molecules are drawn closer, and attractive and repulsive forces between molecules affect molecular motion. When molecules are far apart, attractive forces exist. The magnitude of this attractive force is inversely proportional to the square of the distance between molecules. When molecules collide, they approach a limiting distance of separation when repulsive forces become effective, preventing molecules from directly contacting each other. Molecular contact may cause a chemical reaction, and this occurs only at very high molecular energy levels that exceed the repulsive force. The separation distance between molecules where repulsive forces are effective defines an exclusion zone, which reduces the total volume available for molecules to randomly move. The attractive forces between molecules also restrict molecular motion, and this will have the effect of reducing the quantity and magnitude of impact against the walls of the confining vessel. These attractive forces are referred to as an internal pressure. At low pressures, molecular distance is large, the internal pressure is small, and the excluded volume is small compared with the total volume, and gas molecules obey the ideal gas equation. However, at high pressures, the pressure-volume-temperature relationship deviates from ideality. Gases that deviate from ideal gas behavior are considered real gases. One of the commonly used equations of state for real gases is the van der Waals equation.

Van der Waals proposed corrections to the ideal gas equation based on the excluded volume, nb, and a factor, $n^2a/V^2$, the internal pressure. The fit between experimental P-V-T relationship and calculated values using van der Waals equation of state is very good except in the region of temperature and pressure near the critical point of the gas. For n moles of gas, van der Waals equation of state is:

$$\left(P + \frac{n^2a}{V^2}\right)(V - nb) = nRT \qquad (3.5)$$

Values of the constants a and b in SI units are given in Table 3.1.

**Example 3.7** Calculate the density of air at 150 °C and 5 atm pressure using van der Waals equation of state and the ideal gas equation.

**Table 3.1** Values of van der Waals constants for different gases

| Gas | a Pa $(m^3/kgmole)^2$ | b $m^3/kgmole$ |
|---|---|---|
| Air | $1.348 \times 10^5$ | 0.0366 |
| Ammonia | $4.246 \times 10^5$ | 0.0373 |
| Carbon dioxide | $3.648 \times 10^5$ | 0.0428 |
| Hydrogen | $0.248 \times 10^5$ | 0.0266 |
| Methane | $2.279 \times 10^5$ | 0.0428 |
| Nitrogen | $1.365 \times 10^5$ | 0.0386 |
| Oxygen | $1.378 \times 10^5$ | 0.0319 |
| Water vapor | $5.553 \times 10^5$ | 0.0306 |

*Source*: Calculated from values in the International Critical Tables

**Solution**
Using the ideal gas equation:
When V = 1 m³

$$\text{Density} = W = \frac{PM}{RT};$$

$$= \frac{5(101,325)(29)}{8315(150 + 273)}$$

$$= 4.1766 \, kg/m^3$$

Using van der Waals equation of state: V = 1 m³. Substituting $\frac{W}{M}$ for n in Eq. (3.5)

$$\left(P + \frac{W^2a}{M^2V^2}\right)\left(V - \frac{W}{M}b\right) = \frac{W}{M}RT$$

$$W^3\left(\frac{ab}{M^3V^3}\right) - W^2\left(\frac{a}{M^2V}\right)$$

$$+ W\left(\frac{Pb + RT}{M}\right) - PV = 0$$

Expanding and collecting like terms:
Substituting $a = 1.348 \times 10^5$; $b = 0.0366$; $R = 8315$; $M = 29$; $P = 5$; $T = 150 + 273 = 423$ K; and V = 1 m³.

$$0.2023 \, W^3 - 160.3 \, W^2$$

$$+ 121,923.7 \, W - 506,625 = 0$$

Dividing through by 0.2023:

$$W^3 - 792.3875 \, W^2 + 602,687.6 \, W$$

$$- 2,504,325.3 = 0$$

Solve using the Newton-Raphson iteration technique:

$$f = W^3 - 792.3875 \, W^2$$

$$+ 602,687.6 \, W - 2,504,325.3$$

$$f' = 3 \, W^2 - 1584.775 \, W + 602,687.6$$

Thus, the mass of 1 m³ of air at 150 °C and 5 atm is 4.178 kg = the density in kg/m³. Values calculated using the van der Waals equation of state is more accurate than that obtained using the ideal gas equation. A worksheet in Microsoft Excel to solve this problem is shown in Fig. 3.2.

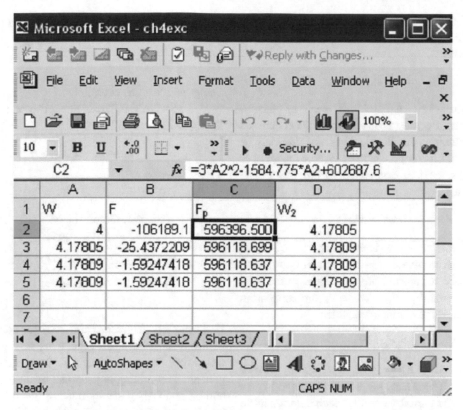

**Fig. 3.2** Spreadsheet for Newton-Raphson method and van der Waals equation to calculate air density

### 3.1.6   Critical Conditions for Gases

When the pressure of a gas is increased by compression, the molecules are drawn closer together and attractive forces between the molecules become strong enough to restrict movement of the molecules. At a certain temperature and pressure, a saturation point of the gas is reached, and an equilibrium condition between gas and liquid can exist. If the energy level in the gas is reduced by removal of the latent heat of vaporization, it will condense into a liquid. The higher the pressure, the higher the saturation temperature of the gas.

If the pressure of the gas is increased as the temperature is maintained following the saturation temperature curve, a point will be reached where gas and liquid become indistinguishable. This particular temperature and pressure is called the critical point. The property of the gas at the critical point is very similar to that of a liquid in terms of dissolving certain solutes, and this property is put into practical use in a process called supercritical fluid extraction, which will be discussed in more detail in Chap. 14.

### 3.1.7   Gas Mixtures

In this section, the concept of partial pressures and partial volumes will be used to elucidate P-V-T relationships of individual components in a gas mixture.

If components of a gas mixture at constant volume are removed one after the other, the drop in pressure accompanying complete removal of one component is the *partial pressure* of that component.

If $P_t$ is the total pressure and $P_a$, $P_b$, $P_c$ ... $P_n$ are partial pressures of the components a, b, c, ... and n, then:

$$P_t = P_a + P_b + Pc + \cdots P_n \qquad (3.6)$$

Equation (3.6) is *Dalton's law of partial pressures*. Because the same volume is occupied by all components, the ideal gas equation may be used on each component to determine the number of moles of that component, from the partial pressure.

$$P_a V = n_a RT \qquad (3.7)$$

The *partial volume* is the change in volume of a gas mixture when each component is removed separately at constant pressure. If $V_t$ is the total volume, and $V_a$, $V_b$, $V$, ... $V_n$ are the partial volumes of the components, then:

$$V_t = V_a + V_b + V_c + \cdots V_n \qquad (3.8)$$

Equation (3.8) is *Amagat's law of partial volumes*. The ideal gas equation may also be used on the partial volume of each component to determine the number of moles of that component in the mixture.

$$PV_a = n_a RT \qquad (3.9)$$

Compositions of gases are expressed as a percentage of each component by volume. It may be seen from Eq. (3.9) that the volume percentage of a gas is numerically the same as the mole percentage.

**Example 3.8** Calculate the quantity of air in the headspace of a can at 20 °C when the vacuum in the can is 10 in. Hg. Atmospheric pressure is 30 in. Hg. The headspace has a volume of 16.4 cm$^3$. The headspace is saturated with water vapor.

**Solution**
The steam tables (Appendix Table A.4) will give the vapor pressure of water at 20 °C of 2336.6 Pa. Let $P_t$ be the absolute pressure inside the can.

$$P_t = (30 - 10)\text{in.Hg}\left(\frac{3386.38 \text{ Pa}}{\text{in.Hg}}\right)$$
$$= 67,728 \text{ Pa}$$

Using the ideal gas equation:
$$V = 16.4 \text{ cm}^3(10^{-6})\text{m}^3/\text{cm}^3 = 1.64 \times 10^{-5} \text{ m}^3;$$
$$T = 20 + 273 = 293 \text{ K}.$$

$$P_{air} = P_t - P_{water}$$

$$n_{air} = \frac{P_{air}V}{RT}$$

$$= \frac{(65,391.4 \text{N/m}^3)(1.64 \times 10^{-5}\text{m}^3)}{8315/(293)}$$

$$= 67728 - 2336.6 = 65391.4 \text{ Pa}$$

$$= 4.40 \times 10^{-7} \text{kgmoles}$$

**Example 3.9** A gas mixture used for controlled atmosphere storage of vegetables contains 5% $CO_2$, 5% $O_2$, and 90% $N_2$. The mixture is generated by mixing appropriate quantities of air and $N_2$ and $CO_2$ gases. 100 m$^3$ of this mixture at 20 °C and 1 atm is needed per hour. Air contains 21% $O_2$ and 79% $N_2$. Calculate the rate at which the component gases must be metered into the system in m$^3$/h at 20 °C and 1 atm.

**Solution**
All percentages are by volume. No volume changes occur on mixing of ideal gases. Because volume percent in gases is the same as mole percent, material balance equations may be made on the basis of volume and volume percentages. Let X = volume $O_2$, Y = volume $CO_2$, and Z = volume $N_2$, fed into the system per hour.

Oxygen balance : 0.21(X) = 100(0.05);
$$X = 23.8 \text{ m}^3$$

$CO_2$ balance : Y = 0.05(100);
$$Y = 5 \text{ m}^3$$

Total volumetric balance :
X + Y + Z = 100
Z = 100 - 23.8 - 5 = 71.2 m$^3$

## 3.2 Thermodynamics

Thermodynamics is a branch of science that deals with energy exchange between components within a system or between a system and its surroundings. A *system* is any matter enclosed within a boundary. The boundary may be real or imaginary and depends solely upon the part of the process under study. Anything outside the boundary is the *surroundings*. The *properties* of a system determine its state just as a state may imply certain properties. Properties may be *extrinsic* (i.e., it is measurable) or it may be intrinsic (i.e., no measurable external manifestations of that property exist). Changes in the intrinsic properties, however, may be measured through changes in energy associated with the change.

*Equilibrium* is a fundamental requirement in thermodynamic transformations. When a system reaches a condition where properties remain constant, a state of equilibrium is attained. The science of thermodynamics deals with a system in equilibrium. Thermodynamics cannot predict how long it will take for equilibrium to occur within a system; it can only predict the final properties at equilibrium.

### 3.2.1 Thermodynamic Variables

Energy involved in thermodynamic transformations is expressed in terms of *heat* (Q), the energy that crosses a system's boundaries due to a difference in temperature, and *work* (W), the energy associated with force displacement. The term "internal energy" (E) is used to define an intrinsic property, which is energy not associated with either work or heat. Internal energy is not measurable, but changes in internal energy can be measured. An intrinsic property *entropy* (S) is also possessed by a system and is a measure of the disorder that exists within a system. The first and second laws of thermodynamics are based on the relationships of the above main thermodynamic variables.

If $\delta Q$ and $\delta W$ are small increments of heat and work energy that crosses a system's boundaries, the accompanying differential change in the internal energy of the system is:

$$dE = \delta Q - \delta W \tag{3.10}$$

Equation (3.10) is also expressed as:

$$\Delta E = Q - W \tag{3.11}$$

Equations (3.10) or (3.11) are expressions for the *first law of thermodynamics*, a fundamental relationship also known as the law of conservation of energy. The symbol $\delta$, called "del," operating on W and Q is not a true differential but rather a finite difference. Although it is considered in mathematical operations as a differential operator, integration produces an increment change in a function, rather than an absolute value. This principle is used when integrating functions such as E and S, which do not have absolute values but for which increment change associated with a process can be calculated.

Entropy cannot be measured, but the change in entropy is defined as the ratio of the reversible energy ($Q_{rev}$) which crossed a system's boundaries and the absolute temperature.

$$dS = \frac{\delta Q_{rev}}{T}; \quad \Delta S = \frac{Q_{rev}}{T} \tag{3.12}$$

The *second law of thermodynamics* states that any process that occurs is accompanied by a positive entropy change, and the entropy change approaches zero for reversible processes. If the process is reversible, any change in entropy in a system is compensated by a change in the entropy of the surroundings such that the net entropy change for the system and its surroundings is zero.

The concept of the entropy or state of disorder of a system has several interpretations. One interpretation is that no truly reversible process is possible, because any process produces a more disordered system than what existed before the change. This change in the state of order requires the application of or a loss of energy. Another interpretation of the concept of entropy involves the spontaneity of processes. Spontaneous processes will occur only when there is enough energy available initially to overcome the requirement for increasing the entropy. Thus, process spontaneity always requires a change from a higher to a lower energy state.

Another intrinsic thermodynamic variable is the enthalpy (H) defined as:

$$H = E + PV \qquad (3.13)$$

In differential form:

$$dH = dE + PdV + VdP \qquad (3.14)$$

Because $PdV = \delta W$; $dE + \delta W = \delta Q$ and:

$$dH = \delta Q + VdP \qquad (3.15)$$

For a constant pressure process, $dP = 0$ and:

$$\delta Q = dH; \delta H = Q \qquad (3.16)$$

A specific heat at constant pressure may be defined as follows:

$$C_p = \left.\frac{dQ}{dT}\right|_p \qquad (3.17)$$

and as:

$$\Delta H = C_p dT \qquad (3.18)$$

Equation (3.18) shows why the enthalpy is referred to as the heat content. In a constant volume process, work is zero, and Eq. (3.10) becomes:

$$dE = \delta Q; \Delta E = Q \qquad (3.19)$$

A specific heat at constant volume may be defined as follows:

$$C_v = \left.\frac{dQ}{dT}\right|_v \qquad (3.20)$$

and as:

$$\Delta E = C_v dT \qquad (3.21)$$

## 3.2.2   The Relationship Between $C_p$ and $C_v$ for Gases

The relationship between $C_p$ and $C_v$ for gases is derived as follows. Substituting $W \cong PdV$ for work in Eq. (3.10) for a constant pressure process:

$$dE = dQ - PdV \qquad (3.22)$$

Taking the derivative with respect to temperature:

$$\frac{dE}{dT} = \left.\frac{dQ}{dT}\right|_p - P\frac{dV}{dT} \qquad (3.23)$$

Equation (3.17) gives $dQ/dT = C_p$ and Eq. (3.21) gives $dE/dT = C_v$. From the ideal gas equation, for 1 mole of gas, $PV = RT$, and $dV/dT$ at constant pressure is $R/P$. Substituting in Eq. (3.23):

$$C_v = C_p - R \qquad (3.24)$$

A useful property in calculating P-V-T and other thermodynamic variables involved in expansion and compression of a gas is the specific heat ratio, $C_p/C_v$, designated by the symbol $\gamma$. The ratio $C_p/R$ is expressed in terms of $\gamma$ as follows:

$$\frac{C_p}{R} = \frac{\gamma}{\gamma - 1} \qquad (3.25)$$

## 3.2.3   P-V-T Relationships for Ideal Gases in Thermodynamic Processes

*Adiabatic processes* are those where no heat is added or removed from the system; therefore, $\delta Q = 0$. Eq. (3.15) then becomes: $dH = V\,dP$. Because for 1 mole of an ideal gas, $V = RT/P$, and because $dH = C_p dT$,

$$C_p dT = RT \left(\frac{dp}{P}\right) \qquad (3.26)$$

$$\ln \left[\frac{P-2}{P_1}\right] = \frac{\gamma}{\gamma-1} \ln \left[\frac{T_2}{T_1}\right]$$

$$\frac{C_p}{R} \int_{T_1}^{T_2} \frac{dT}{T} = \int_{P_1}^{P_2} \frac{dP}{P} \qquad (3.27)$$

Integrating and substituting $C_p/R = \gamma/(\gamma - 1)$.

The ideal gas equation may be used to substitute for P in Eq. (3.27) to obtain an expression for V as a function of T in an adiabatic process. Substitute: $P_1 = nRT_1/V_1$ and $P_2 = nRT_2/V_2$ in Eq. (3.27) and simplifying:

$$\frac{P_2}{P_1} = \left[\frac{T_2}{T_1}\right]^{\frac{\gamma}{\gamma-1}} \qquad (3.28)$$

$$\frac{V_1}{V_2} = \left[\frac{T_2}{T_1}\right]^{\frac{1}{\gamma-1}} \qquad (3.29)$$

Equations (3.28) and (3.29) can be used to derive:

$$\left[\frac{V_1}{V_2}\right]^{\gamma} = \left[\frac{P_2}{P_1}\right] \qquad (3.30)$$

*Isothermal processes* are those where the temperature is maintained constant. Thus, $P_1 V_1 = P_2 V_2$.

*Isobaric processes* are those where the pressure is maintained constant. Thus, $V_1/T_1 = V_2/T_2$.

*Isocratic processes* are those where the volume is maintained constant. Thus, $P_1/T_1 = P_2/T_2$.

### 3.2.4 Changes in Thermodynamic Properties, Work, and Heat Associated with Thermodynamic Processes

Adiabatic:

$$\Delta Q = 0; \Delta S = 0;$$
$$\Delta E = W = \int PdV; \Delta H = \int VdP$$

Isothermal:

$$Q = W = \int PdV; \Delta S = \frac{Q}{T}; \Delta E = 0; \Delta H = O$$

Isobaric:

$$Q = \int C_p dT = \Delta H;$$

$$\Delta S = \int C_p \frac{dT}{T}; W = P\Delta V; \Delta E = Q - W$$

Isocratic:

$$Q = \Delta E = \int C_V dT; \Delta S = \int C_V \frac{dT}{T};$$
$$W = 0; \Delta E = Q$$

### 3.2.5 Work and Enthalpy Change on Adiabatic Expansion or Compression of an Ideal Gas

Work and enthalpy changes are important in determining power input into compressors. Adiabatic compression is what occurs during the compression of a refrigerant in a refrigeration cycle.

In adiabatic compression or expansion:

$$W = \int PdV; \Delta H = \int VdP \qquad (3.31)$$

Equation (3.30) is rearranged, substituting P and V for $P_2$ and $V_2$.

Solving for V:

$$V = \left[(P_1)^{\frac{1}{\gamma}} V_1\right](P)^{\frac{-1}{\gamma}}$$

Differentiating with respect to P:

$$dV = \frac{-1}{\gamma} \left[(P_1)^{\frac{1}{\gamma}} V_1\right][P]^{\frac{-1}{\gamma-1}}$$

Substituting dV in the expression for work and integrating:

$$W = \int_{P_1}^{P_2} -\frac{1}{\gamma} \left[(P_1)^{\frac{1}{\gamma}} V_1\right][P]^{\frac{-1}{\gamma-1+1}}$$

The integral is

$$-\frac{1}{\frac{1}{\gamma}+1}\left[\left[(P_2\right]^{\frac{-1}{\gamma}+1}-\left[P_1\right]^{\frac{-1}{\gamma}+1}\right]$$

$$=-\frac{(P_1)^{\frac{1}{\gamma}}V_1}{\gamma}\int_{P_1}^{P_2}(P)^{\frac{1}{\gamma}}dP$$

Combining the integral with the multiplier and simplifying:

$$W=\frac{P_1V_1}{1-\gamma}\left[\left[\frac{P_2}{P_1}\right]^{\frac{(\gamma-1)}{\gamma}}-1\right] \qquad (3.32)$$

The enthalpy change is determined by evaluating the following integral:

$$\Delta H=V_1\int_{P_1}^{P_2}\left[\frac{P_1}{P}\right]^{\frac{1}{\gamma}}dP \qquad (3.33)$$

### 3.2.6   Work and Enthalpy Change on Isothermal Expansion or Compression of an Ideal Gas

$$\Delta H=\left[\frac{\gamma}{\gamma-1}\right](P_1V_1)\left[\left[\frac{P_2}{P_1}\right]^{\frac{(\gamma-1)}{\gamma}}-1\right] \qquad (3.34)$$

When an ideal gas is subjected to an isothermal process

$$Q=W=\int PdV \, and \, \Delta H=\int PdV+\int VdP$$

Because T is constant, $P=nRT/V$ and $dP=-(nRTV^{-2})\,dV$

$$Q=W=\int_{V_1}^{V_2}nRT\frac{dV}{V}$$

$$Q=nRT\ln\left[\frac{V_2}{V_1}\right] \qquad (3.35)$$

$$\Delta H=W+\int_{V_1}^{V_2}-\left[nRT\frac{dV}{V}\right]$$

The second term in the expression for $\Delta H$ is exactly the same as the expression for W but has a negative sign; therefore, in an isothermal process, $\Delta H=0$.

**Example 3.10** $N_2$ gas trapped inside a cylinder with a movable piston at 80 atm pressure is then allowed to expand adiabatically (no heat added or removed) until the final pressure is 1 atm. The gas is initially at 303 K and the initial volume is l liter. Calculate (a) the work performed by the gas, (b) the entropy change, (c) the enthalpy change, and (d) change in internal energy. The gas has a specific heat ratio, $\gamma$, of 1.41.

**Solution**

(a)  Equation (3.32) will be used to calculate the work. $V_1=0.001\,m^3$; $P_1=80(101325)=8{,}106{,}000$ Pa; and $P_2=101{,}325$ Pa. From Eq. (3.32):

$$W=\frac{(8,106,000Pa)(0.001\,m^3)}{-0.41}$$

$$\left[\left(\frac{101,325}{8,106,0000}\right)^{\frac{0.41}{1.41}}-1\right]$$

$$=-14241.8J$$

(b)  $\Delta S=0$

(c)  From Eq. (3.34):

$$\Delta H=\left[\frac{1.41}{0.41}(8,106,000\,Pa)(0.001\,m^3)\right]$$

$$\left[\left(\frac{101,325}{8,106,0000}\right)^{\frac{0.41}{1.41}}-1\right]=-20,080.9\,J$$

(d)  $\Delta E=W=$ - 14241.7 J

## 3.3   Vapor-Liquid Equilibrium

The molecular attraction that holds liquid molecules together is not strong enough to prevent some molecules from escaping; therefore, some molecules will escape in the gaseous form. If the volume of the system containing the liquid and vapor is held constant, eventually equilibrium will be attained when the rate of escape of molecules from the liquid phase to become vapor equals the rate at which the vapor molecules are recaptured by the liquid phase. This condition of equilibrium exists in all liquids. A liquid will always maintain an envelope of vapor around its surface, and the pressure exerted by that vapor is known as the *vapor pressure*. The pressure exerted by molecules of vapor in equilibrium with a liquid is a function only of temperature. In the absence of any other gas exerting pressure on the liquid surface (e.g., when the liquid is introduced into a container which is under a perfect vacuum), the pressure of the vapor at equilibrium is the vapor pressure. If the total pressure above a liquid is maintained constant at the vapor pressure, heating will not increase the temperature but will cause more liquid molecules to enter the vapor phase. The temperature at a pressure that corresponds to the vapor pressure of a liquid is the *boiling point*, a condition where the whole atmosphere over the liquid surface consists only of gaseous molecules of that liquid. The heat added to the system to generate a unit mass of vapor from liquid at the boiling point is the *heat of vaporization*. When the pressure above the liquid is higher than the vapor pressure, some other gas molecule (e.g., air) would be exerting that pressure in addition to the vapor molecules; therefore, the vapor pressure will be the partial pressure of vapor in the gas mixture surrounding the liquid.

### 3.3.1   The Clausius-Clapeyron Equation

The temperature dependence of the vapor pressure is expressed by Eq. (3.36), the Clausius-Clapeyron equation:

$$\ln\left[\frac{P}{P_1}\right] = \left[\frac{\Delta H_V}{R}\right]\left[\frac{T - T_1}{TT_1}\right] \qquad (3.36)$$

where $\Delta H_v$ is heat of vaporization, P is vapor pressure at temperature T, R is gas constant, and $P_1$ is vapor pressure at temperature $T_1$. Equation (3.36) assumes constant heat of vaporization, but in reality, this quantity changes with temperature. Thus, the most useful function of Eq. (3.36) is in interpolating between values of vapor pressure

listed in abbreviated tables where temperature interval between entries in the table is quite large. The steam tables (Appendix Tables A.3 and A.4) give values of the vapor pressure for water at different temperatures. Because temperature intervals in these tables are rather close, a linear interpolation may be used, instead of a logarithmic function interpolation as suggested by Eq. (3.36).

### 3.3.2   Liquid Condensation from Gas Mixtures

When the partial pressure of component in a gas mixture exceeds the vapor pressure of that component, condensation will occur. If the mixture is in a closed container where the volume is constant, condensation will result in a decrease in total pressure. This principle is responsible for the vacuum that results when a hot liquid is filled into a can and sealed. Replacement of air in the headspace by vapors from the product itself or by steam in a system where steam is flushed over the can headspace prior to sealing results in a reduced pressure after the can cools adequately to condense the vapors in the headspace.

**Example 3.11** A canned food at the time of sealing is at a temperature of 80 °C, and the atmospheric pressure is 758 mm Hg. Calculate the vacuum (in mm Hg) formed inside the can when the contents cools down to 20 °C.

**Solution**
Assume there are no dissolved gases in the product at the time of sealing; therefore, the only gases in the headspace are air and water vapor. From Appendix Table A.3, the vapor pressure of water at 20 °C and 80 °C are 2.3366 and 47.3601 kPa, respectively. In the gas mixture in the headspace, air is assumed to remain at the same quantity in the gaseous phase, while water condenses on cooling.

$$P_{air} = P_t - P_{water}$$

At 80 °C,

$$P_t = 758 \, \text{mm Hg} \left( \frac{1 \, \text{cm Hg}}{10 \, \text{mm Hg}} \right) \left( \frac{1333.33 \, P_a}{\text{cm Hg}} \right)$$
$$= 101,066 P_a$$

Let P = pressure in the headspace at 20 °C.

$$n_{air} = \frac{PV}{RT} = \frac{(101,064 - 47,360.1)V}{8315(80 + 273)}$$
$$= 0.018296 \, V \, \text{kgmole}$$

At 20 °C,

$$n_{air} = \frac{PV}{8315(293)}$$
$$= 4.1046 \times 10^{-7} PV \, \text{kgmole}$$

Because the number of moles of air trapped in the headspace is constant:

$$4.1046 \times 10^{-7} P \, V = 0.018296 \, V$$

P = 44575 Pa absolute = 334  mm Hg absolute

Vacuum = 758 − 334 = 424  mm Hg vacuum

**Example 3.12**  Air at 5 atm pressure is saturated with water vapor at 50 °C. If this air is allowed to expand to 1 atm pressure and the temperature is dropped to 20 °C, calculate the amount of water that will be condensed per m³ of high pressure air at 50 °C.

**Solution**
The vapor pressure of water at 50 °C and 20 °C are 12.3354 and 2.3366 kPa, respectively. Basis: 1 m³ air at 5 atm pressure and 50 °C. The number of moles of air will remain the same on cooling.

$$n_{air} = \frac{[5(101,325) - 12,335.4]1}{8315(50 + 273)}$$
$$= 0.1840 \, \text{kgmole}$$

At 20 °C,

$$n_{air} = \frac{(101,325 - 2336)V}{8315(20 + 273)}$$
$$= 0.04063 \, V$$

Equating: 0.04063 V = 0.1840; V = 4.529 m³ at 20 °C and 1 atm.

At 50 °C,

$$n_{water} = \frac{12,335.4(1)}{8315(323)} = 0.004593$$

At 20 °C,

$$n_{water} = \frac{2336.6(4.529)}{8315(293)} = 0.004344$$

Moles water condensed = 0.004593 − 0.004344 = 0.000249 kg moles.

**Example 3.13**  The partial pressure of water in air at 25 °C and 1 atm is 2.520 kPa. If this air is compressed to 5 atm total pressure to a temperature of 35 °C, calculate the partial pressure of water in the compressed air.

**Solution**
Increasing the total pressure of a gas mixture will proportionately increase the partial pressure of each component. Using the ideal gas equation (Eq. 3.3) for the mixture and for the water vapor, let $V_1$ = the volume of the gas mixture at 25 °C and 1 atm; $P_t$ = total pressure; and $P_w$ = partial pressure of water vapor.

The total number of moles of air and water vapor is:

$$n_t = \frac{P_1 V_1}{RT_1} = \frac{P_2 V_2}{RT_2}$$

The number of moles of water vapor is:

$$n_{water} = \frac{P_{w1} V_1}{RT_1} = \frac{P_{w2} V_2}{RT_2}$$

Assuming no condensation, the ratio $n_t/n_w$ will be the same in the low-pressure and high-pressure air; therefore:

$$\frac{P_1}{P_{w1}} = \frac{P_2}{P_{w2}}; \quad P_{w2} = P_{w1} \frac{P_2}{P_1}$$
$$P_{w2} = \frac{5 \, \text{atm}}{1 \, \text{atm}} (2.520 \, \text{kPa}) = 12.60 \, \text{kPa}$$

The temperature was not used in calculating the final partial pressure of water. The temperature is used in verifying the assumption of no condensation, comparing the calculated final partial pressure with the vapor pressure of water at 35 °C. From the steam tables, Appendix Table A.3, the vapor pressure of water at 35 °C is the pressure of saturated steam at 35 °C, which is 5.6238 kPa. Because the calculated partial pressure of water in the compressed air is greater than the vapor pressure, condensation of water must have occurred and the correct partial pressure of water will be the vapor pressure at 35 °C. Thus, $P_{w2} = 5.6238$ kPa.

## Problems

3.1. Air used for dehydration is heated by burning natural gas and mixing the combustion products directly with air. The gas has a heating value of 1050 BTU/ft$^3$ at 70°F and 1 atm pressure. Assume the gas is 98% methane and 2% nitrogen.
   (a) Calculate the quantity of natural gas in ft$^3$ at 70 EF and 1 atm needed to supply the heating requirements for a dryer that uses 1500 lb dry air per hour at 170°F and 1 atm. Assume the products of combustion will have the same specific heat as dry air of 0.24 BTU/(lb. °F).
   (b) If the air used to mix with the combustion gases is completely dry, what will be the humidity of the air mixture entering the dryer?
3.2. A package having a void volume of 1500 cm$^3$ is to be flushed with nitrogen to displace oxygen prior to sealing. The process used involved drawing a vacuum of 700 mm Hg on the package, breaking the vacuum with nitrogen gas, and drawing another 700 mm Hg vacuum before sealing. The solids in the package prevent total collapse of the package as the vacuum is drawn; therefore, the volume of gases in the package may be assumed to remain constant during the process. If the temperature is maintained constant at 25 °C during the process, calculate the number of gmoles of oxygen left in the package at the completion of the process. Atmospheric pressure is 760 mm Hg.

3.3. Compression of air in a compressor is an adiabatic process. If air at 20°C and 1 atm pressure is compressed to 10 atm pressure, calculate:
   (a) The temperature of the air leaving the compressor.
   (b) The theoretical compressor horsepower required to compress 100 kg of air. The specific heat ratio of air is 1.40; the molecular weight is 29.
3.4. Air at 25 °C and 1 atm that contains water vapor at a partial pressure that is 50% of the vapor pressure at 25 °C (50% relative humidity) is required for a process. This air is generated by saturating room air by passing through water sprays, compressing this saturated air to a certain pressure, P, and cooling the compressed air to 25 °C. The partial pressure of water in the cooled saturated air that leaves the compressor is the vapor pressure of water at 25°C. This air is allowed to expand to 1 atm pressure isothermally. Calculate P such that after expansion, the air will have 50% relative humidity.
3.5. An experiment requires a gas mixture containing 20% $CO_2$, 0.5% $O_2$, and 79.5% $N_2$ at 1 atm and 20 °C. This gas mixture is purchased premixed and comes in a 10 L tank at a pressure of 130 atm gauge. The gas will be used to displace air from packages using a packaging machine that operates by drawing a vacuum completely inside a chamber where the packages are placed, displacing the vacuum with the gas mixture, and sealing the packages. The chamber can hold four packages at a time, and the total void volume chamber with the packages in place is 2500 cm$^3$. How many packages can be treated in this manner before the pressure in the gas tank drops to 1 atm gauge?
3.6. A vacuum pump operates by compressing gases from a closed chamber to atmospheric pressure in order that these gases can be ejected to the atmosphere. A vacuum drier operating at 700 mm Hg vacuum (atmospheric pressure is 760 mm Hg) and 50°C generates 500 g of water vapor per minute by

evaporation from a wet material in the dryer. In addition, the leakage rate for ambient air infiltrating the dryer is estimated to be 1 L/h at 1 atm and 20 °C.

(a) Calculate the total volume of gases that must be removed by the vacuum pump per minute.

(b) If the pump compresses the gas in an adiabatic process, calculate the theoretical horsepower required for the pump. The specific heat ratio for water is 1.30, and that for air is 1.40.

3.7. The mass rate of flow of air (G) used in correlation equations for heat transfer in a dryer is expressed in kg air/m$^2$(h). Use the ideal gas equation to solve for G as a function of the velocity of flow (V, in m/h) of air at temperature T and 1 atm pressure.

3.8. Use van der Waals equation of state to calculate the work done on isothermal expansion of a gas from a volume of 10 to 300 m$^3$ at 80°C. The initial pressure was 10 atm. Calculate the entropy change associated with the process.

3.9. A supercritical $CO_2$ extraction system is being operated at 30.6 MPa and 60 °C in the extraction chamber. The volume of $CO_2$ leaving the system measured at 101.3 kPa and 20 °C is 10 L/min. If the extraction chamber is a tube having a diameter of 50.6 mm and a length of 45 cm, calculate the residence time of the $CO_2$ in the extraction chamber. Residence time = volume of chamber/volumetric rate of flow in the chamber.

## Suggested Reading

Feldner, R. M., & Rosseau, R. W. (1999). *Elementary principles of chemical processes* (2nd ed.). New York: Wiley.

Himmelblau, D. M. (1967). *Basic principles and calculations in chemical engineering* (2nd ed.). Englewood Cliffs: Prentice-Hall.

Martin, M. C. (1986). *Elements of thermodynamics*. New York: Prentice-Hall.

Peters, M. S. (1954). *Elementary chemical engineering*. New York: McGraw-Hill Book Co..

# Energy Balances

"Energy" used to be a term that everybody took for granted. Now, to a layman, energy has been added to the list of the basic necessities of life. Increasing energy costs have forced people to recognize and appreciate the value of energy more than ever before. Energy conservation is being stressed not only in industrial operations but also in almost all aspects of an individual's daily activities.

Energy is not static; it is always in a stage of flux. Even under steady-state conditions, an object absorbs energy from its surroundings and at the same time emits energy to its surroundings at the same rate. When there is an imbalance between the energy absorbed and emitted, the steady state is altered, molecular energy of some parts of the system may increase, new compounds may be formed, or work may be performed. Energy balance calculations can be used to account for the various forms of energy involved in a system.

Energy audits are essential in identifying effectiveness of energy conservation measures and in identifying areas where energy conservation can be done. The technique is also useful in the design of processing systems involving heating or cooling to insure that fluids used for heat exchange are adequately provided and that the equipment is sized adequately to achieve the processing objective at the desired capacity. When energy exchange involves a change in mass due to evaporation or condensation, energy balances can be used during formulation such that after processing, the product will have the desired composition.

## 4.1 General Principles

An energy balance around a system is based on the first law of thermodynamics: the law of conservation of energy. Mechanical (work), electrical, and thermal energy can all be reduced to the same units. Mechanical input into a system to overcome friction, electrical energy, or electromagnetic energy such as microwaves will be manifested by an increase in the heat content of the system. Defining the surroundings of the system on which to make an energy balance is done similar to material balance calculations in Chap. 2. The basic energy balance equation is:

$$\text{Energy in} = \text{Energy out} + \text{Accumulation} \qquad (4.1)$$

If the system is in a steady state, the accumulation term is zero, whereas an unsteady-state system will have the accumulation term as a differential expression. When Eq. (4.1) is used, all energy terms known to change within the system must be accounted for. The heat contents are expressed as enthalpy based on increase in enthalpy from a set reference temperature. Mechanical, electrical, or electromagnetic inputs must all be accounted for if their effects on the

© Springer International Publishing AG, part of Springer Nature 2018
R. T. Toledo et al., *Fundamentals of Food Process Engineering*,
https://doi.org/10.1007/978-3-319-90098-8_4

total heat content are significant. If the system involves only an exchange of energy between two components, the energy balance will be:

Energy gain by component 1
  = Energy loss by component 2    (4.2)

Either Eq. (4.1) or (4.2) will give similar results if Eq. (4.2) is applicable. Equation (4.1) is a general form of the energy balance equation.

## 4.2   Energy Terms

The unit of energy in SI is the joule. Conversion factors in Appendix 1 can be used to convert mechanical and electrical energy units to the SI equivalent. Microwave and radiant energy absorbed by a material are usually expressed as a rate of energy flow, energy/time. Energy from ionizing radiation is expressed as an absorbed dose, the gray (Gy), which has the base units J/kg. Another accepted form for reporting absorbed ionizing radiation is the rad, 100 rd = 1 Gy.

### 4.2.1   Heat

Sensible heat is defined as the energy transferred between two bodies at different temperatures or the energy present in a body by virtue of its temperature. Latent heat is the energy associated with phase transitions, heat of fusion, from solid to liquid, and heat of vaporization, from liquid to vapor.

### 4.2.2   Heat Content, Enthalpy

Enthalpy, as defined in Chap. 3, is an intrinsic property, the absolute value of which cannot be measured directly. However, if a reference state is chosen for all components that enter and leave a system such that at this state the enthalpy is considered to be zero, then the change in enthalpy from the reference state to the current state of a component can be considered as the value of the

absolute enthalpy for the system under consideration. The reference temperature ($T_{ref}$) for determining the enthalpy of water in the steam tables (Appendices 3 and 4) is 32.018 °F or 0.01 °C. The enthalpy of any component of a system that would be equivalent to the enthalpy of water obtained from the steam tables, at any temperature T, is given by:

$$H = C_p(T - T_{ref})    \quad (4.3)$$

$C_p$ in Eq. (4.3) is the specific heat at constant pressure, previously defined in Chap. 3.

### 4.2.3   Specific Heat of Solids and Liquids

The specific heat ($C_p$) is the amount of heat that accompanies a unit change in temperature for a unit mass. The specific heat, which varies with temperature, is more variable for gases compared with liquids or solids. Most solids and liquids have a constant specific heat over a fairly wide temperature range. The enthalpy change of a material with mass m is:

$$q = m \int_{T_1}^{T_2} C_p dT    \quad (4.4)$$

Handbook tables give specific heats averaged over a range of temperature. When average specific heats are given, Eq. (4.4) becomes:

$$q = m \, C_{avg} \, (T_2 - T_1)    \quad (4.5)$$

For solids and liquids, Eqs. (4.3) and (4.5) are valid over the range of temperatures encountered in food processing systems.

Table 4.1 shows the average specific heats of various solids and liquids.

For fat-free fruits and vegetables, purees, and concentrates of plant origin, Siebel (1918) observed that the specific heat varies with moisture content and that the specific heat can be determined as the weighted mean of the specific heat of water and the specific heat of the solids.

**Table 4.1** Specific heat of food products

| Product | % $H_2O$ | $C_{pm}$ |
|---|---|---|
| Dairy products | | |
| Butter | 14 | 2050 |
| Cream, sour | 65 | 2930 |
| Milk, skim | 91 | 4000 |
| Fresh meat, fish, poultry, and eggs | | |
| Codfish | 80 | 3520 |
| Chicken | 74 | 3310 |
| Egg white | 87 | 3850 |
| Egg yolk | 48 | 2810 |
| Pork | 60 | 2850 |
| Fresh fruits, vegetables, and juices | | |
| Apples | 75 | 3370 |
| Apple juice | 88 | 3850 |
| Apple sauce | – | 3730 |
| Beans, fresh | 90 | 3935 |
| Cabbage, white | 91 | 3890 |
| Carrots | 88 | 3890 |
| Corn, sweet, kernels | – | 3320 |
| Cucumber | 97 | 4103 |
| Mango | 93 | 3770 |
| Orange juice, fresh | 87 | 3890 |
| Plums, fresh | 76.5 | 3500 |
| Spinach | 87 | 3800 |
| Strawberries | 91 | 3805 |
| Other products | | |
| Bread, white | 44 | 22,720 |
| Bread, whole wheat | 48.5 | 2850 |
| Flour | 13 | 1800 |

*Source*: Adapted from Polley, S. L. Snyder, O. P., and Kotnour, P. A. Compilation of thermal properties of foods. *Food Technol.* 36(1):76, 1980

For a fat-free plant material with a mass fraction of water M, the specific heat of water above freezing is 1 BTU/(lb . °F) or 4186.8 J/(kg . K), and that of nonfat solids is 0.2 BTU/(lb . °F) or 837.36 J/(kg . K). Because the mass fraction of nonfat solids is (1 − M), the weighted average specific heat for unit mass of material above freezing is:

$$C'_{avg} = 1(M) + 0.2(1 - M)$$
$$= (1 - 0.2)M + 0.2 = 0.8\,M \quad (4.6)$$
$$+0.2 \quad \text{in BTU}/(\text{lb}^{\circ}\text{F})$$

In SI:

$$C_{avg} = 3348\,M + 837.36 \quad \text{in J}/(\text{kg K}) \quad (4.7)$$

Equations (4.6) and (4.7) are different forms of Seibel's equation, which has been used by ASHRAE (1967) in tabulated values for specific heat of fruits and vegetables.

When fat is present, the specific heat above freezing may be estimated from the mass fraction fat (F), mass fraction solids nonfat (SNF), and mass fraction moisture (M), as follows:

$$C_{avg} = 0.4\,F + 0.2\,SNF$$
$$+ M \quad \text{in BTU}/(\text{lb}^{\circ}\text{F}) \quad (4.8)$$

$$C'_{avg} = 1674.72\,F + 837.36\,SNF$$
$$+ 4186.8\,M \quad \text{in J}/(\text{kg K}) \quad (4.9)$$

Below freezing, it is not suitable to use specific heats for the whole mixture because the amount of frozen and unfrozen water varies at different temperatures. It will be necessary to consider the latent heat of fusion of water, and sensible heats of liquid water and ice should be evaluated separately. Refer to the section "Enthalpy Changes in Foods During Freezing" later in this chapter.

Equations (4.8) and (4.9) are general and can be used instead of Eqs. (4.6) and (4.7).

**Example 4.1** Calculate the specific heat of beef roast containing 15% protein, 20% fat, and 65% water.

**Solution**

$$C'_{avg} = 0.15(0.2) + 0.20(0.4) + 0.65(1)$$
$$= 0.76 \text{ BTU}/(\text{lb}^{\circ}\text{F})$$
$$C_{avg} = 0.15(837.36) + 0.2(1674.72)$$
$$+ 0.65(4186.8) = 3182 \text{ J}/(\text{kg K})$$

**Example 4.2** Calculate the specific heat of orange juice concentrate having a solids content of 45%.

**Solution**

Using Siebel's equation, calculating a weighted average specific heat:

$$C'_{avg} = 0.2(0.45) + 1(0.55) = 0.64 \text{ BTU}/(\text{lb}^\circ\text{F})$$

$$C_{avg} = 837.36(0.45) + 4186.8(0.55) = 2679 \text{ J}/(\text{kg K})$$

**Example 4.3** Calculate the heat required to raise the temperature of a 4.535 kg (10 lb) roast containing 15% protein, 20% fat, and 65% water from 4.44 °C (40 °F) to 65.55 °C (150 °F). Express this energy in (a) BTU, (b) joules, and (c) watt-hour.

**Solution**

$C'_{avg}$ from Example 4.1 = 0.76 BTU/(lb · °F)

(a)  $q = mC_{avg}(T_2 - T_1)$

$$= 10 \text{ lb} \frac{0.76 \text{ BTU}}{\text{lb} \times {}^\circ\text{F}}(150 - 40)^\circ\text{F}$$

$$= 836 \text{ BTU}$$

(b)  Use $C_{avg}$ from Example 4.1 = 3182 J/(kg · K)

$q = 4.535$ kg [3182 J/(kg K)] (65.55 − 4.44) K = 0.882 MJ

(c)  $q = 0.882 \text{ MJ} \cdot \dfrac{10^6}{\text{MJ}} \cdot \dfrac{1 \text{ Ws}}{\text{J}} \cdot \dfrac{1 \text{ h}}{3600\text{s}}$

$= 245 \text{ W} \cdot \text{h}$

The specific heat calculated by Siebel's equation is used by the American Society for Heating, Refrigerating, and Air Conditioning Engineers in one of the most comprehensive tabulated values for specific heat of foodstuffs. However, it is overly simplified, and the assumption that all types of nonfat solids have the same specific heat may not always be correct. Furthermore, Siebel's equation for specific heat below the freezing point assumes that all the water is frozen, and this is most inaccurate.

Specific heats of solids and liquids may also be estimated using correlations obtained from Choi and Okos (1987). The procedure is quite unwieldy to do by hand; however, the data may be entered into a spreadsheet to facilitate repetitive calculations. The specific heats, in J/(kg · K), as a function of T (°C) for various components of foods are as follows:

$$\text{Protein}: C_{pp} = 2008.2 + 1208.9 \times 10^{-3}\text{T} - 1312.9 \times 10^{-6}\text{T}^2$$

$$\text{Fat}: C_{pf} = 1984.2 + 1473.3 \times 10^{-3}\text{T} - 4800.8 \times 10^{-6}\text{T}^2$$

$$\text{Carbohydrate}: C_{pc} = 1548.8 + 1962.5 \times 10^{-3}\text{T} - 5939.9 \times 10^{-6}\text{T}^2$$

$$\text{Fiber}: C_{pfi} = 1845.9 + 1930.6 \times 10^{-3}\text{T} - 4650.9 \times 10^{-6}\text{T}^2$$

$$\text{Ash}: C_{pa} = 1092.6 + 1889.6 \times 10^{-3}\text{T} - 3681.7 \times 10^{-6}\text{T}^2$$

Water above freezing :

$$C_{waf} = 4176.2 - 9.0864 \times 10^{-5}\text{T} + 5473.1 \times 10^{-6}\text{T}^2$$

The specific heat of the mixture above freezing is:

$$C_{avg} = P(C_{pp}) + F(C_{pf}) + C(C_{pc}) + Fi(C_{pfi}) + A(C_{pa}) + M(C_{waf}) \quad (4.10)$$

where P, F, Fi, A, C, and M represent the mass fraction of protein, fat, fiber, ash, carbohydrate, and moisture, respectively.

A spreadsheet for calculating specific heat at one temperature using the data given in Example 4.4 is shown in Fig. 4.1.

**Example 4.4** Calculate the specific heat of a formulated food product that contains 15% protein, 20% starch, 1% fiber, 0.5% ash, 20% fat, and 43.5% water at 25 °C.

**Solution**

The calculated values, in J/(kg · K), are $C_{pp} = 2037.6$; $C_{pf} = 2018.0$; $C_{pc} = 1594.1$; $C_{pfi} = 1891.3$; $C_{pa} = 1137.5$; and $C_{waf} = 4179.6$. Substituting in Eq. (4.14):

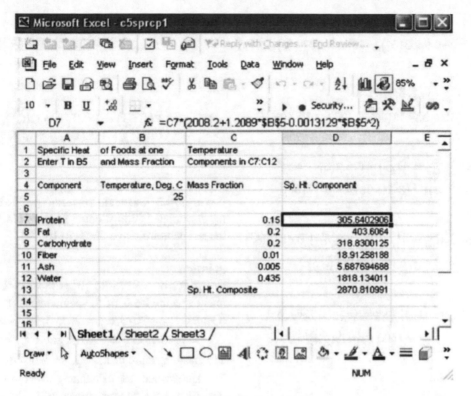

**Fig. 4.1** Spreadsheet for calculating specific heat using data in Example 4.4

$C_{pavg} = 0.15(2037.6) + 0.2(1594.1) + 0.01(1891.3)$

$\qquad + 0.005(1137.5) + 0.2(2018) + 0.435(4179.6)$

$\qquad = 2870.8 \; J/(kg \cong K)$

By comparison, using Siebel's equation:

$C_p = 1674.72(0.2) + 837.36$
$\qquad (0.15 + 0.01 + 0.005 + 0.2)$
$\qquad + 4186.8(0.435)$
$\qquad = 2462 \; J/(kg \; K)$

Values for $C_p$ calculated using Choi and Okos' (1988) correlations are generally higher than those calculated using Seibel's equation at high moisture contents.

Seibel's equations have been found to agree closely with experimental values when M > 0.7 and when no fat is present. Choi and Okos' correlation is more accurate at low moisture contents and for a wider range of product composition

because it is based on published literature values for a wide variety of foods. The simplicity of Siebel's equations however appeals to most users particularly when tolerance for error is not too stringent.

For enthalpy change calculations, Choi and Okos' equations for specific heat must be expressed as an average over the range of temperatures under consideration. The mean specific heat, C*, over a temperature range $T_1$ to $T_2$, where $(T_2 - T_1) = \delta$, $T_2^2 - T_1^2 = \delta^2$, and $T_2^3 - T_1^3 = \delta^3$, is:

$$C^* = \frac{1}{\delta} \int_{T_1}^{T_2} C_p \, dT$$

Thus, the equations for the mean specific heats of the various components over the temperature range $\delta$ become:

Fiber :  $C^*_{pfi} = [1/\delta][1845.9(\delta)$
$$+ 0.9653(\delta^2) - 1550 \times 10^{-6}(\delta^3)]$$

Carbohydrate :  $C^*_{pc} = [1/\delta][1548.8(\delta)$
$$+ 0.9812(\delta^2) - 1980 \times 10^{-6}(\delta^3)]$$

Ash :  $C^*_{pa} = [1/\delta][1092.6$
$$(\delta) + 0.9448(\delta^2) - 1227 \times 10^{-6}(\delta^3)]$$

$$C^*_{avg} = P(C^*_{pp}) + F(C^*_{pf}) + C(C^*_{pc})$$
$$+ Fi(C^*_{pfi}) + A(C^*_{pa}) + M(C^*_{waf})$$

(4.11)

Water :  $C^*_{waf} = [1/\delta][4176.2(\delta) - 4.543$
$$\times 10^{-5}(\delta^2) + 1824 \times 10^{-6}(\delta^3)]$$

**Example 4.5**  Calculate the mean specific heat of the formulated food product in Example 4.4, in the temperature range 25 °C to 100 °C.

Protein :  $C^*_{pp} = [1/\delta][2008.2(\delta) + 0.6045(\delta^2)$
$$- 437.6 \times 10^{-6}(\delta^3)]$$

**Solution**
$\delta = 75\ ^\circ C$; $\delta^2 = 9375\ ^\circ C^2$; $\delta^3 = 984{,}375\ ^\circ C^3$. The mean specific heats of the components in J/(kg .
K) are $C^*_{pp} = 2078$; $C^*_{pf} = 2055.3$; $C^*_{pc} = 1645.4$;

Fat :  $C^*_{pf} = [1/\delta][1984.2(\delta) + 0.7367(\delta^2)$
$$- 1600 \times 10^{-6}(\delta^3)]$$

$C^*_{pfi} = 1948.7$; $C^*_{pa} = 1137.5$; and $C^*_{waf} = 4200$.
The mean specific heat, $C^*_{pavg} = 2904$ J/(kg . K).
The spreadsheet for calculating average specific heat over a temperature range using the data in Example 4.5 is shown in Fig. 4.2.

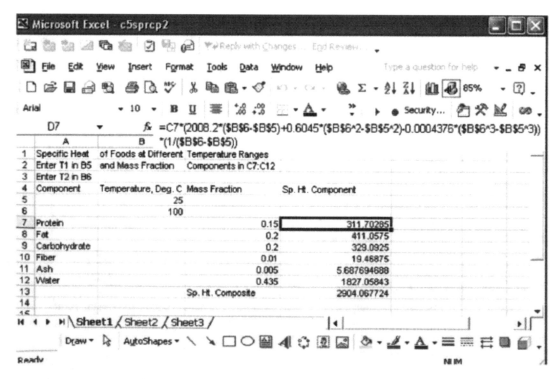

**Fig. 4.2**  Spreadsheet for calculating specific heat as a function of temperature using data in Example 4.5

## 4.3    Enthalpy Changes in Foods During Freezing

### 4.3.1    Correlation Equations Based on Freezing Points of Food Products Unmodified from the Natural State

When considering the heat to be removed during freezing of a food product, a change in phase is involved, and the latent heat of fusion must be considered. Water in a food does not all change into ice at the freezing point. Some unfrozen water exists below the freezing point, and therefore Siebel's equations for specific heat below the freezing point are very inaccurate. The best method for determining the amount of heat that must be removed during freezing, or the heat input for thawing, is by calculating the enthalpy change. One method for calculating enthalpy change below the freezing point (*good only for moisture contents between 73% and 94%*) is the procedure of Chang and Tao (1981). In this correlation, it is assumed that all water is frozen at 227 K (−50 °F).

A reduced temperature ($T_r$) is defined as:

$$T_r = \frac{T - 227.6}{T_f - 227.6} \qquad (4.12)$$

where $T_f$ is the freezing point temperature and T is the temperature at which the enthalpy is being determined. Two parameters, a and b, have been calculated for different products as a function of the mass fraction of moisture in the product, M. The correlation equations are as follows.

Meats:

$$a = 0.316 - 0.247(M - 0.73) \\ - 0.688(M - 0.73)5 \qquad (4.13)$$

$$b = 22.95 + 54.68(a - 0.28) \\ - 5589.03(a - 0.28)5 \qquad (4.14)$$

Vegetables, fruits, juices:

$$a = 0.362 + 0.0498(M - 0.73) \\ - 3.465(M - 0.73)^2 \qquad (4.15)$$

$$b = 27.2 - 129.04(a - 0.23) \\ - 481.46(a - 0.23)^2 \qquad (4.16)$$

The freezing point ($T_f$), in K, is:

$$\text{Meats}: T_f = 271.18 + 1.47\,M \qquad (4.17)$$

$$\text{Fruits and vegetables}: T_f = 287.56 - 49.19\,M \\ + 37.07\,M^2 \qquad (4.18)$$

$$\text{Juices}: T_f \\ = 120.47 + 327.35\,M \\ - 176.49\,M^2 \qquad (4.19)$$

The enthalpy at the freezing point, $H_f$, in J/kg, relative to 227.6 K is:

$$H_f = 9792.46 + 405096\,M \qquad (4.20)$$

The enthalpy at temperature T relative to 227.6 K is determined by:

$$H = H\left[a\,T_r + (1 - a)T_r^{\,b}\right] \qquad (4.21)$$

**Example 4.6** Calculate the freezing point and the amount of heat that must be removed in order to freeze 1 kg of grape juice containing 25% solids from the freezing point to −30 °C.

**Solution**
Y = 0.75.
Using Eq. (4.19) for juices:

$$T_f = 120.47 + 327.35(0.75) \\ - 176.49(0.75)^2 = 266.7 K$$

Using Eq. (4.20):

$$H_f = 9792.46 + 405,096(0.75) = 313,614\ J$$

Using Eq. (4.15):

$$a = 0.362 + 0.0498(0.02) - 3.465(0.02)^2 \\ = 0.3616$$

Using Eq. (4.16):

$$b = 27.2 - 129.04(0.1316) - 481.46(0.1316)^2$$
$$= 1.879$$
$$T_r = (-30 + 273 - 227.6)/(266.7 - 227.6) = 0.394$$

Using Eq. (4.21):

$$H = 313,614\left[(0.3616)0.394 + \left(1 - 0.3616\right)\right.$$
$$\left.(0.394)^{1.879}\right] = 79,457 \text{ J/kg}$$

The enthalpy change from $T_f$ to $-30\,°C$ is

$$\Delta\,H = 313,614 - 79,457 = 234,157 \text{ J/kg}$$

### 4.3.2   Enthalpy Changes During the Freezing of Foods Calculated from Molality of Liquid Water Fraction of the Food

When pure water is frozen, a phase change from liquid water to ice occurs at the freezing point. The cooling curve for water, a plot of temperature against time, will show the temperature sloping toward the freezing point followed by a constant temperature at the freezing point if the rate of cooling is slow enough to permit ice crystals adequate time to develop. The constant temperature is due to the release of energy associated with the phase change; all energy removed from the system at the freezing point serves as an energy sink for the latent heat of fusion. There will be no sensible heat loss (no temperature reduction) until all water has been converted to ice.

In foods, water exists as a solution. All water-soluble compounds in the food contribute to depress the freezing point. Because water crystals consist of pure water, transformation of water to ice is accompanied by an increase in solute concentration in the unfrozen water. Because solutes lower the freezing point, ice formation in foods occurs over a range of temperature. The freezing point is only the start of ice crystal formation. A typical freezing curve of temperature versus time exhibits a section where temperature is almost constant or drops much slower than the rest of the cooling curve. This section of the freezing curve is often referred to as the "ice crystal zone." Generally, a high initial moisture and low heat transfer rate extends the time temperature stays in the ice crystal zone. Food temperature will drop below the freezing point even when liquid water is still present.

### 4.3.3   Freezing Point Depression by Solutes

When the solution is ideal or when the solute concentration is low, the freezing point depression is:

$$\Delta t_f = K_f M$$

where $\Delta\,T_f$ is the freezing point depression relative to the freezing point of water, $0\,°C$; $K_f =$ cryoscopic constant $= 1.86$ for water. $M =$ molality $=$ mole solute/(g water/1000).

Let $n =$ gmoles solute contained in $w$ grams water. If the freezing point of a food is known, $n$ can be calculated from the moisture content and the freezing point depression. Otherwise, the freezing point can be calculated from concentration of solutes and the moisture content. For highly ionized solutes such as salts of sodium and potassium, multiply the actual moles of solute by 2 to obtain $n$.

$$M = \frac{n(1000)}{w}$$

$n$ will be constant during the freezing process, whereas $w$ changes as ice is formed.

Let $T_f =$ freezing point; $\Delta T_f = 0 - T_f = (-T_f)$

$$M = \frac{(-Tf)}{1.86}$$

### 4.3.4   Amount of Liquid Water and Ice at Temperatures Below Freezing

Basis: 1 kg food. Let $w_o$ = original water in the mixture before freezing = weight fraction water $\times$ 1000.

$$n = \left[\frac{(-Tf)}{1.86}\right]\left[\frac{w_o}{1000}\right] = \frac{w_o(-Tf)}{1860}$$

At any temperature below the freezing point T; $\Delta$ Tf = 0 − T = (−T)

$$M = \frac{-(T)}{1.86} = \frac{1000n}{w}$$

$$w = \frac{1000(n)(1.86)}{(-T)}$$
$$= \frac{(1000)(w_o)(-T_f)}{1860}\left(\frac{1.86}{(-T)}\right)$$
$$= \frac{w_o(-T_f)}{(-T)}$$

w = the amount of liquid water in the food at temperature T. The amount of ice = I = $w_o$ − w.

$$I = w_o - w_o\frac{(-T_f)}{(-T)} = w_o\left(1 - \frac{(-T_f)}{(-T)}\right)$$

### 4.3.5   Sensible Heat of Water and Ice at Temperatures Below the Freezing Point

For an increment of temperature dT, water will lose sensible heat according to:

dq = $wC_{pl}dT$

The change in sensible heat for liquid water ($q_{sl}$) from $T_f$ to T is determined by:

$$q_{sl} = \int_{T_f}^{T} C_{pl}\, w\, dT = C_{pl}\, w_o \int_{T_f}^{T} \frac{(-T_f)}{(-T)}\, dT$$

$$= C_{pl}\, w_o(-T_f)Ln\left[\frac{(-T)}{(-T_f)}\right] \quad (4.22)$$

The change in sensible heat of ice ($q_{si}$) from $T_f$ to T is:

$$q_{si} = \int_{T_f}^{T} C_{pi}\, w_o\left[1 - \frac{(-T_f)}{(-T)}\right] dT_o$$

$$= C_{pi}\, w\left[(T_f - T) - (-T_f)\ln\frac{(-T)}{(-T_f)}\right] \quad (4.23)$$

Because the specific heat for water or ice is in J/(kg . K), $w_o$ in the equations for q should be the mass in kilograms.

When considering the energy change associated with a temperature change from any temperature $T_1$ below the freezing point to T, substitute the amount of liquid water at $T_1$ for $w_o$ and $T_1$ for $T_f$.

### 4.3.6   Total Enthalpy Change

The total enthalpy change will consist of sensible heat of fat, sensible heat of nonfat solids, sensible heat of ice, sensible heat of liquid water, and the latent heat of fusion of ice.

The total enthalpy change from the freezing point to any temperature T is:

$$q = \Delta H = F\,C_{pf}(T_f - T)$$
$$+ SNF\,C_{psnf}(T_f - T)$$
$$+ q_{sw} + q_{si} + I\,(334,860)$$

Values for the specific heats of water, ice, fat, and nonfat solids were given earlier in this chapter.

**Example 4.7** Boneless broiler breast meat contains 70.6% water, 24.0% protein, 1.2% ash, and 4.2% fat. The freezing point is −1.2 °C. If this meat is marinated by adding salt solution to obtain a weight gain over the unmarinated meat of 15% and a net salt (NaCl) content of 1.0%, calculate (a) the new freezing point and (b) the enthalpy change as the marinated meat is frozen to −18 °C from the new freezing point per kg of marinated meat.

Basis: 1 kg original meat

$M = 1.2/1.86 = 0.645;$   g water $= 706$
$n = M(w_o)/1000 = .645(706)/1000$
   $= 0.455$ gmoles
For marinated meat:

mass $= 1 + 0.15\,(1) = 1.15$ kg
Salt $= NaCl = 0.01(1.15) = 0.015$ kg
Water $= 1.15 - 0.015 - (1 - 0.706)$
   $= 0.841$ kg $= 841$ g

Consider NaCl to have two moles solute/mole NaCl, the $Na^+$ and $Cl^-$ component.

The molecular weight of NaCl $= 35.5$.

Total gmoles solute in the marinated meat $= 2(0.015)(1000)/35.5 + 0.455 = 1.301$.

The molality of solutes in marinated meat:

$M = 1.301(1000)/841 = 1.547$

(a).   $T_f = 0 - 1.547(1.86) = -2.9°C$
(b).   At $-18°C$:
   $w_o = 841 = 0.841$ kg
   $w = 841(2.9/18) = 135.5g = 0.1355$ kg
   $I = 841 - 135.5 = 705.5g = 0.1355$ kg
   $q_{sl} = 4186.8(0.841)(2.9)\,\ln\,(18/2.9)$
      $= 18,642$ J
      $q_{si} = 2093.4(.841)\left[\{-2.9 - (-18)\}\right.$
   $\left. -2.9\ln\,(18/2.9)\right] = 17,263$ J

Total q $= 18,642 + 17,263 + SNF\,(837.36)$
$(18 - 2.9) + F(1674.72)(18 - 2.9)$
$+I\,(334.860)$

Because the original fat content was 4.2% and there was no fat addition during marination, $F = 0.042$ kg. The solids nonfat will consist of protein (24%), ash (1.2%), and added salt (0.015 kg). SNF $= 0.24 + 0.012 + 0.015 = 0.267$ kg.

Total q $= 18,642 + 17,263 + .267(837.36)(18 - 2.9)$
   $+ .042(1674.72)(18 - 29) + 0.1355(334.860)$
   $= 84,910$ J

Although the original basis is 1 kg of unmarinated meat, note that the amount of water, ice, fat, and solids nonfat are those present in the marinated product; therefore, the calculated q will be the enthalpy change per kg of marinated meat.

### 4.3.7   Specific Heats of Gases and Vapors

From Chap. 3, the specific heat of gases depends upon whether the process is carried out at constant pressure or at constant volume. If a gas is heated by blowing air across heating elements, the process is a constant pressure process. The specific heat is designated by $C_p$, the specific heat at constant pressure. The heat required to raise the temperature of a gas with mass, m, at constant pressure equals the change in enthalpy, $\Delta H$. The enthalpy change associated with a change in temperature from a reference temperature $T_o$ to $T_2$ is:

$$\Delta H = m \int_{T_o}^{T_2} C_p\,dT = mC_{pm}(T_2 - T_o) \quad (4.24)$$

$C_{pm}$ is the mean specific heat over the temperature range $T_o$ to $T_2$, and $C_p$ is the expression for specific heat as a function of T.

If a gas is heated from any temperature $T_1$ to a final temperature $T_2$, the change in enthalpy accompanying the process must be calculated as follows:

$$\begin{aligned}\Delta H &= q \\ &= mC_{pm}(T_2 - T_o) \\ &\quad - mC'_{pm}(T_1 - T_o) \end{aligned} \quad (4.25)$$

where $C'_{pm}$ is mean specific heat from the reference temperature $T_o$ to $T_1$. Tabulated values for the mean specific heat of gases are based on ambient temperature of 77 °F or 25 °C, as the reference temperature. Table 5.2 lists mean specific heats of common gases in the American Engineering System of Units, and Table 4.3 lists specific heats of the same gases in SI. Values for $C_{pm}$ in Tables 4.2 and 4.3 change very little at temperatures ordinarily used in food processes; therefore, $C_{pm}$ based on the temperature range from $T_o$ to $T_2$ can be used for $\Delta H$ between $T_1$ and $T_2$ with very little error in comparison with the use of Eq. (4.25).

**Example 4.8** Calculate the heating requirement for an air drier that uses 2000 $ft^3$/min of air at 1 atm and 170 °F, if ambient air at 70 °F is heated to 170 °F for use in the process.

**Solution**

$q = mC_{pm} (170-70)$

m, the mass of air/min going through the drier, in pounds, will be calculated using the ideal gas equation.

$$PV = \frac{m}{M}RT; M = \text{mol.wt.air} = 29 \text{ lb/lb mole}$$
$$= 126.14 \text{ lb/min}$$

From Table 4.2, $C_{pm}$ is constant at 0.241 BTU/(lb . °F) up to 200 °F; therefore, it will be used for the temperature range 70 °F to 170 °F.

$$= 126.13 \text{ lb/min}$$
$$m = \frac{PVM}{RT} = \frac{14.7(144) \text{ lb/ft}^2 \left[2000 \text{ ft}^3/\text{min}\right][29 \text{ lb/lbmole}]}{[1545 \text{ ft lb}_f/\text{lbmole}^\circ \text{ R}](460 + 170)^\circ \text{ R}}$$

**Table 4.2** Mean heat capacities of various gases from 77 °F to indicated temperature

| Temperature (°F) | Mean specific heat in BTU/lb (°F) | | | | |
| | $O_2$ | $N_2$ | $CO_2$ | $H_2O$ vapor | Air |
| --- | --- | --- | --- | --- | --- |
| 77 | 0.219 | 0.248 | 0.202 | 0.446 | 0.241 |
| 100 | 0.219 | 0.248 | 0.203 | 0.446 | 0.241 |
| 200 | 0.221 | 0.249 | 0.209 | 0.448 | 0.241 |
| 300 | 0.223 | 0.249 | 0.216 | 0.451 | 0.242 |
| 400 | 0.224 | 0.250 | 0.221 | 0.454 | 0.242 |
| 500 | 0.226 | 0.250 | 0.226 | 0.457 | 0.243 |
| 600 | 0.228 | 0.251 | 0.231 | 0.461 | 0.245 |
| 700 | 0.230 | 0.253 | 0.236 | 0.456 | 0.246 |

*Source:* Calculated from Mean molal heat capacity of gases at constant pressure. *In* Harper, J. C. 1976. *Elements of Food Engineering.* AVI Publishing Co., Westport, Conn

**Table 4.3** Mean heat capacities of various gases from 25 °C to indicated temperature

| Temperature (°C) | Mean specific heat J/kgK | | | | | |
| | $O_2$ | $N_2$ | | $CO_2$ | $H_2O$ vapor | Air |
| --- | --- | --- | --- | --- | --- | --- |
| 25 | 916 | 1037 | | 847 | 1863 | 1007 |
| 50 | 919 | 1039 | | 858 | 1868 | 1008 |
| 100 | 926 | 1041 | | 879 | 1880 | 1011 |
| 125 | 930 | 1043 | | 889 | 1886 | 1012 |
| 150 | 934 | 1044 | | 899 | 1891 | 1014 |
| 175 | 936 | 1045 | | 910 | 1897 | 1015 |
| 200 | 938 | 1047 | | 920 | 1903 | 1017 |
| 250 | 945 | 1049 | | 941 | 1914 | 1020 |
| 300 | 952 | 1052 | | 962 | 1925 | 1023 |
| 350 | 960 | 1054 | | 983 | 1937 | 1026 |

*Source:* Calculated from Mean molal heat capacity of gases at constant pressure. *In* Harper, J.C. 1976 *Elements of Food Engineering.* AVI Publishing Co., Westport, Conn

$$q = 126.13 \frac{lb}{min} \times \frac{0.241 \text{ BTU}(170 - 70)^\circ F}{(lb^\circ F)}$$
$$= 3039.8 \text{ BTU/min}$$

**Example 4.9** How much heat would be required to raise the temperature of 10 m³/s of air at 50 °C to 120 °C at 1 atm?

**Solution**
Using Eq. (4.25):

$$C'_{pm} \text{ at } T_1 = 50^\circ C = 1008 \text{ J/(kg.K)}$$
$$C'_{pm} \text{ at } T_2 = 120^\circ C = 1012 \text{ J/(kg.K)}$$

The mean specific heat from $T_o$ to 50 °C and from $T_o$ to 120 °C is different enough to require the use of Eq. (4.25) to determine $\Delta H$. The reference temperature, $T_o$, for $C_{pm}$ in Table 4.3 is 25 °C. The mass is determined using the ideal gas equation.

$$m = \frac{PVM}{RT} = \frac{1 \text{ atm}(10m^3)/s[29 \text{ kg/kg mole}]}{0.08206[m^3 \text{ atm/kg mole K}](50 + 273)(K)}$$
$$= 10.94 \text{ kg/s}$$

$$q = 10.94(1012)(120 - 25) - 10.94(1008)(50 - 25)$$
$$= 776.08 \text{ kJ/s} = 776.08 \text{ kW}$$

## 4.4 Properties of Saturated and Superheated Steam

Steam and water are the two most used heat transfer mediums in food processing. Water is also a major component of food products. The steam tables that list the properties of steam are a very useful reference when determining heat exchange involving a food product and steam or water. At temperatures above the freezing point, water can exist in either of the following forms.

Saturated liquid: liquid water in equilibrium with its vapor. The total pressure above the liquid must be equal to or be higher than the vapor pressure. If the total pressure above the liquid exceeds the vapor pressure, some other gas is present in the atmosphere above the liquid. If the total pressure above a liquid equals the vapor pressure, the liquid is at the boiling point.

Saturated vapor: this is also known as saturated steam and is vapor at the boiling temperature of the liquid. Lowering the temperature of saturated steam at constant pressure by a small increment will cause vapor to condense to liquid. The phase change is accompanied by a release of heat. If heat is removed from the system, temperature and pressure will remain constant until all vapor is converted to liquid. Adding heat to the system will change either temperature or pressure or both.

Vapor-liquid mixtures: steam with less than 100% quality. Temperature and pressure correspond to the boiling point; therefore, water could exist either as saturated liquid or saturated vapor. Addition of heat will not change temperature and pressure until all saturated liquid is converted to vapor. Removing heat from the system will also not change temperature and pressure until all vapor is converted to liquid.

Steam quality: the percentage of a vapor-liquid mixture that is in the form of saturated vapor.

Superheated steam: water vapor at a temperature higher than the boiling point. The number of degrees the temperature exceeds the boiling temperature is the *degrees superheat*. Addition of heat to superheated steam could increase the superheat at constant pressure or change both the pressure and temperature at constant volume. Removing heat will allow the temperature to drop to the boiling temperature where the temperature will remain constant until all the vapor has condensed.

### 4.4.1 The Steam Tables

The steam tables are tabulated values for the properties of saturated and superheated steam.

### 4.4.1.1    The Saturated Steam Table

The saturated steam table consists of entries under the headings of temperature, absolute pressure, specific volume, and enthalpy. A saturated steam table is in the Appendices 3 and 4.

The temperature and absolute pressure correspond to the boiling point, or the temperature and pressure under which steam can be saturated. The absolute pressure at a given temperature is also the vapor pressure.

The rest of the table consists of three general headings, each of which is subdivided into saturated liquid, evaporation, and saturated vapor. The entries under saturated liquid give the properties of liquid water at the indicated temperature. The entries under saturated vapor give the properties of steam at the boiling point. The entries under evaporation are changes due to the phase transformation and are the difference between the properties of saturated vapor and saturated liquid.

Specific volume is the reciprocal of the density. It is the volume in cubic feet occupied by 1 lb of water or steam under the conditions given.

Enthalpy is the heat content of a unit mass of steam or water at the indicated temperature and pressure. Enthalpy values in the steam tables are calculated from a base temperature of 0 °C. The energy change associated with a change in temperature or pressure of steam is the difference in the initial and final enthalpies. The following examples illustrate the use of the steam tables. In these examples, atmospheric pressure is specified and may not be a standard atmosphere. Please refer to the section on "Absolute Pressure and Temperature" in Chap. 3 for a discussion of the difference between atmospheric pressure and a standard atmosphere.

**Example 4.10**    At what vacuum would water boil at 80 °F? Express this in (a) inches of mercury vacuum (given atm pressure = 30 in. Hg) and (b) absolute pressure in Pascals.

**Solution**

From steam tables, Appendix 3, the boiling pressure for water at 80 °F is 0.50683 psia.

(a)    Inches Hg absolute pressure

$$= 0.5068 \frac{lb}{in.^2} \left[ \frac{2.035\ in.Hg}{lb/in.^2} \right]$$
$$= 1.03\ in.Hg\ absolute$$
$$Vacuum = 30\ in.Hg - 1.03\ in.Hg$$
$$= 28.97\ in.Hg\ vacuum$$

(b)    Pressure

$$= \frac{0.50683 lb_f}{in.^2} \left[ \frac{6894.757 Pa}{lb/in.^2} \right]$$
$$= 3.494\ kPa\ absolute\ pressure$$

**Example 4.11**    If 1 lb of water at 100 psig and 252 °F is allowed to expand to 14.7 psia, calculate (a) the resulting temperature after expansion and (b) the quantity of vapor produced.

**Solution**

The absolute pressure = 100 + 14.7 = 114.7 psia. At 252 °F, water will not boil until the pressure is reduced to 30.9 psia. The water therefore is at a temperature much below the boiling point at 114.7 psia, and it would have the properties of liquid water at 252 °F.

(a)    After expansion to 14.7 psia, the boiling point at 14.7 psia is 212 °F. Part of the water will flash to water vapor at 212 °F, and the remaining liquid will also be at 212 °F.

(b)    The enthalpy of water at 252 °F is ($h_f$ at 252 °F) 220.62 BTU/lb. Basis: 1 lb $H_2O$. Heat content = 220.62 BTU. When pressure is reduced to 14.7 psia, some vapor will be formed, but the total heat content of both vapor and liquid at 212 °F and 14.7 psia will still be 220.62 BTU.

If x = wt vapor produced, 1 − x = wt water at 212 °F and 14.7 psia:

$$x(h_g) + (1 - x)(h_f) = 220.62$$

$$h_g = 1150.5 \, BTU/lb; \, h_f = 180.17 \, BTU/lb$$

$$x(1150.5) + (1 - x)(180.17) = 220.62$$

$$x = \frac{220.62 - 180.17}{1150.5 - 180.17}$$

$$= \frac{40.45}{970.33} = 0.0417 \, lb \, H_2O$$

**Example 4.12** If water at 70 °F is introduced into an evacuated vessel, initially at 0 psia, what would be the pressure inside the vessel when equilibrium is finally attained at 70 °F?

**Solution**
Because the vessel is completely evacuated, the gaseous phase after introduction of water will be 100% water vapor. Upon introduction, water will vaporize until the pressure of water in the space above the liquid equals the vapor pressure.

Pressure = vapor pressure of water at 70 °F
From  steam  tables,  pressure  at
    70 °F = 0.36292 psia.

**Example 4.13** If the vessel in Example 4.11 had initially 14.7 psia absolute pressure and contained completely dry air, what would be the absolute pressure after introducing the water, assuming that none of the original air had escaped during the process?

**Solution**

$$\begin{aligned} Pressure &= partial \, pressure \, of \, air + partial \, pressure \\ &\quad of \, water(volume \, is \, constant) \\ &= original \, pressure \, of \, dry \, air + vapor \\ &\quad pressure \, of \, water \\ &= 14.7 + 0.36292 = 15.063 \, psia \end{aligned}$$

**Example 4.14** How much heat would be given off by cooling steam at 252 °F and 30.883 psia to 248 °F, at the same pressure?

**Solution**
First, check the state of water at 30.883 psia and 252 °F and 248 °F. From steam tables, the boiling point of water at 30.883 psia is 252 °F. Therefore, steam at 252 °F and 30.883 psia is saturated vapor. At 30.883 psia and 248 °F, water will be in the liquid state, because 248 °F is below the boiling temperature at 30.883 psia.

Heat given off = q = $h_g$ at 252 °F − $h_f$ at 248 °F
From steam tables, Appendix 4

$h_g$ at 252 °F = 1164.78 BTU/lb
$h_f$ at 248 °F = 216.56 BTU/lb
q = 1164.78–216.56 = 948.22 BTU/lb

Saturated steam is a very efficient heat transfer medium. Note that for only a 2 °F change in temperature, 948 BTU/lb of steam is given off. The heat content of saturated vapors comes primarily from the latent heat of vaporization, and it is possible to extract this heat simply by causing a phase change at constant temperature and pressure.

### 4.4.1.2  The Superheated Steam Tables
A superheated steam table is in Appendix 2. Both temperature and absolute pressure must be specified to accurately define the degree of superheat. From the temperature and absolute pressure, the specific volume v in $ft^3/lb$ and the enthalpy h in BTU/lb can be read from the table. Example problems on the use of the superheated steam tables are as follows.

**Example 4.15** How much heat is required to convert 1 lb of water at 70 °F to steam at 14.696 psia and 250 °F?

**Solution**
First determine the state of steam at 14.696 psia and 250 °F. At 14.696 psia, the boiling point is 212 °F. Steam at 250 °F and 14.696 psia is superheated steam. From the superheated steam table, h at 250 °F is 1168.8 BTU/lb.

Heat required $= h_g$ at $250°F$ and

$$14.696\text{ psia} - h_f \text{ at } 70°F$$
$$= 1168.8\text{ BTU/lb} - 38.05\text{ BTU/lb}$$
$$= 1130.75\text{ BTU/lb}$$

**Example 4.16**  How much heat would be given off by cooling superheated steam at 14.696 psia and 500 °F to 250 °F at the same pressure?

**Solution**
Basis: 1 lb of steam.

Heat given off $= q = h$ at 14.696 psia and $500°F - h_g$
at 14.696 psia and $250°F$
$$= 1287.4 - 1168.8 = 118.6\text{ BTU/lb}$$

Superheated steam is not a very efficient heating medium. Note that a 250 °F change in temperature is accompanied by the extraction of only 118.6 BTUs of heat.

### 4.4.1.3  Double Interpolation from Superheated Steam Tables

Because the entries in the table are not close enough to cover all conditions, it may be necessary to interpolate between entries to obtain the properties under a given set of conditions. In the case of superheated steam where both temperature and pressure are necessary to define the state of the system, a double interpolation is sometimes necessary. The following example shows how the double interpolation is carried out.

**Example 4.17**  Calculate the enthalpy of superheated steam at 320 °F and 17 psia.

**Solution**
Entries in Appendix 2 show enthalpies at 15 and 20 psia and 300 °F and 350 °F. The tabular entries and the need for interpolation are as follows:

| P (psia) | Enthalpy 300 °F | 350 °F |
|---|---|---|
| 15 | 1192.5 | 1216.2 |
| 17 | ? | ? |
| 20 | 1191.4 | 1215.4 |

Enthalpies at 300 °F and 350 °F at 17 psia are obtained by interpolating between 15 and 20 psia at each temperature.

At 300 °F: P = 15, h = 1192.5; P = 20, h = 1191.4; at P = 17, h =?

$$h_{(300°F \wedge 17\text{ psia})} = 1191.4 + \frac{(1192.5 - 1191.4)}{(20 - 15)}(20 - 17)$$

$$= 1192.06\text{ BTU/lb}$$

At 350 °F: P = 15, h = 1216.2; P = 20, h = 1215.4; at P = 17, h =?

$$h_{(350°F \wedge 17\text{ psia})} = 1215.4 + \frac{(1216.2 - 1215.4)}{(20 - 15)}(20 - 17)$$

$$= 1215.88\ \text{BTU/lb}$$

Now it is possible to interpolate between 300 °F and 350 °F to obtain the enthalpy at 17 psia and 320 °F.

$$h = 1215.88 + \frac{(1215.88 - 1192.06)}{(350 - 300)}(350 - 320)$$

$$= 1215.88 - 14.29 = 1201.59\text{ BTU/lb}$$

### 4.4.2    Properties of Steam Having Less Than 100% Quality

If steam is not 100% vapor, the properties can be determined on the basis of the individual properties of the component.

If x = % quality
$$v = xv_g + (1 - x)v_f$$
$$h = xh_g + (1 - x)h_f$$

**Example 4.18**  Calculate the enthalpy of steam at 252 °F having 80% quality.

**Solution**
From the saturated steam tables, Appendix 3, at 252 °F, saturated steam or water has the following properties:

$h_f = 220.62$ BTU/lb
$h_g = 1164.78$ BTU/lb
$h = 1164.78(0.8) + 220.62(0.2) = 931.82$
    $+ 44.12 = 975.94$ BTU/lb

Note that only the temperature of steam is given in this problem. If either the temperature or pressure is given, but not both, steam is at the boiling point.

## 4.5    Heat Balances

Heat balance calculations are treated in the same manner as material balances. The amount of heat entering a system must equal the amount of heat leaving a system, or:

$$\text{Heat in} = \text{heat out} + \text{accumulation}$$

At a steady state, the accumulation term is zero and heat entering the system must equal what leaves the system.

Heat balance problems are facilitated by using diagrams that show process streams bringing heat and taking heat out of a system.

**Example 4.19** Calculate the amount of water that must be supplied to a heat exchanger that cools 100 kg/h of tomato paste from 90 °C to 20 °C. The tomato paste contains 40% solids. The increase in water temperature should not exceed 10 °C while passing through the heat exchanger. There is no mixing of water and tomato paste in the heat exchanger.

**Solution**
The diagram of the system is shown in Fig. 4.3.

This problem may be solved by assuming a datum from which enthalpy calculations are made. This datum temperature is the lowest of the process stream temperatures. Let $T_1 =$ inlet water temperature $= 20$ °C, and $T_2 =$ exit water temperature $= 20 + 10 = 30$ °C. Let 20 °C be the reference temperature for enthalpy calculations. The specific heat of water $= 4187$ J/(kg $\cong$ K) and that of tomato paste is obtained, using Eq. (5.7).

$$C_{avg} = 3349(0.6) + 837.36 = 2846.76 \text{ J/(kg.K)}$$

Heat content of entering tomato paste:

$$q_1 = (100 \text{ kg}) \, [2846.76 \text{ J/(kg.K)}](90 - 20) \text{ K} = 19.927 \text{ MJ}$$

Heat content of tomato paste leaving system:

$$q_2 = 100 \text{kg}[2846.76 \text{ J/(kg.K)}](20 - 20) \text{ K} = 0$$

Let W $=$ kg water entering the system

$$q_3 = \text{W kg}[4187 \text{ J/(kg.K)}](20 - 20) \text{ K} = 0$$
$$q_4 = \text{heat content of water leaving the system}$$
$$= \text{W kg } [4187 \text{ J/(kg.K)}](30 - 20) \text{ K}$$
$$= 41,870(\text{W}) \text{ J}$$

The heat balance is:

$$q_1 + q_3 = q_2 + q_4$$

Because $q_2$ and $q_3 = 0$, $q_1 = q_4$, and:

$$(19.927 \text{ MJ}) \frac{10^6 \text{J}}{\text{MJ}} = 41,871(\text{W}) \text{ J}$$
$$\text{W} = \frac{19.927 \times 10^6}{41870} = 475.9 \text{ kg}$$

The heat balance may also be expressed as follows:

Heat gain by water $=$ heat loss by the tomato paste

$$100 \text{kg} \left( 2846.76 \frac{\text{J}}{\text{kg.K}} \right) (90 - 20) \text{k}$$
$$= \text{W} \left( 4187 \frac{\text{J}}{\text{kg.K}} \right) \cdot (T_1 + 10 - T_1) \text{K}$$
$$41870\text{W} = 100(2846.76)(70); \quad \text{W} = 475.9 \text{kg}$$

If there is no mixing and exchange of mass between process streams, calculation is simplified by equating heat gain by one process stream to heat loss by the other. However, when mixing and material transfer occurs, a heat balance based on heat content of each stream entering and leaving the system will simplify analysis of the problem.

**Fig. 4.3** Diagram of heat exchange during cooling of tomato paste in Example 5.19

**Fig. 4.4** Diagram of material and energy balance in apple dehydration

**Example 4.20** Calculate the amount of saturated steam at 121.1 °C that must be supplied to a dehydrator per hour. Steam condenses in the heater, which heats the drying air from steam to water at 121.1 °C. The dehydrator is operated as follows: Apples at 21.1 °C enter the dehydrator with 80% moisture and leave the dehydrator at 37.7 °C and 10% moisture. One hundred pounds per hour of fresh apples enter the drier. Fresh air at 21.1 °C and a humidity of 0.002 kg $H_2O$/kg dry air enter the drier, mixes with recycled hot air until the humidity is 0.026 kg $H_2O$/kg dry air, and is heated to 76.7 °C using steam in a finned heat exchanger. Hot air leaves the drier at 43.3 °C and a humidity of 0.04 kg $H_2O$/kg dry air.

**Solution**

The diagram for the process is shown in Fig. 4.4. Use as a basis 1 h of operation.

The system looks complicated with the hot air recycling, but if the boundary of the system is enlarged as shown in Fig. 4.4, solving the problem is considerably simplified.

The problem will be solved by separating air into dry air and water vapor and apples into dry matter and water, and heat contents of each component entering and leaving the system will be calculated. The lowest temperature, 21.1 °C, may be used as the base temperature for heat content calculations as in the preceding example, but because the steam tables will be used to determine the enthalpy of water and steam, it will be more consistent to use the reference temperature for the steam tables of 0 °C.

The following enthalpies are to be calculated:

Heat input = q

$q_1$ = enthalpy of $H_2O$ in entering air (vapor at 21.1 °C)

$q_2$ = enthalpy of dry air entering at 21.1 °C

$q_3$ = enthalpy of $H_2O$ in apples entering (liquid at 21.1 °C)

$q_4$ = enthalpy of dry matter in apples entering at 21.1 °C

$q_5$ = enthalpy of $H_2O$ in exit air (vapor at 43.3 °C)

$q_6$ = enthalpy of dry air leaving at 43.3 °C

$q_7$ = enthalpy of $H_2O$ in apples leaving (liquid at 37.7 °C)

$q_8$ = enthalpy of dry matter in apples leaving at 37.3 °C

The heat balance is:

$$q + q_1 + q_2 + q_3 + q_4 = q_5 + q_6 + q_7 + q_8$$

$q_2$ and $q_6$ are calculated using the quantity of dry air that enters the system per hour. The latter is obtained by performing a material balance on water.

Amount of water lost by apples
= amount of water gained by air

Solids balance for apples:

$$x = \frac{45.4(0.2)}{0.9} = 10.09 \text{ kg/h}$$

$45.4(0.2) = x(0.9); x$ = wt dried apples

$\dfrac{\text{Water gained by air}}{\text{kg dry air}} = (0.04 - 0.002) = 0.038$

Water lost by apples = 45.4–10.09 = 35.31 kg/h

Let w = mass of dry air:

Total water gained by air = 0.038 w
Material balance on water: 0.038 w = 35.31
w = 929.21 kg dry air/h
The mean specific heat of air is obtained from Table 4.3. These values are:

$25^\circ C : C_{pm} = 1008$ J/(kg.$E$)
$50^\circ C : C_{pm} = 1007$ J/(kg.K)

These mean specific heat will be used for determining enthalpies of air at 21.1 °C and 43.3 °C.

From the section "Specific Heat of Solids and Liquids":

$C_p$ for solids non − fat in apples
= 837.36 J/(kg.K)

Use steam tables for heat contents of water and steam. Calculating the enthalpies,

$q = m\, C_{pm}\, (T - 0)$ because 0 °C is used as the reference temperature.

$$q_1 = (929.21 \text{ kg dry air})\frac{0.002 \text{ kg } H_2O}{\text{kg dry air}} (h_g \text{ at } 21.1^\circ C)$$

From Appendix 4, $h_g$ at 21.1 °C by interpolation = 2.54017 MJ/kg.

$q_1 = 929.21(0.002)(2.54017 \times 10^6)$
$\quad = 4.7207$ MJ

$q_2 = (929.21)$ kg dry air$(1008)\dfrac{J}{kg} \cdot K(21.1 - 0)$
$\quad = 19.7632$ MJ

$q_3 = 45.4(0.8)(h_f \text{ at } 21.1^\circ C)$

From Appendix 4, $h_f$ at 21.1 °C = 0.08999 MJ/kg.

$q_3 = 45.4(0.8)(0.08999 \times 10^6) = 3.2684$ MJ

$q_4 = (45.4)(0.2)$ kg dry solids$\left(\dfrac{837.36 J}{kg \cdot K}\right)$
$\quad (21.1 - 0)^\circ C = 0.16043$ MJ

$q_5 = (929.21 \text{ kg dry air})\left(0.04\dfrac{\text{kg } H_2O}{\text{kg dry air}}\right)$
$\quad (h_g \text{ at } 43.3^\circ C)$

From Appendix 4, $h_g$ at 43.3 °C = 2.5802 M/kg.

$q_5 = (929.21)(0.04)(2.5802 \times 10^6) = 95.9019$ MJ

$q_6 = (929.21 \text{ kg dry air})\dfrac{1007 J}{kg \cdot K}(43.3 - 0)$
$\quad = (929.21)(1007)(43.3) = 40.5164$ MJ

$q_7 = (10.09)(0.1) \text{ kg } H_2O(h_f \text{ at } 37.7^\circ C)$

From Appendix 4, $h_f$ at 37.7 °C = 0.15845 MJ/kg.

$$q_7 = (10.09)(0.1)(0.15845)$$
$$= 0.15987 \text{ MJ}$$
$$q_8 = (10.09)\,(0.9)\text{ kg dry matter}\left(\frac{837.36\,\text{J}}{\text{kg F}}\right)$$
$$(37.7 - 0)$$
$$= (10.09)(0.9)(837.36)(37.7)$$
$$= 0.28667 \text{ MJ}$$

The heat balance with q in MJ is:

$$q+ \ [4.7207 + 19.7632 + 3.2684 + 0.16043]$$
$$= [95.9019 + 40.5164 + 0.15988 + 0.28667]$$
$$q = 136.8648 - 27.91273 = 108.952 \text{ MJ}$$

Because the basis is 1 h of operation, the heating requirement for the process is 108.952 MJ/h.

From Appendix 4, the heat of vaporization of steam at 121.1 °C is 2.199144 MJ/kg.

$$\text{The amount of steam supplied/h} = \frac{108.952\,\text{MJ/h}}{2.199144\,\text{MJ/kg}}$$
$$= 49.543 \text{ kg/h}$$

**Alternative Solution**
The problem will be solved using a balance of heat sinks and heat source in the system. Consider the system as a whole, enclosing the recycle stream within the system boundary. Entering the system are air and apples, which both gain sensible heat through the process. Also entering the system is steam, which is the only heat source. Incoming air with the water vapor may be combined, and using an average specific heat, the sensible heat gain is easily calculated. Additional moisture in the air leaving the system is accounted for by moisture evaporation from apples, which may be assumed to occur at the temperature the apples leave the system. Thus energy gain by vaporized water is the difference between the enthalpy of the vapor at 43.3 °C and the enthalpy of liquid at 37.7 °C. The sensible

heat gain by apples will be assessed with the initial weight and specific heat. In equation form, the energy balance is $m_{ma}\ C_{pm}$ (43.3–21.1) + $m_{ap}\ C_p$ (37.7–21.1) + $m_v$ ($h_{g@43.3}$ c − $h_{f@37.7}$ c) = $m_s$($h_{fg@121.1}$ c) where $m_{ma}$, $m_{ap}$, and $m_s$ are the mass of moist air, wet apples, and steam, respectively. $C_{pm}$ is the average specific heat of the moist air, and $C_p$ is the specific heat of the wet apples. The specific heat of apples = 0.8 (4186.8) + 0.2(837.36) = 3516.9 J/(kg . K). Mass of vapor is the same as the amount removed from apples and is determined by a material balance:

$$m_v = 45.4 - \frac{45.4(0.2)}{0.9} = 35.31\,\frac{\text{kg}}{\text{h}}$$

Air will gain (0.04–0.002) or 0.038 kg water/kg dry air. Thus, the mass of dry air entering the system is 35.31/0.038 = 929.21 kg. The mass of moist air is:

$$m_{ma} = \left[\frac{929.21\,\text{kg dry air}}{\text{h}}\right]\left[\frac{1.002\,\text{kg moist air}}{\text{kg dry air}}\right]$$
$$= 931.07\frac{\text{kg}}{\text{h}}$$

The specific heat of air and water, respectively, at 50 °C are 1008 and 1868 J/(kg . K). Thus, the mean specific heat for moist air is $C_{pm}$ = [1008 (1) + 1868 (0.002)](1/1.002) = 1009.7 J/(kg . K).

The enthalpy of vapor at 43.3 °C($h_{g@43.3}$ c) from Appendix 4 by interpolation between 42.5 °C and 45 °C is:

$$2.5788+ [(2.5832 - 2.5788)/2.5](43.3 - 42.5)$$
$$= 2.58021 \text{ MJ/kg.}$$

The enthalpy of liquid at 37.7 °C = 0.15816 MJ/kg. The latent heat of evaporation at 121.1 °C ($h_{fg@121.1C}$) from Appendix 4 by interpolation is:

$$2.20225- [(2.20225 - 2.19519)/(122.5 - 120)](121.1 - 120)$$
$$= 2.19914 \text{ J/kg.}$$

Solving for $m_s$:

$$m_s = \frac{(931.07)(1009.7)(22.2) + (45.4)(3516.9)(16.6) + (35.31)(2.42236 \times 10^6)}{2.19914 \times 10^6}$$

$$= 49.58 \frac{kg}{h}$$

**Example 4.21** Calculate the amount of steam at 121.1 °C (250 °F) that must be added to 100 kg of a food product with a specific heat of 3559 J/(kg . K) to heat the product from 4.44 °C (40 °F) to 82.2 °C (180 °F) by direct steam injection.

**Solution**
The diagram of the system is shown in Fig. 4.5. Let x = kg of steam required. From steam table, the enthalpies of water and steam are as follows:

For steam at 121.1 °C or 250 °F, $h_g$ from Table A.3 = 1164.1 BTU/lb = 2.70705 MJ/kg.

$$\frac{\text{Heat loss steam}}{\text{kg}} = h_g \text{ at } 121.1^\circ C - h_f \text{ at } 82.2^\circ C$$

For water at 82.2 °C or 180 °F, $h_f$ = 148.00 BTU/lb = 0.34417 MJ/kg = 2.70705–0.34417– = 2.36288 MJ/kg.

Total heat loss steam = x(2.36288) MJ
Heat gain by product = 100kg[3559 J/(kg ≅ K)]

$$(82.2 - 4.44) \text{ K}$$
$$= 27.67478 \text{ MJ}$$
$$x (2.36288) = 27.67478$$
$$x = 11.71 \text{ kg steam required}$$

## Problems

4.1. What would be the pressure generated when milk is heated to 135 °C in a closed system? If the system is not pressurized, can this temperature be attained?

4.2. A method for heating food with saturated steam at temperatures below the boiling point of water is by carrying out the process under a vacuum. At what vacuum should a system be operated to heat a material with saturated steam at 150 °F?

4.3. If a retort indicates a pressure of 15 psig but mercury in glass thermometer registers only 248 °F, what does this indicate?

**Fig. 4.5** Diagram of a direct steam injection process for heating a food product

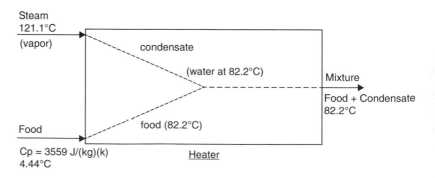

Assume that both instruments are in working order and have been properly calibrated.

4.4. An evaporator is operated at 15 in. Hg vacuum. What would be the temperature of the product inside the evaporator? Assume product has the same boiling point as water.

4.5. How much heat is required to convert 1 kg water at 20 °C to steam at 120 °C?

4.6. How much heat must be removed to convert 1 lb steam at 220 °F to (a) water at 220 °F and (b) water at 120 °F?

4.7. One pound of steam at 260 °F contains 80% steam and 20% liquid water. How much heat will be released by this steam when it is allowed to condense to water at 200 °F?

4.8. At what temperature would water be expected to boil at 10 in. Hg vacuum? Atm pressure = 14.696 psia.

4.9. How much steam at 250 °F would be required to heat 10 lb of water from 70 °F to 210 °F in a direct steam injection heater?

4.10. How much heat will be required to convert steam at 14.696 psig to superheated steam at 600 °F at the same pressure?

4.11. Ten pounds of water at a pressure of 20 psig is heated to a temperature of 250 °F. If this water is allowed to empty into an open vessel at atmospheric pressure, how much of the water will remain in the liquid phase?

4.12. (a) If water at 70 °F is introduced into an evacuated vessel where the original pressure is 0 psia, what would be the pressure inside the vessel at equilibrium? Assume no change in the temperature of water. (b) If the original pressure is 14.696 psia, what will be the final pressure?

4.13. Determine the heat content in BTU/lb for water (it could be liquid, saturated steam, or superheated steam) under the following conditions: (a) 180 °F and 14.696 psia pressure, (b) 300 °F and 14.696 psia pressure, and (c) 212.01 °F and 14.696 psia pressure.

4.14. In the formulation of a pudding mix, it is desired that the solids content of the product would be 20%. The product leaving the batch tank has a temperature of 26.67 °C (80 °F), and this is preheated to 90.56 °C (195 °F) by direct steam injection using culinary steam (saturated) at 104.4 °C (220 °F) followed by heating in a closed system to sterilizing temperatures. There is no further loss or gain of moisture in the rest of the process. What should be the solids content of the formulation in the batch tank such that after the direct steam injection heating the final solids content of the product will be 20%? Use Siebel's equation for calculating the specific heat of the product.

4.15. A fruit juice at 190 °F is allowed to flash into an essence recovery system maintained at a vacuum of 29 in. of mercury. Atmospheric pressure is 29.9 in. The vapors that flash off are rectified to produce an essence concentrate, and the juice after being stripped of the aromatic constituents is sent to an evaporator for concentration. Assuming sufficient resident time for the juice in the system to allow equilibrium in temperature between the liquid and the vapor, calculate:

(a) The temperature of the juice leaving the essence recovery system.

(b) The solids content of the juice leaving the system if the original solids content is 10%. Assume no additional heat input and the latent heat for vaporization is derived from the loss in sensible heat of the liquid. The specific heat of the solids is 0.2 BTU/(lb ≅ °F).

(c) The quantity of water vaporized per 100 lb of juice entering the system.

4.16. An evaporator has a heat transfer surface area that would allow the transfer of heat at the rate of 100,000 BTU/h. If this evaporator is concentrating apple juice from 10% to 45% solids, under a vacuum of 25 in. Hg (atmospheric pressure is 30 in. of mercury), how much apple juice can be processed per hour?

4.17. Orange juice concentrate at 45% total solids leaves the evaporator at 50 °C. This is frozen into slush in swept surface heat exchangers until half of the water is in the form of ice crystals prior to filling into cans, and the cans are frozen at −25 °C. Assume that the sugars are all hexose sugars and that the freezing point reduction can be determined using $\Delta T_f = K_f m$, where $K_f$ = the cryoscopic constant = 1.86 and m = the molality. Calculate:

(a) The total heat that must be removed from the concentrate in the swept surface heat exchangers per kg of concentrate processed

(b) The amount of heat that must be further removed from the concentrate in frozen storage

(c) The amount of water still in the liquid phase at −25 °C

Note: The moisture content is beyond the range where Chang and Tao's correlation is applicable. Determine the freezing point by calculating the freezing point depression: $\Delta T_b = K_f m$. The specific heat of the solids is the same below and above freezing. The specific heat of ice is 2093.4 J/(kg$K). The heat of fusion of ice is 334,860 J/kg. The juice contains 42.75% soluble solids.

4.18. In a falling film evaporator, fluid is pumped to the top of a column and the fluid falls down as a film along the heated wall of the column increasing in temperature as it drops. When the fluid emerges from the column, it is discharged into a vacuum chamber where the fluid drops in temperature by flash evaporation until it reaches the boiling temperature at the particular vacuum employed. If juice containing 15% solids is being concentrated to 18% solids in one pass through the heated column, and the vacuum in the receiving vessel is maintained at 25 in. Hg, calculate the temperature of the fluid as it leaves the column such that when flashing occurs the desired solids content will be obtained.

4.19. When sterilizing foods containing particulate solids in the Jupiter system, solids are heated separately from the fluid components by tumbling the solids in a double cone processing vessel with saturated steam contacting the solids. The fluid component of the food is heated, held until sterile, and cooled using conventional fluid heating and cooling equipment. The cooled sterile liquid is pumped into the double cone processing vessel containing the hot solids, which cools the latter and drops the pressure to atmospheric. After allowing the mixture to cool by cooling the walls of the processing vessel, the sterile mixture is filled aseptically into sterile containers

(a) Meat and gravy sauce is being prepared. Beef cubes containing 15% solids nonfat, 22% fat, and 63% water are heated from 4 °C to 135 °C, during which time condensate accumulates within the processing vessel with the meat. Saturated steam at 135 °C is used for heating. Calculate the total amount of meat and condensate at 135 °C.

(b) The gravy mix is of equal weight as the raw meat processed and consists of 85% water and 15% solids nonfat. Calculate the temperature of the mixture after equilibration if the gravy mix is at 20 °C when it is pumped into the processing chamber containing the meat at 135 °C.

4.20. The chillers in a poultry processing plant cool broilers by contacting the broilers with a mixture of water and ice. Broilers enter at 38 °C and leave the chillers at 4 °C. USDA requires an overflow of 0.5 gallons of water per broiler processed, and this must be replaced with fresh water to maintain the liquid level in the chiller. Melted ice is part of this overflow requirement. If a plant processes 7000 broilers/h and the broilers average 0.98 kg with a composition of 17% fat, 18% solids nonfat, and 65% water, calculate the ratio by weight of ice to fresh water that must be added into the chiller to meet the overflow requirement and the cooling load. Fresh water is at 15 °C, and the overflow is at 1.5 °C.

4.21. Saturated steam at 280 °F is allowed to expand to a pressure of 14.696 psia without a loss of enthalpy. Calculate (a) the temperature and (b) the weight of high-pressure steam needed to produce 100 m$^3$/min of low-pressure steam at 14.696 psia and the temperature calculated in (a).

4.22. In one of the systems for ultrahigh-temperature sterilization, milk enters a chamber maintained at 60 psia and 800 °F in an atmosphere of superheated steam where it discharges from a plenum into vertical tubes where it falls down in a thin film while exposed to the steam. The milk will be at the boiling temperature at 60 psia on reaching the bottom of the heating chamber. After a sterilizing hold time at constant temperature, the milk is discharged into a vacuum chamber for rapid cooling. If the vacuum chamber is at 15 in. Hg vacuum, calculate:

(a) The temperature of the milk leaving the flash chamber.

(b) The total solids content. Raw milk enters the heater at 2 °C and contains 89% water, 2% fat, and 9% solids nonfat. Given: the enthalpy of superheated steam at 60 psia and 800 °F is 1431.3 BTU/lb. Saturated steam temperature at 60 psia is 292.7 °F. The enthalpy of saturated liquid ($h_f$) at 292.7 °F is 260.7 BTU/lb.

4.23. (a) Calculate the freezing point of fresh strawberries that contain 8.5% soluble solids, 1% insoluble solids, and 90.5% water. Assume that the average molecular weight of the soluble solids is 261.

(b) If sucrose is added to the above strawberries in the ratio 1 part of sugar to 3 parts strawberries by weight, calculate the new freezing point.

(c) Calculate the change in enthalpy of the sugared strawberries from 20 °C to −20 °C.

4.24. A food mix containing 80% solids nonfat and 20% fat on a dry basis is to be extruded. Water is added continuously along with the product into the extruder, and the mixture temperature increases to 135 °C at a pressure of 600 kPa at the die entrance. Extrudate leaves the die and immediately expands to atmospheric pressure, releasing vapor as it exits the die. Assume that the temperature of the extrudate is 100 °C immediately after leaving the die. It is desired that the moisture content of the extrudate will be 18% after the pressure reduction. Calculate the rate of moisture addition to the extruder if the solid feed originally contains 10% water and is fed at the rate of 30 kg/h.

4.25. A food product that contains 15% solids nonfat, 2% fat, and 83% water is to be pasteurized by heating to 95 °C by direct steam infusion using 90% quality steam at 15 psig. After heating to 95 °C, the product is held in a holding tube and cooled in a heat exchanger; therefore, there is no opportunity for removing added water from the steam condensate later in the process. Calculate the moisture content of the raw product entering the heater such that the desired moisture content will be obtained after heating and cooling.

## Suggested Reading

American Society for Heating, Refrigerating and Air Conditioning Engineers. (1967). *Guide and data book, applications for 1966 and 1967*. New York: ASHRAE.

American Society of Mechanical Engineers. (1967). *American Society of Mechanical Engineers 1967 steam tables. Properties of saturated and superheated steam from 0.08865 to 15,500 lb per sq in. absolute pressure.* Windsor: Combustion Engineering Inc.

Chang, H. D., & Tao, L. C. (1981). Correlation of enthalpy of food systems. *Journal of Food Science, 46,* 1493.

Choi, Y., & Okos, M. R. (1987). Effects of temperature and composition on thermal properties of foods. In M. Le Maguer & P. Jelen (Eds.), *Food engineering and process applications* (Vol. I, pp. 93–102). New York: Elsevier.

Charm, S. E. (1971). *Fundamentals of food engineering* (2nd ed.). Westport: AVI Publishing.

Felder, R. M., & Rousseau, R. W. (1999). *Elementary principles of chemical processes* (2nd ed.). New York: Wiley.

Himmelblau, D. M. (1967). *Basic principles and calculations in chemical engineering* (2nd ed.). Englewood Cliffs: Prentice-Hall.

Hougen, O. A., & Watson, K. M. (1946). *Chemical process principles. Part I. Material and energy balances.* New York: Wiley.

McCabe, W. L., Smith, J. C., & Harriott, P. (1985). *Unit operations of chemical engineering* (4th ed.). New York: McGraw-Hill Book.

Seibel, J. E. (1918). *Compend of mechanical refrigeration and engineering* (9th ed.). Chicago: Nickerson and Collins.

Sinnott, R. K. (1996). *Chemical engineering* (Vol. 6, 2nd ed.). Oxford: Butterworth-Heinemann.

Watson, E. L., & Harper, J. C. (1988). *Elements of food engineering* (2nd ed.). New York: Van Nostrand Reinhold.

# Flow of Fluids

**5**

Fluids are substances that flow without disintegration when pressure is applied. This definition of a fluid includes gases, liquids, and certain solids. A number of foods are fluids. In addition, gases such as compressed air and steam are also used in food processing, and they exhibit resistance to flow just like liquids. In this chapter, the subject of fluid flow will be discussed from two standpoints: the resistance to flow and its implications in the design of a fluid handling system and evaluation of rheological properties of fluid foods.

## 5.1 The Concept of Viscosity

Viscosity is a measure of resistance to flow of a fluid. Although molecules of a fluid are in constant random motion, the net velocity in a particular direction is zero unless some force is applied to cause the fluid to flow. The magnitude of the force needed to induce flow at a certain velocity is related to the viscosity of a fluid. Flow occurs when fluid molecules slip past one another in a particular direction on any given plane. Thus, there must be a difference in velocity, *a velocity gradient*, between adjacent molecules. In any particular plane parallel to the direction of flow, molecules above and below that plane exert a resistance to the force that propels one molecule to move faster than the other. This resistance of a material to flow or deformation is known as

stress. *Shear stress* ($\tau$) is the term given to the stress induced when molecules slip past one another along a defined plane. The velocity gradient ($-dV/dr$ or $\gamma$) is a measure of how rapidly one molecule is slipping past another; therefore, it is also referred to as the *rate of shear*. The position from which distance is measured in determining the shear rate is the point in the flow stream where velocity is maximum; therefore, as distance r increases from this point of reference, V decreases and the velocity gradient is a negative quantity. Because shear stress is always positive, expressing the shear rate as $-dV/dr$ satisfies the equality in the equation of shear stress as a function of shear rate. A plot of shear stress against shear rate for various fluids is shown in Fig. 5.1.

Fluids that exhibit a linear increase in the shear stress with the rate of shear (Eq. 5.1) are called *Newtonian fluids*. The proportionality constant ($\mu$) is called the *viscosity*:

$$\tau = \mu \left( -\frac{dV}{dr} \right) \qquad (5.1)$$

Newtonian fluids are those that exhibit a linear relationship between the shear stress and the rate of shear. The slope $\mu$ is constant; therefore, the viscosity of a Newtonian fluid is independent of the rate of shear. The term viscosity is appropriate to use only for Newtonian fluids. Fluids with characteristics deviating from Eq. (5.1) are called *non-Newtonian* fluids. These fluids exhibit either

© Springer International Publishing AG, part of Springer Nature 2018
R. T. Toledo et al., *Fundamentals of Food Process Engineering*,
https://doi.org/10.1007/978-3-319-90098-8_5

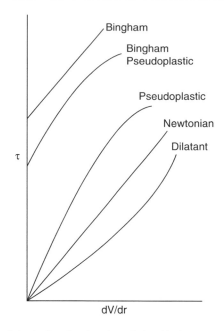

**Fig. 5.1** A plot showing the relationship between shear stress and shear rate for different types of fluids

shear-thinning or shear-thickening behavior, and some exhibit a yield stress (i.e., a threshold stress that must be overcome before the fluid starts to flow). The two most commonly used equations for characterizing non-Newtonian fluids are the power law model (Eq. 5.2) and the Herschel-Bulkley model for fluids (Eq. 5.3):

$$\tau = K(\gamma)^n \qquad (5.2)$$

$$\tau = \tau_o + K(\gamma)^n \qquad (5.3)$$

Equation (5.2) can fit the shear stress versus shear rate relationships of a wide variety of foods. Equation (5.2) can also be used for fluids which exhibit a yield stress as represented by Eq. (5.3), if $\tau_o$ is very small compared to $\tau$ or if the shear rate is very high. n is the *flow behavior index* and is dimensionless. K is the *consistency index* and will have units of pressure multiplied by time raised to the *n*th power.

Equations (5.2) and (5.3) can be rearranged as follows:

$$\tau = \left[K(\gamma)^{n-1}\right]\gamma \qquad (5.4)$$

$$\tau - \tau_o = \left[K(\gamma)^{n-1}\right]\gamma \qquad (5.5)$$

Equations (5.4) and (5.5) are similar to Eq. (5.1), and the factor that is the multiplier of $\gamma$ is the apparent viscosity defined as follows:

$$\mu_{app} = K(\gamma)^{n-1} \qquad (5.6)$$

For fluids having characteristics which fit Eqs. 5.4 and 5.5, the apparent viscosity can also be expressed as

$$\mu_{app} = \tau/\gamma \qquad (5.7)$$

$$\mu_{app} = (\tau - \tau_o)/\gamma \qquad (5.8)$$

Substitution of Eq. (5.2) for $\tau$ in Eq. (5.7) also yields Eq. (5.6).

The apparent viscosity has the same units as viscosity, but the value varies with the rate of shear. Thus, when reporting an apparent viscosity, the shear rate under which it was determined must also be specified.

Equation (5.6) shows that the apparent viscosity will be decreasing with increasing shear rate if the flow behavior index is less than 1. This type of fluids exhibits "shear thinning" behavior and are referred to as *shear thinning fluids* or *pseudoplastic fluids*.

When the flow behavior index of a fluid is greater than 1, $\mu_{app}$ increases with increasing rates of shear, and the fluid is referred to as *shear-thickening* or *dilatant* fluids. Shear thinning behavior is exhibited by emulsions and suspensions where the dispersed phase has a tendency to aggregate following the path of least resistance in the flow stream, or the particles align themselves in a position that presents the least resistance to flow. Shear-thickening behavior will be exhibited when the dispersed phase swells or changes shape when subjected to a shearing action or when the molecules are so long that they tend to cross-link with each other, trapping molecules of the dispersion medium. Shear-thickening behavior is seldom observed in foods, although it has been reported in suspensions of clay and high molecular weight organic polymers.

Fluids that exhibit a yield stress and a flow behavior index of 1 are referred to as *Bingham plastics*. Some foods exhibit a yield stress and a flow behavior index less than 1, and there is no general term used for these fluids, although the terms *Herschel-Bulkley* and *Bingham pseudoplastic* have been used.

## 5.2  Rheology

Rheology is the science of flow and deformation. When a material is stressed, it deforms and the rate and nature of the deformation that occurs characterizes its rheological properties. The science of rheology can be used on solids or fluids. Some food materials exhibit both fluid and solid behaviors, and these materials are called *visco-elastic*. In general, *flow* indicates the existence of a velocity gradient within the material, and this characteristic is exhibited only by fluids. The subject of rheology in this section covers only flow properties of fluids. In foods, rheology is useful in defining a set of parameters that can be used to correlate with a quality attribute. These parameters can also be used to predict how the fluid will behave in a process and in determining energy requirements for transporting the fluid from one point in a processing plant to another.

### 5.2.1  Viscometry

Instruments used for measuring flow properties of fluids are called viscometers. Newtonian viscosity can be easily measured since only one shear rate needs to be used, and therefore viscometers for this purpose are relatively simple compared to those for evaluating non-Newtonian fluids. Viscometers require a mechanism for inducing flow that should be measurable, a mechanism for measuring the applied force, and the geometry of the system in which flow is occurring must be simple in design such that the force and the flow can be translated easily into a shear stress and shear rate and graphed as shown in Fig. 5.1.

#### 5.2.1.1  Viscometers Based on Fluid Flow Through a Cylinder

These viscometers are called capillary or tube viscometers, depending on the inside diameter. The principle of operation is based on the Poiseuille equation, if the fluid is Newtonian. The Rabinowitsch-Mooney equation applies when the fluid is non-Newtonian. The Poiseuille equation will be derived based on the Newtonian flow equation (Eq. 5.1), and the Rabinowitsch-Mooney equation will be derived based on the power law equation (Eq. 5.2).

#### 5.2.1.2  Derivation of the Poiseuille Equation

Figure 5.2 shows a tube of length L and radius R. A pressure $P_1$ exists at the entrance and $P_2$ exists at the end of the tube. At a distance r from the center, a ring of thickness dr is isolated, and at this point, the fluid velocity is V, the point velocity. The shear stress at this point is force resisting flow/area of the fluid undergoing shear. The force resisting flow on the fluid occupying the area of the ring is the pressure drop times the area of the ring, and the area of the plane where the fluid is sheared is the circumferential area of the cylinder formed when the ring is projected through length, L:

**Fig. 5.2** Differential control element for analysis of fluid flow through a tube

$$\tau = \frac{(P_1 - P_2)(\pi r^2)}{2\pi r L} = \frac{(P_1 - P_2)r}{2L} \qquad (5.9)$$

Substituting Eqs. (5.9) in (5.1):

$$\frac{(P_1 - P_2)r}{2L} = \mu\left(-\frac{dV}{dr}\right)$$

Separating variables and integrating:

$$\int dV = \frac{(P_1 - P_2)}{2L\mu}\int -r dr$$

$$V = \frac{(P_1 - P_2)}{2L\mu}\left(\frac{-r^2}{2}\right) + C$$

The constant of integration can be determined by applying the boundary condition: at $r = R$, $V = 0$. $C = (P_1 - P_2)(R^2)/4L\mu$:

$$V = \frac{(P_1 - P_2)}{4L\mu}\left(R^2 - r^2\right) = \frac{\Delta P}{4L\mu}\left(R^2 - r^2\right)$$
$$(5.10)$$

Equation (5.10) gives the point velocity in a flow stream for any fluid flowing within a cylindrical tube. The point velocity is not very easily measured. However, an average velocity defined as the volumetric flow rate/area can be easily measured. Equation (5.10) will be expressed in terms of the average velocity. Consider a pipe of radius R and a control volume that is the walls of a hollow cylinder of thickness dr within the pipe, shown in Fig. 5.3.

The cross-sectional area of the ring of thickness dr is $dA = \pi[(r + dr)^2 - r^2)] = \pi(r^2 + 2rdr + (dr)^2 - r^2) = 2\pi r dr + (dr)^2$. Because dr is small and $(dr)^2$ is negligible, $dA = 2\pi r dr$. The volumetric rate of flow (volume/time) through the control volume is $VdA = 2\pi r V dr$. The total volume going through the pipe will be the integral of the volumetric rate of flow through the control volume from $r = 0$ to $r = R$. Substituting Eq. (5.10) for V:

$$\bar{V}\left(\pi R^2\right) = \frac{(P_1 - P_2)}{4L\mu}(2\pi)\int_0^R \left(R^2 - r^2\right)r\, dr$$

Rearranging and integrating:

$$\bar{V} = \frac{(P_1 - P_2)}{2L\mu R^2}\left[\frac{R^2 r^2}{2} - \frac{r^4}{4}\right]\Big|_0^R$$

Substituting limits, combining terms, and substituting $\Delta P$ for $(P_1 - P_2)$:

$$\bar{V} = \frac{(P_1 - P_2)R^2}{8L\mu} = \frac{\Delta P R^2}{8L\mu} \qquad (5.11)$$

Equation (5.11) is the Poiseuille equation and can be used to determine the viscosity of a Newtonian fluid from pressure drop data when the fluid is allowed to flow through a tube or a capillary.

When Eq. (5.11) is used to determine the viscosity of non-Newtonian fluids, the viscosity obtained will be an apparent viscosity. Thus a

**Fig. 5.3** Pipe showing ring of thickness dr used as a control element for analysis of fluid flow

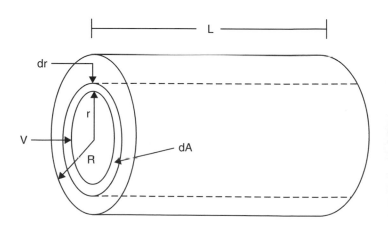

viscosity obtained from measurements using a single rate of flow is

$$\mu_{app} = \frac{\Delta PR^2}{8L\bar{V}}$$

Equations (5.10) and (5.11) may be combined to obtain an expression for $V/\bar{V}$ which can be differentiated to obtain a rate of shear as a function of the average velocity.

$$V = \bar{V}\left[\frac{\Delta P}{4L\mu}\right](R^2 - r^2)\left[\frac{8L\mu}{\Delta PR^2}\right]$$

Simplifying:

$$V = 2\bar{V}\left[1 - \left(\frac{r}{R}\right)^2\right] \qquad (5.12)$$

Equation (5.12) represents the velocity profile of a Newtonian fluid flowing through a tube expressed in terms of the average velocity. The equation represents a parabola where the maximum velocity is $2\bar{V}$ at the center of the tube ($r = 0$). Differentiating Eq. (5.12):

$$\frac{dV}{dr} = 2\bar{V}\left[\frac{2r}{R^2}\right]$$

The shear rate at the wall ($r = R$) for a Newtonian fluid is

$$-\frac{dV}{dr}\Big|w = \frac{4\bar{V}}{R} \qquad (5.13)$$

### 5.2.1.3 Velocity Profile and Shear Rate for a Power Law Fluid

Equations (5.1) and (5.9) can be combined to give

$$\frac{(P_1 - P_2)r}{2L} = K\left(-\frac{dV}{dr}\right)^n$$

Rearranging and using $\Delta P$ for ($P_1 - P_2$):

$$\frac{dV}{dr} = \left[\frac{\Delta P}{2LK}\right]^{1/n}(r)^{1/n}$$

Integrating and substituting the boundary condition, $V = 0$ at $r = R$:

$$V = \left[\frac{\Delta P}{2LK}\right]^{1/n}\left[\frac{1}{(1/n) + 1}\right]\left[R^{(1/n)+1} - r^{(1/n)+1}\right] \qquad (5.14)$$

Equation (5.14) represents the velocity profile of a power law fluid. The velocity profile equation will be more convenient to use if it is expressed in terms of the average velocity. Using a procedure similar to that used in the section "Derivation of the Poiseuille Equation," the following expression for the average velocity can be derived:

$$\bar{V}(\pi R^2)$$
$$= \int_0^R 2\pi r\left[\frac{\Delta P}{2LK}\right]^{1/n}\left[\frac{n}{n+1}\right]\left[R^{(1/n)+1} - r^{(1/n)+1}\right]dr$$

Integrating and substituting limits:

$$\bar{V} = \left[\frac{\Delta P}{2LK}\right]^{1/n}[R]^{(n+1)/n}\left[\frac{n}{3n+1}\right] \qquad (5.15)$$

The velocity profile in terms of the average velocity is

$$V = \bar{V}\left[\frac{3n+1}{n+1}\right]\left[1 - \left[\frac{r}{R}\right]^{(n+1)/n}\right] \qquad (5.16)$$

Figure 5.4 shows $V/\bar{V}$ as a function of r/R for various values of n. When $n = 1$, the velocity profile is parabolic, and Eqs. (5.12) and (5.16) give the same results. The velocity profile flattens out near the center of the pipe as n decreases.

Differentiating Eq. (5.16) and substituting $r = R$ for the shear rate at the wall:

$$-\frac{dV}{dr}\Big|w = \bar{V}\left[\frac{3n+1}{n+1}\right]\left[\frac{n+1}{n}\right][R]^{-(n+1)/n}[R]^{1/n}$$

Simplifying:

$$-\frac{dV}{dr}\Big|w = \bar{V}\left(\frac{3n+1}{n}\right)\left(\frac{1}{R}\right) \qquad (5.17)$$

Equation (5.17) is a form of the Rabinowitsch-Mooney equation used to calculate shear rates for non-Newtonian fluids flowing through tubes. Converting into a form similar to Eq. (5.13):

**Fig. 5.4** Plot of $V/\bar{V}$ as a function of position in the pipe for fluids with different values of the flow behavior index n

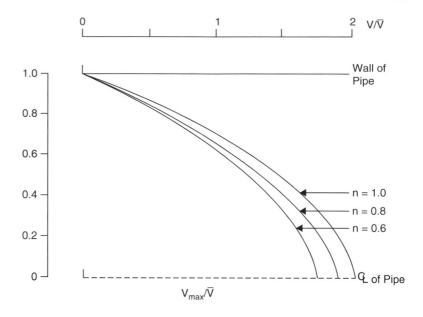

$$-\frac{dV}{dr}\Big|_w = \frac{4\bar{V}}{R}\left[\frac{3}{4}+\frac{1}{4n}\right] \qquad (5.18)$$

Equation (5.18) shows that the shear rate at the wall for a non-Newtonian fluid is similar to that for a Newtonian fluid except for the multiplying factor (0.75 + 0.25/n).

### 5.2.1.4   Glass Capillary Viscometers

The simplest viscometer operates on gravity-induced flow and is commercially available in glass. Figure 5.5 shows a Cannon-Fenske-type viscometer. These viscometers are available with varying size of the capillary. The size of capillary is chosen to minimize the time of efflux for viscous fluids. The Poiseuille equation (Eq. 5.11) is used to determine the viscosity in these viscometers. The pressure drop needed to induce flow is

$$\Delta P = \rho g h$$

where $\rho$ is the density of the fluid and h is the height available for free fall in the viscometer (Fig. 5.5). The viscosity is then calculated as

**Fig. 5.5** Diagram of a glass capillary viscometer showing the components and parameters used in fitting viscometer data to the Poiseuille equation using the above expression for $\Delta P$ in Eq. (5.13) as follows:

$$\frac{\mu}{\rho} = \frac{ghR^2}{8L\overline{V}} \qquad (5.19)$$

The ratio $\mu/\rho$ is the *kinematic viscosity*.

The viscosity is obtained from the time of efflux of the fluid through the viscometer. To operate the viscometer, fluid is pipetted into the large leg of the viscometer until the lowermost bulb is about half full. Fluid is then drawn into the small leg to about half of the uppermost bulb. When the suction on the small leg is released, fluid will flow, and timing is started when the fluid meniscus passes through the first mark. When the fluid meniscus passes the second mark, timing is stopped and the efflux time is recorded. The lower portion of the small leg of the viscometer is a capillary of radius R and length L. If $t_e$ is the time of efflux, the average velocity $\overline{V}$ will be $L/t_e$. Substituting in Eq. (5.18):

$$\frac{\mu}{\rho} = \frac{ghR^2}{8L} t_e \qquad (5.20)$$

For each viscometer, the length and diameter of capillary, and height available for free fall, are specific; therefore, these factors can be grouped into a constant, $k_v$, for a particular viscometer. The kinematic viscosity can then be expressed as

$$\frac{\mu}{\rho} = k_v t_e \qquad (5.21)$$

The viscometer constant is determined from the efflux time of a fluid of known viscosity and density.

**Example 5.1** A glass capillary viscometer when used on a fluid with a viscosity of 10 centipoises allowed the fluid to efflux in 1.5 min. This same viscometer used on another fluid allowed an efflux time of 2.5 min. If the densities of the two fluids are the same, calculate the viscosity of the second fluid.

**Solution**

Using Eq. (5.21) to solve for the viscometer constant:

$$k_v = \frac{\mu}{\rho t_e} = \frac{10}{1.5\rho}$$

For the second fluid:

$$\mu = \rho \left[\frac{10}{1.5\rho}\right](2.5) = 16.67 \, \text{centipoises}$$

### 5.2.1.5   Forced Flow Tube or Capillary Viscometry

For very viscous fluids, gravity-induced flow is not sufficient to allow measurement of viscosity. Forced flow viscometers are used for these fluids. In order to obtain varying flow rates in viscometers, some means must be provided to force a fluid through the viscometer at a constant rate. This may be obtained by using a constant pressure and measuring the flow rate that develops or by using a constant flow rate and measuring the pressure drop over a length of test section. Flow rates through glass capillary viscometers may also be varied by applying pressure instead of relying solely on gravity flow.

The flow properties of a fluid are the constants in Eqs. (5.1), (5.2), and (5.3), which can be used to characterize the relationship between the shear stress and the rate of shear. These are the viscosity, if the fluid is Newtonian, the flow behavior index, the consistency index, and the yield stress. Equation (5.3) is most general. If a fluid has a yield stress, the test is carried out at very high rates of shear in order that $\tau_o$ is very small in comparison to $\tau$ and Eq. (5.3) simplifies to Eq. (5.2). A fluid with properties that fits Eq. (5.1) is a special case of a general class of fluids which fits Eq. (5.2), when $n = 1$. Evaluation of $\tau_o$ in Eq. (5.3), from data at low shear rates, is simple, once n is established from data at high shear rates. The shear rate at the wall may be determined using Eqs. (5.13) and (5.18). These equations show that the shear rate is the product of some factor which is independent of the rate of flow and another factor which is a function of the rate of flow, i.e., the average velocity, $\bar{V}$, the volumetric rate of flow Q, or even the speed of a piston $V_p$ which delivers fluid to the tube. In equation form:

$$\gamma_w = F_1 \bar{V} = F_2 Q = F_3 V_P$$

The shear stress at the wall is determined by substituting R for r in Eq. (5.9):

$$\tau_w = \frac{\Delta P R}{2L} \qquad (5.22)$$

Equation (5.22) shows that $\tau_w$ is a product of a factor which is independent of the rate of flow and the pressure drop. In equation form: $\tau_w = F_{p1}$ $(\Delta P) = F_{p2}$ (height of manometer fluid) $= F_{p3}$ (transducer output). Substitution of any of the above expressions for $\tau_w$ and $\Delta_w$ into Eq. (5.1):

$$PF_{p1} = \left[\bar{V}F_1\right]^n$$

P and $\bar{V}$ are, respectively, measurements from which pressure or velocity can be calculated. Taking logarithms of both sides:

$$\log P + \log F_{p1} = n \log (\bar{V}) + n \log F_1$$

Rearranging:

$$\log P = n \log \bar{V} + \left(n \log F_1 - \log F_{p1}\right)$$

Thus, a log-log plot of any measure of pressure against any measure of velocity can give the flow behavior index n for a slope.

The shear rate can then be calculated using either Eq. (5.13) or (5.18), and the shear stress can be calculated using Eq. (5.22). Taking the logarithm of Eq. (5.2): $\log \tau_w = \log K + n \log \gamma_w$.

The intercept of a log-log plot of $\tau_w$ against $\gamma_w$ will give K.

### Example 5.2

A tube viscometer having an inside diameter of 1.27 cm and a length of 1.219 m is used to determine the flow properties of a fluid having a density of 1.09 g/cm³. The following data were collected for the pressure drop at various flow rates measured as the weight of fluid discharged from the tube. Calculate the flow behavior and consistency indices for the fluid.

| Data with pressure drop (in kPa) | |
| --- | --- |
| $(P_1 - P_2)$ | Flow rate (g/s) |
| 19.197 | 17.53 |
| 23.497 | 26.29 |
| 27.144 | 35.05 |
| 30.350 | 43.81 |
| 42.925 | 87.65 |

### Solution

Figure 5.6 shows a plot of log $\Delta P$ against log (mass rate of flow). The slope, n = 0.5, indicates that the fluid is non-Newtonian. Equation (5.18) will be used to solve for $\gamma_w$ and Eq. (5.22) for $\tau_w$.

Solving for $\tau_w$: R = 0.5(1.27 cm)(0.01 m/cm) = 0.00635 m; L = 1.219 m:

$$\tau_w = [0.00635(0.5)/1.219]\ \Delta P$$
$$= 0.002605\ \Delta P\ \text{Pa}$$

The average velocity $\bar{V}$ in m/s can be calculated by dividing the mass rate of flow by the density and the cross-sectional area of the tube. Let q = mass rate of flow in g/s:

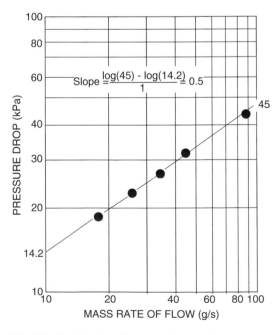

**Fig. 5.6** Log-log plot of pressure drop against mass rate of flow to obtain the flow behavior index from the slope

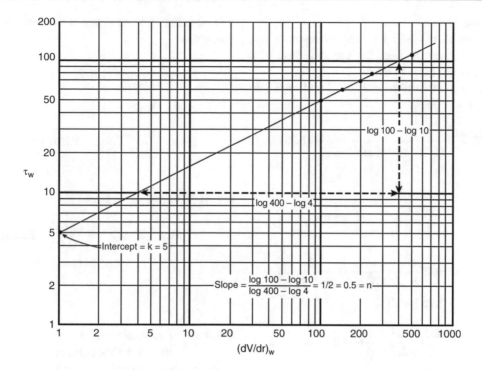

**Fig. 5.7** Log-log plot of shear stress against shear rate to obtain the consistency index from the intercept

$$\bar{V} = q \frac{g}{s} \frac{cm^3}{1.09g} \frac{m^3}{(100)^3 cm^3} \frac{1}{\pi(0.00635)^2 m^2}$$
$$= 0.007242 \, q \, m/s$$

Equation (5.18) is used to calculate the shear rate at the wall. $n = 0.5$:

$$\gamma_w = \frac{4\bar{V}}{R}\left(0.75 + \frac{0.25}{n}\right) = \frac{4(0.007242 \, q)}{0.00635}(1.25)$$
$$= 5.7047 \, q$$

The shear stress and shear rates are the following:

| $\tau_w$ (Pa) | $\gamma_w$ (1/s) |
|---|---|
| 50.008 | 99.966 |
| 61.209 | 149.92 |
| 70.710 | 199.87 |
| 79.062 | 249.83 |
| 111.82 | 499.83 |

Equation (5.2) will be used to determine K. The yield stress $\tau_o$ is assumed to be much smaller than the smallest value for $\tau_w$ measured. Thus, Eq. (5.3) reduces to the same form as Eq. (5.2), the logarithm of which is as follows:

$$\log \tau_w = \log K + n \log \gamma_w$$

A log-log plot of shear stress against the shear rate is shown in Fig. 5.7. The intercept is the consistency index:

$$k = 5 \, Pa \cdot s^n.$$

**Example 5.3** A fluid induces a pressure drop of 700 Pa as it flows through a tube having an inside diameter of 0.75 cm and a length of 30 cm at a flow rate of 50 cm$^3$/s.

(a) Calculate the apparent viscosity defined as the viscosity of a Newtonian fluid that would exhibit the same pressure drop as the fluid at the same rate of flow through this tube.

(b) This same fluid flowing at the rate of 100 cm$^3$/s through a tube 20 cm long and 0.75 cm in inside diameter induces a pressure drop of 800 Pa. Calculate the flow behavior and consistency indices for this fluid.

What would be the shear rates at 50 and 100 cm$^3$/s?

**Solution**

Equation (5.11) will be used to calculate the apparent viscosity from the pressure drop of a fluid flowing through a tube.

(a)   $R = (0.0075)(0.5) = 0.00375$ m; $L = 0.3$ m; $\Delta P = 700$ Pa:

$$\bar{V} = 50 \frac{cm^3}{s} \frac{1\,m^3}{(100)^3 cm^3} \frac{1}{\pi(0.00375)^2} = 1.1318\,m/s$$

$$\mu_{app} = \frac{\Delta P R^2}{8 L \bar{V}}$$

$$\mu_{app} = \frac{700(0.00375)^2}{8(0.3)(1.1318)} = 0.003624\,Pa \cdot s$$

(b)   $\bar{V} = 100 \frac{cm^3}{s} \frac{1\,m^3}{(100)^3 cm^3} \frac{1}{\pi(0.00375)^2}$

$$= 2.2635\,m/s$$

$$\mu_{app} = \frac{800(0.00375)^2}{8(0.2)(2.2635)} = 0.003106\,Pa \cdot s$$

Equation (5.6) will be used to solve for K and n.

Let $\varphi = 0.75 + 0.25/n$:

Equation (6.18) : $\gamma_w = \dfrac{4\bar{V}}{R}(\phi)$

Equation (6.6) : $\mu_{app} = K(\gamma)^{n-1}$

Substituting $\phi$ : $\mu_{app} = K\left[\frac{4\phi}{R}\right]^{n-1}\bar{V}^{(n-1)}$

Using the subscripts 1 and 2 for the velocity and apparent viscosity at 50 and 100 cm³/s, respectively:

$$\frac{\mu_{app}1}{\mu_{app}2} = \frac{\left(K(4\phi/R)^{n-1}\bar{V}_1\right)^{n-1}}{\left(K(4\phi/R)^{n-1}\bar{V}_2\right)^{n-1}} = \left(\frac{V_1}{V_2}\right)^{n-1}$$

$$n = 1 + \frac{\log(1.16674)}{\log(0.5009)}$$

$$\log(1.16674) = (n-1)\log(0.5009)$$

$$\log\left(\frac{0.003624}{0.003106}\right) = (n-1)$$

$$\times \log\left(\frac{1.1318}{2.2635}\right) = 0.777$$

$$\phi = 0.75 + \frac{0.25}{0.777} = 1.072$$

Using Eq. (5.6) on either of the two apparent viscosities:

$$\gamma_w = \frac{4(1.1317)}{0.00375}(1.072) = 1294\,s^{-1}$$

Using Eq. (5.6):

$$K = \frac{0.003624}{(1294)^{0.777-1}} = 0.0178\,Pa \cdot s^n$$

(c)   At 50 cm³/s, $\gamma_w = 1294$ s⁻¹

At 100 cm³/s, $\gamma_w = \dfrac{4(2.2635)}{0.00375}(1.072)$

$$= 2414\,s^{-1}$$

### 5.2.1.6   Evaluation of Wall Effects in Tube Viscometry

The equations derived in the previous sections for evaluating flow properties of fluids by tube or capillary viscometry are based on the assumption that there is zero slip at the wall. This condition may not always be true for all fluids, particularly suspensions.

To determine if slip exists at the wall, it will be necessary to conduct experiments on the same fluid using viscometers of different radius. Kokini and Plotchok (1987) defined a parameter, $ß_c$, which can be used to correct for slip at the tube wall. $ß_c$ is the slope of a plot of $Q/(\pi R^3 \tau_w)$ vs $1/R^2$ obtained on the same fluid using different viscometers of varying radius at flow rates which would give a constant shear stress at the wall, $\tau_w$. Q is the volumetric rate of flow, and R is the viscometer radius. With $ß_c$ known, a corrected volumetric flow rate $Q_c$ is calculated as follows:

$$Q_c = Q - \pi R \tau_w ß_c \qquad (5.23)$$

The corrected flow rate $Q_c$ is then used instead of the actual flow rate in determining the mean velocity and shear rates in the determination of the flow behavior index and the consistency index of fluids.

The experiments will involve a determination of pressure drop at different rates of flow on at least four tubes, each with different radius. The value of ß$_c$ may vary at different values of $\tau_w$; therefore, for each tube, data are plotted as log Q against log $\tau_w$. The four linear plots representing data obtained on the four tubes are drawn on the same graph, and at least four values of $\tau_w$ are selected. Points on each line at a constant value of $\tau_w$ are then selected. At each value of $\tau_w$, a plot is made of Q/($\pi$ R$^3$$\tau_w$) against 1/R$^2$. The plot will be linear with a slope, ß$_c$. The value of ß$_c$ may be different at different $\tau_w$; therefore, plots at four different values of $\tau_w$ are needed. After determining ß$_c$ values at each $\tau_w$, Eq. (5.23) is used to determine Q$_c$, which is then used to determine Q$_w$. An example is given in Kokini and Plotchok's (1987) article.

### 5.2.1.7  Glass Capillary Viscometer Used as a Forced Flow Viscometer

Fluids that exhibit non-Newtonian behavior but have low consistency indices are usually difficult to study using tube viscometers because of the low pressure drop. A glass capillary viscometer may be used as a forced flow viscometer by attaching a constant pressure source to the small leg of the viscometer. A simple setup is shown in Fig. 5.8.

The pressure source is a water column of movable height. To operate, the fluid is pipetted into the large leg of the viscometer. The fluid level in the u-tube is made level with the fluid in the large bulb of the viscometer by adjusting the height of the dropping funnel, and the stopcock is shut off. The viscometer is disconnected from the pressure source, and fluid is drawn to the top of the second bulb on the small viscometer leg by suction. After reconnecting the viscometer to the pressure source, the height of the dropping funnel is raised, and pressure is applied by opening the stopcock. The time of efflux is then measured as the time for the fluid meniscus to pass through the first and second mark on the viscometer. Pressure is discontinued by closing the stopcock, and the difference in fluid level between the dropping funnel and the large bulb of the viscometer is measured. The flow behavior and consistency indices will be calculated from the efflux times at different levels of fluid in the pressure source.

**Fig. 5.8** Diagram of a system used to adapt a glass capillary viscometer for forced flow viscometry

In a capillary viscometer, the volume of fluid which passes through the capillary is the volume which fills the section between the two marks on the viscometer. Two other critical factors are the length of the capillary (L) and the height available for free fall (H). These measurements are made on the viscometer as indicated in Fig. 5.8. The radius of the capillary may be calculated from the viscometer constant using Eq. (5.19). When used as a forced flow viscometer, the pressure drop will be the pressure equivalent to the height of the forcing fluid as shown in Fig. 5.8 and the pressure equivalent to the height available for free fall of the fluid in the viscometer. The calculations are shown in the following example.

**Example 5.4** A Cannon-Fenske-type glass capillary viscometer has a height available for free fall of 8.81 cm and a length of capillary of 7.62 cm. When used on distilled water at 24 °C, an efflux time of 36 s was measured. The volume of fluid which drains between the two marks on the viscometer was 3.2 cm$^3$.

(a)   Calculate the radius of the capillary.

(b)   When used as a forced flow viscometer on a concentrated acid whey containing 18.5% total solids at 24 °C, the following data were collected:

| Height of ($H_2$) forcing fluid (cm) | Efflux time ($t_e$) (s) |
|---|---|
| 0 | 92 |
| 7.1 | 29 |
| 20.8 | 15 |
| 34.3 | 9 |

The whey has a density of 1.0763 g/cm³. Calculate the flow behavior and consistency indices.

**Solution**

(a)   At 24 °C, the density and viscosity of water are 947 kg/m³ and 0.00092 Pa · s, respectively.

Using Eq. (5.11):

$$\mu = \frac{\Delta P R^2}{8 L \bar{V}}; \quad \Delta P = \rho g H$$

$$\mu = \frac{\rho g H R^2 \pi R^2 t_e}{3.2 \times 10^{-6}(8)L} \quad R^4 = \frac{\left(3.2 \times 10^{-6}\right) 8 L \mu}{\rho g H \pi t_e}$$

$$\bar{V} = \frac{\text{Volume}}{\pi} R^2 t_e$$

Substituting known values and solving for R:

$$R = \left[ \frac{0.00092\left(3.2 \times 10^{-6}\right)(8)(0.0762)}{997\,(9.8)(0.0881)(\pi)(36)} \right]^{0.25}$$

$$= 3.685 \times 10^{-4} \text{m}$$

(a)   The total pressure forcing the fluid to flow is

$$\Delta P = g(\rho_1 H + \rho_2 H_2)$$

$$\Delta P = 9.8\,[(1076.3)(0.0881) + (997)(H_2)]$$
$$= 9.8(94.822 + 997 H_2)\ \text{Pa}$$

Let q = volumetric rate of flow:

$$q = \frac{3.2}{t_e} \text{cm}^3/\text{s}$$

The pressure-inducing flow and the volumetric rate of flow calculated using the above equations are as follows:

| $\Delta P$ (Pa) | q (cm³/s) |
|---|---|
| 929 | 0.0564 |
| 1623 | 0.110 |
| 2961 | 0.213 |
| 4280 | 0.355 |

A log-log plot of $\Delta P$ against q shown in Fig. 5.9 gives n = 0.84.

The consistency index will be calculated from the rates of shear and the shear stress:

**Fig. 5.9** Log-log plot of pressure drop against the rate of flow to determine the flow behavior index, n, from the slope

$$\bar{V} + \frac{q \times 10^{-6}}{\pi\left(R^2\right)} = 2.3466\,q$$

$$\gamma_w = \frac{4\bar{V}}{R}\left[0.75 + \frac{0.25}{n}\right] = 26699\,q$$

$$= \frac{4(2.3466)}{3.683 \times 10^{-4}}\left[0.75 + \frac{0.25}{0.84}\right]$$

The shear stress is calculated from $\Delta P$. Substituting known values in Eq. (5.22):

$$\tau_w = \frac{\Delta PR}{2L} = \frac{\Delta P\left(3.683 \times 10^{-4}\right)}{2(0.00762)}$$

$$= 24.1666 \times 10^{-4}\Delta P$$

The shear stress and shear rates are the following:

| $\tau_w$ (Pa) | $\gamma_w$ (1/s) |
| --- | --- |
| 2.245 | 1505 |
| 3.922 | 2936 |
| 7.155 | 5687 |
| 10.343 | 9478 |

Figure 5.10 shows a log-log plot of shear stress against shear rate. A point on the line can be used to calculate K. The first data point (2.245,1505) falls directly on the line:

$$K = \frac{\tau_w}{\left(\gamma_w\right)^n} = \frac{2.245}{(1505)^{0.84}} = 0.0048\,\text{Pa} \cdot s^n$$

## 5.2.2  Effect of Temperature on Rheological Properties

Temperature has a strong influence on the resistance to flow of a fluid. It is very important that temperatures be maintained constant when making rheological measurements. The flow behavior index, n, is relatively constant with temperature, unless components of the fluid undergo chemical changes at certain temperatures. The viscosity and the consistency index, on the other hand, are highly temperature dependent.

The temperature dependence of the viscosity and the consistency index can be expressed in terms of the Arrhenius equation:

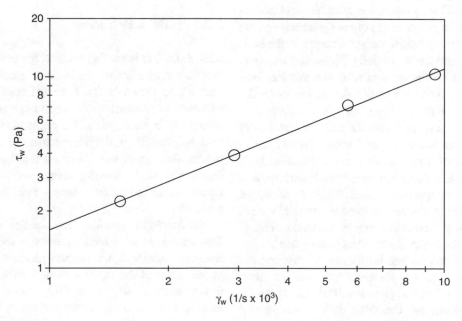

**Fig. 5.10** Log-log plot of shear stress against shear rate to determine a point on the line that can be used to calculate the consistency index, K

$$\ln\left(\frac{\mu}{\mu_1}\right) = \frac{E_a}{R}\left(\frac{1}{T} - \frac{1}{T_1}\right) \quad (5.24)$$

where $\mu$ is the viscosity at absolute temperature T; $\mu_1$ is the viscosity at temperature $T_1$; $E_a$ is the activation energy, J/gmole; R is the gas constant, 8.314 J/(gmole K); and T is the absolute temperature.

Equation (5.24) is useful in interpolating between values of $\mu$ at two temperatures. When data at different temperatures are available, Eq. (5.14) may be expressed as

$$\ln \mu = A + \frac{B}{T} \quad (5.25)$$

where the constants B and A are slope and intercept, respectively, of the plot of $\ln(\mu)$ against 1/T. The same expressions may be used for the temperature dependence of the consistency index, K.

There are other expressions for the dependence of the viscosity or consistency index on temperature such as the Williams-Landel-Ferry (WLF) equation and the Fulcher equation (Rao et al. 1987), but these also have limitations for general use, and determination of the parameters from experimental data requires specialized curve-fitting computer programs.

The WLF equation, however, has implications in the flow behavior of polymer solutions and the viscoelastic properties of gels; therefore, the basis for the model will be briefly discussed. The model is based on the concept of a free volume. Any polymer, by virtue of the size of the molecule, contains a free volume. The free volume may include the space within the helix in a protein or starch molecule or the space formed when sections of a random coil mesh. The magnitude of the free volume is temperature dependent. At a specific temperature called the glass transition temperature, the free volume is essentially zero, and this will be manifested by a drastic change in the thermophysical properties of the material. The free volume is best manifested by the specific volume. A plot of the specific volume with temperature above the glass transition temperature is continuous, and the slope defines the thermal expansion coefficient. The point in the curve

where a discontinuity occurs is the glass transition temperature. The WLF model relates the ratio, $\mu/\mu_g$, where $\mu$ is the viscosity at temperature T and $\mu_g$ is the viscosity at the glass transition temperature, $T_g$, to a function of $T - Tg$. The Arrhenius equation is the simplest to use and is widely applied in expressing the temperature dependence of a number of factors.

**Example 5.5** The flow behavior index of applesauce containing 11% solids is 11.6 and 9.0 Pa · s$^n$, at 30 °C and 82 °C, respectively. Calculate the flow behavior index at 50 °C.

**Solution**

Using Eq. (6.24), T = 303 K; $T_1$ = 355 K:

$$\frac{E_a}{R} = \frac{\ln(11.6/9.0)}{1/303 - 1/355} = 524.96K$$

$$= 11.6[e]^{524.96(1/323-1/303)}$$

Using Eq. (5.14) with $k_1$ = 11.6, $T_1$ = 303 K, and T = 323 K:

$$K = 11.6(0.898) = 10.42 \text{ Pa} \cdot \text{s}^n$$

### 5.2.3 Back Extrusion

Back extrusion is another method for evaluating the flow properties of fluids. It is particularly useful with materials that have the consistency of a paste or suspensions that have large particles suspended in them, since the suspended solids tend to intensify wall effects when the fluid is flowing in a small tube. When a rotational viscometer is used, pulsating torque readings are usually exhibited when a lumpy fluid is being evaluated.

The key to the accurate determination of fluid flow properties using back extrusion is the assurance of annular flow, i.e., the plunger must remain in the center of the larger stationary cylinder all throughout the test. Figure 5.11 shows a back extrusion device designed for use on an Instron universal testing machine. A stanchion positioned

TO LOAD
CELL

ORIGINAL
FLUID LEVEL

$L_p$

$R_s$

$R_P$    TO BASE
$R_C$    PLATE

**Fig. 5.11** Diagram of a back extrusion cell suitable for measuring flow properties of fluids with lumpy or paste-like consistency

at the center of the large cylinder which is machined to a close tolerance to fit an opening at the center of the plunger guides the movement of the plunger and ensures concentric positioning of the plunger and outer cylinder at all times during the test.

The equations that govern flow during back extrusion have been derived by Osorio and Steffe (1987). Let $F_t$ be the force recorded at the maximum distance of penetration of the plunger $L_p$. The height of the column of fluid which flowed through the annulus, L, is

$$L = \frac{A_p L_p}{A_a}$$

where $A_p$ is the area of the plunger which displaces fluid and $A_a$ is the cross-sectional area of the annulus:

$$A_p = \pi\left(R_p^2 - R_s^2\right); \quad A_a = \pi\left(R_c^2 - R_p^2\right)$$

where $R_c$ is the radius of the cylinder, $R_p$ is the radius of the plunger, and $R_s$ is the radius of the stanchion which positions the plunger in the

center of the cylinder. $L_p$ cannot easily be measured on the plunger and cylinder assembly itself since the fluid level in the cylinder is not visible outside, but it can easily be read from the force tracing on the Instron chart, by measuring the distance from the initiation of the rise in force and the point of maximum force on the chart. Figure 5.12 shows a typical back extrusion force-displacement curve.

Part of the total force $F_t$ counteracts the force of gravity on the mass of fluid in the annulus.

Force of gravity on annular fluid $= \rho g \, L\pi$ $(R_c^2 - R_p^2)$

The net force, $F_n$, which is exerted by the fluid to resist flow through the annulus, is the total force corrected for the force of gravity:

$$F_n = F_t - \rho g L\pi\left(R_c^2 - R_p^2\right)$$

where $\rho$ = density of the fluid.

The flow behavior index, n, is the slope of a log-log plot of $F_n/L$, against the velocity of the plunger, $V_p$.

Determination of the shear rate at the plunger wall and the shear stress at this point involves a rather complex set of equations which were derived by Osorio and Steffe (1987). The derivation is based on a dimensionless radius, $\lambda$, which is the ratio $r_m/R_c$, where $r_m$ is the radius of a point in the flow stream where velocity is maximum. The deferential momentum and mass balance equations were solved simultaneously to obtain values of $\lambda$ for different values for the flow behavior index of fluids and different annular gaps. The authors presented results of the calculations as a table of $\lambda$ for different n and annular gap size expressed as $\sigma$, the ratio of plunger to cylinder radius ($R_p/R_c$). Table 5.1 shows these values. Figure 5.13 related the value of $\sigma$ to a parameter $\theta$ for fluids with different n. The parameter $\theta$ is used to calculate the shear rate at the wall from the plunger velocity, $V_p$.

Let $\sigma = R_p/R_c$; $\sigma_s = R_s/R_c$; $\lambda = r_m/R_c$

The parameter $\theta$ in Fig. 5.12 is a dimensionless parameter which satisfies the expressions for the integral of the dimensionless point velocity of fluid through the annulus and the quantity of fluid displaced by the plunger. The terms $V_p$, P,

**Fig. 5.12** Typical force
deformation curve obtained
for back extrusion showing
how total force, plunger
displacement, and yield
force are obtained

**Table 5.1**  Values of $\lambda$ for different values of $\sigma$ and $n$

|  | $n$ | | | | | | | | | |
|---|---|---|---|---|---|---|---|---|---|---|
| $\sigma$ | 0.1 | 0.2 | 0.3 | 0.4 | 0.5 | 0.6 | 0.7 | 0.8 | 0.9 | 1.0 |
| 0.1 | 0.4065 | 0.4889 | 0.5539 | 0.6009 | 0.6344 | 0.6586 | 0.6768 | 0.6907 | 0.7017 | 0.7106 |
| 0.2 | 0.5140 | 0.5680 | 0.6092 | 0.6398 | 0.6628 | 0.6803 | 0.6940 | 0.7049 | 0.7138 | 0.7211 |
| 0.3 | 0.5951 | 0.6313 | 0.6587 | 0.6794 | 0.6953 | 0.7078 | 0.7177 | 0.7259 | 0.7326 | 0.7382 |
| 0.4 | 0.6647 | 0.6887 | 0.7068 | 0.7206 | 0.7313 | 0.7399 | 0.7469 | 0.7527 | 0.7575 | 0.7616 |
| 0.5 | 0.7280 | 0.7433 | 0.7547 | 0.7636 | 0.7705 | 0.7761 | 0.7807 | 0.7846 | 0.7878 | 0.7906 |
| 0.6 | 0.7871 | 0.7962 | 0.8030 | 0.8082 | 0.8124 | 0.8158 | 0.8186 | 0.8209 | 0.8229 | 0.8246 |
| 0.7 | 0.8433 | 0.8480 | 0.8516 | 0.8544 | 0.8566 | 0.8584 | 0.8599 | 0.8611 | 0.8622 | 0.8631 |
| 0.8 | 0.8972 | 0.8992 | 0.9007 | 0.9019 | 0.9028 | 0.9035 | 0.9042 | 0.9047 | 0.9052 | 0.9055 |
| 0.9 | 0.9493 | 0.9498 | 0.9502 | 0.9504 | 0.9507 | 0.9508 | 0.9510 | 0.9511 | 0.9512 | 0.9513 |

From: Osorio and Steffe (1987); used with permission

and K in the expression of $\theta$ in Fig. 5.13 are $V_p$
for plunger velocity, P for pressure drop per unit
distance traveled by the fluid, and k for the fluid
consistency index.

Knowing the flow behavior index of the fluid
and the dimensionless plunger radius, $\sigma$, a value
for $\lambda$ is obtained from Table 5.1. Figure 5.13 is
then used to obtain a value for $\theta$. The shear rate at
the plunger wall is calculated using Eq. (5.26):

$$-\left.\frac{dV}{dr}\right|_{R_p} = \frac{V_p}{R_c\theta}\left(\sigma^2 - \sigma_s^2\right)\left[\frac{\lambda^2}{\sigma} - \sigma\right]^{\frac{1}{n}} \quad (5.26)$$

The shear stress at the plunger wall is calculated from the net force, $F_n$, exerted on the base of
the column of fluid to force it up the annulus.

The shear stress at the wall, $\tau_w$, is calculated
using Eq. (5.27). Equations (5.26) and (5.27) are
derived to account for the fact that part of the

**Fig. 5.13** Graph of the parameter θ used to determine the shear rate from the plunger velocity. (Source: Osorio and Steffe 1987. Used with permission)

plunger area being occupied by the stanchion which positions the plunger in the center of the concentric cylinder does not displace fluid up through the annular gap:

$$\tau_w = \frac{F_n}{2\pi L R_p}\left(\frac{\lambda^2 - \sigma^2}{\lambda^2 - \sigma^2}\right) \qquad (5.27)$$

The consistency index of the fluid can then be determined from the intercept of a log-log plot of shear stress and shear rate.

The yield stress can be calculated from the residual force after the plunger has stopped, $F_{ne}$, corrected for the hydrostatic pressure of the height of fluid in the annulus:

$$\tau_o = \frac{(F_{ne})}{2\pi(R_p + R_c)L} \qquad (5.28)$$

where $F_{ne} = F_{te} - \rho g L \pi\left(R_c^2 - R_p^2\right)$ and $F_{te}$ is the residual force after the plunger has stopped.

**Example 5.6**  The following data were collected in back extrusion of a fluid through a device similar to that shown in Fig. 5.8. The diameters were 2.54 cm for the stanchion, 6.64 cm for the plunger, and 7.62 cm for the outer diameter. The fluid had a density of 1.016 g/cm$^3$. The total force, the distance penetrated by the plunger, and the plunger speed at each of the tests are as follows:

| Plunger speed, $V_p$ (mm/min) | Penetration depth, $L_p$ (cm) | Total force, $F_p$ (N) |
|---|---|---|
| 50 | 15.2 | 11.44 |
| 100 | 13.9 | 12.16 |
| 150 | 16.0 | 15.36 |
| 200 | 13.2 | 13.60 |

Calculate the flow behavior and consistency indices for this fluid.

**Solution**
The flow behavior index is calculated by determining the ratio $F_n/L$ and plotting against $V_p$:

$$L = L_p \frac{\left[(0.0332)^2 - (0.0127)^2\right]}{\left[(0.0381)^2 - (0.0332)^2\right]}$$

$$= 2.6905 L_p$$

$$F_n = F_t - 1016(9.8)(2.6905 L_p)$$

$$(\pi)\left[(0.0381)^2 - (0.0332)^2\right]$$

$$= F_t - 29.4028\ L_p$$

Values of $F_n$, $L$, and $F_n/L$ at different $V_p$ are as follows:

| $V_p$ (mm/min) | $L$ (m) | $F_n$ (N) | $F_n/L$ (N/m) |
|---|---|---|---|
| 50 | 0.408 | 6.98 | 17.1 |
| 100 | 0.374 | 8.08 | 21.59 |
| 150 | 0.430 | 10.66 | 24.79 |
| 200 | 0.355 | 9.72 | 27.37 |

A log-log plot of $F_n/L$ against $V_p$ is shown in Fig. 5.14. The slope is 0.33, the flow behavior index, n. The value of $\lambda$ for $\sigma = 0.871$, n = 0.33 is obtained by interpolation from Table 5.1.

The value of $\lambda$ for n = 3, $\sigma = 0.871$ is

$$\lambda = 0.9007 + \frac{(0.9504 - 0.9019)(0.071)}{0.1}$$

$$= 0.9351$$

The value of $\lambda$ for n = 0.4, $\sigma = 0.871$ is

$$\lambda = 0.9019 + \frac{(0.9504 - 0.9019)(0.071)}{0.1}$$

$$= 0.9363$$

The value of $\lambda$ for n = 0.33 is obtained by interpolation:

$$\lambda = \frac{0.9351 + (0.9363 - 0.9351)(0.03)}{0.01}$$

$$= 0.9354$$

The value of $\theta$ is obtained from Fig. 5.13. The curves for n = 0.3 and n = 0.4 do not intercept $\sigma = 0.871$ in Fig. 5.13. However, the curves are linear at low values of $\theta$ and may be represented by the following equations:

**Fig. 5.14** Log-log plot of corrected force to extrude fluid through an annular gap, against piston velocity, to obtain the flow behavior index, n, from the slope

$$n = 0.3 : \log\theta = 1.48353 - 7.5767\sigma$$
$$n = 0.4 : \log\theta = 1.9928 - 7.4980\sigma$$

Thus, at $n = 0.3$, $\theta = 7.665 \times 10^{-6}$; at $n = 0.4$, $\theta = 2.8976 \times 10^{-5}$. At $n = 0.33$, $\theta = 1.4058 \times 10^{-5}$ by interpolation. Knowing the value of n, $\theta$, and $\lambda$, $\tau_w$ and $\gamma_w$ can be calculated. Using Eq. (5.26):

$$\sigma = 0.871, \lambda = 0.9354, \sigma_s = 0.3333, 1/n = 3.$$
$$\gamma_w = \frac{1}{0.381\left(1.4058 \times 10^{-5}\right)}\left[(0.871)^2\right.$$
$$\left. -(0.3333)^2\right]\left[\frac{(0.9354)^2}{0.087}\right]^3\left(\frac{1}{60,000}\right)V_p$$
$$= 0.04796\ V_p;$$

where $V_p$ is in mm/min.
Using Eq. (5.27):

$$\tau_w = \frac{(0.9354)^2 - (0.871)^2}{2\pi(0.0332)\left[(0.9354)^2 - (0.3333)^2\right]}\left(\frac{F_n}{L}\right)$$
$$= 0.73(F_n/L)$$

The values of the shear stress and shear rates are the following:

| $\tau_w$ (Pa) | $\lambda_w$ |
|---|---|
| 12.48 | 2.398 |
| 15.76 | 4.796 |
| 18.09 | 7.194 |
| 19.98 | 9.592 |

The consistency index, K, is determined from the intercept of the log-log plot in Fig. 5.15:

$$K = 9.2\ \text{Pa} \cdot \text{s}^n$$

### 5.2.4 Determination of Rheological Properties of Fluids Using Rotational Viscometers

The most common rotational viscometer used in the food industry is the concentric cylindrical viscometer. The viscometer shown in Fig. 5.16

**Fig. 5.15** Log-log plot of
shear stress against shear
rate to obtain the
consistency index, K, from
the intercept

**Fig. 5.16** Schematic diagram of a rotational viscometer

consists of concentric cylinders with fluid in the
annular space. This system is also referred to as a
Couette. The torque needed to rotate one of the
cylinders is measured. This torque is proportional
to the drag offered by the fluid to the rotation of
the cylinder.

If the outer cup is stationary and if the torque
measured is A, the force acting on the surface of
the inner cylinder to overcome the resistance to
rotation will be A/R.

The shear stress $\tau_w$ at the wall is

$$\tau_w = \frac{A}{R}\frac{1}{2\pi RL} = \frac{A}{R^2(2\pi L)} \quad (5.29)$$

For a Newtonian fluid in narrow gap
viscometers where the rotating cylinder has a
radius, R, and the cup radius is $R_c$, the shear rate
at the wall at a rotational speed of N rev/s is

$$\gamma_w = \frac{4\pi N R_c}{\left(R_c^2 - R^2\right)} \quad (5.30)$$

If the fluid is non-Newtonian with a flow
behavior index, n, the shear rate at the wall
has been derived by Kreiger and Maron (1954)
for a Couette-type viscometer ($R_c/R < 1.2$) as
follows:

$$\gamma_w$$

$$= N\left[\frac{4\pi R_c^2}{(R_c^2 - R^2)} + 2\pi\left[1 + \frac{2}{3}\ln\left(\frac{R_c}{R}\right)\right]\left(\frac{1-n}{n}\right) + \frac{2\pi}{3}\ln\left(\frac{R_c}{R}\right)\left[\frac{1-n}{n}\right]^2\right]$$

$$(5.30a)$$

Equation (5.30) is a special case of Eq. (5.30a) when the flow behavior index is equal to 1. The factor in brackets which is a multiplier for N in Eq. (5.30a) is easily calculated using a spreadsheet program.

**Example 5.7** A rotational viscometer with a spring constant equivalent to 673.7 dyne $\cong$ cm full scale is used on a narrow gap viscometer with a rotating cylinder diameter of 29.5 mm and a cup diameter of 32 mm. The cylinders are 44.5 mm high. Assume end effects are negligible. This Couette-type viscometer is used to measure the flow properties of whole eggs, and the data are shown in the spreadsheet (Fig. 5.17) with N in rev/min entered in cells A4–A7 and torque in % of full scale entered in cells B4–B7. Calculate the flow behavior and consistency indices in SI units. The term in brackets to multiply N to obtain the shear rate in Eq. (5.30a) has a value of 87.2 when n is equal to 0.66 and is entered in cell A12 in Fig. 5.17.

**Solution**
From the spreadsheet (Fig. 5.17), the slope of the regression of log(torque) versus log(RPM) is 0.66. This is the value of n used to calculate the multiplying factor for N to obtain the shear rate from the rotational speed. The consistency index is the antilog of the intercept of a regression of log $(\tau)$ against log$(\gamma)$. The intercept is $-1.246$ and the value of $K = 10^{-1.246} = 0.057$ Pa $\cdot$ s$^n$.

### 5.2.4.1 Wide-Gap Rotational Viscometer

A wide-gap rotational viscometer consists of a cylinder or a disk-shaped bob on a spindle, which rotates in a pool of liquid. A torque transducer attached to the base of the rotating spindle measures the drag offered by the fluid to the rotation of the bob. Different size spindles may be used. Based on the spindle size and the rotational speed, the indicated torque can be converted to an apparent viscosity by using an appropriate conversion factor specific for the size of spindle and rotational speed. For a shear-thinning fluid, the apparent viscosity increases as the rotational speed decreases. A log-log plot of the apparent viscosity against rotational speed will have a slope equivalent to n − 1. If one spindle is used at several rotational speeds, the unconverted torque reading is directly proportional to the shear stress, and the rpm is directly proportional to the shear rate. A log-log plot of torque reading against rpm will have a slope equivalent to the flow behavior index n.

**Example 5.8** A Brookfield viscometer model RVF was used to evaluate the apparent viscosity of tomato catsup. One spindle (No. 4) gave readings within the measuring scale of the instrument at four rotational speeds. The viscometer constant was 7187 dyne $\cong$ cm full scale. The torque readings in percent full scale at various rotational speeds in revolutions/min (rpm), respectively, are as follows: (53.5, 2), (67, 4), (80.5,10), and (97, 20).

**Solution**
n is the slope of a log-log plot of torque against rpm.

| Speed (rpm) | Indicator reading (% full scale) | Torque (dyne-cm) |
| --- | --- | --- |
| 2 | 53.5 | 3845 |
| 4 | 67 | 4815 |
| 10 | 80.5 | 5786 |
| 20 | 97 | 6971 |

The data plotted in Fig. 5.18 shows a slope of 0.25, which is the value of n.

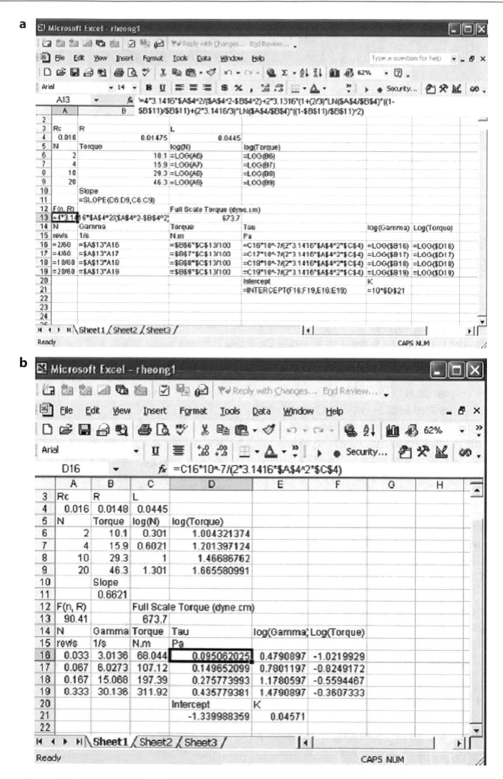

**Fig. 5.17** (**a**) Spreadsheet program in Microsoft Excel showing data entered and calculated values of shear stress; shear rate; flow behavior index, n; and consistency index, K. (**b**) Spreadsheet program in Microsoft Excel showing formulas in the cells used to calculate values in Fig. 5.17a

**Fig. 5.18** Log-log plot of torque against rotational speed to obtain the flow behavior index from the slope

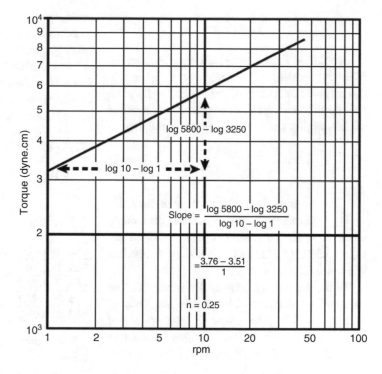

## 5.2.4.2   Wide-Gap Viscometer with Cylindrical Spindles

An important characteristic of a wide-gap viscometer is that the fluid rotates in the proximate vicinity of the surface of the rotating spindle at the same velocity as the spindle (zero slip) and the fluid velocity reaches zero at some point before the wall of the stationary outer cylinder. If these characteristics are met in a system, Kreiger and Maron (1952) derived an expression for the shear rate at the rotating inner cylinder wall as follows:

Let $\omega$ be the angular velocity at a distance $r$ from the center of the rotating spindle. $\gamma_w = (-dV/dr)$. Because $V = \omega r$, then $dV = rd\omega$:

$$-\frac{dV}{dr} = rd\frac{\omega}{dr}$$

The shear stress "$\tau$" at a point in the fluid a distance "$r$" from the center of the rotating spindle is

$$\tau = \frac{\text{Torque}}{2\pi r^2 h} \qquad r^2 = \frac{\text{Torque}}{2\pi\tau h}$$

Differentiating the expression for shear stress:

$$d\tau = \frac{\text{Torque}}{2\pi h} - 2\frac{dr}{r^3}$$

Substituting $r^2$:

$$d\tau = \frac{\text{Torque}}{2\pi h}\left(-2\frac{dr}{r}\right)\left(\frac{2\pi h}{\text{Torque}}\right)d\tau$$
$$= \frac{-2\tau dr}{r}$$

Solving for $dr/r$, $dr/r = (-1/2)d\tau/\tau$.

Because $(-dV/dr) = F(\tau)$, then $r\,(d\omega/dr) = F(\tau)$, and $d\omega = (dr/r)F(\tau)$.

Substituting $(dr/r) = d\tau/2\tau$, $d\omega = (d\tau/2\tau)F(\tau)$ and $d\omega = (1/2)F(\tau)\,d\ln(\tau)$.

At the surface of the spindle, the shear stress is $\tau_w$ and the angular velocity is $2\pi N$. Thus, $d(2\pi N) = (1/2)\,F(\tau_w)\,d\ln(\tau_w)$. Because $(-dV/dr)_w = F(\tau_w)$, then $2\pi dN = (1/2)(-dV/dr)\,d\ln(\tau_w)$ and $-(dV/dr)_w = 4\pi dN/d\ln(\tau_w)$.

Because $d\ln(N) = dN/N$, then $dN = N\,d\ln(N)$.

$$-(dV/dr)_w = 4\pi N/[d\ln(\tau_w)/d\ln(N)]$$

The denominator is the slope of the log-log plot of torque versus $N$ and has the same value as the flow behavior index $n$:

$$\gamma_w = \frac{4\pi N}{n} \qquad (5.31)$$

Brookfield Engineering Laboratories gives the following equation for shear rate at a point x distant from the center of the rotating cylindrical spindle in a wide-gap viscometer. R is the radius of the spindle, and Rc is the radius of the cup:

$$\gamma = \frac{4\pi N Rc^2 R^2}{x^2(Rc^2 - R^2)} \qquad (5.31a)$$

If Rc is much bigger than R, then $Rc^2/(Rc^2 - R^2) \sim 1$; therefore, at the surface of the spindle where $x = R$, the shear rate is $\gamma = 4\pi N$. Thus Eqs. (5.30), (5.31), and (5.31a) are equivalent when $n = 1$.

To use Eq. (5.31), a value of n is needed. The value of n is determined by plotting log(Torque) against log(N). The slope is the value of n. Knowing n, Eq. (5.31) can then be used to calculate $\gamma_w$. $\tau_w$ is calculated using Eq. (5.29). A log-log plot of $\tau_w$ against $\gamma_w$ will have K for an intercept.

**Example 5.9** The following data were obtained in the determination of the flow behavior of a fluid using a rotational viscometer. The viscometer spring constant is 7197 dyne $\cong$ cm full scale. The spindle has a diameter of 0.960 cm and a height of 4.66 cm.

| N (rev/min) | Torque (fraction of full scale) |
|---|---|
| 2 | 0.155, 0.160, 0.157 |
| 4 | 0.213, 0.195, 0.204 |
| 10 | 0.315, 0.275, 0.204 |
| 20 | 0.375, 0.355, 0.365 |

Calculate the consistency and flow behavior indices for this fluid.

**Solution**
Equation (5.29) will be used to calculate $\tau_w$ and Eq. (5.31) for $\gamma_w$.

It will be necessary to determine n before Eq. (5.31) can be used. The slope of a log-log plot of torque against the rotational speed will be the value of n. This plot shown in Fig. 5.19 shows a slope of 0.36, which is the value of n. The

multiplying factor to convert the rotational speed in rev/min to the shear rate at the wall of the rotating cylinder, $\gamma_w$, in s$^{-1}$ units is as follows: $4(3.1416)/[0.36(60)] = 0.5818$. The multiplying factor to convert the indicated torque as a fraction of viscometer full-scale torque to the shear stress at the rotating cylinder wall, $\tau_w$, in Pascal is

$$7187(10^{-7})/\left[2(3.1416)(0.0096/2)^2(0.0466)\right]$$
$$= 106.5367.$$

The values for $\tau_w$ and $Q_w$ are the following:

| Shear stress ($\tau_w$) (Pa) | Shear rate ($\gamma_w$) (1/s) |
|---|---|
| 16.5, 17.0, 16.7 | 1.16 |
| 22.7, 20.8, 21.7 | 2.33 |
| 33.6, 29.3, 21.7 | 5.82 |
| 39.9, 37.8, 38.9 | 11.64 |

The plot of $\tau_w$ against $\gamma_w$ shown in Fig. 5.20 shows that $\tau_w = 15.2$ Pa at $\gamma_w = 1$; thus, $K = 15.2$ Pa Xs$^n$.

## 5.3   Continuous Viscosity Monitoring and Control

In the formulation of foods where viscosity or consistency is a quality attribute, continuous monitoring of flow properties and automatic control of the feeding of ingredients is important. Two good examples of these processes are blending of tomato paste with sugar, vinegar, and flavoring in manufacturing of catsup and blending of flour, water, and flavoring ingredients to make batter for battered and breaded fried fish, poultry, or vegetables. These processes are rather simple because the level of only one ingredient in the formulation controls the consistency of the blend and therefore will be very suitable for automatic control of flow properties. The principles for continuous viscosity monitoring are similar to those used in rheology. Some means must be used to divert a constant flow of fluid from the main pipeline to the measuring device. There are several configurations of viscosity measuring devices suitable for continuous monitoring in flowing

**Fig. 5.19** Log-log plot of indicated torque against the rotational speed to obtain the flow behavior index, n, from the slope

systems. Three of the easiest to adapt on a system are described below.

### 5.3.1   Capillary Viscometer

A capillary viscometer may be used for continuous monitoring of consistency by tapping into the main piping, using a metering pump to deliver a constant flow through a level capillary, and returning the fluid to the main pipe downstream from the intake point. The metering pump must be a positive displacement pump to prevent flow fluctuations due to changes in the main pipeline pressure. A differential pressure transducer mounted at the entrance and exit of the capillary is used to measure the pressure drop. Poiseuille equation (Eq. 5.11) is used to calculate the apparent viscosity from the pressure difference measured and the flow delivered by the metering pump.

### 5.3.2   Rotational Viscometer

Rotational viscometers could be used for continuous monitoring of viscosity in open vessels. Provision must be allowed for changes in fluid levels to ensure that the immersion depth of the spindle is appropriate. Mounting the instrument on a float device resting on the fluid surface will ensure the same spindle immersion depth regardless of fluid level. Some models of rotational viscometers can remotely display the torque; others give a digital readout of the torque which can be easily read on the instrument.

For fluids flowing through pipes, problems may be encountered with high fluid velocities. Semicontinuous measurement may be made by periodically withdrawing fluid from the pipe through a sampling valve and depositing this sample into a cup where the viscosity is measured by a rotational viscometer. This type of measurement will have a longer response time than a truly in-line measurement.

**Fig. 5.20** Log-log plot of shear stress against shear rate to obtain the consistency index, K, from the intercept

### 5.3.3    Viscosity-Sensitive Rotameter

This device is similar to a rotameter and operates on the principle that the drag offered by an obstruction in a field of flow is proportional to the viscosity and the rate of flow. A small float positioned within a vertical tapered cylinder will assume a particular position when fluid a fixed flow   rate and with a particular viscosity is flowing through the tube. This device needs a constant rate of flow to be effective, and usually, a metering pump draws out fluid from a tap on a pipe, delivers the fluid through the viscosity sensitive rotameter, and returns the fluid to the pipe through a downstream tap.

### 5.4    Flow of Falling Films

Fluid films are formed when fluids flow over a vertical or inclined surface or when fluid is withdrawn from a tank at a rapid rate. The differential equation for the shear stress and shear rate relationship for the fluid coupled with a differential force balance will allow a resolution of the thickness of the film as a function of fluid flow properties.

### 5.4.1    Films of Constant Thickness

When the rate of flow of a fluid over a vertical or inclined plane is constant, the film thickness is also constant. Figure 5.21 shows a fluid film flowing down a plane inclined at an angle $\alpha$ with the vertical plane. If the width of the plane is W, a force balance on a macroscopic section of the film of length L and of thickness x, measured from the film surface, is as follows:

Mass of the differential increment $= LW\rho x$
Force due to gravity on mass $= LW\rho gx$
Component of gravitational force along inclined
   plane $= LW\rho gx \cos(\alpha)$

**Fig. 5.21** Diagram of a control element used in analysis of fluid film flowing down an inclined plane

Fluid resistance against the flow $= \tau WL$

If the fluid follows the power law equation (Eq. 5.2), a force balance results in:

$$WLK\left[-\frac{dV}{dr}\right]^n = LW\rho gx \, \cos{(\alpha)}$$

Rearranging:

$$-\frac{dV}{dx} = \left[\frac{\rho \, gx \, \cos{(\alpha)}}{K}\right]^{1/n}$$

Integrating:

$$\int dV = -\left[\frac{\rho \, g \, \cos{(\alpha)}}{K}\right]^{1/n}\int [x]^{1/n}dx$$

The integrated expression is

$$V = -\left[\frac{\rho \, g \, \cos{(\alpha)}}{K}\right]^{1/n}\left[\frac{n}{n+1}\right][x]^{(n+1)/n} + C$$

The boundary condition, at $x = \delta$, $V = 0$, when substituted in the above expression will give the value of the integration constant, C. $\delta$ is the film thickness, and since x is measured from the surface of the film, the position $x = \delta$ is the surface of the

incline along which the film is flowing. The final expression for the fluid velocity profile is

$$V = \left[\frac{\rho \, g \, \cos{(\alpha)}}{K}\right]^{1/n}\left[\frac{n}{n+1}\right]$$
$$[\delta]^{(n+1)/n}\left[1 - \left[\tfrac{x}{\delta}\right]^{(n+1)/n}\right]$$

The thickness of the film $\delta$ can be expressed in terms of an average velocity of flow, $\bar{V}$. The expression is derived as follows:

$$W\delta\bar{V} = \int_0^\delta VW \, dx = W\left[\frac{\rho g \, \cos{(\alpha)}}{K}\right]^{1/n}\left[\frac{n}{n+1}\right]$$

$$[\delta]^{(n+1)/n}\int_0^\delta\left[1 - \left[\frac{x}{\delta}\right]^{(n+1)/n}\right]dx\bar{V}$$

$$= \left[\frac{\rho g \, \cos{(\alpha)}}{K}\right]^{1/n}\left[\frac{n}{2n+1}\right][\delta]^{(n+1)/n}$$

$$(5.32)$$

The film thickness is

$$\delta = \left[\frac{\bar{V}(2n+1)K^{1/n}}{n[\rho \, g \, \cos{(\alpha)}]^{1/n}}\right]^{n/(n+1)} \qquad (5.33)$$

**Example 5.10** Applesauce at 80 °C has flow properties of K = 9 Pa · s$^n$; n = 0.33, $\rho$ = 1030 kg/m$^3$. This product is to be deaerated by allowing to flow as a film down the vertical walls of a tank 1.5 m in diameter. If the film thickness desired is 5 mm, calculate the mass rate of flow of applesauce to be fed into the deaerator. At a film thickness of 5 mm and 80 °C, this film must be exposed 15 s to a vacuum of 25 in. Hg to reduce dissolved oxygen content to a level needed for product shelf stability. Calculate the height of the tank needed for the film to flow over.

**Solution**

Equation (5.32) will be used to calculate the average velocity of the film needed to give a film thickness of 5 mm. Since the side is vertical, $\alpha = 0$ and $\cos \alpha = 1$:

$$\bar{V} = \left[\frac{(1030)(9.8)}{9.0}\right]^{1/0.33}\left[\frac{(0.005)^{4.03}}{5.03}\right]$$

$$= 0.1845\,\text{m/s}$$

The mass rate of flow, $\bar{m}$, is

$$\bar{m} = \frac{0.1845\,\text{m}}{\text{s}}(\pi)\left[(0.75)^2 - (0.75 - .005)^2\right]$$

$$\text{m}^2\left(\frac{1030\,\text{kg}}{\text{m}^3}\right) = 4.48\,\text{kg/s}$$

The height of the vessel needed to provide the 15 s required exposure for deaeration is

$$h = \frac{0.1845\,\text{m}}{\text{s}}(15\,\text{s}) = 2.77\,\text{m}$$

### 5.4.2    Time-Dependent Film Thickness

Problems of this type are encountered when fluids cling as a film on the sides of a storage vessel when the vessel is emptied (drainage) or when a solid is passed through a pool of fluid and emerges from the fluid coated with a fluid film (withdrawal). The latter process may be encountered when applying batter over a food product prior to breading and frying. The solution to problems of withdrawal and drainage is the same.

Figure 5.22 shows a fluid film with width W falling down a vertical surface. A control segment of the film having a thickness $\delta$ located a distance z from the top, and a length dz. along the surface on which it flows, is shown. The solution will involve solving partial differential equations which involves a mass balance across the control volume.

If $\bar{V}$ is the average velocity of the fluid in the film at any point z distant from the top, the mass of fluid entering the control volume from the top is

$$M_i = \bar{V}\delta W\rho$$

The mass leaving the control volume from the lower part of the control volume is

$$M_e = \left[\bar{V}\delta + \frac{\partial}{\partial z}(\bar{V}\delta)dz\right]W\rho$$

The accumulation term due to the difference between what entered and left the control volume is

$$\frac{\partial\delta}{\partial z}\rho\,Wdz$$

The mass balance across the control volume is

$$\bar{V}\delta W\rho = \left[\bar{V}\delta + \frac{\partial}{\partial z}(\bar{V}\delta)dz\right]\rho W + \frac{\partial\delta}{\partial z}\rho\,W\,dz$$

Simplification will produce the partial differential equation which describes the system.

$$\frac{\partial}{\partial z}(\bar{V}\delta) = -\frac{\partial\delta}{\partial t} \qquad (5.34)$$

The solution to this partial differential equation will be the product of two functions, G(t) which is a function only of t and F(z) which is a function only of z:

$$\delta = F(z)G(t) \qquad (5.35)$$

Partial differentiation of Eq. (5.35) gives

$$\frac{\partial\delta}{\partial t} = F(z)\frac{\partial G(t)}{\partial t} \qquad (5.36)$$

$$\frac{\partial\delta}{\partial z} = G(t)\frac{\partial F(z)}{\partial z} \qquad (5.37)$$

Equation (5.32) represents $\bar{V}$ for a fluid film flowing down a vertical surface ($\cos\alpha = 1$). Partial differentiation of $\bar{V}\delta$ with respect to z gives

$$\frac{\partial}{\partial z}(\bar{V}\delta) = \frac{\partial}{\partial z}\left[\left[\frac{\rho g}{K}\right]^{1/n}\left[\frac{n}{2n+1}\right][\delta]^{(n+1)/n}\delta\right]$$

$$= \left[\frac{\rho g}{K}\right]^{1/n}[\delta]^{(n+1)/n}\frac{\partial\delta}{\partial z}$$

$$(5.38)$$

Substituting Eq. (5.34) for M/Mz $(\bar{V}\delta)$, Eq. (5.36) for Mδ/Mt, and Eq. (5.37) for Mδ/Mz in Eq. (5.38):

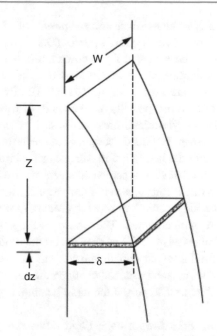

**Fig. 5.22** Diagram of a control element used in analysis of time dependent flow of fluid film down a surface

$$-F(z)\frac{\partial G(t)}{\partial t} = \left[\frac{\rho}{K}\right]^{1/n}[F(z)G(t)]^{(n+1)/n}G(t)$$
$$\times \frac{FF(z)}{\partial z}$$

Each side of the partial differential equation contains only one variable; therefore, the equation can be separated and each side integrated independently by equating it to a constant. Let the constant $= \lambda$.

The function with the time variable is obtained by integrating the left hand side of the equation, setting it equal to $\lambda$. The function with the z variable is obtained the same way, using the right side of the equation:

$$F(z) = \left[\lambda\left[\frac{K}{\rho g}\right]^{1/n}\left[\frac{n}{n+1}\right]z\right]^{n/(n+1)} + C_2$$

$$G(t) = \left[\lambda\left[\frac{n+1}{n}\right]t\right]^{-n/(n+1)} + C_1$$

The boundary conditions, $\delta = 0$ at $z = 0$ and $t = 4$, will be satisfied when the integration constants $C_1$ and $C_2$ are zero. When expressions for $F(z)$ and $G(t)$ are combined according to Eq. (5.35), the constant $\lambda$ cancels out. The final

expression for the film thickness as a function of time and position is

$$\delta = \left[\left[\frac{K}{\rho g}\right]^{1/n}\left[\frac{z}{t}\right]\right]^{n/(n+!)} \tag{5.39}$$

**Example 5.11** Calculate the thickness of batter that would be attached to the sides of a brick-shaped food product, at the midpoint of its height, if the product is immersed in the batter, removed, and allowed to drain for 5 s before breading. The batter has a flow behavior index of 0.75, a density of 1003 kg/m$^3$, and a consistency index of 15 Pa · s$^n$. The vertical side of the product is 2 cm high.

**Solution**
$z = 0.001$ m; $t = 5$ s. Using Eq. (5.38):

$$\delta = \left[\left[\frac{15}{1003(9.8)}\right]^{1/0.75}\left(\frac{0.001}{5}\right)\right]^{0.75/1.75}$$

$$= 0.000599\,\text{m}$$

**Example 5.12** Applesauce is rapidly drained from a drum 75 cm in diameter and 120 cm high. At the end of 60 s after the fluid is drained from the drum, calculate the mass of applesauce adhering to the walls of the drum. The applesauce has a flow behavior index of 0.34, a consistency index of 11.6 Pa · s$^n$, and a density of 1008 kg/m$^3$

**Solution**
Equation (5.39) will be used. However, the thickness of the film will vary with the height; therefore, it will be necessary to integrate the quantity of fluid adhering at the different positions over the height of the vessel. At any given time, t, the increment volume of fluid dq, over an increment of height dz, will be

$$dq = \pi\left[r^2 - (r - \delta)^2\right]dz$$

The total volume of film covering the height of the drum will be

$$q = \int_0^z \pi(2r\delta - \delta_2)\, dz$$

Substituting the expression for $\delta$ (Eq. 5.39) and integrating over z, holding time constant:

$$q = 2\pi r\left[\frac{2}{\rho g}\right]^{1/(n+1)}\left[\frac{1}{t}\right]^{n/(n+1)}\left[\frac{n+1}{2n+1}\right]$$
$$\times [z]^{(2n+1)/(n+1)}$$
$$-\pi\left[\frac{K}{\rho}\right]^{2/(n+1)}\left[\frac{1}{t}\right]^{2n/(n+1)}\left[\frac{n+1}{3n+1}\right]$$
$$\times [z]^{(3n+1)/(n+1)}$$

Substituting values of known quantities:

$$K/\rho g = 11.6/(1008)(9.8)$$
$$= 0.0011742/(n+1) = 1.515;$$

$$2n/(n+1) = 0.507; (n+1)/(3n+1)$$
$$= 0.66331;$$

$$1/(n+1) = 0.7463; n/(n+1)$$
$$= 0.254; (n+1)/2n+1)$$
$$= 0.7976$$

$$V = 2\pi(0.375)(0.001174)^{0.7463}(1/60)^{0.254}(0.7976)$$
$$= (1.2)^{1/0.7976} - \pi(0.001174)^{1.515}(1/60)^{0.507}$$
$$= (0.6633)(1.2)^{1/0.6633}$$
$$= 0.0054289 - 0.0000125101$$
$$= 0.005416\ m^3$$

The mass of applesauce left adhering to the walls of drum is

$$0.005416 m^3 (1008)kg/m^3 = 5.46\ kg$$

### 5.4.3  Processes Dependent on Fluid Film Thicknesses

Fluid films are involved in processes that require rapid rates of mass and/or heat transfer. Examples are as follows:

*Ultrahigh-Temperature Sterilization of Fluid Milk and Cream* One system (DaSi) rapidly heats a fluid film as it flows down multiple tubes in a superheated steam atmosphere. The steam and fluid are introduced in parallel. The fluid is pumped to a specially constructed distributor inside a cylindrical pressure vessel where it flows down the outside periphery of the vertically oriented tube in a thin film. Heat transfer into the thin film from the superheated steam with which it is in direct contact is very rapid resulting in very short residence time to heat the product to sterilization temperatures. The short heating time minimizes adverse flavor changes on the product. The fluid flow rate and the outside surface area of the tubes must be matched to obtain the necessary film thickness to permit the rapid heating.

*Falling Film Evaporators* Fluid flows as a film down the surface of a heated tube(s) and discharges into a plenum where a vacuum allows liquid to vaporize by flash evaporation. To maintain efficiency of heating, fluid flow must be at a rate that would permit the desired thickness of fluid film to develop without flooding the whole inside cross-sectional area of the tube.

*Deaeration of Food Fluids* Viscous fluids at elevated temperatures are exposed to a vacuum while flowing in a thin film to facilitate removal of dissolved oxygen. One system (FMC) sprays the fluid at the top of a tall column allowing the fluid to it the walls and flow down as a falling fluid film. The column dimensions needed to successfully deaerate the product may be calculated as shown in the example in the previous section "Films of Constant Thickness."

*Stripping of Volatile Flavor Component from Foods* A system developed by Flavourtech called a "Spinning Cone Column" consists of a rotating shaft to which multiple cones arranged at even intervals along the height of the shaft are attached. The cone faces upwards with the base radiating outward from the shaft and the apex of the cone is pierced by the rotating shaft to which the cone is fastened. Another stationary cone concentric with the rotating cone but open at the apex and attached at the periphery of the base to the wall of the cylindrical column creates a channel to guide the fluid film as it travels from one cone element to the next. Volatile flavor compounds are stripped from the fluid using an inert gas flowing countercurrent to the fluid or by vapors generated by flash evaporation when the fluid is introduced into the column at a high temperature and a vacuum is maintained in the column. The top cone is a rotating cone and fluid entering the column is deposited at the apex of this cone. The fluid film flows upward on the wall of the rotating cone due to centrifugal force of rotation and spills over the base periphery where it falls into the top of a stationary cone. Fluid film flows by gravity over the stationary cone toward the central shaft where it drops into the apex of another rotating cone. The process repeats between the alternating rotating and stationary cones. The system has been shown to be effective in stripping volatile flavors from an aqueous slurry of tea, recovery of volatile flavors from fruit and vegetable juices and purees, deaeration of fruit juice and purees, removing alcohol from wine or beer, and recovery of flavors from exhaust air streams such as onion dehydrator or fruit dehydrator. The mass transfer from a large area of thin fluid film makes this system very effective compared to other types of contractors used in extraction operations.

## 5.5    Transportation of Fluids

The basic equations characterizing flow and forces acting on flowing fluids are derived using the laws of conservation of mass, conservation of momentum, and the conservation of energy.

### 5.5.1    Momentum Balance

Momentum is the product of mass and velocity and has units of $kg \cdot m/s$. When mass is expressed as a mass rate of flow, the product with velocity s the rate of flow of momentum and will have units of $kg \cdot m/s^2$, the same units as force. When a velocity gradient exists in a flowing fluid, momentum transfer occurs across streamlines, and the rate of momentum transfer per unit area, $d(mV/A)/dt$, the momentum flux, has units of $kg/m \cdot s^2$. The units of momentum flux are the same as that of stress or pressure. The pressure drop in fluids flowing through a pipe which is attributable to fluid resistance is the result of a momentum flux between streamlines in a direction perpendicular to the direction of flow. The momentum balance may be expressed as

$$\text{Rate of momentum flow in} + \sum F$$
$$= \text{Rate of momentum flow out}$$
$$+ \text{Accumulation}$$

$\sum F$ is the sum of external forces acting on the system (e.g., atmospheric pressure, stress on the confining vessel) or forces exerted by restraints on a nozzle discharging fluid. Because momentum is a function of velocity, which is a vector quantity (i.e., it has both magnitude and direction), all the terms in the above equation are vector quantities. The form of the equation given above implies that a positive sign on EF indicates a force applied in a direction entering the system. When making a momentum balance, the component of the force or velocity acting in a single direction (e.g., x, y, or z component) is used in the equation. Thus, in three-dimensional space, a momentum balance may be made for the x, y, and z directions.

The force balance used to derive the Poiseuille equation (Eq. 5.9) is an example of a momentum balance over a control volume, a cylindrical shell of fluid flowing within a cylindrical conduit. Equation (5.9) shows that the momentum flow (momentum flux multiplied by the surface area of the control volume) equals the net force (pressure drop multiplied by the cross-sectional area of the control volume) due to the pressure acting on the control element.

When the control volume is the whole pipe (i.e., the wall of the pipe is the boundary surrounding the system under consideration), it will be possible to determine the forces acting on the pipe or its restraints. These forces are significant in instances where a fluid changes direction as in a bend, when velocity changes as in a converging pipe section, when a fluid discharges out of a nozzle, or when a fluid impacts a stationary surface.

Momentum flow (rate of change of momentum, $\bar{M}$) has the same units as force and is calculated in terms of the mass rate of flow, $\bar{m}$; kg/s; the velocity, V; or the volumetric rate of flow, q, as follows:

$$\bar{M} = \bar{m}V = AV^2\rho = qV\rho$$

A is the area of the flow stream perpendicular to the direction of flow and is the fluid density. The momentum flux (rate of change of momentum per unit area) is the quotient, $\bar{M}/A$. A force balance over a control volume is the same as a balance of momentum flow.

**Example 5.13** Calculate the force acting on the restraints of a nozzle from which fluid is discharging to the atmosphere at the rate of 5 kg/s. The fluid has a density of 998 kg/m³. It enters the nozzle at a pressure of 238.3 kPa above atmospheric pressure. The nozzle has a diameter of 6 cm at the inlet and 2 cm at the discharge.

**Solution**
Assume that the velocity of the fluid through the nozzle is uniform (i.e., there is no velocity variation across the pipe diameter). The pipe wall may be used as a system boundary for making the momentum balance, in order to include the force acting on the restraint. The momentum balance with subscripts 1 and 4 referring to the inlet and discharge of the nozzle, respectively, is

$$\bar{m}V_1 + P_1A_1 + F_x = \bar{m}\,V_4 + P_4A_4$$

Atmospheric pressure ($P_a$) acts on the inlet to the nozzle entering the control volume, and this is also the pressure at the discharge. Because $P_1$ is pressure above atmospheric at the inlet, the momentum balance becomes:

$$\bar{m}\,V_1 + P_aA_1 + P_1A_1 + F_x = \bar{m}V_4 + P_aA_4$$

Solving for $F_x$:

$$F_x = \bar{m}\,(V_4 - V_1) - P_a(A_1 - A_4) - P_1A_1$$

Substituting known quantities and solving for $V_1$ and $V_4$:

$$A_1 = 0.002827 \text{ m}^2;$$
$$A_4 = 0.001257$$

$$V_1 = \frac{5\,\text{kg}}{\text{s}}\frac{\text{m}^3}{998\,\text{kg}}\frac{1}{0.002827\,\text{m}^2} = 1.77\,\text{m/s}$$

$$V_4 = \frac{5\,\text{kg}}{\text{s}}\frac{\text{m}^3}{998\,\text{kg}}\frac{1}{0.001257\,\text{m}^2} = 3.99\,\text{m/s}$$

$$\begin{aligned}F_X &= 5(3.99 - 1.77) - 438000(0.002827)\\ &\quad - 101,300(0.002827 - 0.001257)\\ &= 11.1 - 1238.2 - 159 = -1386\,\text{N}\end{aligned}$$

The negative sign indicates that this force is acting in a direction opposite the momentum flow out of the system; thus, the nozzle is being pushed away from the direction of flow.

## 5.5.2  The Continuity Principle

The principle of conservation of mass in fluid dynamics is referred to as the continuity principle. This principle is applied whenever there is a change of velocity; a change in diameter of the conduit, split flow; or a change in density of compressible fluids such as gases. The continuity principle for flow in one direction, expressed in equation form, is as follows:

$$\rho_1\,V_1A_1 = \rho_2V_2A_2 + A\frac{\partial}{\partial t}(\rho V) \qquad (5.40)$$

If velocity is changing in all directions, a three-dimensional mass balance should be made. The above equation is a simplified form suitable for use in fluid transport systems where velocity is directed in one direction. At a steady state, the time derivative is zero. If the fluid is incompressible, the density is constant, and the continuity equation reduces to:

$$V_1A_1 = V_2A_2 \qquad (5.41)$$

**Example 5.14** A fluid having a density of 1005 kg/m3 is being drawn out of a storage tank 3.5 m in diameter through a tap on the side at the lowest point in the tank. The tap consists of a short length of 1.5 in. nominal sanitary pipe (ID = 0.03561 m) with a gate valve. If fluid flow out of the tap at the rate of 40 L/min, calculate the velocity of the fluid in the pipe and the velocity at which the fluid level recedes inside the tank.

**Solution**

The mass rate of flow, $\bar{m}$, is constant. The mass rate of flow is the product of the volumetric rate of flow and the density, $\bar{m} = q\rho$:

$$\bar{m} = \left[\frac{40\,L}{min} \cdot \frac{0.001\,m^3}{L} \cdot \frac{1\,min}{60\,s}\right] \cdot \left[\frac{1005\,kg}{m^3}\right]$$

$$= 0.67\,kg/s$$

The velocity of fluid in the pipe is

$$V = \frac{q}{A} = \frac{\bar{m}}{\rho A}$$

$$V_{pipe} = \left[\frac{0.67\,kg}{s}\right] \cdot \frac{1\,m^3}{1005\,kg}$$

$$\cdot \frac{1}{\pi(0.01781)^2 m^2} = 0.669\,m/s$$

From the continuity principle:

$$A_{pipe}V_{pipe} = A_{Tank}$$

$$V_{Tank}V_{tank} = V_{pipe} \cdot \frac{A_{pipe}}{A_{tank}} = 0.669\frac{(0.03561)^2}{(3.5)^2}$$

$$= 6.92 \times 10^{-5}\frac{m}{s}$$

## 5.6    Fluid Flow Regimes

Equation (5.10) indicates that a Newtonian fluid flowing inside a tube has a parabolic velocity profile. Fluid molecules in the center of the tube flow at the maximum velocity. The equation holds only when each molecule of the fluid remains in the same radial position as the fluid traverses the length of the tube. The path of any molecule follows a well-defined streamline, which can be readily shown by injecting a dye into the flow stream. When injected in a very fine stream, the dye will trace a straight line parallel to the direction of flow. This type of flow is called streamline or laminar flow. Figure 5.13 shows streamline flow that is observed when a dye is continuously injected into a flow stream (A), and the development of the parabolic velocity profile (D) that can be observed when a viscous dye solution is injected into a tube filled with a very viscous fluid at rest (B) and flow is allowed to develop slowly (C).

As rate of flow increases, molecular collisions occur at a more frequent rate and cross-overs of molecules across streamlines occur. Eddy currents develop in the flow stream. This condition of flow is turbulent, and the velocity profile predicted by Eq. (5.10) no longer holds. Figure 5.14 shows what can be observed when a dye is continuously injected into a fluid in turbulent flow. Swirling and curling of the dye can be observed. A certain amount of mixing also occurs. Thus, a particle originally introduced at the center of the tube may traverse back and forth between the wall and the center as the fluid travels the length of the tube. The velocity profile in turbulent flow is flat compared to the parabolic profile in laminar flow. The ratio of average to maximum velocity in laminar flow is 0.5 as opposed to an average ratio of approximately 0.8 in fully developed turbulent flow.

### 5.6.1    The Reynolds Number

The Reynolds number is a dimensionless quantity that can be used as an index of laminar or turbulent flow. The Reynolds number, represented by Re, is a function of the tube diameter, D, the average velocity $\bar{V}$, the density of the fluid $\rho$, and the viscosity $\mu$:

**Fig. 5.23** Development of laminar velocity profile during flow of a viscous fluid through a pipe

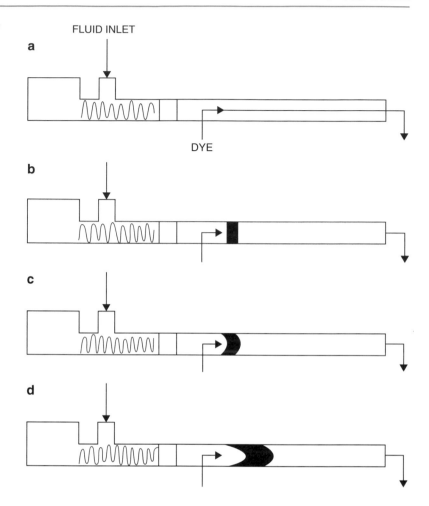

$$Re = \frac{D\bar{V}\rho}{\mu} \qquad (5.42)$$

For values of the Reynolds number below 2100, flow is laminar and the pressure drop per unit length of pipe can be determined using Eq. (5.11) for a Newtonian fluid and Eq. (5.5) for power law fluids. For Reynolds numbers above 2100, flow is turbulent, and pressure drops are obtained using an empirically derived friction factor chart (Figs. 5.23 and 5.24).

### 5.6.2   Pipes and Tubes

Tubes are thin walled cylindrical conduits whose nominal sizes are based on the outside diameter.

Pipes are thicker walled than tubes, and the nominal size is based on the inside diameter. The term "sanitary pipe" used in the food industry is used for stainless steel tubing where the nominal designation is based on the outside diameter. Table 5.2 shows dimensions for steel and sanitary pipes and heat exchanger tubes.

### 5.6.3   Frictional Resistance to Flow of Newtonian Fluids

Flow through a tubular conduit is accompanied by a drop in pressure. The drop in pressure is equivalent to the stress that must be applied on the fluid to induce flow. This stress can be considered as a frictional resistance to flow. Working

**Fig. 5.24** Path traced by a dye injected into a fluid flowing through a tube in turbulent flow

**Table 5.2** Pipe and heat exchanger tube dimensions (numbers in parentheses represent the dimension in meters)

| Nominal size(in.) | Steel pipe (Sch 40) | | Sanitary pipe | | Heat exchanger tube (18 Ga) | |
| | ID in./(m) | OD in./(m) | ID in./(m) | OD in./(m) | ID in./(m) | OD in./(m) |
|---|---|---|---|---|---|---|
| 0.5 | 0.622 | 0.840 | – | – | 0.402 | 0.50 |
| | (0.01579) | (0.02134) | | | (0.01021) | (0.0217) |
| 0.75 | 0.824 | 1.050 | – | – | 0.652 | 0.75 |
| | (0.02093) | (0.02667) | | | (0.01656) | (0.01905) |
| 1 | 1.049 | 1.315 | 0.902 | 1.00 | 0.902 | 1.00 |
| | (0.02664) | (0.03340) | (0.02291) | (0.0254) | (0.02291) | (0.0254) |
| 1.5 | 1.610 | 1.900 | 1.402 | 1.50 | 1.402 | 1.50 |
| | (0.04089) | (0.04826) | (0.03561) | (0.0381) | (0.03561) | (0.0381) |
| 2.0 | 2.067 | 2.375 | 1.870 | 2.00 | – | – |
| | (0.05250) | (0.06033) | (0.04749) | (0.0508) | | |
| 2.5 | 2.469 | 2.875 | 2.370 | 2.5 | – | – |
| | (0.06271) | (0.07302) | (0.06019) | (0.0635) | | |
| 3.0 | 3.068 | 3.500 | 2.870 | 3.0 | – | – |
| | (0.07793) | (0.08890) | (0.07289) | (0.0762) | | |
| 4.0 | 4.026 | 4.500 | 3.834 | 4.0 | – | – |
| | (0.10226) | (0.11430) | (0.09739) | (0.1016) | | |

*ID* inside diameter
*OD* outside diameter

against this stress requires the application of power in transporting fluids. An expression for the pressure drop as a function of the fluid properties and the dimension of the conduit would be useful in predicting power requirements for inducing flow over a given distance or for determining the maximum distance a fluid will flow with the given pressure differential between two points. For Newtonian fluids in laminar flow, the Poiseuille equation (Eq. 5.11) may be expressed in terms of the inside diameter (D) of a conduit, as follows:

$$\frac{\Delta P}{L} = \frac{32\bar{V}\mu}{D^2} \qquad (5.43)$$

Equation (5.43) can be expressed in terms of the Reynolds number:

$$\frac{\Delta P}{L} = \frac{32\bar{V}\mu}{D^2} \cdot \frac{(D\bar{V}\rho)/\mu}{Re}$$

$$\frac{\Delta P}{L} = \frac{2\left(\frac{16}{Re}\right)(\bar{V})^2(\rho)}{D} \qquad (5.44)$$

Equation (5.44) is similar to the Fanning equation derived by dimensional analysis where the friction factor f is 16/Re for laminar flow. The Fanning equation, which is applicable in both laminar and turbulent flow, is

$$\frac{\Delta P}{\rho} = \frac{2f(\bar{V})^2 L}{D} \qquad (5.45)$$

For turbulent flow, the Fanning friction factor f may be determined for a given Reynolds number and a given roughness factor from a friction factor chart as shown in Fig. 5.25. For smooth tubes, a plot of log f obtained from the lowest curve in Fig. 5.25 against the log of Reynolds number would give the following relationships between f and Re:

$$f = 0.048\,(Re)^{-0.20}\quad 10^4 < Re < 10^6 \quad (5.46)$$

$$f = 0.193\,(Re)^{-0.35}\quad 3 \times 10^3 < Re < 10^4 \tag{5.47}$$

**Example 5.15** What pressure must be generated at the discharge of a pump that delivers 100 L/min of a fluid having a specific gravity of 1.02 and a viscosity of 100 centipoises? The fluid flows through a 1.5 in. (nominal) sanitary pipe, 50 m long. The pipe is straight and level, and the discharge end of the pipe is at atmospheric pressure.

**Solution**

The Reynolds number is determined to determine if flow is laminar or turbulent:

$$Re = \frac{D\bar{V}\rho}{\mu}$$

Substituting units of terms in the equation for Re:

$$Re = \frac{D(m)\bar{V}(m/s)\rho(kg/m^3)}{\mu(Pa \cdot s)}$$

[Note that the Reynolds number will be dimensionless if the base units for Pa · s, kg/(m s), are substituted in the dimensional equation.]

Converting values of variables into SI units:

$$\rho = 1.02 \cdot \frac{1000\,kg}{m^3} = 1020\,\frac{kg}{m^3}$$

$$D = 1.402\,in. \cdot \frac{0.0254\,m}{in.} = 0.0356\,m$$

$$q = \frac{100\,L}{min} \cdot \frac{0.001\,m^3}{L} \cdot \frac{1\,min}{60\,s} = 0.00167\,\frac{m^3}{s}$$

$$\bar{V} = \frac{q}{area} = \frac{0.00167\,m^3}{s} \cdot \frac{1}{\left(\frac{\pi}{4}\right) \cdot (0.0356)^2 m^2}$$

$$= 1.677\,m/s$$

$$\mu = 100\,cP \cdot \frac{0.001\,Pa \cdot s}{cP} = 0.1\,Pa \cdot s$$

Calculating the Reynolds number:

$$Re = \frac{D\bar{V}\rho}{\mu} = \frac{(0.0356)(1.677)(1020)}{0.1} = 609$$

Flow is laminar. Using Eq. (5.44):

$$\begin{aligned}\Delta P &= \frac{2(16/Re)(\bar{V})^2(\rho)(L)}{D}\\ &= \frac{2(16)(1.677)^2(1020)(50)}{609(0.0356)}\\ &= 211,699\,Pa\,gauge\end{aligned}$$

Because the pressure at the end of the pipe is 0 gauge pressure (atmospheric pressure), the pressure at the pump discharge will be

$$P = 211,699\,Pa\,guage$$

**Example 5.16** Milk with a viscosity of 2 centipoises and a specific gravity of 1.01 is being pumped through a 1 in. (nominal) sanitary pipe at the rate of 3 gallons per minute. Calculate the pressure drop in $lb_f/in^2$ per foot length of level straight pipe.

**Solution**

Use SI units and Eqs. (5.40) or (5.41) to calculate $\Delta P$ and convert to $lb_f/in^2$. From Table 5.2, a 1 in (nominal) sanitary pipe will have an inside diameter of 0.02291 m:

$$\begin{aligned}q &= 3\frac{gal}{min} \cdot \frac{0.00378541\,m^3}{gal} \cdot \frac{1\,min}{60\,s}\\ &= 0.00018927\,m^3/s\end{aligned}$$

$$\bar{V} = \frac{q}{A} = 0.00018927\frac{m^3}{s} \cdot \frac{1}{\pi/4(0.02291)^2 m^3}$$

$$- 0.459\,m/s$$

$$\rho = 1.01 \cdot \frac{1000\,kg}{m^3} = 1010\,kg/m^3$$

$$Re = \frac{D\bar{V}\rho}{\mu} = \frac{0.02291(0.459)(1010)}{0.002} = 5310$$

Because Re > 2100 flow is turbulent, use Eq. (5.45) for calculating $\Delta P$ and Eq. (5.47) for f at Re < $10^4$. Sanitary pipe is considered a smooth pipe:

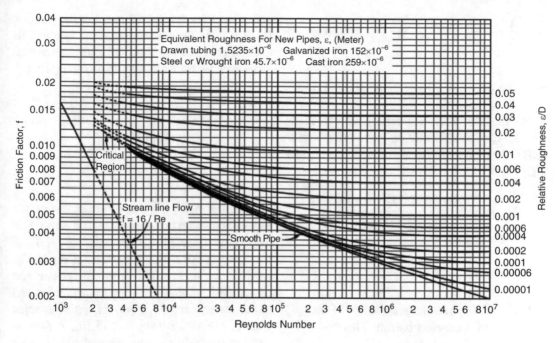

**Fig. 5.25**  The Moody diagram for the Fanning friction factor (based on Moody, L. F. Friction factors for pipe flow. Trans. ASME. 66:671, 1944)

$$f = 0.193(\text{Re})^{-0.35} = 0.193(5310)^{-0.35}$$
$$= 0.0095$$

f = 0.0095 can also be obtained from Fig. 5.25. The pressure drop per unit length of pipe may be calculated as $\Delta P/L$. Because the units used in the calculations are in SI, the result will be Pa/m. This can then be converted to the desired units of (lb $f_f$/in.$^2$)/ft. It may also be solved by substituting a value for L in Eq. (5.45):

$$L = 1\text{ft} \cdot \frac{0.3048\text{m}}{\text{ft}} = 0.3048\text{m}$$

$$\Delta P = \frac{2f\ V^2 L\rho}{D}$$

$$\Delta P = \frac{2(0.0095)(0.459)^2(0.3048)(1010)}{0.02291}$$

$$\Delta P = 53.79\,\text{Pa}$$

$$\Delta P = 53.79\,\text{Pa} \cdot \frac{1\dfrac{\text{lb}_f}{\text{in}^2}}{6894.757\,\text{Pa}} = 0.0078\,\dfrac{\dfrac{\text{lb}_f}{\text{in}^2}}{\text{ft of pipe}}$$

**Example 5.17**  Calculate the pressure drop for water flowing at the rate of 10 gal./min through 100 m of level straight wrought iron pipe having an inside diameter of 0.3579 m. Use a density for water of 62.4 lb/ft$^3$ and a viscosity of 0.98 centipoises.

**Solution**
This problem is different from Examples 5.15 and 5.16 in that the pipe is not smooth and therefore Eqs. (5.46) or (5.47) cannot be used for calculating f. After calculating the Reynolds number, f can be determined from a Moody diagram (Fig. 5.25) if flow is turbulent, and $\Delta P$ can be calculated using Eq. (5.45).

From Table 5.2, D = 0.03579 m:

$$q = \frac{10\,\text{gal}}{\text{min}} \cdot \frac{0.00378541\,\text{m}^3}{\text{gal}} \cdot \frac{1\,\text{min}}{60\,\text{s}}$$
$$= 0.0006309\,\text{m}^3/\text{s}$$

$$\text{Re} = \frac{D\bar{V}\rho}{\mu} = \frac{0.03579(0.6274)(999.7)}{0.00098}$$
$$= 22,906$$

$$\rho = 62.4\frac{\text{lb}_m}{\text{ft}^3} \cdot \frac{(3.281)^3\text{ft}^3}{\text{m}^3} \cdot \frac{0.45359\,\text{kg}}{\text{lb}_m}$$
$$= 999.7\,\text{kg/m}^{3-}$$

$$V = \frac{q}{A} = \frac{0.0006309 \, \text{m}^3/\text{s}}{(\pi/4)(0.03579)^2(\text{m}^2)} = 0.6274 \, \text{m/s}$$

$$\mu = 0.98 \, \text{cP} \cdot \frac{0.001 \, \text{Pa} \cdot \text{s}}{\text{cP}} = 0.00098 \, \text{Pa} \cdot \text{s}$$

To use Fig. 5.25, first determine ε/D. From Fig. 5.24, ε = 0.0000457 m. ε/D = 0.0000457/0.03579 = 0.00128. For Re = 2.29 × 10⁴, and ε/D = 0.00128, f = 0.0069.

$$\Delta P = \frac{2(0.0069)(0.6274)^2(100)(999.7)}{0.03579}$$

$$= 15.17 \, \text{kPa}$$

### 5.6.4    Frictional Resistance to Flow of Non-Newtonian Fluids

For fluids whose flow behavior can be expressed by Eq. (5.2) (pseudoplastic or power law fluids), the Rabinowitsch-Mooney equation (Eq. 5.18) for the shear rate at the wall was derived in the section "Velocity Profile and Shear Rate for a Power Law Fluid." Substituting the shear stress at the wall (Eq. 5.22) for $\tau_w$ and the shear rate at the wall, $\gamma_w$ (Eq. 5.18), in Eq. (5.2):

$$\frac{\Delta PR}{2L} = K \left[ \frac{\bar{V}}{R} \cdot \frac{3n+1}{n} \right]^n$$

$$\Delta P = \frac{2LK(\bar{V})^n}{(R)^{n+1}} \left[ \frac{3n+1}{n} \right]^n \qquad (5.48)$$

Equation (5.44) for the pressure drop of a fluid in laminar flow through a tubular conduit may be written in terms of the radius of the tube and the Reynolds number as follows:

$$\Delta P = \frac{2(16/\text{Re})(\bar{V})^2 \rho L}{2R} \qquad (5.49)$$

Equation (5.49) is simply another form of Eq. (5.44). These equations are very general and can be used for any fluid. In laminar flow, the friction factor, expressed as 16/Re, may be used as a means of relating the pressure drop of a

non-Newtonian fluid to a dimensionless quantity, Re. Thus, Re for a non-Newtonian fluid can be determined from the flow behavior and consistency indices. Equating Eqs. (5.48) and (5.49):

$$\frac{2LK(\bar{V})^n}{(R)^{n+1}} \left[ \frac{3n+1}{n} \right]^n = \frac{2(16/\text{Re})(\bar{V})^2 L\rho}{2R}$$

$$\text{Re} = \frac{8(\bar{V})^{2-n}(R)^n \rho}{K\left[ \frac{3n+1}{n} \right]^n}$$

$$(5.50)$$

Equation (5.50) is a general expression for the Reynolds number that applies for both Newtonian (n = 1) and non-Newtonian liquids (n ≠ 1). If n = 1, Eq. (5.50) is exactly the same as Eq. (5.42), when K = μ. Pressure drops for power law non-Newtonian fluids flowing through tubes can be calculated using Eq. (5.49) using the Reynolds number calculated using Eq. (5.50), if flow is laminar. If flow is turbulent, the Reynolds number can be calculated using Eq. (5.50) and the pressure drop calculated using Eq. (5.45) with the friction factor f determined from this Reynolds number using Eqs. (5.46) or (5.47). The Moody diagram (Fig. 5.25) can also be used to determine "f" from the Reynolds number.

**Example 5.18** Tube viscometry of a sample of tomato catsup shows that flow behavior follows the power law equation with K = 125 dyne As/cm² and n = 0.45. Calculate the pressure drop per meter length of level pipe if this fluid is pumped through a 1-in. (nominal) sanitary pipe at the rate of 5 gallons (US)/min. The catsup has a density of 1.13 g/cm³.

**Solution**
Convert given data to SI units. n = 0.45 (dimensionless):

$$K = 125 \frac{\text{dyne s}^n}{\text{cm}^2} \cdot \frac{1 \times 10^{-5} \text{N}}{\text{dyne}} \cdot \frac{(100)^2 \text{cm}^2}{\text{m}^2}$$

$$= 12.5 \, \text{Pa} \cdot \text{s}^n$$

From Table 5.1, D = 0.02291 m; R = 0.01146 m:

$$q = \frac{5 \text{ gal}}{\text{min}} \cdot \frac{0.003478541 \, \text{m}^2}{\text{gal}} \cdot \frac{1 \text{ min}}{60 \text{ s}}$$

$$q = 0.00031545 \, \text{m}^3/\text{s}$$

$$\bar{V} = \frac{q}{A} = \frac{0.00031545 \, \text{m}^3/\text{s}}{\left(\frac{\pi}{4}\right) \cdot (0.02291)^2 \cdot \text{m}^2} = 0.7652 \, \text{m/s}$$

$$\rho = \frac{1.13 \text{g}}{\text{cm}^3} \cdot \frac{(100)^3 \text{cm}^3}{\text{m}^3} \cdot \frac{1 \text{ kg}}{1000 \text{ g}}$$

$$\rho = 1130 \, \text{kg/m}^3$$

$$L = 1 \, \text{m}$$

The Reynolds number is calculated using Eq. (5.50):

$$Re = \frac{8(0.7652)^{2-0.45}(0.01146)^{0.45}(1130)}{12.5 \left[\frac{3(.45)+1}{0.45}\right]^{0.45}}$$

$$= 30.39$$

Flow is laminar and pressure drop is calculated using Eq. (5.45):

$$\Delta P = \frac{2\left(\frac{16}{30.39}\right)(0.7652^2)(1130)(1)}{0.02291} = 30.41 \, \text{kPa}$$

**Example 5.19** Peach puree having a solids content of 11.9% has a consistency index K of 72 dyne As$^n$/cm$^2$ and a flow behavior index of 0.35. If this fluid is pumped through a 1-in. (nominal) sanitary pipe at 50 gallons (US)/min, calculate the pressure drop per meter of level straight pipe. Peach puree has a density of 1.07 g/cm$^3$.

**Solution**

$$D = 0.02291 \text{ (from Table 6.2);}$$
$$R = 0.01146 \, \text{m}; n = 0.35$$

$$K = 72 \frac{\text{dyne s}^n}{\text{cm}^2} \cdot \frac{(10)^{-5}}{\text{dyne}}$$

$$\cdot \frac{(100)^2 \text{cm}^2}{\text{m}^2} = 7.2 \, \text{Pa} \cdot \text{s}^n$$

$$\rho = 1070 \, \text{kg/m}^3 q$$

$$= 50 \frac{\text{gal}}{\text{min}} \frac{0.00378541 \, \text{m}^3}{\text{gal}} \frac{1 \text{ min}}{60 \text{ s}}$$

$$= 0.0031545 \, \text{m}^3/\text{s}$$

$$\bar{V} = \frac{q}{A} = \frac{0.00378541 \, \text{m}^3}{(\pi/4)(0.02291)^2} = 7.6523 \, \text{m/s}$$

Using Eq. (5.50):

$$Re = \frac{8(7.6523)^{2-0.35}(0.01146)^{0.35}(1070)}{7.2 \left[\frac{3(0.35)+1}{0.35}\right]^{0.35}}$$

$$= 3850$$

Flow is turbulent. The sanitary pipe is a smooth pipe, so Eq. (5.47) will be used to calculate f:

$$f = 0.193(\text{Re})^{-0.35} = 0.193(3850)^{-0.35}$$
$$= 0.010731$$

P can now be calculated using Eq. 5.42. L = 1 m:

$$\Delta P = \frac{2(0.01073)(7.6523)^2(1)(1070)}{0.02291}$$

$$= 58.691 \, \text{kPa/m}$$

### 5.6.5  Frictional Resistance Offered by Pipe Fittings to Fluid Flow

The resistance of pipe fittings to flow can be evaluated in terms of an equivalent length of straight pipe. Each type of pipe fitting has its specific flow resistance expressed as a ratio of equivalent length of straight pipe (L=) over its diameter. Table 5.3 lists the specific resistance of various pipe fittings. The equivalent length of a fitting, which is the product of L=/D obtained from Table 5.3 and the pipe diameter, is added to the length of straight pipe within the piping system to determine the total drop pressure drop across the system.

**Example 5.20** Calculate the pressure drop due to fluid friction across 50 m of a 1-in. (nominal) sanitary pipe that includes five 90-degree elbows in the piping system. Tomato catsup with properties described in Example 5.18 in the section "Frictional Resistance to Flow for non-

**Table 5.3** Specific resistance of various pipe fittings expressed as an equivalent length of straight pipe to pipe diameter ratio

| Fitting | L'/D (dimensionless) |
|---|---|
| 90° elbow std. | 35 |
| 45° elbow std. | 15 |
| Tee (used as a coupling), branch plugged | 20 |
| Tee (used as an elbow) entering the branch | 70 |
| Tee (used as an elbow) entering the tee run | 60 |
| Tee branching flow | 45 |
| Gate valve, fully open | 10 |
| Globe valve, fully open | 290 |
| Diaphragm valve, fully open | 105 |
| Couplings and unions | Negligible |

**Table 5.4** Energy terms involved in the mechanical energy balance for fluid flow in a piping system, the formulas for calculating them, and their units

| Energy term | Formula | Dimensional expression | Formula (basis: 1 kg) | Unit |
|---|---|---|---|---|
| Potential energy pressure | $m\left(\dfrac{P}{\rho}\right)$ | $\dfrac{\text{kg}\left(N\cdot m^{-2}\right)}{\text{kgm}^{-3}}$ | $\dfrac{P}{\rho}$ | joule/kg |
| Elevation | $mgh$ | $\text{kg}(m\cdot s^{-2})(m)$ | $gh$ | joule/kg |
| Kinetic energy | $\frac{1}{2}mV^2$ | $\text{kg}(m\cdot s^{-1})^2$ | $\dfrac{V^2}{2}$ | joule/kg |
| Work input (from pump) | $W$ | | $W$ | joule/kg |
| Frictional resistance | $\dfrac{m\Delta P_f}{\rho}$ | $\dfrac{\text{kg}\left(N\cdot m^{-2}\right)}{\text{kg}\cdot m^{-3}}$ | $\dfrac{\Delta P_f}{\rho}$ | joule/kg |

Newtonian Fluids" flows through the pipe at the rate of 5 gallons (US)/min.

**Solution**

In Example 5.18 of the section referred to above, the tomato catsup pumped under these conditions exhibited a pressure drop of 30.41 kPa/m of level pipe. The equivalent length of the five elbows is determined as follows:

L/D from Table 5.3 for a 90-degree elbow is 35. The diameter of the pipe is 0.02291

Total equivalent length of straight pipe = 50 m + 5(35)(0.02291) = 54 m:

$$\Delta P = (30.41 \text{ kPa/m})(54 \text{ m}) = 1642 \text{ kPa}$$

## 5.7  Mechanical Energy Balance: The Bernoulli Equation

When fluids are transferred from one point to another, a piping system is used. Because fluids exhibit a resistance to flow, an energy loss occurs as the fluid travels downstream along the pipe. If the initial energy level is higher than the energy at any point downstream, the fluid will flow spontaneously. On the other hand, if the energy change needed to take the fluid to a desired point downstream exceeds the initial energy level, energy must be applied to propel the fluid through the system. This energy is an input provided by a pump.

An energy balance across a piping system is similar to the energy balance made in this chapter. Table 5.4 lists the energy terms involved in fluid flow, their units, and the formulas for calculating them. All energy entering the system including the energy input must equal that leaving the system and the energy loss due to fluid friction. The boundaries of the system must be carefully defined to identify the input and exit energies. As in previous problems in material and energy balances, the boundaries may be moved around to isolate the unknown quantities being asked in the problem. The mechanical energies involved are primarily kinetic and potential energy and mechanical energy at the pump.

The potential and kinetic energy could have finite values on both the source and discharge points of a system. Work input should appear on the side of the equation opposite the frictional resistance term. A balance of the intake and exit energies in any given system including the work input and resistance terms based on a unit mass of fluid is

$$\frac{P_1}{\rho} + gh_1 + \frac{V_1^2}{2} + W_s = \frac{P_2}{\rho} + gh_2 + \frac{V_2^2}{2} + \frac{\Delta P_f}{\rho}$$
(5.51)

Equation (5.51) is the Bernoulli equation.

**Example 5.21** A pump is used to draw tomato catsup from the bottom of a deaerator. The fluid level in the deaerator is 10 m above the level of the pump. The deaerator is being operated at a vacuum of 0.5 $kg_f/cm^2$ (kilogram force/cm$^2$). The pipe connecting the pump to the deaerator is a 2.5-in. (nominal) stainless steel sanitary pipe, 8 m long with one 90-degree elbow. The catsup has a density of 1130 kg/m$^3$, a consistency index K of 10.5 Pa · s$^n$, and a flow behavior index n of 0.45 (dimensionless). If the rate of flow is 40 L/min, calculate the pressure at the intake side of the pump to induce the required rate of flow.

**Solution**
Figure 5.26 is a diagram of the system. The problem asks for the pressure at a point in the system before work is applied by the pump to the fluid. Therefore, the term $W_s$ in Eq. (5.51) is zero. The following must be done before quantities can be substituted into Eq. (5.51).

Assume atmospheric pressure is 101 kPa. $P_1$ must be converted to absolute pressure. The volumetric rate of flow must be converted to velocity (V). The frictional resistance to flow ($\Delta P_f/\rho$ must be calculated:

$$\text{Vacuum} = 0.5 \frac{kg_f}{cm^2} \cdot \frac{0.8 \, m \cdot kg}{kg_f \cdot s^2} \cdot \frac{(100)^2 cm^2}{m^2}$$

$$= 49 \, kPa$$

$P_1 = $ Atmospheric pressure − vacuum
$= 101 - 49 = 52 \, kPa.$

$P_1 = 0.5 \, kf/cm^2$ (vacuum)
$V_1 = 0$
$h_1 = 10 \, m$

10 m

$h_2 = 0$
$q = 40 \, l/min$
$P_2 = ?$

**Fig. 5.26** Diagram of piping system for catsup from a deaerator to a pump

$$q = 40\frac{L}{min} \cdot \frac{0.001 \, m^3}{L} \cdot \frac{1 \, min}{60 \, s}$$
$$= 0.0006666 \, m^3/s$$

Cross-sectional area, $A = \frac{\pi}{4}(0.06019)^2$
$$= 0.002845 \, m^2$$

From Table 5.2, D = 0.06019; R = 0.0300:

$$\bar{V} = \frac{q}{A} = \frac{0.0006666 \ m^3/s}{0.002845 \ m^2} = 0.2343 \ m/s$$

The frictional resistance to flow can be calculated using Eq. (5.45) after the Reynolds number has been calculated using Eq. (5.50):

$$Re = \frac{8(0.2343)^{2-0.45}(0.03009)^{0.45}(1130)}{10.5\left[\frac{3(0.45)+1}{0.45}\right]}$$

$$= 8.921$$

Flow is laminar and using Eq. (5.45):

$$\frac{\Delta P_f}{L\rho} = \frac{2(16/8.920)(0.2343)^2}{0.060019}$$
$$= 3.272 \ J/(kg \cdot m)$$

where L is the length of straight pipe + equivalent length of fitting.

From Table 5.3, the equivalent length of a 90-degree elbow is 35 pipe diameters:

$$\frac{\Delta P_f}{\rho} = 3.272(10.1) = 33.0 \text{ J/kg}$$
$$L = 8 + 1(35)(0.06019) = 10.1 \text{ m}$$

Substituting all terms in Eq. (5.51):

$$P_2 = \left[\frac{52, \ 00}{1130} + 98 - 33.0 - 0.03\right](1130)$$
$$\frac{52, \ 000}{1130} + 9.8(10) + 0 = \frac{P_2}{1130}$$
$$+ 0 + \frac{(0.2343)^2}{2} + 33$$
$$= (46.02 + 9833.00.03) \times (1130)$$
$$= 125.4 \text{ kPa absolute}$$

This problem illustrates one of the most important factors often overlooked in the design of a pumping system, that is, to provide sufficient suction head to allow fluid to flow into the suction side of the pump. As an exercise, the reader can repeat these same calculations using a pipe size of 1.0 in. at the given volumetric flow rate. The calculated absolute pressure at the intake side of the pump will be negative, a physical impossibility.

Another factor that must be considered when evaluating the suction side of pumps is the vapor pressure of the fluid being pumped. If the suction pressure necessary to induce the required rate of flow is lower than the vapor pressure of the fluid at the temperature it is being pumped, boiling will occur, vapor lock develops, and no fluid will flow into the intake side of the pump. For example, suppose the pressure at the pump intake is 69.9 kPa. From Appendix Table A.3, the vapor pressure of water at 90 °C is 70 kPa. If the height of the fluid level is reduced to 4.8 m, the pressure at the pump intake will be 69 kPa; therefore, under these conditions, if the temperature of the catsup is 90 °C and above, it would not be possible to draw the liquid into the intake side of the pump.

**Example 5.22** The pump in Example 5.21 delivers the catsup to a heat exchanger and then to a filler. The discharge side of the pump consists of 1.5-in. stainless steel sanitary pipe 12 m long with two elbows. Joined to this pipe is the heat exchanger, which is 20 m of steam-jacketed 1-in. stainless steel sanitary pipe with two U-bends. (Assume the resistance of a U-bend is double that of a 90-degree elbow.) Joined to the heat exchanger is another 20 m of 1.5 in. pipe with two 90-degree elbows, and the catsup is discharged into the filler bowl at atmospheric pressure. The pipe elevation at the point of discharge is 3 m from the level of the pump. The piping system before the pump and the rates of flow are the same as that in Example 5.21. Calculate the power requirement for the pump. The diagram of the system is shown in Fig. 5.27.

**Solution**

From Example 5.21, $h_1 = 10$ m; $V_1 = 0$; $P_1 = 52$ kPa. $\Delta P_f/\rho$ for the section prior to the pump $= 33$ J/kg. $\Delta P_f/\rho$ for the section at the discharge side of the pump will be calculated in two steps: for the 1.5-in. pipe and the 1-in. pipe at the heat exchanger. $D = 1.5$-in. pipe $= 0.03561$ m, $R = 0.0178$.

Equivalent length of four 90-degree elbows $= 4(35)(0.03561) = 4.9854$ m.

$$\text{Total length} = 20 + 12 + 4.99$$
$$= 36.99 \text{ m.}$$

$$A = \frac{\pi}{4}(0.03561)^2 = 0.00099594 \text{ m}^2$$

$$\frac{\Delta P_f}{\rho} = \frac{2(16/35.82)(0.6693)^2(36.99)}{0.03561} = 415.7 \text{ J/kg}$$

$$\bar{V} = \frac{q}{A} = \frac{0.0006666 \text{ m}^3/\text{s}}{0.0009954 \text{ m}^2} = 0.6693 \text{ m/s}$$

$$Re = \frac{8(0.6693)^{2-0.45}(0.0178)^{0.45}(1130)}{10.5\left[\frac{3(0.45)+1}{0.45}\right]^{0.45}}$$

$$= 35.87$$

**Fig. 5.27** Diagram of a piping system for catsup from a de-aerator to a pump and heat exchanger and filler

D of 1-in. pipe $= 0.02291$; R $= 0.01146$ m:

A $\quad = \pi(0.01146)^2 = 0.0004126$ m$^2$

$V = \dfrac{q}{A} = \dfrac{0.0006666 \text{ m}^3/\text{s}}{0.0004126 \text{ m}^2} = 1.616 \text{ m/s}$

$\dfrac{\Delta P_f}{\rho} = \dfrac{2(16/115.2)(1.616)^2(L)}{0.02291} = 31.64(L) \text{ J/kg}$

Re $\quad = \dfrac{8(1.616)^{2-0.45}(0.01146)^{0.45}(1130)}{10.5\left[\frac{3(0.45)+1}{0.45}\right]^{0.45}} = 115.3$

The equivalent length of two U-bends $= 2(2)$ $(35)(D)$: L $= 20 + 2(2)(35)(0.02291) = 23.2$ m:

$$\frac{\Delta Pf}{\rho} = 31.64(23.27) = 734.0 \text{ J/kg}$$

Total resistance to flow $= 33.0 + 415.7 +$ $734.0 = 1182.7$ J/kg.

Substituting in Eq. (5.51), $P_2 = 101{,}000$ Pa; $h_2 = 3$ m; and $V_2 = 0.6697$ m/s:

$\dfrac{52{,}000}{1130} + 9.8(10) + 0 + W_s$

$= \dfrac{101{,}000}{1130} + 9.8(3)$

$+ \dfrac{(0.6697)^2}{2} + 1183.2$

$W_s = \dfrac{101{,}000}{1130} + 9.8(3) + \dfrac{(0.6697)^2}{2}$

$\quad + 1183.2 - \dfrac{52{,}000}{1130} - 9.8(10)$

$= 89.4 + 29.4 + 0.2 + 1182.7$

$\quad -46 - 98 = 1157.7$ J/kg

Mass flow rate $= q\,(\rho)$

$\quad\quad\quad\quad\quad\quad = 0.0006666(1130)$

$\quad\quad\quad\quad\quad\quad = 0.7532$ kg/s

Power

$= (1157.7 \text{ J/kg})(0.7532 \text{ kg/s})$

$= 872$ watts

## 5.8    Pumps

### 5.8.1    Types of Pumps and Their Characteristics

Pumps for transporting liquids may be classified into two general classes: *positive displacement* and *centrifugal*. Positive displacement pumps can operate effectively over a range of relatively slow-drive shaft rotational speeds, and they deliver a fixed volume per revolution. These pumps can handle viscous liquids, generate high discharge pressure, are self-priming, and have some suction lift capability. Because of the relatively flat flow delivery within a moderate range of discharge pressure, positive displacement pumps are useful as metering pumps in processes where flow rate is a critical parameter for success of a process. Positive displacement pump flow must not be throttled; otherwise, high pressures will develop and break the pipes, destroy bearing seals in the pump drive shaft, or burn up the drive motor because of the excessive load. In contrast, centrifugal pumps depend on an impeller rotating at high rotational speeds to increase the kinetic energy of fluid at the periphery of the impeller propelling the fluid at high velocity toward the pump discharge. Centrifugal pumps must be primed before they can generate any flow, or the source must be elevated above the pump to permit natural flooding of the pump inlet by gravity flow. Pressures generated are low, and these pumps are not capable of pumping very viscous fluids. Fluid aeration, foaming, and fluid temperature elevation in the pump casing are other undesirable characteristics of centrifugal pumps particularly when operating at head pressures close to the maximum for the particular pump. However, when pressures against which the pump must work are low, and the fluid has relatively low viscosity, centrifugal pumps are the least expensive pump to use.

Positive displacement pumps may be classified as reciprocating plunger or piston, gear, lobe, diaphragm, or progressing cavity pumps. Plunger or piston pumps generate the highest pressure, but they deliver pulsating flow and develop pressure hammers, which might be detrimental to

equipment downstream or to the piping system. Diaphragm pumps have the pumped fluid within a cavity formed by the diaphragm and the pump head cover. The diaphragm repeatedly moves forward and retracts. With each retraction of the diaphragm, the enlarged volume of the fluid cavity permits fluid to enter the cavity, while forward motion of the diaphragm forces the fluid from the cavity toward the discharge. Check valves control the intake and discharge of fluid from the diaphragm cavity. A major advantage of diaphragm pumps is the complete isolation of the fluid from the surroundings while in the diaphragm cavity preventing contamination or aeration of the product. Diaphragm pumps are also available with an air drive rather than electric and are a good feature in areas that are constantly wet. All other positive displacement pumps have moving shafts that turn the pump rotor or gear, and these shafts must be adequately sealed to prevent fluid from leaking around the shaft. Progressing cavity pumps have the advantage of a gradual buildup of pressure as the fluid moves forward from the fluid intake at the opposite end of the pump from the discharge. Thus, the shaft seal for progressing cavity pumps only works against a fraction of the pressure generated at the pump discharge.

### 5.8.2    Factors to Be Considered in Pump Selection

*Flow and Discharge Pressure* Pumps are sized according to the amount of fluid they deliver at designated discharge pressures. The latter is commonly called the *head*, which is defined as the height of a column of fluid that generates a pressure at its base equivalent to the pressure registered by a gauge placed at the discharge port of the pump. A graph of flow as a function of the head is called the *performance curve* of a pump and is available from pump manufacturers. Each make, model, and size of pump has a performance curve, which should be consulted to verify if a given pump can satisfactorily perform the required application.

*Fluid Viscosity* The viscosity of the fluid not only determines the head against which the pump must work but also the net positive suction head available at the suction inlet to the pump. For example, centrifugal pumps require a *net positive suction head* (NPSH) to be available, in the application under consideration, which is greater than that specified for a particular make and model of the pump. The NPSH is discussed in more details later in this section. Generally, the limitation of centrifugal pumps in effectively pumping a fluid of high viscosity is the relatively high available NPSH they require compared to positive displacement pumps. However, even with positive displacement pumps, fluids with very high viscosity that would exhibit a very high pressure drop due to fluid friction in a short length of pipe connecting the suction to the fluid reservoir could prevent acceptable functioning of the pump for the desired application.

*Fluid temperature* affects the available NPSH. High fluid vapor pressure results in flashing of vapor to negate the suction developed by the pump and result in a condition called *vapor lock*. Fluid is prevented from entering the pump resulting in zero flow. To prevent vapor lock when pumping high-temperature fluids, the fluid source must be elevated above the pump level in order that gravity flow will permit flooding of the pump inlet at the desired rate of flow.

*Fragility of Suspended Particles* Fluids that contain fragile suspended particles require pumps that have large cavities to hold the particles without compressing them, and there should be a minimum of restrictions, sharp turns, and areas of low fluid velocities in the flow stream to prevent solids from bridging while flowing out of the pump.

*Ease of Cleaning* Fluids that are good substrates for microbiological growth will require a pump that can be easily cleaned. Pumps should be capable of being cleaned in place. Otherwise, the pump should be capable of being cleaned by simply removing the pump head cover without

having to disconnect the pump from the connecting pipes.

*Abrasiveness of Suspended Solids in the Fluid* Suspended solids that are abrasive will require a pump where there is a minimum likelihood of having particles trapped between a stationary and moving part. Centrifugal and diaphragm pumps are ideal for this type of application.

*Corrosiveness of Fluid* Material of construction of any part of the pump that contacts an abrasive fluid must be of the type that will resist chemical attack of the fluid.

### 5.8.3  Performance Curves of Pumps

The performance curve of a volute-type centrifugal pump is shown in Fig. 5.28. A characteristic of centrifugal pumps is the low head that they can work against. This particular pump was effectively shut off at a pressure equivalent to 5.3 m of fluid (51.9 kPa ga). Another characteristic is the required NPSH. For example, at the maximum pump efficiency, flow is 3.5 kg/s and the head is 2.8 m. This will require a NPSH of 1.5 m. The NPSH is defined as follows:

$$NPSH \text{ available} = \left(\frac{1}{\rho g}\right)(P_{atm} \pm \rho g h - \Delta P_f - P_v)$$

where $\rho$, g, h, and $\Delta P_f$ are as previously defined. $P_{atm}$ is the atmospheric pressure, and $P_v$ is the fluid vapor pressure.

**Example 5.23** A fluid with a density of 1004 kg/m$^3$ and a viscosity of 0.002 Pa s is to be pumped using the pump with a performance curve shown in Fig. 5.18. If the inlet pipe is 1.5 in. sanitary pipe, the suction pipe section is 1.2 m long with one 90-degree elbow, fluid source is 1 m below pump level, and the fluid vapor pressure is the same as that of water at 35 °C (5.6238 kPa), calculate the NPSH available when operating at a head of 2.8 m, and determine if the pump will be

**Fig. 5.28** Performance curve of a volute type centrifugal pump

suitable for operating under the given conditions. The atmospheric pressure is 101,000 Pa.

**Solution**

The pipe diameter is 0.03561 m. The equivalent length of pipe L = 1.2 + 1(35)(0.03561) = 2.45 m. At a head of 2.8 m, the flow rate is 3.5 kg/s:

$$\bar{V} = 3.5\frac{kg}{s} \cdot \frac{1\,m^3}{1004\,kg} \cdot \frac{1}{\pi\left(\frac{.03561}{2}\right)^2} = 3.5\frac{m}{s}$$

Using Poiseuille equation:

$$\Delta P_f = \frac{8L\mu V}{R^2} = \frac{8(2.446)(0.002)(3.5)}{\left(\frac{0.03561}{2}\right)^2} = 432\,Pa$$

The NPSH available is greater than the 1.5 m required by the pump; therefore it will be possible to operate the pump under the specified conditions.

$$NPSH\ available$$
$$= \frac{1}{(1004)(9.8)}[101000 - 9.8(1000)(1)$$
$$-432 - 5624] = 8.65\ m$$

The performance curve of a positive displacement pump shown in Fig. 5.29 depicts a flow rate that depends on the speed of the rotor. Because a fixed volume is delivered with each rotation of the rotor, the rate of flow is proportional to the rotational speed. Note that the rate of flow at 400 rpm is almost double that at 200 rpm. High discharge pressures generated are one order of magnitude greater than those generated by centrifugal pumps. The flow reduction with an increase in head is gradual. Flow reduction with increase in head depends on the rotor design. A tight clearance between the rotor and the casing minimizes backflow of fluid at high pressures thus maintaining a fairly flat flow rate with pressure. Positive displacement pumps generally develop more constancy of flow with increasing head as the viscosity of the fluid increases because backflow is reduced as fluid viscosity increases.

Attaining high pressures and maintaining constancy of flow with increasing pressures, however, is not desirable if these conditions are not required in an application. The reduction in backflow requires tight clearances between the rotor and casing, and to generate high pressures, there must be a tight shaft seal increasing friction thus reducing the pump efficiency.

**Fig. 5.29** Performance curve for a two-lobe rotary positive displacement pump

The *pump efficiency* is defined as the ratio of the hydraulic horsepower to the brake horsepower. Let H be the pressure at pump discharge expressed as height of fluid in meters, g the acceleration due to gravity in m/s$^2$, and $\bar{m}$ the mass rate of flow in kg/s:

$$\text{Hydraulic Horsepower} = \frac{H \cdot g \cdot \bar{m}}{745.7}$$

The numerator is powered in watts and the denominator is 745.7 W/HP.

The brake horsepower is what the pump manufacturer specifies for the motor drive of a pump to deliver a specified rate of flow against a specified head. Efficiencies of centrifugal pumps and positive displacement pumps are rather low as shown in Figs. 5.28 and 5.29. Centrifugal pump efficiencies peak at 28% for the centrifugal pump shown in Fig. 5.28, and the maximum efficiency occurs over a very narrow flow rate-head combination. In contrast, positive displacement pump efficiency (Fig. 5.29) remains relatively constant over a wide range of head. Higher rotor speeds induce higher efficiency than at lower speeds. Efficiencies for this particular pump were 50% or 60% at 200 and 400 rpm, respectively, at the optimum head and flow rate. Energy losses in pumps are attributable primarily to backflow and mechanical friction.

## Problems

5.1. The following data were obtained when tomato catsup was passed through a tube having an inside diameter of 1.384 cm and a length of 1.22 m.

| Flow rate (cm$^3$/s) | P (dynes/cm$^2$) |
|---|---|
| 107.5 | $50.99 \times 10^4$ |
| 67.83 | $42.03 \times 10^4$ |
| 50.89 | $33.07 \times 10^4$ |
| 40.31 | $29.62 \times 10^4$ |
| 10.10 | $15.56 \times 10^4$ |
| 8.80 | $14.49 \times 10^4$ |
| 33.77 | $31.00 \times 10^4$ |
| 53.36 | $35.14 \times 10^4$ |
| 104.41 | $46.85 \times 10^4$ |

Determine the fluid consistency index K and the flow behavior index n of this fluid.

5.2. Figure 5.30 shows a water tower and the piping system for a small manufacturing plant. If water flows through the system at 40 L/min, what would be the pressure at point B? The pipes are wrought iron pipes. The water has a density of 998 kg/m$^3$ and a viscosity of 0.8 centipoise. The pipe is 1.5-in. (nominal) wrought iron pipe.

5.3. The catsup in Problem 1 (n = 0.45, K = 6.61 Pa $\cdot$ s$^n$) is to be heated in a shell

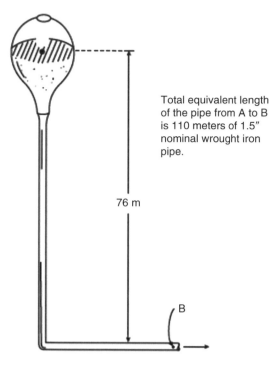

Total equivalent length of the pipe from A to B is 110 meters of 1.5" nominal wrought iron pipe.

76 m

B

**Fig. 5.30** Water tower and piping system for Problem 2

and tube heat exchanger. The exchanger has a total of 20 tubes 7 m long arranged parallel inside a shell. Each tube is a 3/4-in. outside diameter, 18-gauge heat exchanger tube. (From a table of thickness of sheet metal and tubes, an 18-gauge wall is 0.049 in.) It is possible to arrange the fluid flow pattern by the appropriate selection of heads for the shell, such that number of passes and the number of tubes per pass can be varied. Calculate the pressure drop across the heat exchanger (the pressure at the heat exchanger inlet necessary to push the product through the heat exchanger) if a flow rate of 40 L of product per min (density = 1013 kg/m$^3$) is going through the system for (a) two-pass (10 tubes/pass) and (b) five-pass system (four tubes per pass). Consider only the tube resistance, and neglect the resistance at the heat exchanger heads.

5.4. Figure 5.31 shows a deaerator operated at 381-mm-Hg vacuum. (Atmospheric pressure is 762 mm Hg.) It is desired to allow

a positive suction flow into the pump. If the fluid has a density of 1040 kg/m$^3$, the flow rate is 40 L/min, and the pipe is 1.5-in. sanitary pipe, calculate the height "h"; the bottom of the de-aerator must be set above the pump level in order that the pressure at the pump intake is at least 5 kPa above atmospheric pressure. The fluid is Newtonian and has a viscosity of 100 centipoises. [Note: Total length of straight pipe from first elbow below the pump to the entrance to the pump = 4 m. Distance from pump level to horizontal pipe = 0.5 m.]

5.5. Calculate the total equivalent length of 1-in. wrought iron pipe that would produce a pressure drop of 70 kPa due to fluid friction, for a fluid flowing at the rate of 50 L/min. The fluid has a density of 988 kg/m$^3$ and a viscosity of two centipoises.

5.6. Calculate the horsepower required to pump a fluid having a density of 1040 kg/m$^3$ at the rate of 40 L/min through the system shown in Fig. 5.32. $\Delta P_f/\rho$ calculated for the system is 120 J/kg. Atmospheric pressure is 101 kPa.

5.7. Calculate the average and maximum velocities of a fluid flowing at the rate of 20 L/min through a 1.5-in. sanitary pipe. The fluid has a density of 2030 kg/m$^3$ and a viscosity of 50 centipoises. Is the flow laminar or turbulent?

5.8. Determine the inside diameter of a tube that could be used in a high-temperature, short-time heater-sterilizer such that orange juice with a viscosity of 3.75 centipoises and a density of 1005 kg/m$^3$ would flow at a volumetric flow rate of 4 L/min and have a Reynolds number of 2000 while going through the tube.

5.9. Calculate the pressure generated at the discharge of a pump that delivers a pudding mix ($\rho$ = 995 kg/m$^3$; K = 1.0 Pa · s$^n$, n = 0.6) at the rate of 50 L/min through 50 m of a 1.5-in. straight, level 1.5-in. stainless steel sanitary pipe. What would be the equivalent viscosity of a Newtonian fluid that would give the same pressure drop?

**Fig. 5.31** Diagram of Problem 4

**Fig. 5.32** Diagram for Problem 6

5.10. What pipe diameter will give a rate of flow of 4 ft./min for a fluid delivered at the rate of 2 gal/min?

5.11. Calculate the viscosity of a fluid that would allow a pressure drop of 35 kPa over a 5 m length of $^3/_4$-in. stainless steel sanitary pipe if the fluid is flowing at 2 L/min and has a density of 1010 kg/m$^3$. Assume laminar flow.

5.12. A fluid is evaluated for its viscosity using a Brookfield viscometer. Collected data of rotational speed in rev/min and corresponding apparent viscosity in centipoises, respectively, are 20, 7230; 10, 12,060; 4, 25,200; and 2, 39,500. Is the fluid Newtonian or non-Newtonian? Calculate the flow behavior index, n, for this fluid.

5.13. A fluid having a viscosity of 0.05 lb$_{mass}$/ft. (s) requires 30 s to drain through a capillary viscometer. If this same viscometer is used to determine the viscosity of another fluid and it takes 20 s to drain, calculate the viscosity of this fluid. Assume the fluids have the same densities.

5.14. Figure 5.33 shows a storage tank for sugar syrup that has a viscosity of 15.2 centipoises at 25 °C and a density of 1008 kg/m$^3$. Friction loss includes an entrance loss to the drain pipe that equals the kinetic energy gain of the fluid and resistance to flow through the short section of the drain pipe.

(a) Formulate an energy balance equation for the system.

(b) Formulate the continuity equation that represents the increment change in fluid level in the tank as a function of the fluid velocity through the drain pipe.

(c) Solve simultaneously equations formulated in (a) and (b) to calculate the time to drain the tank to a residual level 1 m from the bottom of the tank.

(d) Calculate the amount of residual fluid in the tank including the film adhering to the side of the tank.

DRAIN PIPE
4 in SANITARY
5 m LONG
1 GATE VALVE

**Fig. 5.33** Diagram for Problem 14

5.15. A fluid tested on a tube viscometer 0.75 cm in diameter and 30 cm long exhibited a pressure drop of 1200 Pa when the flow rate was 50 cm$^3$/s.

(a) Calculate the apparent viscosity and the apparent rate of shear under this condition of flow.

(b) If the same fluid flowing at the rate of 100 cm$^3$/s through a viscometer tube 0.75 cm in diameter and 20 cm long exhibits a pressure drop of 1300 Pa, calculate the flow behavior and consistency indices. Assume wall effects are negligible.

5.16. Apparent viscosities in centipoises (cP) of 4000, 2500, 1250, and 850 were reported on a fluid at rotational speeds of 2, 4, 10, and 20 rev/min. This same fluid was reported to require a torque of 900 dyne cm to rotate a cylindrical spindle 1 cm in diameter and 5 cm high within the fluid at 20 rev/min.

(a) Calculate the flow behavior and consistency indices for this fluid.

(b) When exhibiting an apparent viscosity of 4000 cP, what would have been the shear rate under which the measurement was made?

5.17. The flow behavior and consistency indices of whey at a solids content of 24% have been reported to be 0.94 and $4.46 \times 10^{-3}$ Pa · s$^n$, respectively. If a rotational viscometer having a full-scale torque of 673.7 dyne Acm is to be used for testing the flow behavior of this fluid, determine the diameter and height of a cylindrical spindle to be used such that at the slowest speed of 2 rev/min. The minimum torque will be 10% of the full-scale reading of the instrument. Assume a length to diameter ratio of 3 for the spindle. Would this same spindle induce a torque within the range of the instrument at 20 rev/min?

5.18. Egg whites having a consistency index of 2.2 Pa · s$^n$ and a flow behavior index of 0.62 must be pumped through a 1.5-in. sanitary pipe at a flow rate, which would induce a shear rate at the wall of 150/s. Calculate the rate of flow in L/min and the pressure drop due to fluid flow resistance under these conditions.

5.19. The system shown in Fig. 5.34 is used to control the viscosity of a batter formulation used on a breading machine. A float indicates the fluid level in the reservoir, and by attaching the float to an appropriate transducer, addition of dry ingredients and water into the mixing tank may be regulated to maintain the batter consistency. The appropriate fluid level in the reservoir may be maintained when a different consistency of the batter is required, by changing the length of the pipe draining the reservoir. If the fluid is Newtonian with a viscosity of 100 cP and if fluid density is 1004 kg/m$^3$, calculate the feed rate that must be metered into the reservoir to maintain the level shown. Assume entrance loss

**Fig. 5.34** Diagram for Problem 19

is negligible compared to fluid resistance through the drain pipe.

5.20. The system shown in Fig. 5.35 has been reported to be used for disintegrating wood chips in the pulp industry, after digestion. It is desired to test the feasibility of using the same system on a starchy root crop such as cassava or sweet potatoes, to disrupt starch granules for easier hydrolysis with enzymes to produce sugars for alcoholic fermentation. In simulating the system on a small scale, two parameters are of importance: the shear rate of the slurry as it passes through the discharge pipe and the impact force of the fluid against the plate. Assume that entrance loss from the tank to the discharge pipe is negligible. Under the conditions shown, calculate the shear rate through the drain pipe and the impact force against the plate at the time the drain pipe is

P = +696 kPa gage

H = 6 m

DRAIN PIPE
30 cm ID
20 cm LONG

GATE
VALVE

NOZZLE
DISCHARGE
15 cm ID

IMPACT PLATE

**Fig. 5.35** Diagram for Problem 20

viscometer using a cylindrical spindle 1 cm in diameter and 5 cm long are as follows, with speed in rev/min and torque in % of full scale, respectively: 2, 11; 4, 18; 10, 34; and 20, 56. The viscometer constant is 7187 dyne cm. Calculate the apparent viscosity of Product A and Product B at 0.5 rev/min. At this rotational speed, did the apparent viscosity of Product B match that of Product A?

5.23. An FMC deaerator 2.97 m high and 96.5 cm in diameter is rated to deaerate from 4.2 to 8.4 kg/s of product. If the product has a density of 1008 kg/m$^3$ and has a flow behavior index of 0.44 and a consistency index of 8.1 Pa s$^n$, calculate the film thickness and film velocity to achieve the mid-range (6.3 kg/s) of the specified capacity. If half the deaerator height is to be covered by the fluid film, calculate the time available for any gas bubbles to leave the film into the vapor space in the chamber.

first opened. The slurry has flow behavior and consistency indices of 0.7 and 0.8 Pa · s$^n$ and a density of 1042 kg/m$^3$.

5.21. In a falling film direct-contact steam heater for sterilization, milk is pumped into a header that distributes the liquid to several vertical pipes, and the liquid flows as a film in laminar flow down the pipe. If the fluid has a density of 998 kg/m$^3$ and a viscosity of 1.5 cP, calculate the flow rate down the outside surface of each of 3.7-cm outside diameter pipes in order that the fluid film will flow at a Reynolds number of 500. Calculate the fluid film thickness when flow develops at this Reynolds number.

5.22. A sauce product is being formulated to match a reference product (Product A) that has a consistency index of 12 Pa s$^n$ and a flow behavior index of 0.55. Rheological measurements of the formulated product (Product B) on a wide-gap rotational

## Suggested Reading

Bennet, C. O., & Myers, J. E. (1962). *Momentum, heat, and mass transport*. New York: McGraw-Hill Book Co.

Charm, S. E. (1971). *Fundamentals of food engineering* (2nd ed.). Westport: AVI Publishing Co.

Felder, R. M., & Rousseau, R. W. (1999). *Elementary principles of chemical processes* (2nd ed.). New York: Wiley.

Foust, A. S., Wenzel, L. A., Clump, C. W., Maus, L., & Andersen, L. B. (1960). *Principles of unit operations*. New York: Wiley.

Geankoplis, C. J. (1993). *Transport processes and unit operations* (3rd ed.). Englewood Cliffs: Prentice-Hall.

Heldman, D. R., & Singh, R. P. (1981). *Food process engineering*. Westport: AVI Publishing Co..

Kokini, J. L., & Plutchok, G. J. (1987). Viscoelastic properties of semisolid foods and their biopolymeric components. *Food Technology, 41*(3), 89.

Kreiger, I. M., & Maron, S. H. (1952). Direct determination of flow curves of non-Newtonian fluids. *Journal of Applied Physics, 33*, 147.

Kreiger, I. M., & Maron, S. H. (1954). Direct determination of the flow curves of non-Newtonian fluids. III. Standardized treatment of viscometric data. *Journal of Applied Physics, 25*, 72.

Leniger, H. A., & Beverloo, W. A. (1975). *Food process engineering*. Boston: D. Riedel Publishing Co.

McCabe, W. L., & Smith, J. C. (1967). *Unit operations of chemical engineering* (2nd ed.). New York: McGraw-Hill Book Co.

Osorio, F. A., & Steffe, J. F. (1987). Back extrusion of power flow fluids. *Journal of Texture Studies, 18*, 43.

Rao, M. A., Shallenberger, R. S., & Cooley, H. J. (1987). Effect of temperature on viscosity of fluid foods with high sugar content. In M. Le Maguer & P. Jelen (Eds.), *Food engineering and process applications* (Vol. I, pp. 23–38). New York: Elsevier.

# Heat Transfer

Heat transfer is the movement of energy from one point to another by virtue of a difference in temperature. Heating and cooling are manifestations of this phenomenon, which is used in industrial operations and in domestic activities. Increasing energy costs and in some cases inadequate availability of energy will require peak efficiency in heating and cooling operations. An understanding of the mechanisms of heat transport is needed in order to recognize limitations of heating and cooling systems, which can then lead to adoption of practices that circumvent these limitations. In industrial and domestic heating and cooling, energy use audits can be used to determine total energy use and energy use distribution within the process, to identify areas of high energy use, and to target these areas for energy conservation measures.

## 6.1 Mechanisms of Heat Transfer

Heat will be transferred from one material to another when there is a difference in their temperature. The temperature difference is the driving force which establishes the rate of heat transfer.

### 6.1.1 Heat Transfer by Conduction

When heat is transferred between adjacent molecules, the process is called conduction. This is the mechanism of heat transfer in solids.

### 6.1.2 Fourier's First Law of Heat Transfer

According to Fourier's first law, the heat flux, in conduction heat transfer, is proportional to the temperature gradient:

$$\frac{q}{A} = -k\frac{dT}{dx} \qquad (6.1)$$

In Eq. (6.1), q is the rate of heat flow, and A is the area through which heat is transferred. A is the area perpendicular to the direction of heat flow. The expression q/A, the rate of heat transfer per unit area, is called the heat flux. The derivative dT/dx is the temperature gradient. The negative sign in Eq. (6.1) indicates that positive heat flow will occur in the direction of decreasing temperature. The parameter k in Eq. (6.1), the thermal conductivity, is a physical property of a material. Values of the thermal conductivity of common materials of construction and insulating materials and of food products are given in Appendices 9 and 10, respectively.

### 6.1.3   Estimation of Thermal Conductivity of Food Products

The thermal conductivity of materials varies with the composition and, in some cases, the physical orientation of components. Foods, being of biological origin, possess highly variable composition and structure; therefore, k of foods presented in the tables is not always the same for all foods in the category listed. The effect of variations in the composition of a material on values of the thermal conductivity has been reported by Choi and Okos (1987). Their procedure may be used to estimate k from the composition. k is calculated from the thermal conductivity of the pure component $k_i$ and the volume fraction of each component, $X_{vi}$. An important assumption used in this estimation procedure is that the contribution of each component to the composite thermal conductivity is proportional to the component volume fraction as follows:

$$k = \sum (k_i X_{vi}) \qquad (6.2)$$

The thermal conductivity in W/(m · K) of pure water ($k_w$), ice ($k_{ic}$), protein ($k_p$), fat ($k_f$), carbohydrate ($k_c$), fiber ($k_{fi}$), and ash ($k_a$) is calculated at the temperature, T in °C, using Eqs. (6.3, 6.4, 6.5, 6.6, 6.7, 6.8, and 6.9), respectively:

$$k_w = 0.57109 + 0.0017625\,T - 6.7306$$
$$\times\, 10^{-6}\,T^2 \qquad (6.3)$$

$$k_{ic} = 2.2196 - 0.0062489\,T + 1.0154$$
$$\times\, 10^{-4}\,T^2 \qquad (6.4)$$

$$k_p = 0.1788 + 0.0011958\,T - 2.7178$$
$$\times\, 10^{-6}\,T^2 \qquad (6.5)$$

$$k_f = 0.1807 - 0.0027604\,T - 1.7749$$
$$\times\, 10^{-7}\,T^2 \qquad (6.6)$$

$$k_c = 0.2014 + 0.0013874\,T - 4.3312$$
$$\times\, 10^{-6}\,T^2 \qquad (6.7)$$

$$k_{fi} = 0.18331 + 0.0012497\,T - 3.1683$$
$$\times\, 10^{-6}\,T^2 \qquad (6.8)$$

$$k_a = 0.3296 + 0.001401\,T - 2.9069$$
$$\times\, 10^{-6}\,T^2 \qquad (6.9)$$

The volume fraction $X_{vi}$ of each component is determined from the mass fraction $X_i$, the individual densities $\rho_i$, and the composite density, $\rho$ as follows:

$$X_{vi} = \frac{X_i \rho}{\rho_i} \qquad (6.10)$$

The individual densities, in kg/m³, are obtained using Eqs. (6.12, 6.13, 6.14, 6.15, 6.16, 6.17 and 6.18), respectively, for water ($\rho_w$), ice ($\rho_{ic}$), protein ($\rho_p$), fat ($\rho_f$), carbohydrate ($\rho_c$), fiber ($\rho_{fi}$), and ash ($\rho_a$):

$$\rho = \frac{1}{[\sum (X_i/\rho_i)]} \qquad (6.11)$$

$$\rho_w = 997.18 + 0.0031439\,T$$
$$- 0.0037574\,T^2 \qquad (6.12)$$

$$\rho_{ic} = 916.89 - 0.13071\,T \qquad (6.13)$$

$$\rho_p = 1329.9 - 0.51814\,T \qquad (6.14)$$

$$\rho_f = 925.59 - 0.41757\,T \qquad (6.15)$$

$$\rho_c = 1599.1 - 0.31046\,T \qquad (6.16)$$

$$\rho_{fi} = 1311.5 - 0.36589\,T \qquad (6.17)$$

$$\rho_a = 2423.8 - 0.28063\,T \qquad (6.18)$$

A spreadsheet program in Excel shown in Fig. 6.1 may be used to calculate the thermal conductivity from the composition of a material.

**Fig. 6.1** Spreadsheet program in Excel used to calculate the thermal conductivity of milk

**Fig. 6.2** Spreadsheet program in Excel used to calculate the thermal conductivity of lean pork

To use this program, enter the temperature in cell A2 and the mass fraction of each specified component in B4–B10. Excel will calculate required intermediate values and display the composite thermal conductivity in cell G13. Enter zero for a component whose mass fraction is not given.

**Example 6.1** Calculate the thermal conductivity of lean pork containing 7.8% fat, 1.5% ash, 19% protein, and 71.7% water, at 19 °C.

**Solution**
The spreadsheet showing values of the component mass fraction in B4–B10 and the composite thermal conductivity is shown in Fig. 6.2. The composite thermal conductivity is 0.4970 W/m · AK displayed in cell G13 in Fig. 6.2.

**Example 6.2** Calculate the thermal conductivity of milk which contains 87.5% water, 3.7% protein, 3.7% fat, 4.6% lactose, and 0.5% ash, at 10 °C.

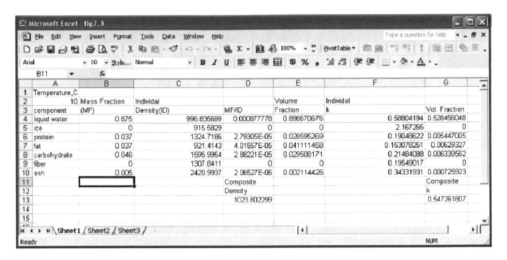

**Fig. 6.3**  Excel spreadsheet showing the solution for Example 6.2

## Solution

The spreadsheet program (Fig. 6.1) is used with the mass fraction of components and the temperature entered in the corresponding cells. Results are shown in Fig. 6.3. The composite thermal conductivity is shown in cell G13 to be 0.5473 W/m · K.

Data on average composition of foods from USDA Handbook 8, which can be used to estimate thermophysical properties, is given in Appendix 8.

### 6.1.4  Fourier's Second Law of Heat Transfer

When the rate of heat transfer across a solid is not uniform (i.e., there is a difference in the rate at which energy enters and leaves a control volume), this difference will be manifested as a rate of change of temperature with time, within the control volume. This type of problem is called unsteady-state heat transfer. Figure 6.4 shows a control volume for analyzing heat transfer into a cube with sides dx, dy, and dz. The following are heat balance equations across the control volume in the x, y, and z directions. Partial differential equations are used since in each case only one direction is being considered:

**Fig. 6.4**  Control volume for analysis of heat transfer into a cube

$$q = q_x + q_y + q_z = \rho \, dx \, dy \, dz \, C_p \frac{\delta T}{\delta t}$$

$$q_x = k \, dy \, dz \left[ \left. \frac{\partial T}{\partial x} \right|_1 - \left. \frac{\partial T}{\partial x} \right|_2 \right]$$

$$q_y = k \, dx \, dz \left[ \left. \frac{\partial T}{\partial y} \right|_1 - \left. \frac{\partial T}{\partial y} \right|_2 \right]$$

$$q_z = k \, dx \, dz \left[ \left. \frac{\partial T}{\partial z} \right|_1 - \left. \frac{\partial T}{\partial z} \right|_2 \right]$$

Combining:

$$\rho C_p \frac{\partial T}{\partial t} = k \left[ \frac{\left. \frac{\partial T}{\partial x} \right|_1 - \left. \frac{\partial T}{\partial x} \right|_2}{dx} + \frac{\left. \frac{\partial T}{\partial y} \right|_1 - \left. \frac{\partial T}{\partial y} \right|_2}{dy} \right]$$

$$+ k \left[ \frac{\left. \frac{\partial T}{\partial z} \right|_1 - \left. \frac{\partial T}{\partial z} \right|_2}{dz} \right]$$

The difference in the first derivatives divided by dx, dy, or dz is a second derivative, therefore:

$$\frac{\partial T}{\partial t} = \frac{k}{\rho C_p}\left[\frac{\partial^2 T}{\partial x^2} + \frac{\partial^2 T}{\partial y^2} + \frac{\partial^2 T}{\partial z^2}\right] \quad (6.19)$$

Equation (6.19) represents Fourier's second law of heat transfer. The rate of change of temperature with time at any position within a solid conducting heat is proportional to the second derivative of the temperature with respect to distance at that particular point. The ratio $k/(\rho \cdot C_p)$ is $\alpha$, the thermal diffusivity. The spreadsheet program for k (Fig. 6.1) and $C_p$ (Chap. 2, Fig. 2.1) can be used to estimate the thermal diffusivity of foods from their composition. An Excel program for calculating $\alpha$ determined from food composition is given in Appendix 11.

### 6.1.5 Temperature Profile for Unidirectional Heat Transfer Through a Slab

If heat is transferred under steady-state conditions, and A is constant along the distance, x, the temperature gradient, dT/dx, will be constant, and integration of Eq. (6.1) will result in an expression for temperature as a linear function of x. Substituting the boundary conditions, $T = T_1$ at $x = x_1$, in Eq. (6.20):

$$T = -\frac{q/A}{k} \cdot x + C \quad (6.20)$$

Substituting the boundary condition, $T = T_1$ at $x = x_1$, in Eq. (6.20):

$$T = -\frac{q/A}{k} \cdot (x_1 - x) + T_1 \quad (6.21)$$

Equation (6.21) is the expression for the steady-state temperature profile in a slab where A is constant in the direction of x. The temperature gradient, which is constant in a slab transferring heat at a steady state, is equal to the ratio of the heat flux to the thermal conductivity.

If the temperature at two different points in the solid is known (i.e., $T = T_1$ at $x = x_1$ and $T = T_2$ at $x = x_2$), Eq. (6.21) becomes:

$$T_1 - T_2 = -\frac{q/A}{k}(x_1 - x_2)$$
$$\frac{q}{A} = -k \cdot \frac{\Delta T}{\Delta x} \quad (6.22)$$

Substituting Eq. (6.22) in Eq. (6.21):

$$T = -\frac{\Delta T}{\Delta x} \cdot (x_1 - x) + T_1 \quad (6.23)$$

To keep the convention on the signs, and ensure that proper temperatures are calculated at any position within the solid, $\Delta T = T_2 - T_1$ with increasing subscripts in the direction of heat flow, and $\Delta x = x_2 - x_1$ is positively increasing in the direction of heat flow. Thus, if a temperature drop $\Delta T$ across any two points in a solid separated by distance $\Delta x$ is known, it will be possible to determine the temperature at any other point within that solid using Eq. (6.23). Equations (6.22) and (6.23) also show that when heat is transferred in a steady state in a slab, the product, $k(\Delta T/\Delta x)$, is a constant, and:

$$k\left[\frac{\Delta T}{\Delta x}\right]_1 = k\left[\frac{\Delta T}{\Delta x}\right]_2 = \cdots \quad (6.24)$$

If the solid consists of two or more layers having different thermal conductivity, the heat flux through each layer is constant; therefore, the temperature profile will exhibit a change in slope at each junction between layers. A composite slab containing three materials having thermal conductivity $k_1$, $k_2$, and $k_3$ is shown in Fig. 6.5. The temperature profile shows a different temperature gradient within each layer. For a composite slab, Eq. (6.24) becomes:

$$k_1\left(\frac{\Delta T_1}{\Delta x_1}\right) = k_2\left(\frac{\Delta T_2}{\Delta x_2}\right) = k_3\left(\frac{\Delta T_3}{\Delta x_3}\right) \quad (6.25)$$

Equation (6.25) can be used to experimentally determine the thermal conductivity of a solid by placing it between two layers of a second solid and determining the temperature at any two points within each layer of solid after a steady state has been achieved. The existence of a steady state and unidimensional heat flow can be proved by equality of the first and third terms. Two of the

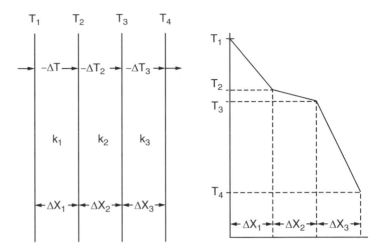

**Fig. 6.5** Diagram of composite slab and temperature drop at each layer

terms involving a known thermal conductivity $k_1$ and the unknown thermal conductivity $k_2$ can be used to determine $k_2$.

**Example 6.3** Thermocouples embedded at two points within a steel bar, 1 and 2 mm from the surface, indicate temperatures of 100 °C and 98 °C, respectively. Assuming no heat transfer occurring from the sides, calculate the surface temperature.

**Solution**

Because the thermal conductivity is constant, either Eqs. (6.23) or (6.25) can be used. Following the convention on signs for the temperature and distance differences, $T_2 = 98$, $T_1 = 100$, $x_2 = 2$ mm, and $x_1 = 1$ mm. The temperature gradient $\Delta T/\Delta x = (T_2 - T_1)/(x_2 - x_1) = (98-100)/0.001(2-1) = -2000$. Equation (6.23) becomes:

$$T = (-2000)(x_1 - x) + T_1$$

At the surface, $x = 0$, and at point $x_1 = 0.001$, $T_1 = 100$. Thus:

$$T = 2000(0.001) + 100 = 102\overset{\circ}{}C$$

**Example 6.4** A cylindrical sample of beef 5 cm thick and 3.75 cm in diameter is positioned

between two 5-cm-thick acrylic cylinders of exactly the same diameter as the meat sample. The assembly is positioned inside an insulated container such that the bottom of the lower acrylic cylinder contacts a heated surface maintained at 50 °C and the top of the upper cylinder contacts a cool plate maintained at 0 °C. Two thermocouples each are embedded in the acrylic cylinders, positioned 0.5 cm and 1.5 cm from the sample-acrylic interface. If the acrylic has a thermal conductivity of 1.5 W/(m · K), and the temperatures recorded at steady state are, respectively, 45 °C, 43 °C, 15 °C, and 13 °C, calculate the thermal conductivity of the meat sample.

**Solution**

Equation (6.25) will be used. Proceeding along the direction of heat flow, let $T_1 = 45$, $T_2 = 43$, $T_3 =$ temperature at the meat and bottom acrylic cylinder interface, $T_4 =$ temperature at the meat and top acrylic cylinder interface, $T_5 = 15$, and $T_6 = 13$ °C. Let $x_1 = 0$ at the lowermost thermocouple location, $x_2 = 1$, $x_3 = 1.5$, $x_4 = 6.5$, $x_5 = 7$, and $x_6 = 8$ cm. Because the temperatures have reached a steady state, the rate of heat transfer across any part of the system is equal; therefore, the temperature differences indicated by thermocouples within the acrylic should be equal if they are separated the same distances apart. Thus: $(T_2 - T_1)/(x2 - x1) = (T6 - T5)/(x6 - x5) = -2/0.01 = -200$. The heat flux through the system is:

$$\frac{q}{A} = -k\left(\frac{\Delta T}{\Delta x}\right) = -1.5(-200) = 300\,W$$

Using Eq. (6.23), the temperatures at the acrylic and meat sample interfaces are:

$$T_3 = -(-200)(0.01 - 0.015) + 43 = 42\,^{\circ}C$$
$$T_4 = -(-200)(0.07 - 0.065) + 15 = 16\,^{\circ}C$$

Using Eq. (6.25), with $\Delta T$ as the temperature drop across the meat sample $(T_3 - T_4)$, and $\Delta x =$ the thickness of the meat sample, the thermal conductivity of the meat is:

$$k = \frac{q/A}{\Delta T/\Delta x} = \frac{300}{26/0.05} = 0.5769\,W/(m \cdot K)$$

### 6.1.6 Conduction Heat Transfer Through Walls of a Cylinder

When heat flows through the walls of a cylinder, the area perpendicular to the direction of heat flow changes with position. A in Eq. (6.1) at any position within the wall is the surface area of a cylinder of radius r, $A = 2\pi rL$. Equation 6.1 may be expressed as:

$$q = -k(2\pi rL)\frac{dT}{dr}$$

When heat transfer occurs in a steady state, q is constant:

$$-\frac{dr}{r} = \frac{2\pi Lk}{q}\,dT \tag{6.26}$$

After integration and substitution of the boundary conditions, at $r = r_1$, $T = T_1$ and at $r = r_2$, $T = T_2$, the resulting equation for the rate of heat transfer expressed in Eq. (6.27) is obtained:

$$q = \frac{T_1 - T_2}{(\ln(r_2/r_1))/2\pi Lk} \tag{6.27}$$

Equation (6.27) follows the convention established in Eq. (6.24), that q is positive in the direction of decreasing temperature. Increasing subscripts represent positions proceeding farther away from the center of the cylinder.

### 6.1.7 The Temperature Profile in the Walls of a Cylinder in Steady-State Heat Transfer

At any point r, the temperature T may be obtained from Eq. (6.27) by substituting r for $r_2$ and T for $T_2$:

$$T = \frac{\ln(r/r_1)q}{2\pi Lk} + T_1$$

Substituting Eq. (6.27) for q:

$$T = \frac{\ln(r/r_1)}{\ln(r_2/r_1)}(T_1 - T_2) + T_1 \tag{6.28}$$

or:

$$\frac{(T - T_1)}{\ln(r/r_1)} = \frac{(T_1 - T_2)}{\ln(r_2/r_1)} \tag{6.29}$$

If the wall of the cylinder consists of layers having different thermal conductivities, the temperature profile may be obtained using the same procedure used in deriving Eqs. (6.28) and (6.29), except that the thermal conductivity of each layer is used.

$$\frac{(T - T')}{\frac{1}{k_2}\ln\left(\frac{r}{r'}\right)} = \frac{(T_1 - T_2)}{\frac{1}{k_1}\ln\left(\frac{r_2}{r_1}\right)} = \frac{q}{2\pi L} \tag{6.30}$$

In Eq. (6.30), the temperatures $T_1$ and $T_2$ must transect a layer bounded by $r_1$ and $r_2$, which has a uniform thermal conductivity $k_1$. Similarly, the layer bounded by r and $r'$ where the temperatures are T and $T'$ must also have a uniform thermal conductivity, $k_2$. Each of the terms in Eq. (6.30) is proportional to the rate of heat transfer. Equations (6.30) and (6.25) represent a fundamental relationship in steady-state heat transfer by conduction (i.e., the rate of heat flow through each layer in a multilayered solid is equal).

**Example 6.5** A ¾-in. steel pipe is insulated with 2-cm-thick fiberglass insulation. If the inside wall of the pipe is at 150 °C, and the temperature at the outside surface of the insulation is 35 °C, (a) calculate the temperature at the interface between the pipe and the insulation, and (b) calculate the rate of heat loss for 1 m length of pipe.

**Solution**
From Table 6.2 on pipe dimensions, the inside diameter of a ¾-in. steel pipe is 0.02093 m, and the outside diameter is 0.02667 m. From Appendix 9, the thermal conductivity of steel is 45 W/(m · K) and that of fiberglass is 0.035 W/(m · K). From the pipe dimensions, $r_1$ = 0.01047 m, $r_2$=0.01334, and $r_3$=(0.01334+0.020)=0.03334 m. $T_1$ = 150. $T_2$ = the temperature at the pipe-insulation interface. $T_3$ = 35 °C. $T_1$ and $T_2$ transect the metal wall of the pipe, with k = 45 W/(m · K). From Eq. (7.30): T = $T_2$, T' = $T_s$, r = $r_3$ and r' = $r_2$, $k_1$ = 0.035 W/(m · K), and $k_2$ = 45 W/(m · K). Substituting in Eq. (6.30):

$$\frac{(T_2 - T_s)k_1}{\ln(r_3/r_2)} = \frac{(T_1 - T_2)k_2}{\ln(r_2/r_1)} 0.038210\, T_2$$

$$-1.33735 = 185.75(150)$$

$$-185.75\, T_2 \frac{(T_2 - 35)(0.035)}{\ln(0.03334/0.01334)}$$

$$= \frac{(150 - T_2)(45)}{\ln(0.01334/0.01047)}$$

$$T_2(185.75 + 0.03821)$$

$$= 27,864\, T_2 = 149.97^\circ C$$

### 6.1.8    Heat Transfer by Convection

This mechanism transfers heat when molecules move from one point to another and exchanges energy with another molecule in the other location. Bulk molecular motion is involved in convection heat transfer. Bulk molecular motion is induced by density changes associated with difference in fluid temperature at different points in the fluid, condensation, or vaporization (natural

convection) or when a fluid is forced to flow past a surface by mechanical means (forced convection). Heat transfer by convection is evaluated as the rate of heat exchange at the interface between a fluid and a solid. The rate of heat transfer by convection is proportional to the temperature difference and is expressed as:

$$q = hA(T_m - T_s) = hA\,\Delta T \qquad (6.31)$$

where h is the heat transfer coefficient, A is the area of the fluid-solid interface where heat is being transferred, and $\Delta T$, the driving force for heat transfer, is the difference in fluid temperature, $T_m$, and the solid surface temperature, $T_s$. Convection heat transfer is often represented as heat transfer through a thin layer of fluid that possesses a temperature gradient, at the fluid-surface interface. The temperature, which is assumed to be uniform at $T_m$ in the fluid bulk, gradually changes through the fluid film until it assumes the solid surface temperature past the film. Thus, the fluid film may be considered as an insulating layer that resists heat flow between the fluid and the solid. The fluid film is actually a boundary layer that has different properties and different velocity from the bulk of the fluid. The magnitude of the heat transfer coefficient varies in an inverse proportion to the thickness of the boundary layer. Conditions that result in a reduction of the thickness of this boundary layer will promote heat transfer by increasing the value of the heat transfer coefficient.

#### 6.1.8.1    Natural Convection
Natural convection depends on gravity and density and viscosity changes associated with temperature differences in the fluid to induce convective currents. The degree of agitation produced by the convective currents depends on the temperature gradient between the fluid and the solid surface. When the $\Delta T$ is small, convective currents are not too vigorous, and the process of heat transfer is referred to as free convection. The magnitude of the heat transfer coefficient in free convection is very low, of the order 60 W/(m² · K) for air and 60–3000 W/(m² · K) for water. When the surface is in contact with a liquid, and the surface temperature exceeds the boiling point of

the liquid, bubbles of superheated vapor are produced at the solid-liquid interface. As these bubbles leave the surface, the boundary layer is agitated resulting in very high heat transfer coefficients. This process of heat transfer is called nucleate boiling, and the magnitude of the heat transfer coefficient for water is of the order 5000–50,000 W/(m$^2$· K). When the $\Delta T$ is very high, excessive generation of vapor at the interface produces an insulating layer of vapor that hinders heat transfer. This process of heat transfer is called film boiling, and the heat transfer coefficient is much lower than that in nucleate boiling.

Another form of natural convection is the transfer of heat from condensing vapors. Condensing vapors release a large amount of energy on condensation; therefore, heat transfer coefficients are very high. When the vapors condense as droplets, which eventually coalesce and slide down the surface, the vapor is always in direct contact with a clean surface, and therefore, heat transfer coefficients are very high. This type of heat transfer is called dropwise condensation. Heat transfer coefficients of the order 10,000 W/(m$^2$· K) are common in dropwise condensation. When vapors condense as a film of liquid on a surface, the liquid film forms a barrier to heat transfer, and heat transfer coefficients are lower. This process of heat transfer is called filmwise condensation, and the magnitude of the heat transfer coefficient may be in the order of 5000 W/(m$^2$· K).

### 6.1.8.2   Forced Convection

In forced convection heat transfer, heat transfer coefficients depend on the velocity of the fluid, its thermophysical properties, and the geometry of the surface. In general, heat transfer coefficients for noncondensing gases are about two orders of magnitude lower than that for liquids. Techniques for calculating heat transfer coefficients are discussed in the section "Local Heat Transfer Coefficients."

## 6.1.9   Heat Transfer by Radiation

Heat transfer by radiation is independent of and additive to that transferred by convection. Electromagnetic waves traveling through space may be intercepted by a suitable surface and absorbed, raising the energy level of the intercepting surface. When the electromagnetic waves are of the frequency of light, the phenomenon is referred to as radiation. All bodies at temperatures higher than absolute zero emit energy in proportion to the fourth power of their temperatures. In a closed system, bodies exchange energy by radiation until their temperatures equalize. Radiation heat transfer, like convection, is a surface phenomenon; therefore, the conditions at the surface determine the rate of heat transfer. Thermal radiation includes the spectrum ranging from the high ultraviolet (0.1 μm) through the visible spectrum (0.4–0.7 μm) to infrared (0.7–100 μm). Surfaces emit thermal radiation in a range of wavelengths; therefore, radiation may be expressed as a *spectral intensity*, which is the intensity at each wavelength, or a *total intensity*, which is the integral of the energy emitted over a range of wavelengths. Surfaces also receive energy from the surroundings. The total energy received by a surface is called *irradiation*, which could also be considered as *spectral* or total. The irradiation received by a surface may be absorbed or reflected. The total energy leaving a surface is the sum of emitted energy by virtue of its temperature and the reflected energy and is called the *radiosity*. The fraction of incident energy absorbed by a surface is called the *absorptivity* ($\alpha$), and the fraction reflected is called the *reflectivity* ($\rho$). Some of the energy may be transmitted across the surface, and the fraction of incident energy transmitted is the *transmissivity* ($\tau$). Thus, $\alpha + \rho + \tau = 1$.

### 6.1.9.1   Types of Surfaces

Surfaces may be classified according to their ability to absorb radiation, as *black* or *gray* bodies. A black body is one that absorbs all incident radiation. An example of a black body is the interior of a hollow sphere that has a small opening to admit radiation. All the energy that enters the small opening is reflected back and forth within the inside of the sphere, and, eventually, it will be totally absorbed. Emissivity ($\varepsilon$) is a property that is the fraction of radiation emitted or absorbed by a black body at a given temperature that is actually emitted or absorbed by a surface at the same

**Table 6.1** Emissivity of various materials

| Material | Temp.(°C) | Emissivity |
|---|---|---|
| Aluminum, bright | 170 | 0.04 |
| Aluminum paint | 100 | 0.3 |
| Chrome, polished | 150 | 0.06 |
| Iron, hot rolled | 20 | 0.77 |
| Brick, mortar, plaster | 20 | 0.77 |
| Glass | 90 | 0.94 |
| Oil paints, all colors | 212 | 0.92–0.96 |
| Paper | 95 | 0.92 |
| Porcelain | 20 | 0.93 |
| Stainless steel, polished | 20 | 0.24 |
| Wood | 45 | 0.82–0.93 |
| Water | 32–212 | 0.96 |

*Source:* Bolz; R. E., and Tuve, G. L. (eds.) 1970 *Handbook of tables for Applied Engineering Science.* CRC Press, Cleveland, Ohio

temperature. Thus, black bodies have $\varepsilon = 1$. If a surface absorbs a fraction ($\varepsilon < 1$) of incident radiation equally at all wavelengths, the surface is considered gray. Table 6.1 shows emissivity of some surfaces.

#### 6.1.9.1.1 Kirchhoff's Law

A body at constant temperature is in equilibrium with its surroundings, and the amount of energy absorbed by radiation will be exactly the same as that emitted. Thus, the absorptivity of a surface ($\alpha$) is exactly the same as the emissivity ($\varepsilon$), and these two properties may be used interchangeably.

#### 6.1.9.1.2 Stefan-Boltzmann Law

The energy flux emitted by a black body is directly proportional to the fourth power of the absolute temperature:

$$\frac{q}{A} = \sigma T^4 \qquad (6.32)$$

$\sigma$ is the Stefan-Boltzmann constant which has a value of $5.6732 \times 10^{-8}$ W/(m$^2$ · K$^4$). For gray bodies:

$$q = A\sigma\varepsilon T^4 \qquad (6.33)$$

The intensity of energy flux radiating from a black surface is a function of the wavelength,

and a wavelength exists where the intensity of radiation is maximum. The energy flux from a black surface at an absolute temperature T as a function of the wavelength $\lambda$ is:

$$\frac{q}{A} = \frac{C_1}{\lambda^5} \cdot \frac{1}{[e^{C_2/(\lambda T)} - 1]} \qquad (6.34)$$

$C_1$ and $C_2$ are constants. The total flux over the entire spectrum will be:

$$\frac{q}{A} = \int_{-\infty}^{\infty} \left[ \frac{C_1 \lambda^{-5}}{e^{C_2/(\lambda T)}} - 1 \right] d\lambda \qquad (6.35)$$

Equation (6.34) can be used to show that there will be a wavelength where the energy flux is maximum. The total energy flux over the whole spectrum is the energy radiated by a body at a temperature, T; thus, Eq. (6.35) is equivalent to Eq. (6.32).

#### 6.1.9.1.3 Wein's Displacement Law

The wavelength for maximum energy flux from a body shifts with a change in temperature. The product of the wavelength for maximum flux intensity and absolute temperature is a constant. $\lambda_{max} \cdot T = 2.884 \times 10^{-3}$ m · K.

#### 6.1.9.1.4 Lambert's Law

This is also called the cosine law. The energy flux over a solid angle $\omega$ in a direction $\varphi$ from a normal drawn toward the surface is a function of cos $\varphi$:

$$\frac{dq}{dA \, d\omega} = \frac{\varepsilon\sigma T^4}{\pi} \cos\phi \qquad (6.36)$$

Equation (6.36) is the basis for the derivation of view factors used in calculating the rate of radiation heat transfer between two bodies having the same emissivities. Figure 6.6 shows two area increments transferring heat by radiation and the representation of the cosine law according to Eq. (6.36). The solid angle $d\omega$ with which the area element $dA_1$ is viewed from $dA_2$ is:

$$d\omega = \frac{\cos\phi_2 dA_2}{r^2}$$

Integrating Eq. (6.36) after substituting the expression for the solid angle $(d\omega)$ will give the rate of heat transfer $(q_{1-2})$ from area 1 to area 2:

$$q_{1-2} = \varepsilon_1\sigma\left(T_1^4 - T_2^4\right) \int_{A_1} \int_{A_2} \frac{\cos\phi_2 dA_2}{\pi r^2}$$
$$\cdot dA_1 \cos\phi_1$$

The double integral times $1/A_1$ is the view factor $(F_{1-2})$ used to calculate the energy transferred by area 1, and when the multiplier is $1/A_2$, it becomes the view factor $(F_{2-1})$ for area 2. The equations for the rate of heat transfer become:

$$q_{2-1} = (F_{2-1})A_2\varepsilon\sigma\left(T_1^4 - T_2^4\right) \qquad (6.37)$$

$$q_{1-2} = (F_{1-2})A_1\varepsilon\sigma\left(T_1^4 - T_2^4\right) \qquad (6.38)$$

Because in a steady state the rate of heat transfer from $A_1$ to $A_2$ is the same as that from $A_2$ to $A_1$, the product of $A_1$ and the view factor for $A_1$ are equal to the product of $A_2$ and the view factor for $A_2$. If the two surfaces have different emissivities, the effective view factor $(\bar{F})$ derived by Jacob and Hawkins (1957) is:

$$\frac{1}{\bar{F}_{2-1}A_2} = \frac{1}{A_2F_{2-1}} + \frac{1}{A_1}\left(\frac{1}{\varepsilon_1} - 1\right) \\ + \frac{1}{A_2}\left(\frac{1}{\varepsilon_2} - 1\right) \qquad (6.39)$$

$$\frac{1}{\bar{F}_{1-2}A_1} = \frac{1}{A_1F_{1-2}} + \frac{1}{A_1}\left(\frac{1}{\varepsilon_1} - 1\right) \\ + \frac{1}{A_2}\left(\frac{1}{\varepsilon_2} - 1\right) \qquad (6.40)$$

The rate of heat transfer based on area $A_1$ or $A_2$ using the effective view factor is:

$$q = A_1\bar{F}_{1-2}\sigma\left(T_1^4 - T_2^4\right) \\ = A_2\bar{F}_{2-1}\sigma\left(T_1^4 - T_2^4\right) \qquad (6.41)$$

View factors for some geometries and procedures for applying view factor algebra are available in the literature (e.g., Rohsenow and Hartnett 1973). View factors for simple geometries are given in the succeeding section on "Radiant Energy Exchange."

### 6.1.9.2 Effect of Distance Between Objects on Heat Transfer

The total energy intercepted by an area from a point on another area is dependent on the solid angle with which the point views the area. The surface area seen from a point over a solid angle $\omega$ may be considered as the base of a cone of distance $r$ from the point and has an area of $\omega r^2$. Because the energy leaving the point toward this viewed area is q, the energy flux is $q/\omega r^2$, indicating that flux is inversely proportional to the square of the distance from the source.

**Fig. 6.6** Two area elements and representation of the cosine law for heat transfer by radiation

Radiant energy flux from a source weakens as a body moves away from a source.

### 6.1.9.3    Radiant Energy Exchange

View factors for simple geometries:

1. Small object ($A_1$) surrounded by a large object:

$$\bar{F}_{1-2} = \varepsilon_1; \quad F_{1-2} = 1$$

2. Large parallel planes with equal areas:

$$\frac{1}{\bar{F}_{1-2}} = \frac{1}{\varepsilon_1} + \frac{1}{\varepsilon_2} - 1 \qquad (6.42)$$

3. Two parallel disks with centers directly in line (from Rohsenow and Hartnett 1973): a = diameter of one disk and b = diameter of the other; c = distance between disks. A1 is the area of the larger disk with diameter b (Fig. 6.7):

$$\bar{F}_{1-2} = 0.5\left[Z - \sqrt{(Z^2 - 4X^2Y^2)}\right] \qquad (6.43)$$

4. Two parallel long cylinders of equal diameters, b separated by distance 2a (from Roshenow and Hartnett 1973) (see Fig. 6.7):

$$\left(\bar{F}\right)_{1-2}$$
$$= \frac{2}{\pi}\left[\sqrt{(X^2 - 1)} - X + \frac{\pi}{2} - \arccos\left(1/x\right)\right] \qquad (6.44)$$

where X = 1+ a/b.

A spreadsheet program in Excel for solving the view factors represented by Eqs. (6.44) is shown in Fig. 6.8.

**Example 6.6** Glass bottles may be prevented from breaking on filling with hot pasteurized juice when their temperature is close to that of the juice being filled. The bottles are rapidly heated by passing through a chamber that has top, bottom, and side walls heated by natural gas. The glass bottles may be considered as an object completely surrounded by a radiating surface. The glass bottles have an emissivity of 0.94,

a mass of 155 g each, a specific heat of 1256 J/(kg · K), and a surface area of 0.0219 m². If the glass bottles are to be heated from 15.5 °C to 51.6 °C in 1 min, calculate the temperature of the walls of the chamber to achieve this average heating rate when the glass is at the midpoint of the temperature range (33.6 °C).

**Solution**

The required heat transfer rate is m $C_p$ $\Delta T$:

$$q = \frac{0.155(1256)(51.6 - 15.5)}{60} = 117.1\,\text{W}$$

Equations (6.39) and (6.41) may be used to calculate $F_{1-2}$ and $q_{1-2}$. Because $A_2$ in Eq. (6.29) is large, the third term on the right of the equation is zero, and because $F_{1-2} = 1$, $\bar{F}_{1-2} = \varepsilon$. The heat transfer by radiation may also be calculated using Eq. (6.38):

$$F_{1-2} = 1; \quad T_2 = 33.6 + 273 = 303.6\text{kq}$$

$$= A\varepsilon\sigma\left(T_1^4 - T_2^4\right)$$

$$q = 0.0219(0.94)(5.6732e - 0.8)\left(T^4 - (306.6)^4\right)$$

$$T^4 = \frac{117.1}{0.0219(0.94)(5.6732e - 0.8)} + (306.6)^4$$

$$T = \left(1.091034 \times 10^{11}\right)^{0.25} = 574.7\,\text{K}$$

**Example 6.7** Cookies traveling on a conveyor inside a continuous baking oven occupy most of the area on the surface of the conveyor. The top wall of the oven directly above the conveyor has an emissivity of 0.92, and the cookies have an emissivity of 0.8. If the top wall of the oven has a temperature of 175 °C, calculate the average rate of heat transfer by radiation between the cookies per unit area on the side that faces the top wall of the oven when the cookie surface temperature is 70 °C.

**Solution**

Assume the layer of cookies on the conveyor and the top oven wall constitute a set of long parallel plates. The view factor $\bar{F}_{1-2}$ is calculated using Eq. (6.42):

**Fig. 6.7** Figure of two
parallel disks and two
parallel long cylinders
exchanging energy by
radiation

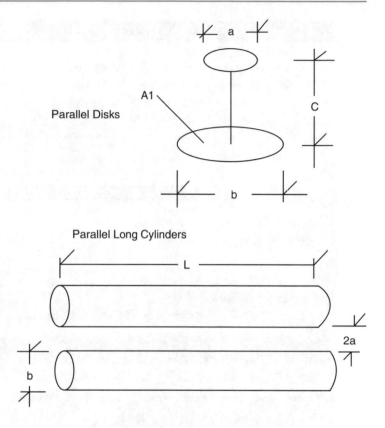

Using Eq. (6.41):

$$\frac{q}{A} = 0.748(5.6732 \times 10^{-8})(448^4 - 343^4)$$
$$= 1122 \text{ W/m}^2$$

$$\frac{1}{\overline{F}_{1-2}} = \frac{1}{0.92} + \frac{1}{0.80} - 1 = 1.3369$$

$$\overline{F}_{1-2} = \frac{1}{1.3369} = 0.748$$

$$T_1 = 70 + 273 = 343\text{K};$$

$$T_2 = 175 + 273 = 448\text{K}$$

**Example 6.8** The temperature registered by a thermocouple in a gas stream is often the result of steady-state heat transfer by convection from the gas to the thermocouple junction. However, when radiation heat transfer from a surrounding wall is significant, the actual temperature indicated by the thermocouple will be higher than the gas temperature. Consider a superheater for steam that consists of a gas-heated pipe through which steam is flowing. The pipe wall temperature is 500 °C, and the heat transfer coefficient between the thermocouple junction and the superheated air is 342 W/m² K. If the thermocouple reads 200 °C, what is the actual temperature of the superheated steam? Assume the thermocouple junction is a black body.

**Solution**

The thermocouple junction may be considered as an area totally enclosed by another; therefore, $F_{1-2} = 1$. Because the temperature of the thermocouple junction is higher than that of steam, energy will be lost by the junction due to convection. To maintain a constant temperature, the heat transfer by radiation will equal that lost by convection and the heat balance becomes:

$$hA(T - T_s) = \sigma A(T_w^4 - T^4)$$

Solving for $T_s$:

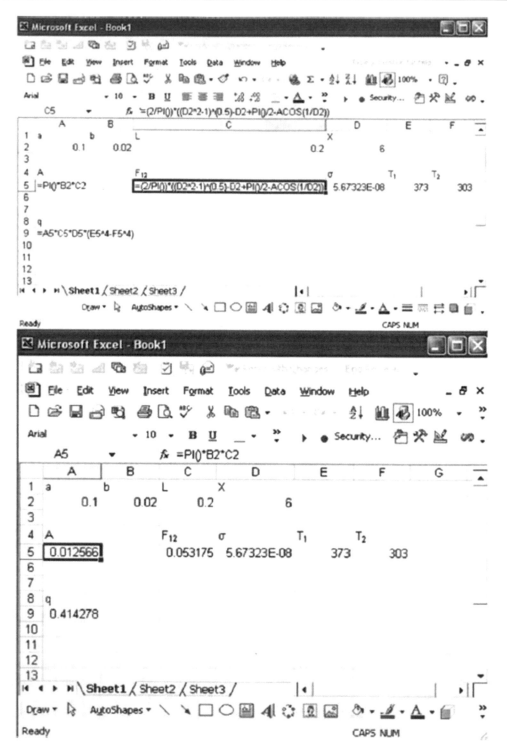

**Fig. 6.8** Spreadsheet program in excel to solve view factors Eqs. 6.43 and 6.44

$$T_s = (1/h)\left[hT - \sigma\left(T_w{}^4 - T^4\right)\right]$$

$T_w$ and $T$ in the expression for radiation heat transfer are in absolute temperature of 773 and 473 K, respectively; therefore, for consistency $T$ and $T_s$ in the expression for convection heat transfer should also be expressed in Kelvin. Solving for $T_s$ in Kelvin:

$$T_s$$
$$= (1/342)\left[342A\,473 - 5.6732 \times 10^{-8}\left\{(773)^4 - (473)^4\right\}\right]$$
$$= 422 \text{ K or } 149\ ^\circ C$$

Note that the actual steam temperature $T_s$ will approach the indicated temperature $T$ as $h$ approaches infinity or if $T_w$ approaches $T$.

**Example 6.9** A vacuum belt dryer consists of a belt that conveys the product. This belt travels over a heated plate. Another heated plate is positioned over this belt and extends the whole length of the lower plate. Heat transfer from the heated top plate to the product on the conveyor is by radiation, and the view factor may be considered as that for long parallel plates. The top plate is sand blasted stainless steel with $\varepsilon = 0.52$. Apple slices $\varepsilon = 0.85$ are being dried, and the desired product temperature during drying equals the boiling temperature of water at an absolute pressure of 7 mmHg, the pressure inside the vacuum dryer. If the surface area of sliced apples at 86% water is 0.731 $cm^2/g$, calculate the rate of evaporation of water from the apple surface due to heat transfer by radiation, expressed in g water evaporated/(h Ag dry apple solids) if the top plate temperature is maintained at 80 °C.

**Solution**
At 7 mm Hg absolute pressure, the pressure in Pascals is:

$$P = 0.7 \text{ cm Hg} \times 1333.33 \text{ Pa/cm Hg}$$
$$= 933.33 \text{ Pa}$$

From Appendix 4, $T = 5 + [2.5/(1.0365–0.8724)]$
$(0.93333–0.8724) = 5.9\ °C$
Using Eq. (6.42): $\bar{F}_{1-2} = 1/[1/0.52 + 1/0.85–1] = 0.4763$
Using Eq. (6.41): $q/A_1 = 0.4763(5.6732 \times 10^{-8})$
$[(353)^4 - (278.9)^4] = 256.07 \text{ W/m}^2$

From Appendix 4, the latent heat of evaporation of water at .9333 kPa is:

$$h_{fg} = 2.4897$$
$$-[(2.4897 - 2.4839)/(1.0365 - 0.8724)]$$
$$(0.9333 - 0.8724) = 2.4918 \text{ MJ/kg}$$

Drying rate:

$$\frac{dW}{dt} = \left[\frac{256.07 \text{ J}}{(s \cdot m^2)}\right] \cdot \left[\frac{kg}{2491800}\right] \cdot \left[\frac{1000g}{kg}\right] \cdot \left[\frac{3600 \text{ s}}{h}\right]$$

$$= 360.95 \frac{g\,water}{(m^2 \cdot h)}$$

The apple surface area in $m^2/g$ dry matter:

$$= [0.731 \text{ cm}^2/(1 - 0.86) \text{ g dry matter}]$$
$$[1 \text{ m}^2/10000 \text{ cm}^2] = 0.000522$$

$$dW/dt = [369.95 \text{ g water}/(m^2 A\,h)]$$
$$\times [0.000522 \text{ m}^2/\text{g dry matter}]$$
$$= 0.193 \text{ g water}/(h\,A\,g\,dry\,apple\,solids)$$
evaporation from the
radiant heat transfer alone.

## 6.1.10    Microwave and Dielectric Heating

Microwaves like light are also electromagnetic vibration. Heat transfer is dependent on the degree of excitability of molecules in the absorbing medium and the frequency of the field to which the medium is exposed. Dielectric heating is the term used when relatively low frequencies are used and the material is placed between two electrodes to which an electric current is passed. Frequencies from 60 Hz to 100 MHz may be used for dielectric heating. Microwave heating refers to the use of electromagnetic waves of very high frequency making it possible to transmit the energy through space. The most common frequencies used for microwave heating are 2450 MHz and 915 MHz. Domestic microwave ovens operate at 2450 MHz. The equations that govern heat transfer by microwave and dielectric systems are the same.

### 6.1.10.1  Energy Absorption by Foods in a Microwave Field

The energy absorbed by a body is:

$$\frac{q}{V} = 0.556\left(10^{-12}\right) f E^2 e \tan(\delta) \qquad (6.45)$$

where q/V = energy absorbed, W/cm$^3$; f = frequency, Hz; e = dielectric constant, an index of the rate at which energy penetrates a solid; dimensionless tan($\delta$) = dielectric loss factor, an index of the extent to which energy entering the solid is converted to heat; and dimensionless E = field strength in volts/cm$^2$. e and tan($\delta$) are properties of the material and are functions of composition and temperature. f and E are set by the type of microwave generator used. Table 6.2 shows the dielectric constant and the dielectric loss factor for foods, food components, and some packaging materials. Metal containers are opaque to microwave (i.e., microwaves are reflected from the surface therefore none passes across to food contained inside). However, an electrically conductive metal finite electrical resistance will heat up in the same manner as an electrical resistance wire will heat up when an electric current is passed through it. Similarly, electrically conductive wires will heat very rapidly in a microwave field. A continuous metal sheet with very low electrical resistance will not heat up in a microwave field. However, a discontinuous metal sheet such as metallized plastic contains many small areas of metal that presents a large resistance to electrical current flow; therefore, intense heating occurs in these materials. Such materials called absorbers or intensifiers are used in microwavable packages of frozen breaded fried foods or pizza to ensure a crispy crust when heated in a microwave. Glass and plastic are practically transparent to microwaves, i.e., they transmit microwaves, and very little energy is absorbed.

The frequency of microwave power generated by a microwave generator is declared on the nameplate of the unit. The power output is also supplied by the manufacturer for each unit. The coupling efficiency of a microwave unit is expressed as the ratio of power actually supplied to the unit and the actual power absorbed by the material heated. When the quantity of material being heated is large, the power generated by the system limits the power absorbed, rather than the value predicted by Eq. (6.45). The time it takes to heat a large quantity of material can be used to determine the microwave power output of a unit.

If a maximum q/v is determined by varying the quantity of food heated until further reduction results in no further increase in q/v, this value will be the limiting power absorption and will be dependent on the dielectric loss properties of the material according to Eq. (6.45). If the dielectric constant and loss tangent of the material are known, it will be possible to determine the electromagnetic field strength which exists, and differences in heating rates of different components in the food mixture can be predicted using Eq. (6.45).

**Table 6.2** Dielectric properties of food and other materials

| Material | Temperature °C | e'' | tan($\delta$) |
|---|---|---|---|
| Beef (raw) | −15 | 5.0 | 0.15 |
| Beef (raw) | 25 | 40 | 0.30 |
| Beef (roast) | 23 | 28 | 0.20 |
| Peas (boiled) | −15 | 2.5 | 0.20 |
| Peas (boiled) | 23 | 9.0 | 0.50 |
| Pork (raw) | −15 | 6.8 | 1.20 |
| Pork (roast) | 35 | 23.0 | 2.40 |
| Potatoes (boiled) | −15 | 4.5 | 0.20 |
| Potatoes (boiled) | 23 | 38.0 | 0.30 |
| Spinach (boiled) | −15 | 13.0 | 0.50 |
| Spinach (boiled) | 23 | 34.0 | 0.80 |
| Suet | 25 | 2.50 | 0.07 |
| Porridge | −15 | 5.0 | 0.30 |
| Porridge | 23 | 47.0 | 0.41 |
| Pyrex | 25 | 4.80 | 0.0054 |
| Water | 1.5 | 80.5 | 0.31 |
| Water | 25 | 76.7 | 0.15 |
| 0.1 M NaCl | 25 | 75.5 | 0.24 |

*Sources:* Copson, 1971. *Microwave Heating. AVI* Publishing Co. Westport, Conn.; Schmidt, W. 1960. Phillips *Tech. Rev* 3.89

### 6.1.10.2  Relative Heating Rates of Food Components

When the power output of a microwave unit limits the rate of energy absorption by the food, components having different dielectric properties will have different heating rates. Using subscripts 1 and 2 to represent component 1 and 2, $e''$ to represent the product of e and $\tan(\delta)$, and C to represent the constant, Eq. (6.45) becomes:

$$q_1 = \frac{m_1}{\rho_1} CfE^2 e_1''$$

$$q_2 = \frac{m_2}{\rho_2} CfE^2 e_2''$$

Because $P = q1 + q2$ and $q = mC_p \, dT/dt$:

$$Cf \; E^2 = \frac{P}{(m_1/\rho_1)e_1'' + (m_2/\rho_2)e_2''}$$

$$\frac{dT_1}{dt} = \frac{\rho_2 e_1'' P}{C_{p1}\left(\rho_2 e_1'' m_1 + \rho_1 e_2'' m_2\right)}$$

(6.46)

$$\frac{dT_2}{dt} = \frac{\rho_1 e_2'' P}{C_{p2}\left(\rho_2 e_1'' m_1 + \rho_1 e_2'' m_2\right)} \qquad (6.47)$$

The relative rate of heating is:

$$\frac{dT_1}{dT_2} = \frac{\rho_2 e_1'' C_{p2}}{\rho_1 e_2'' C_{p1}} \qquad (6.48)$$

Similar expressions may be derived for more than two components.

**Example 6.10**   The dielectric constant of beef at 23 °C and 2450 MHz is 28 and the loss tangent is 0.2. The density is 1004 kg/m$^3$ and the specific heat is 3250 J/(kg · K). Potato at 23 °C and 2450 MHz has a dielectric constant of 38 and a loss tangent of 0.3. The density is 1010 kg/m$^3$ and the specific heat is 3720 J/(kg · K).

(a)   A microwave oven has a rated output of 600 W. When 0.25 kg of potatoes were placed inside the oven, the temperature rise after 1 min of heating was 38.5 °C. When 60 g of potato was heated in the oven, a temperature rise of 40 °C was observed after 20 s. Calculate the average power output of the oven and the mass of potatoes that must be present such that power output of the oven is limiting the rate of power absorption rather than the capacity of the material to absorb the microwave energy.

(b)   If potatoes and beef are heated simultaneously, what would be the relative rate of heating?

**Solution**

(a)   Assume that 0.25 kg mass of product is sufficient to make microwave power availability the rate limiting factor for microwave absorption:

$$P = 0.25\,\text{kg} \left[\frac{3720 \text{ J}}{\text{kg} \cdot \text{K}}\right] \left[\frac{38.5 \text{ K}}{1 \text{ min}}\right] \left[\frac{1 \text{ min}}{60 \text{ s}}\right]$$

$$= 596.75 \text{ W}$$

When a small amount of material is heated:

$$P = 0.06\,\text{kg} \left[\frac{3720 \text{ J}}{\text{kg} \cdot \text{K}}\right] (40 \text{ K}) \left[\frac{1}{20 \text{ s}}\right] = 446.4 \text{ W}$$

The amount absorbed with the small mass in the oven is much smaller than when a larger mass was present; therefore, it may be assumed that power absorption by the material limits the rate of heating:

$$\frac{q}{V} = \left[\frac{446.4\text{W}}{0.06 \text{ kg}\dfrac{1\text{m}^3}{1010\text{kg}}}\right] \cdot \frac{1\text{m}^3}{10^{-6}\text{cm}^3}$$

$$= 7.5144 \text{ W/cm}^3$$

Because with a small load in the oven, the power absorption is only 7.5177 W/cm$^3$, this may be assumed to be $(q/V)_{\text{lim}}$, the maximum rate of power absorption by the material. If the power output of the oven as calculated above is 596.75 W, the mass present where the rate of power absorption equals the power output of the oven is:

$$\left.\frac{q}{V}\right|_{\text{lim}} = \frac{P(\rho)10^{-6}}{m}$$

$$m = \frac{P(\rho)10^{-6}}{(q/V)_{\text{lim}}}$$

$$= \frac{596.75\left(1010 \times 10^{-6}\right)}{7.5177}$$

$$= 0.08 \text{ kg}$$

Thus, any mass greater than 0.08 kg will heat inside this microwave oven at the rate determined by the power output of the oven.

(b)   Using Eq. (6.48): Let subscript 1 refer to beef and subscript 2 refer to the potato:

$$\frac{dT_1}{dT_2} = \frac{1010(28)(0.2)(3720)}{1004(38)(0.3)(3250)} = 0.565$$

Thus, the beef will be heating slower than the potatoes.

## 6.2   Temperature Measuring Devices

Temperature is defined as the degree of thermal agitation of molecules. Changes in molecular motion of a gas or liquid will change the volume or pressure of that fluid, and a solid will undergo a dimensional change such as expansion or contraction. Certain metals will lose electrons when molecules are thermally excited, and when paired with another metal whose molecules could receive these displaced electrons, an electromotive force will be generated. These responses of materials to changes in temperature are utilized in the design of thermometers, electronic and mechanical temperature measuring devices. The calibration of all temperature measuring devices is based on conditions of use. Some electronic instruments have ambient temperature compensation integrated in the circuitry, and others do not. Similarly, fluid-filled thermometers are calibrated for certain immersion depths; therefore, partial immersion of a thermometer calibrated for total immersion will lead to errors in measurement. Thus, conditions of use that correspond to the calibration of the instrument must be known, for accurate measurements. For other conditions, the instrument will need recalibration. Thermometers filled with fluids other than mercury may have to be recalibrated after a certain time of use to ensure that breakdown of the fluid, which might change its thermal expansion characteristics, has not occurred. In the food industry, a common method of calibration involves connecting the thermometers to a manifold which contains saturated steam. The pressure of the saturated steam can then be used to determine the temperature from steam tables and compared to the thermometer readings. Recalibration of fluid-filled thermometers is normally not done. Thermometers used in critical control points in processes must be replaced when they lose accuracy. Electronic temperature measuring devices must be recalibrated frequently against a fluid-filled thermometer.

The temperature indicated by a measuring device represents the temperature of the measuring element itself, rather than the temperature of the medium in contact with the element. The accuracy of the measurement would depend on how heat is transferred to the measuring element. The temperature registered by the instrument will be that of the measuring element after heat exchange approaches equilibrium.

Thermometers will not detect oscillating temperatures if the mass of the temperature measuring element is such that the lag time for heat transfer equilibration is large in relation to the period of the oscillation. Response time of temperature measuring devices is directly proportional to the mass of the measuring element; the smaller the mass, the more accurate the reading.

Measurement of the surface temperature of a solid in contact with either a liquid or a gas is not very accurate if the measuring element is simply laid on the solid surface. Depending upon the thickness of the measuring element, the temperature indication will be intermediate between that of the surface and the fluid temperature at the interface. Accurate solid surface temperatures can only be determined by embedding two thermocouples a known distance from the surface and extrapolating the temperature readings toward the surface to obtain the true surface temperature.

Measurements of temperatures of gases can be influenced by radiation. As shown in the previous section on Radiation Heat Exchange, an unshielded thermometer or thermocouple in a gas stream surrounded by surfaces of a different temperature the gas will read a temperature different from the true gas temperature because of radiation. Shielding of measuring elements is necessary for accurate measurements of gas temperatures.

### 6.2.1   Liquid-in-Glass Thermometers

Liquid-in-glass thermometers most commonly use mercury for general use and mineral spirits, ethanol, or toluene for low-temperature use. When properly calibrated, any temperature stable liquid may be used, and non-mercury-filled thermometers are now available for use as temperature indicators in food processing facilities.

### 6.2.2   Fluid-Filled Thermometers

This temperature measuring device consists of a metal bulb with a long capillary tube attached to it. The end of the capillary is attached to a device that causes a definite movement with pressure transmitted to it from the capillary. The capillary and bulb are filled with fluid that changes in pressure with changes in temperature. Capillary length affects the calibration and so does the ambient temperature surrounding the capillary. Thus, temperature measuring devices of this type must be calibrated in the field. Figure 6.9B shows a typical fluid-filled thermometer with a spiral at the end of the capillary. The movement of the spiral is transmitted mechanically through a rack and pinion arrangement or a lever to a needle which moves to indicate the temperature on a dial. Some stainless steel-cased dial-type thermometers are of this type. Proper performance of these instruments requires total immersion of the bulb.

### 6.2.3   Bimetallic Strip Thermometers

When two metals having dissimilar thermal expansion characteristics are joined, a change in temperature will cause a change in shape. For example, thermostats for domestic space cooling and heating utilize a thin strip of metal fixed at both ends to a thicker strip of another metal. An increase in temperature will make the thinner metal, which has a higher thermal coefficient of expansion, to bend, and the change in position may open or close properly positioned electrical contacts, which then activates heating or cooling systems. In some configurations, the bimetallic strip may be wound in a helix or a spiral. The strip is fixed at one end, and the free end would move with changes in temperature to position an indicator needle on a temperature scale. This type of thermometer requires total immersion of the bimetallic strip for proper performance.

**Fig. 6.9** Schematic diagram of temperature-measuring devices. (**a**) RTD circuit and (**b**) fluid-filled thermometer

**Fig. 6.10** Thermocouple circuit showing ice water as the reference

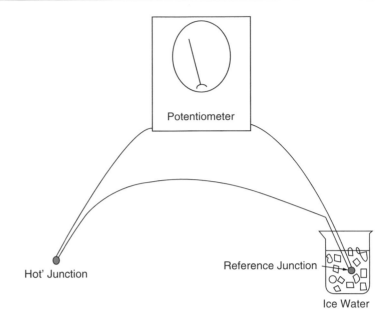

### 6.2.4    Resistance Temperature Devices (RTDs)

The principle of RTDs is based on the change in resistance of a material with changes in temperature. Figure 6.9a shows an RTD circuit. The resistance strip may be an insulated coil of resistance wire encased in a metal tube (resistance bulb) or a strip of metal such as latinum. Other materials such as semiconductors exhibit large drop in resistance with increasing temperatures. The whole instrument consists of the resistance or semiconductor element for sensing the temperature, electronic circuitry for providing the excitation voltage, and a measuring circuit for the change in resistance. Response is usually measured as a change in voltage or current in the circuit. RTDs vary in size and shape depending upon the use. Total immersion of the sensing element is necessary for accurate results.

### 6.2.5    Thermocouples

A thermocouple is a system of two separate dissimilar metals with the ends fused together forming two junctions. When the two junctions are at different temperatures, an electromotive force is generated. This electromotive force is proportional to the difference in temperature between the two junctions. A thermocouple circuit is shown in Fig. 6.10. One of the junctions, the reference junction, is immersed in a constant temperature bath which is usually ice water. Electromotive force values at various temperatures with ice water as the reference are tabulated and available in handbooks.

### 6.2.6    Radiation Pyrometers

This type of temperature sensing device focuses the energy received from a surface through a lens on a device such as a photovoltaic cell or a thermopile. This energy stimulates the production of an electromotive force which can then be measured. Since the amount of energy received per unit area by a receiving surface from a source varies inversely as the distance, these types of measuring devices have distance compensation features. Distance compensation is achieved by focusing the energy received on the lens of the instrument into a smaller area having an energy sensitive surface. Radiation pyrometers are useful for sensing very high temperatures such as in furnaces, flames, and red hot or molten metals. Some, which have been calibrated for operation at near ambient temperatures, are very useful for non-contact monitoring of surface temperatures of processing equipment.

### 6.2.7   Accurate Temperature Measurements

*Fluid-in-Glass Thermometers* The most common of these is the mercury in glass thermometer, but there are other fluids used. The temperature indicated is based on thermal expansion of the fluid in the thermometer. The temperature markings on the stem are constructed based on the volume of the bulb, the diameter of the glass capillary into which the fluid emerges from the bulb, and the coefficient of thermal expansion of the fluid. The temperature scale in the stem is marked based on a specified *minimum immersion depth*. A *total immersion* thermometer must have the bulb and the whole stem immersed in the fluid to be measured.

*Partial Immersion* Thermometers have the minimum immersion depth marked on the stem below the temperature scale. The minimum immersion depth for partial immersion thermometers is usually 76 mm from the tip of the bulb. Errors are introduced when the immersion depth is below the minimum. The magnitude of the error will depend on the length of the exposed mercury column and the ambient temperature around the exposed mercury column.

*Thermocouples* Errors in temperature measurement with thermocouples come from improper contact between thermocouple wires at the measuring junction; conduction of heat along the thermocouple wires, which would make the temperature at the measuring junction different from the medium measured; interference with convection heat transfer at the measuring junction; radiation heat transfer between the measuring junction and the surroundings; resistance to current flow by the thermocouple wire; inaccurate reference junction compensation; and accuracy of the instrument that reads the electromotive force developed within the thermocouple circuit. One of the most common problems in temperature measurement is in the selection of the size of thermocouple wire. Although a very thin wire thermocouple eliminates the effect of heat conduction, a long thermocouple lead with thin wires would present a high resistance in the circuit resulting in a significant drop in the measured electromotive force. Another common problem is the joining of the thermocouple wires at the measuring junction. The two wires must have good electrical contact, and the joint must contribute minimum electrical resistance in the thermocouple circuit. With thin wires, the bare ends are twisted at least three turns before soldering or brazing, while with thicker wires, the two ends may simply be butted together and brazed or welded.

## 6.3   Steady-State Heat Transfer

### 6.3.1   The Concept of Resistance to Heat Transfer

Equation (6.25), derived from Fourier's first law, shows that in a steady-state system the quantity of heat passing through any part of the system must equal that passing through any other part and equal the total passing through the system. The problem of heat transfer through multiple layers can be analyzed as a problem involving a series of resistance to heat transfer. The transfer of heat can be considered as analogous to the transfer of electrical energy through a conductor. $\Delta T$ is the driving force equivalent to the voltage E in electrical circuits. The heat flux q is equivalent to the current, I.

Ohm's law for electrical circuits is:

$$I = \frac{E}{R} \tag{6.49}$$

For heat transfer through a slab:

$$\frac{q}{A} = \frac{\Delta T}{[\Delta X/k]} = \frac{\Delta T}{R} \quad (6.50)$$

Thus, in comparing Eqs. (6.48) and (6.49), R is equivalent to $\Delta X/k$. The resistance to heat transfer is $\Delta X/k$ and is the "R" rating used in the insulation industry to rate the effectiveness of insulating materials. For multilayered materials in the geometry of a slab where A is constant in the direction of increasing x, the overall resistance to heat transfer is the sum of the individual resistance in series, and:

$$R = R1 + R2 + R3 + \ldots \ldots Rn$$

Thus:

$$R = \frac{\Delta x_1}{k_1} + \frac{\Delta x_2}{k_2} + \ldots \ldots \frac{\Delta x_n}{k_n}$$

and:

$$q = \frac{\Delta T}{R}$$

$$= \frac{\Delta T}{[\ln (r_2/r_1)/2\pi Lk_1 + \ln (r_3/r_2)/2\pi Lk_2 + \ldots \ln (r_{n+1}/r_n)/2\pi Lk_n]}$$

and:

$$\frac{\Delta T_1}{[\ln (r_2/r_1)2\pi Lk_1]} = \frac{\Delta T_2}{[\ln (r_3/r_2)/2\pi Lk_2]}$$
$$= \ldots \frac{\Delta T_n}{[\ln (r_{n+1}/r_n)/2\pi Lk_n]} \quad (6.55)$$

For convection heat transfer:

$$q = h A \Delta T = \frac{\Delta T}{(1/hA)}$$
$$R = \frac{1}{hA} \quad (6.56)$$

### 6.3.2   Combined Convection and Conduction: The Overall Heat Transfer Coefficient

Most problems encountered in practice involve heat transfer by combined convection and conduction. Usually, the temperatures of fluids on

$$\frac{q}{A} = \frac{\Delta T}{[\Delta x_1/k_1 + \Delta x_2/k_2 \ldots \ldots \ldots \Delta x_n/k_n]} \quad (6.51)$$

Equation (6.50) can also be written as:

$$\frac{q}{A} = \frac{\Delta T_1}{R_1} = \frac{\Delta T_2}{R_2} = \frac{\Delta T_3}{R_3} = \ldots \ldots \frac{\Delta T_n}{R_n} = \frac{\Delta T}{R} \quad (6.52)$$

Equation (6.52) is similar to Eq. (6.25) previously derived using Fourier's first law.

For heat transfer through a cylinder, the heat transfer rate in Eq. (6.27) results in an expression for the heat transfer resistance as:

$$R = \frac{\ln (r_2/r_1)}{2\pi LK} \quad (6.53)$$

For resistances in series:

$$(6.54)$$

both sides of a solid are known, and the rate of heat transfer across the solid is to be determined. Heat transfer involves convective heat transfer between a fluid on one surface, conductive heat transfer through the solid, and convective heat transfer again at the opposite surface to the other fluid. The rate of heat transfer may be expressed in terms of U, the overall heat transfer coefficient, or in terms of R, an overall resistance.

Consider a series of resistances involving n layers of solids and n fluid to surface interfaces. The thermal conductivity of the solids are $k_1 \ldots k_2 \ldots k_3 \ldots k_n$, and the heat transfer coefficients are $h_1 \ldots h_2 \ldots h_n$ with subscript n increasing along the direction of heat flow:

$$q = UA \Delta T = \frac{\Delta T}{R} \quad (6.57)$$

For a slab:

Because A is the same across the thickness of a slab:

$$R = \frac{1}{UA}$$

$$R = \sum \left[\frac{1}{h_n A}\right] + \sum \left[\frac{X_n}{k_n A}\right] = \frac{1}{UA} \quad (6.58)$$

$$\frac{1}{U} = \frac{1}{h_1} + \frac{\Delta x_1}{k_1} + \frac{\Delta x_2}{k_2} + \cdots \cdots \frac{\Delta x_n}{k_n}$$
$$+ \cdots \cdots \frac{1}{h_n} \quad (6.59)$$

or:

$$U = \frac{1}{[\sum (1/h_n) + \sum (x_n/k_n)]} \quad (6.60)$$

For a cylinder:

$$R = \sum \left[\frac{\ln (r_{n+1}/r_n)}{2\pi L k_n}\right] + \sum \left[\frac{1}{h_n A_n}\right] = \frac{1}{UA}$$

If the A used as a multiplier for U is the outside area, then $U = U_o$ the overall heat transfer coefficient based on the outside area. $U_i = $ overall heat transfer coefficient based on the inside area. Using $h_i = $ inside heat transfer coefficient and $h_o = $ outside heat transfer coefficient, $r_i$ and $r_o$ are inside and outside radius of the cylinder, respectively:

$$(2\pi r_o L)U_o$$
$$= \frac{1}{(1/2\pi L) \sum [\ln (r_{n+1}/r_n)/k_n] + (1/2\pi L)[1/h_o r_o + 1/h_{ir_i}]}$$

or:

$$U_o$$
$$= \frac{1}{r_o \sum [\ln (r_n/r_{n-1})/k_n] + [r_o/r_i h_i] + [1/h_o]U_o} = \frac{1}{\frac{r_o}{r_i h_i} + \frac{r_o \ln (r_2/r_1)}{k_1} + \frac{r_o \ln (r_3/r_2)}{k_2}}$$
$$+ \cdots \frac{r_o \ln (r_n/r_{n-1})}{k_n} + \frac{1}{h_o} U_i$$
$$= \frac{1}{r_i \sum [\ln (r_n/r_{n-1})/k_n] + [r_i/r_o h_o] + [1/h_i]} \quad (6.61)$$

or:

$$\frac{1}{U_i} = \frac{1}{h_i} + \frac{r_i \ln (r_2/r_1)}{k_1} + \frac{r_i \ln (r_3/r_2)}{k_2}$$
$$+ \cdots \cdots \frac{r_i \ln (r_n/r_{n-1})}{k_n} + \frac{r_i}{r_o h_o} \quad (6.62)$$

**Example 6.11** Calculate the rate of heat transfer across a glass pane that consists of two 1.6-mm-thick glass separated by 0.8-mm layer of air. The heat transfer coefficient on one side that is at 21 °C is 2.84 W/(m². K) and on the opposite side that is at −15 °C is 11.4 W/(m². K). The thermal conductivity of glass is 0.52 W/(m · K) and that of air is 0.031 W/(m · K).

**Solution**

When stagnant air is trapped between two layers of glass, convective heat transfer is minimal, and the stagnant air layer will transfer heat by conduction. There are five resistances to heat transfer. $R_1$ is the convective resistance at one surface exposed to air, $R_2$ is the conductive resistance of one 1.6-mm-thick layer of glass, $R_3$ is the conductive resistance of the air layer between the glass, $R_4$ is the conductive resistance of the second 1.6-mm-thick layer of glass, and $R_5$ is the convective resistance of the opposite surface exposed to air. Using Eq. (6.9):

$$\frac{1}{U} = \frac{1}{h_1} + \frac{x_1}{k_1} + \frac{x_2}{k_2} + \frac{x_3}{k_3} + \frac{1}{h_2}$$

$$= \frac{1}{2.8} + \frac{1.6 \times 10^{-3}}{0.52} + \frac{0.8 \times 10^{-3}}{0.031}$$

$$+ \frac{1.6 \times 10^{-3}}{0.52} + \frac{1}{11.4} = 0.352$$

$$+ 0.0031 + 0.0258 + 0.0031 + 0.0877 = 0.4718$$

$$U = 2.12 \frac{W}{m^2 \cdot K}$$

$$\frac{q}{A} + U\Delta T = 2.12(21 - (-15)) = 76.32 \frac{W}{m^2}$$

**Example 6.12** (a) Calculate the overall heat transfer coefficient for a 1-in. (nominal) 16-gauge heat exchanger tube when the heat transfer coefficient is 568 W/(m$^2$ K) inside and 5678 W/(m$^2\cdot$ K) on the outside. The tube wall has a thermal conductivity of 55.6 W/(m $\cdot$ K). The tube has an inside diameter of 2.21 cm and a 1.65-mm wall thickness. (b) If the temperature of the fluid inside the tube is 80 °C and 120 °C on the outside, what is the inside wall temperature?

**Solution**

$r_i = 2.21/2 = 1.105$ cm

$r_o = 1.105 + 0.1651 = 1.2701$ cm

(a) Using Eq. (6.61):

$$\frac{1}{U_O} = \frac{1}{5678} + \frac{1.2701 \times 10^2 \ \ln{(1.2701/1.105)}}{55.6}$$

$$+ \frac{1.2701 \times 10^{-2}}{(1.105 \times 10^{-2})(568)}$$

$$= 1.76 \times 10^{-4} + 0.318 \times 10^{-4} + 20.236 \times 10^{-4}$$

$$= 22.315 \times 10^{-4}; U_o = 448 \text{W/m}^2 \cdot \text{K}$$

(b) Temperature on the inside wall:

q(overall) = q(across inside convective heat transfer coefficient).

Let $T_f$ = temperature of fluid inside tube and $T_w$ = temperature of the wall:

$$U_o A_o \Delta T = h_i A_i (T_w - T_f)$$
$$448(2\pi r_o L)(120 - 80) = 568(2\pi r_i L)(T_w - 80)$$
$$(T_w - 80) = \frac{448(r_o)(40)}{568(r_i)}$$
$$= 80 + \frac{448(1.2701)(40)}{568(1.105)}$$
$$= 80 + 36.3 = 116.3\overset{\circ}{}C$$

## 6.4 Heat Exchange Equipment

Heat exchangers are equipment for transferring heat from one fluid to another. In its simplest form, a heat exchanger could take the form of a copper tube exposed to air, where fluid flowing inside the tube is cooled by transferring heat to ambient air. When used for transferring large quantities of heat per unit time, heat exchangers take several forms to provide for efficient utilization of the heat contents of both fluids exchanging heat and to allow for compactness of the equipment. The simplest heat exchanger to make is one of tubular design. Heat exchangers commonly used in the food industry include the following.

*Swept Surface Heat Exchangers* The food product passes through an inner cylinder, and the heating or cooling medium passes through the annular space between the inner cylinder and an outer cylinder or jacket. A rotating blade running the whole length of the cylinder continuously agitates the food as it passes through the heat exchanger and, at the same time, continuously scrapes the walls through which heat is being transferred (the heat transfer surface). This type of heat exchanger can be used to heat, cool, or provide heat to concentrate viscous food products. Figure 6.11 shows a swept surface heat exchanger.

*Double-Pipe Heat Exchanger* This heat exchanger consists of one pipe inside another. The walls of the inner pipe form the heat transfer surface. This type of heat exchanger is usually built and installed in the field. A major disadvantage is the relatively large space it occupies for the quantity of heat exchanged, compared to other types of heat exchangers. Figure 6.12 shows a double-pipe heat exchanger.

*Shell-and-Tube Heat Exchanger* This heat exchanger consists of a bundle of tubes enclosed by a shell. The type of head arrangement allows for one tube pass or multiple passes for the product. In a one-pass arrangement, the product enters at one end and leaves at the opposite end. In a multipass arrangement, the product may travel back and forth through different tubes with each pass before finally leaving the heat exchanger. The heat exchange medium on the outside of the tubes is usually distributed using a system of baffles. Figure 6.13 shows two types of shell-and-tube heat exchangers.

**Fig. 6.11** Diagram of a swept surface heat exchanger

**Fig. 6.12** Diagram of a double-pipe heat exchanger

**Fig. 6.13** Diagram of a single-pass and a multiple-pass shell-and-tube heat exchanger

*Plate Heat Exchanger* This type of heat exchanger was developed for the dairy industry. It consists of a series of plates clamped together on a frame. Channels are formed between each plate. The product and heat transfer medium flow through alternate channels. Because of the narrow channel between the plates, the fluid flows at high velocities and in a thin layer resulting in very high heat transfer rates per unit heat transfer surface area. The plate heat exchanger is mostly used for heating fluids to temperatures below the boiling point of water at atmospheric pressure. However, units designed for high-temperature service are commercially available. Plate heat exchangers are now used in virtually any application where tubular heat exchangers were previously commonly used. Newer designs have strength to withstand moderate pressure or vacuum. A major limitation is the inability to handle viscous liquids. Figure 6.14 shows a plate heat exchanger.

**Fig. 6.14** Diagram of a plate heat exchanger showing alternating paths of processed fluid and heat exchange medium

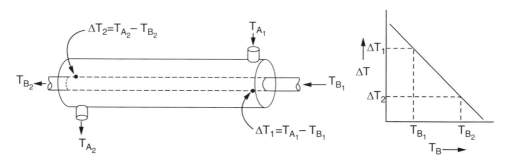

**Fig. 6.15** Diagram of the temperature profiles of fluid along the length of a heat exchanger tube

### 6.4.1   Heat Transfer in Heat Exchangers

The overall heat transfer coefficient in tubular exchangers is calculated using Eqs. (6.61) or (6.62). Because in heat exchangers the $\Delta T$ can change from one end of the tube to the other, a mean $\Delta T$ must be determined for use in Eq. (6.57) to calculate q.

### 6.4.2   The Logarithmic Mean Temperature Difference

Refer to Fig. 6.15 for the diagram and the representation of the symbols used in the following derivation.

If the $\Delta T$ changes linearly along the length of the heat exchanger relative to the temperature of one of the fluids, the slope of the line representing $\Delta T$ vs. $T_B$ is:

$$\text{Slope} = \frac{d(\Delta T)}{dT_B} + \frac{\Delta T_2 - \Delta T_1}{T_{B2} - T_{B1}}$$

Rearranging:

$$dT_B = d\Delta T \frac{T_{B2} - T_{B1}}{\Delta T_2 - \Delta T_1}$$

The amount of heat transferred across any point in the exchanger is:

$$q = UdA\Delta T$$

This heat transferred will cause a rise in product temperature, $dT_B$, and can be also expressed as:

$$q = m \, C_p \, dT_B$$

Equating:

$$U \, dA \, \Delta T = m \, C_p \, dT_B$$

Substituting the expression for $dT_B$:

$$U \, dA \, \Delta T = mC_p \left[ \frac{d \, \Delta T (T_{B2} - T_{B1})}{(\Delta \, T_2 - \Delta \, T_1)} \right]$$

Rearranging:

$$mC_p(T_{B2} - T_{B1}) \frac{d\Delta T}{\Delta T} = U(\Delta T_2 - \Delta T_1)(dA)$$

Integrating:

$$mC_p(T_{B2} - T_{B1}) \int_{\Delta T_1}^{\Delta T_2} \frac{d\Delta T}{\Delta T} = U(\Delta T_2 - \Delta T_1) \int_0^A dA$$

The limits are such that at the entrance of fluid B into the heat exchanger, $\Delta T = \Delta T_1$ and $A = 0$ at this point. When fluid B leaves the heat exchanger, $\Delta T = \Delta T_2$ and $A = A$ the total heat transfer area for the heat exchanger:

$$mC_p(T_{B2} - T_{B1}) \ln \left( \frac{\Delta T_2}{\Delta T_1} \right)$$

$$= UA(\Delta T_2 - \Delta T_1)$$

$mC_p(T_{B2} - T_{B1}) = q$, the total amount of heat absorbed by fluid B. Thus:

$$q = UA \frac{\Delta T_2 - \Delta T_1}{\ln (\Delta T_2 / \Delta T_1)} = UA \, \Delta T_L$$

where $\Delta T_L$ is the logarithmic mean $\Delta T$ expressed as:

$$\Delta T_L = \frac{\Delta T_2 - \Delta T_1}{\ln \left( \frac{\Delta T_2}{\Delta T_1} \right)} \qquad (6.63)$$

When heat is exchanged between two liquids or between liquid and gas, the temperature of both fluids will be changing as they travel through the heat exchanger. If both fluids enter on the same end of the unit, flow will be in the same direction, and this type of flow is cocurrent. Countercurrent flow exists when the fluids flow in opposite direction through the unit. True countercurrent or cocurrent flow would occur in double-pipe, plate, and multiple-tube single-pass shell-and-tube heat exchangers. When true cocurrent or countercurrent flow exists, the rate of heat transfer can be calculated by:

$$q = U_i A_i \Delta T_L \qquad \text{or;}$$
$$q = U_o A_o \Delta T_L \qquad\qquad (6.64)$$

The log mean temperature difference is evaluated between the inlet and exit $\Delta T$ values.

In multipass heat exchangers, the fluid inside the tubes will be traveling back and forth across zones of alternating high and low temperatures. For this type of heat exchangers, a correction factor is used on the $\Delta T_L$. The correction factor depends upon the manner in which shell fluid and tube fluid pass through the heat exchanger:

$$q = U_i A_i F \Delta T_L \qquad \text{or}$$
$$q = U_o A_o F \Delta T_L \qquad\qquad (6.65)$$

where F is the correction factor. The reader is referred to a heat transfer textbook or handbook for graphs showing correction factors to $\Delta T$ for multipass shell-and-tube heat exchangers.

**Example 6.13** Applesauce is being cooled from 80 °C to 20 °C in a swept surface heat exchanger. The overall coefficient of heat transfer based on the inside surface area is 568 W/m$^2 \cdot$ K. The applesauce has a specific heat of 3187 $\cdot$ J/kg K and is being cooled at the rate of 50 kg/h. Cooling water enters in countercurrent flow at 10 °C and leaves the heat exchanger at 17 °C. Calculate (a) the quantity of cooling water required and (b) the required heat transfer surface area for the heat exchanger.

**Solution**
The diagram of the heat exchanger in countercurrent flow is shown in Fig. 6.16.

**Fig. 6.16** Diagram of a heat exchanger in a counterflow configuration for the fluids exchanging heat

(a)   The rate of heat transfer:

$$50\frac{kg}{h} \cdot 3187\frac{J}{kg \cdot K} \cdot (80 - 20)^{\circ}K = 9,561,000$$

$$\frac{J}{h} = 2.656 \text{ kW}$$

Let x = quantity of water used. The specific heat of water is 4186 J/kg·K:

$$x(4186)(17 - 10) = 9561000$$

$$x = \frac{9,561,000}{4186(7)} = 326 \text{ kg/h}$$

(b)   $\Delta T_2 = 20 - 10 = 10^{\circ}C$
$\Delta T_1 = 80 - 17 = 63^{\circ}C$
Using Eq. (6.63):

$$\Delta T_L = \frac{63 - 10}{\ln 63/10} = \frac{53}{\ln 6.3} = 28.8$$

The rate of heat transfer is calculated using Eq. (6.64):

$$q = U_i A_i \Delta T_L = 568 \ A_i \ (28.8)$$

$$(568)(Ai)(28.8) = 2656$$

$$A_i = \frac{2656}{568(28.8)} = 0.162 \text{ m}^2$$

## 6.5   Local Heat Transfer Coefficients

A major problem in heat transfer is the estimation of heat transfer coefficients to be used for design purposes. These values of h must be determined from the properties of the fluid and the geometry of the system.

### 6.5.1   Dimensionless Quantities

The use of dimensionless quantities arises from the principle of similarity. Equations that describe different systems having similar characteristics can be superimposed on each other to form a single expression suitable for all systems. Thus, if the physical characteristics of a fluid and the conditions that exist in an experiment are expressed in terms of dimensionless quantities, it will be possible to extrapolate results of an experiment to other fluids and other conditions. The principle of similarity makes it unnecessary to experimentally establish equations for heat transfer to each fluid. A general correlation equation will be suitable for all fluids. The determination of the different dimensionless quantities involved in relationships between variables describing a system is done by a method called dimensional analysis. In dimensional analysis, an equation relating the various variables is first assumed, and by performing an analysis of the base dimensions of the variables to make the equation dimensionally consistent, specific groupings of the variables are formed. The following dimensionless quantities have been identified and used in correlations involving the heat transfer coefficient.

*Nusselt Number (Nu)*  This expression involves the heat transfer coefficient (h), the characteristic dimension of the system (d), and the thermal conductivity of the fluid (k). This dimensionless expression may be considered as the ratio of the characteristic dimension of a system and the thickness of the boundary layer of fluid that would transmit heat by conduction at the same rate as that calculated using the heat transfer coefficient:

$$Nu = h\frac{d}{k} \qquad (6.66)$$

*Reynolds Number (Re)*  This expression involves the characteristic dimension of the system (d), the velocity of the fluid (V), the density ($\rho$), and the viscosity ($\mu$). It may be considered as the ratio of inertial forces to the frictional force. The Reynolds number has been discussed in Chap. 5.

*Prandtl Number (Pr)*  This expression involves the specific heat ($C_p$), the viscosity ($\mu$), and the thermal conductivity (k). It may be considered as the ratio of rate of momentum exchange between molecules and the rate of energy exchange between molecules that lead to the transfer of heat:

$$Pr = \frac{C_p\mu}{k} \qquad (6.67)$$

*Grashof Number (Gr)*  This quantity involves the characteristic dimension of a system (d), the acceleration due to gravity (g), the thermal expansion coefficient ($\beta$), the density of the fluid ($\rho$), the

viscosity ($\mu$), and the temperature difference $\Delta T$ between a surface and the fluid temperature beyond the boundary layer. This number may be considered as a ratio of the force of gravity to buoyant forces that arise as a result of a change in temperature of a fluid:

$$Gr = \frac{d^3\ g\ \beta\ \rho^2\ \Delta T}{\mu^2} \qquad (6.68)$$

*Peclet Number (Pe)*  This dimensionless number is the product of the Reynolds number and the Prandtl number:

$$Pe = Re\,Pr = \frac{\rho\,VC_p\ d}{k} \qquad (6.69)$$

*Rayleigh Number (Ra)*  This dimensionless number is the product of the Grashof number and the Prandtl number:

$$Ra = Gr\,Pr = \frac{d^3\ g\ \beta\ C_p\rho^2\ \Delta T}{\mu\ k} \qquad (6.70)$$

*Graetz Number (Gz)*  This is similar to the Peclet number. It was derived from an analytical solution to the equations of heat transfer from a surface to a fluid flowing along that surface in laminar flow. The Graetz number is:

$$Gz = \frac{\pi}{4}\left[Re\cdot Pr\cdot\left(\frac{d}{L}\right)\right] = \frac{\dot{m}\ C_p}{kL} \qquad (6.71)$$

where $\dot{m}$ is the mass rate of flow, kg/s.

## 6.5.2   Equations for Calculating Heat Transfer Coefficients

These equations generally take the form:

$$Nu = f\left(Re, Pr, \left(\frac{L}{d}\right), Gr, \left(\frac{\mu}{\mu_w}\right)\right) \qquad (6.72)$$

The Grashof number is associated with free convection, and the length to diameter ratio (L/d) appears when flow is laminar. When calculating these dimensionless quantities, the thermophysical properties of fluids at the arithmetic mean temperature at the inlet and exit are used. The viscosity of fluid at the wall ($\mu_w$) affects heat transfer and is included in the correlation equation to account for the difference in cooling and heating processes. Most of the English literature on correlation equations for heat transfer are based on the general expression:

$$\text{Nu} = \alpha(\text{Re})^\beta (\text{Pr})^\gamma \left[\frac{L}{d}\right]^\delta \qquad (6.73)$$

where $\alpha$, $\beta$, $\gamma$, and $\delta$ are constants obtained from correlation analysis of experimental data.

A number of correlation equations suitable for use under various conditions have been published. A summary of equations suitable for use under commonly encountered conditions in food processing is given in Appendix 12. In order to illustrate the use of these equations in the design of a heating or cooling system, some equations are also given in this chapter.

### 6.5.2.1  Simplified Equations for Natural Convection to Air or Water

For natural convection to air or water, Table 6.3 lists simplified equations based on those given by McAdams (1954).

### 6.5.2.2  Fluids in Laminar and Turbulent Flow Inside Tubes

Equation (6.74), used for laminar flow, was originally derived by Leveque and is discussed in a number of heat transfer textbooks. Equation

(6.75) is another form of Eq. (6.74). Equation (6.76), commonly referred to as the Dittus-Boelter equation, is used for turbulent flow:

$$\text{Nu} = 1.615 \left[\text{RePr}\left(\frac{d}{L}\right)\right]^{0.33} \left(\frac{\mu}{\mu_w}\right)^{0.14} \qquad (6.74)$$

Sieder and Tate (1936) introduced the use of the ratio of viscosity of fluid bulk and that at the wall as a multiplying factor to account for the difference in the heat transfer coefficient between a fluid and a heated or a cooled surface:

$$\text{Nu} = 1.75 (\text{Gz})^{0.33} \left(\frac{\mu}{\mu_w}\right)^{0.14} \qquad (6.75)$$

$$\text{Nu} = 0.023 (\text{Re})^{0.8} (\text{Pr})^{0.33} \left(\frac{\mu}{\mu_w}\right)^{0.14} \qquad (6.76)$$

### 6.5.2.3  Heat Transfer to Non-Newtonian Fluids in Laminar Flow

Metzner et al. (1957) used Leveque's solution to heat transfer to a moving fluid in laminar flow as a basis for non-Newtonian flow heat transfer. They defined the delta function, a factor $\Delta^{0.33}$, as the ratio $\text{Nu}_n/\text{Nu}$ (non-Newtonian Nusselt number/ Newtonian Nusselt number).

Using the delta function as a multiplying factor for the Newtonian Nusselt number, Eq. (6.74) can be used for non-Newtonian fluids as follows:

$$\text{Nu} = 1.75 \Delta^{0.33} \text{Gz}^{0.33} \left(\frac{\mu}{\mu_w}\right)^{0.14} \qquad (6.77)$$

**Table 6.3** Equations for calculating heat transfer coefficients in free convection from water or air

| Surface conditions | Equation | Value of C for: Air | Water |
|---|---|---|---|
| Horizontal cylinder heated or cooled | $h = C\left(\frac{\Delta T}{D}\right)^{0.25}$ | 1.396 | 291.1 |
| Fluid below heated horizontal plate | $h = C(\Delta T)^{0.25}$ | 2.4493 | |
| Fluid below heated horizontal plate | $h = C(\Delta T)^{0.25}$ | 1.3154 | |
| Fluid above cooled horizontal plate | $h = C(\Delta T)^{0.25}$ | 1.3154 | |
| Fluid below cooled horizontal plate | $h = C(\Delta T)^{0.25}$ | 2.4493 | |
| Vertical cylinder heated or cooled | $h = C\left(\frac{\Delta T}{D}\right)^{0.25}$ | 1.3683 | 127.1 |
| Vertical plate heated or cooled | $h = C\left[\frac{\Delta T}{L}\right]^{0.25}$ | 1.3683 | 127.1 |

*Source:* Calculated from values reported in McAdams, W. H. 1954. *Heat Transmission Co.,* New York. 3rd ed. McGraw-Hill Book Co., New York.6

When the flow behavior index is greater than 0.4 for all values of Gz, $\Delta^{0.33} = [(3n + 1)/4n]^{0.33}$. When n < 0.4, $\Delta^{0.33}$ is determined as follows:

$$\Delta^{0.33} = -0.24\,n + 1.18 \text{ when Gz} = 5.0$$
$$(6.78)$$

$$\Delta^{0.33} = -0.60\,n + 1.30 \text{ when Gz} = 10.0$$
$$(6.79)$$

$$\Delta^{0.33} = -0.72\,n + 1.40 \text{ when Gz} = 15.0$$
$$(6.80)$$

$$\Delta^{0.33} = -0.35\,n + 1.57 \text{ when Gz} = 25.0$$
$$(6.81)$$

Equation (6.81) can also be used when Gz > 25, 0.1 < n < 0.4.

For Gz > 25 and n < 0.1, the relationship between $\Delta^{0.33}$ and n is nonlinear, and the reader is referred to the paper of Metzner et al. (1957) for determination of the delta function. Most food fluids will have n > 0.1; therefore, the equations for the delta function given above should be adequate to cover most problems encountered in food processing.

### 6.5.2.4   Adapting Equations for Heat Transfer Coefficients to Non-Newtonian Fluids

The approach of Metzner et al. (1957) to adapting correlation equations for heat transfer coefficients derived for Newtonian fluids to non-Newtonian fluids will allow the use of practically any of the extensive correlations derived for Newtonian fluids for systems involving non-Newtonian fluids. For non-Newtonian fluids, the viscosity is not constant in the radial direction of the pipe. It will be possible, however, to use the equations derived for Newtonian fluids by using an equivalent viscosity for the non-Newtonian fluid that will cause the same pressure drop at the flow rate under consideration. A similar approach was used by Metzner et al. (1957) for heat transfer to non-Newtonian fluids in turbulent flow.

For a non-Newtonian fluid that follows the power law equation:

$$\tau = K\,(\gamma)^n$$

where $\tau$ = the shear stress, $\gamma$ = the shear rate, K = the consistency index, and n = the flow behavior index.

The Reynolds number at an average rate of flow V is:

$$Re = \frac{8(V)^{2-n}R^n\rho}{K[(3n+1)/n]^n}$$

The viscosity to be used in the Prandtl number and the Sieder-Tate viscosity correction term for the fluid bulk is determined as follows.

For a Newtonian fluid:

$$Re = \frac{DV\rho}{\mu}$$

Substituting the Reynolds number of the non-Newtonian fluid:

$$\mu = \frac{DV\rho}{\mu}$$
$$= K\left(\tfrac{3n+1}{n}\right)^n\left[\frac{2RV\rho}{8V^{2-n}R^n\rho}\right]$$
$$\mu = \frac{DV\rho}{8(V)^{2-n}R^n\rho/[K(3+1/n)^n]}$$
$$\mu = \frac{K}{4}\left[\frac{3n+1}{n}\right]^n(R)^{1-n}(V)^{n-1}$$
$$(6.82)$$

At the wall, the viscosity $\mu_w$ used in the Sieder-Tate viscosity correction term is determined from the apparent viscosity (Eq. 5.6, Chap. 5) as follows:

$$\mu_w = K(\gamma_w)^{n-1}$$

The shear rate at the wall, $\gamma_w$, is calculated using the Rabinowitsch-Mooney equation (Eq. 5.18, Chap. 5):

$$\gamma_w = \frac{8V}{D}\left[\frac{3n+1}{4n}\right] = \frac{2V}{D}\left[\frac{3n+1}{n}\right]$$
$$\mu_w = K\left(\tfrac{2V}{D}\right)^{n-1}\left[\tfrac{3n+1}{n}\right]^{n-1}$$
$$(6.83)$$

Equations (6.82) and (6.83) are the viscosity terms for non-Newtonian fluids used in the correlation equations for heat transfer coefficients to Newtonian fluids.

As an example, Eq. (6.74) will be used to determine an equivalent equation for non-Newtonian fluids. Because Leveque's derivation of Eq. (6.74) was based on the fluid velocity profile at the boundary layer near the pipe wall, the viscosity term in the Reynolds number should be the apparent viscosity at the shear rate existing at the pipe wall. The viscosity term in the Prandtl number is based on viscous dissipation and conduction in the bulk fluid and should be the apparent viscosity based on the average velocity. Substituting Eq. (6.83) for the $\mu$ term in the equation for the Reynolds number $\Phi$ and Eq. (6.82) for the $\Phi$ term in the equation for the Prandtl number, the Nusselt number expression of Eq. (6.74) becomes:

$$Nu = 1.615 \left[ \left[ \frac{DV\rho}{K[2V/D]^{n-1}[(3n+1)/n]^{n-1}} \right. \right.$$
$$\left. \left. \times \left[ \frac{C_p(K/4)[(3n+1)/n]^n[D/2]^{1-n}V^{n-1}}{} \left[ \frac{D}{L} \right] \right] \right]^{0.33} \left[ \frac{\mu}{\mu_w} \right]^{0.14} \right.$$

Simplifying:

$$Nu = 1.615 \left[ \frac{3n+1}{4n} \right]^{0.33} \left[ \frac{D^2 V\rho C_p}{kL} \right]^{0.33} \left[ \frac{\mu}{\mu_w} \right]^{0.14}$$

Substituting the mass rate of flow calculated from the volumetric rate of flow and the density:

$$\dot{m} = \frac{\pi D^2}{4} \cdot V \cdot \rho \qquad D^2 V\rho = \frac{4}{\pi}\dot{m}$$
$$Nu = 1.615 \left[ \frac{4}{\pi} \right]^{0.33} \left[ \frac{\dot{m} C_p}{kL} \right]^{0.33} \left[ \frac{3n+1}{4n} \right]^{0.33} \left[ \frac{\mu}{\mu_w} \right]^{0.14}$$

This expression is similar to Eq. (6.77) except for the value of the constant.

**Example 6.14** Calculate the rate of heat loss from a 1.524 m inside diameter horizontal retort, 9.144 m long. Steam at 121 °C is inside the retort. Ambient air is at 25 °C. The retort is made out of steel ($k = 42$ W/m · K) and has a wall thickness of 0.635 cm.

**Solution**

The heat transfer coefficient from the outside surface to ambient air will be controlling the rate

of heat transfer; therefore, variations in the steam side heat transfer coefficient will have little effect on the answer. Assume the steam side heat transfer coefficient is 6000 W/(m² · K).

The system is a horizontal cylinder with a vertical plate at each end. The outside heat transfer coefficient can be calculated using the appropriate equations from Table 6.3.

For a horizontal cylinder to air:

$$h = 1.3196 \left[ \frac{\Delta T}{D_o} \right]^{0.25}$$

For a vertical plate to air:

$$h = 1.3683 \left[ \frac{\Delta T}{L} \right]^{0.25}$$

$\Delta T$ is evaluated from the outside wall temperature. The problem can be solved using a trial and error procedure by first assuming a heat transfer coefficient, calculating a wall temperature, and recalculating the heat transfer coefficient until the assumed values and calculated values converge.

The magnitude of the heat transfer coefficient in free convection to air is of the order 5 W/m² · K. Equation (6.62) can be used to calculate an overall heat transfer coefficient for the cylindrical surface:

$$\frac{1}{U_i} = \frac{r_i}{r_o h_o} + \frac{r_i \ln r_o/r_i}{k} + \frac{1}{h_i} = 0.1984$$
$$+ 0.00015 + 0.000167 = \frac{0.762}{0.76835(5)}$$
$$+ \frac{0.762 \ln(0.76835/0.762)}{42} + \frac{1}{6000}$$
$$U_i = 5.03 \, \text{W}/(\text{m}^2\text{AK})$$
$$U_i A_i \Delta T = h_o A_o \Delta t_w$$
$$U_i (2\pi r_i L)\Delta T = h_o (2\pi r_o L)\Delta t_w$$
$$U_i r_i \Delta T = h_o r_o \ \Delta t_w$$
$$\Delta T_w = \frac{U_i r_i \Delta T}{h_o r_o} = \frac{5.03(0.762)(121 - 25)}{5(0.76835)}$$
$$= 95.78 °\text{C}$$
$$h_o = 1.3196 \left( \frac{\Delta T_w}{D_o} \right)^{0.25} = 1.3196 \left( \frac{95.78}{1.5367} \right)^{0.25}$$
$$= 3.71 \, \text{W}/(\text{m}^2\text{K})$$

The calculated value is less than the assumed $h_o$. Use this calculated value to recalculate $U_i$. All the terms in the expression for $1/U_i$, previously used, are the same except for the first term. Assume $h_o = 3.71$ W/(m$^2 \cdot$ K):

$$\frac{1}{U_i} = \frac{0.762}{0.76835(3.71)} + 0.00015 + 0.000167$$

$$U_i = 3.7365 \ W/(m^2 \, K)$$

$$\Delta T_w = \frac{3.7365(0.762)(121 - 25)}{3.71(0.76835)} = 95.89°C$$

$$h_o = 1.3196 \left(\tfrac{95.89}{1.5367}\right)^{0.25} = 3.71 \, W/(m^2 \, K)$$

Because the assumed and calculated values are the same, the correct $h_o$ must be $h_o = 3.71$ W/(m$^2 \cdot$ K) and:

$$q = U_i A_i$$
$$\Delta T = 3.7365(\pi)(1.524)(9.144)(121 - 25)$$
$$= 15,696 \ W$$

For the ends, examination of the equation for the heat transfer coefficient reveals that h with vertical plates is approximately 4% higher than h with horizontal cylinders for the same $\Delta T$ if L is approximately the same as D. Assume $h = 3.71$ $(1.04) = 3.86$ W/(m$^2 \cdot$ K). Because the end is a vertical flat plate, U can be calculated using Eq. (6.59):

$$\frac{1}{U} = \frac{1}{3.86} + \frac{0.00635}{42} + \frac{1}{6000}$$
$$U = 3.855 \, W/(m^2 \cdot K)$$

$$UA\Delta T = h_o A \Delta T_w$$

$$\Delta T_w = \frac{U\Delta T}{h_o} = \frac{3.855(121 - 25)}{3.86} = 95.88$$

$$h_o = 1.3683 \left[\tfrac{95.87}{1.5367}\right]^{0.25} = 3.846 \, W/(m^2 \cdot K)$$

The assumed and calculated values are almost the same. Use $h_o = 3.846$ W/(m$^2 \cdot$ K):

$$\frac{1}{U} = \frac{1}{3.846} + 0.000156 + 0.000167$$
$$U = 3.835 \, W/(m^2 \cdot K)$$

Because there are two sides, the area will be:
$$A = 2\pi R^2$$

$$q = UA\Delta T$$
$$= (3.835)(\pi)(1.5367)^2(2)(121 - 25)$$
$$= 5463W$$

Total heat loss $= 15,704 + 5463 = 21,167$ W.

**Example 6.15** Calculate the overall heat transfer coefficient for applesauce heated from 20°C to 80 °C in a stainless steel tube 5 m long with an inside diameter of 1.034 cm and a wall thickness of 2.77 mm. Steam at 120 °C is outside the tube. Assume a steam side heat transfer coefficient of 6000 W/(m$^2 \cdot$ K). The rate of flow is 0.1 m/s. Applesauce has a density of 995 kg/m$^3$, and this density is assumed to be constant with temperature. The value for n is 0.34 and the values for K are 11.6 at 30 °C and 9.0 at 82 °C in Pa$\cdot$ s units. Assume K changes with temperature according to an Arrhenius type relationship as follows:

$$\log K = A + \frac{B}{T}$$

where A and B are constants and T is the absolute temperature. The thermal conductivity may be assumed to be constant with temperature at 0.606 W/(m $\cdot$ K). The specific heat is 3817 J/(kg $\cdot$ K). The thermal conductivity of the tube wall is 17.3 W/(m $\cdot$ K).

First, determine the temperature dependence of K. B is the slope of a plot of log K against 1/T:

$$B = \frac{\log 11.6 - \log 9.0}{(1/303 - 1/355)} = \frac{1.064458 - 0.954243}{0.0033 - 0.00282}$$

$$= 227.99 A = \log 11.6 - \frac{227.99}{303}$$

$$= 1.064458 - 0.752433 = 0.3120 \log K$$

$$= 0.3143 + \frac{227.99}{T}$$

The arithmetic mean temperature for the fluid is $(20 + 80)/2 = 50$ °C. At this temperature, $K = \log^{-1}(0.3143 + 227.3/323) = 10.42$:

$$G = \frac{0.1 \ m}{s} \frac{1}{\pi (0.00517)^2 \ m^2}$$
$$= 1190.88 \ kg/(s \cdot m^2)$$

The equivalent Newtonian viscosity is calculated using Eq. (6.82):

$$\mu = \frac{10.42}{4} \left(\frac{3(0.34) + 1}{0.34}\right)^{0.34} (0.00517)^{1-0.34}(0.1)^{0.34-1}$$
$$= \frac{10.42}{4}(1.8328)(0.030967)(4.5709) = 0.676 \ Pa \cdot s$$

$$Re = \frac{0.01034(0.1)(995)}{0.676} = 1.521; \quad \text{flow is laminar}$$

Equation (6.77) must be used, because n < 0.4 and the delta function deviates from $\delta = (3n + 1)/4n$. Solving for the Graetz number:

$$\dot{m} = \pi\left(\frac{D^2}{4}\right)(V)(\rho) = \pi(0.00517)^2(0.1)(995)$$

$$= 0.008355$$

Using Eq. (6.79) for n < 0.4 and Gz = 10:

$$Gz = \frac{0.008355(3817)}{[(0.606)(5)]} = 10.52$$

$$\Delta^{0.33} = -0.6(0.34) + 1.30 = 1.096$$

The Nusselt number is:

$$Nu = 1.75(1.096)(10.525)^{0.33}\left(\frac{\mu}{\mu_w}\right)^{0.14}$$

$$= 4.170\left(\frac{\mu}{\mu_w}\right)^{0.14} = \frac{hD}{k}$$

Solving for h:

$$h = \frac{4.17[\mu/\mu_w]^{0.14}(0.606)}{0.01034} = 244.39\left[\frac{\mu}{\mu_w}\right]^{0.14}$$

Assume most of the temperature drop occurs across the fluid side resistance. Assume that $T_w = 116\,°C$.

K at 116 °C = $\log^{-1}(0.3126 + 227.99/389) = 7.91$ Pa · s.

Using Eq. (6.83):

$$\mu_w = 7.91\left[\frac{(2)(0.1)}{0.01034}\right]^{0.34-1}\left[\frac{3(0.34)+1}{0.34}\right]^{0.34-1}$$

$$= 7.91(0.141548)(0.30849) = 0.3454h_i$$

$$= 244.39\left[\frac{0.676}{0.3454}\right]^{0.14} = 268.48 \ W/(m^2 \cdot \leq K)$$

Using Eq. (6.62) for the overall heat transfer coefficient:

$$\frac{1}{U_i} = \frac{r_i}{r_o h_o} + \frac{r_i \ln r_o/r_i}{k} = \frac{1}{h_i}$$

$$r_i = 0.00517; \quad r_o = 0.00517$$

$$+0.00277 = 0.00794$$

$$\frac{1}{U_i} = \frac{0.00517}{0.00794(6000)}$$

$$+\frac{0.00517 \ \ln(0.00794/0.00517)}{17.3} + \frac{1}{268.48}$$

$$= 0.000109 + 0.000128$$

$$+0.003725 = 0.003962;$$

$$U_i = 252.4$$

Check if wall temperature is close to assumed temperature:

$$U_iA_i \ \Delta T = hA_i \ \Delta T_w \Delta T_w = \frac{U_iA_i \ \Delta T}{h_iA_i}$$

$$= \frac{252.42}{268.48}(120 - 50) = 65.8$$

$$T_w = 50 + 65.8 = 115.8$$

This is close to the assumed value of 116 °C; therefore:

$$h_i = 268.48 \ W/(m^2 \cdot K) \text{ and } U_i$$

$$= 252.42 \ W/(m^2 \cdot K)$$

**Example 6.16** The applesauce in Example 6.14 is being cooled at the same rate from 80 °C to 120 °C in a double-pipe heat exchanger with water flowing in the annular space outside the tube at a velocity of 0.5 m/s. The inside diameter of the outer jacket is 0.04588 m. Water enters at 8 °C and leaves at 16 °C. Calculate the food side and the water side heat transfer coefficient and the overall heat transfer coefficient.

**Solution**

First, calculate the water side heat transfer coefficient. At a mean temperature of 12 °C, water has the following properties: k = 0.58 W/(m · K); $\mu = 1.256 \times 10^{-3}$ Pa · s; $\rho = 1000$ kg/m$^3$; $C_p = 4186$ J/(kg · K); and $D_o = 0.01588$ m. For an annulus, the characteristic length for the Reynolds number is 4 (cross-sectional area/wetted perimeter):

$$D = \frac{4(\pi)(d_2^2 - d_1^2)/4}{(\pi)(d_1 + d_2)} = d_2 - d_1 = 0.03$$

$$Re = \frac{D_o\bar{V}\rho}{\mu} = \frac{(0.030)(0.5)(1000)}{1.256 \times 10^{-3}} = 11,942$$

$$Pr = \frac{C_p\mu}{k} = \frac{4186(1.256 \times 10^{-3})}{0.58} = 9.06$$

For water in turbulent flow in an annulus, Monod and Allen's equation from Appendix 12 with the viscosity factor is:

$$\frac{hD_o}{k} = 0.02 \ Re^{0.8}Pr^{0.33}\left[\frac{d_2}{d_1}\right]^{0.53}\left[\frac{\mu}{\mu_w}\right]^{0.14}$$

$$= 0.02(11942)^{0.8}(9.06)^{0.33}\left[\frac{0.04588}{0.01588}\right]^{0.53}\left[\frac{\mu}{\mu_w}\right]^{0.14}$$

$$= 2565\left[\frac{\mu}{\mu_w}\right]^{0.14}$$

Assume that $T_w = 2 \ °C$ higher than the arithmetic mean fluid temperature. $T_w = 15 \ °C$; $\mu_w = 1.215 \times 10^{-3} \ Pa \cdot s$.

Now, calculate the food fluid side heat transfer coefficient. Since the fluid arithmetic mean

$$h_o = \frac{0.58}{0.03}(0.02)(1827)(2.069)(1.754)\left[\frac{\mu}{\mu_w}\right]^{0.14}$$

$$h_o = 2565\left[\frac{1.256}{1.215}\right]^{0.14} = 2577 \ W/(m^2 \cdot K)$$

temperature is the same as in Example 6.14:

$$h_i = 256.7\left[\frac{\mu}{\mu_w}\right]^{0.14}$$

The overall $\Delta T$ between the water and the applesauce based on the mean temperature is:

$$50 - 12 = 38°C. \ T_w = 15°C.K_w$$
$$= \log^{-1}\left(\frac{0.3126 + 227.99}{288}\right)$$
$$= 12.69 \ Pa \cdot s$$

From Example 6.14, $\mu = 0.676 \ Pa \cdot s$. Using Eq. (6.83):

$$\mu_w = 12.69\left[\frac{2(0.1)}{0.01034}\right]^{0.34-1}\left[\frac{3(0.34)+1}{0.34}\right]^{0.34-1}$$

$$= 0.554h_i = 256.7\left[\frac{0.676}{0.554}\right]^{0.14}$$

$$= 263.95 \ W/(m^2 \cdot K)$$

Calculating for $U_i$ using Eq. (6.62):

$$\frac{1}{U_i} = \frac{0.00517}{0.00794(2577)}$$
$$+ \frac{0.00517 \ \ln(0.00794/0.00517)}{17.3} + \frac{1}{263.95}$$
$$= 0.000253 + 0.000128 + 0.003789$$
$$= 0.004170 U_i = 239.83 \ W/(m^2 \cdot K)$$

Solving for the wall temperature:

$$\Delta T_w = \frac{U_i \Delta T}{h_i} = \frac{(239.83)(38)}{263.95} = 34.5°C$$

$$T_w = 50 - 34.5 = 15.5°C$$

The small difference in the calculated and assumed wall temperatures will not alter the calculated $h_i$ and $h_o$ values significantly; therefore, a second iteration is not necessary. The values for the heat transfer coefficients are:

Water side: $h_o = 2577 \ W/(m^2 \cdot K)$
Food fluid side: $h_i = 263.95 \ W/(m^2 \cdot K)$
Overall heat transfer coefficient: $U_i = 239.83 \ W/(m^2 \cdot K)$

## 6.6 Unsteady-State Heat Transfer

Unsteady-state heat transfer occurs when food is heated or cooled under conditions where the temperature at any point within the food or the temperature of the heat transfer medium changes with time. In this section, procedures for calculating temperature distribution within a solid during a heating or cooling process will be discussed.

### 6.6.1 Heating of Solids Having Infinite Thermal Conductivity

Solids with very high thermal conductivity will have a uniform temperature within the solid. A heat balance between the rate of heat transfer and the increase in sensible heat content of the solid gives:

$$mC_P\frac{dT}{dt} = hA(T_m - T)$$

Solving for $T$ and using the initial condition, $T = T_o$ at $t = 0$:

$$\ln\left(\frac{T_m - T}{T_m - T_o}\right) = \frac{hA}{mC_P}t \qquad (6.84)$$

Equation (6.84) shows that a semilogarithmic plot of $\theta = (T_m - T)/(T_m - T_o)$ against time will be linear with a slope $-hA/mC_p$. If a solid is heated or cooled, the dimensionless temperature ratio, $\theta$, plotted semilogarithmically against time, will show an initial lag before becoming linear. If the extent of the lag time is defined such that the point of intersection of the linear portion of the heating curve with the ordinate is defined, it will be possible to calculate point temperature changes in a solid for long heating times, from an empirical value of the slope of the heating curve. This principle is used in thermal process calculations for foods in Chap. 8.

Fluids within a well-stirred vessel exchanging heat with another fluid that contacts the vessel walls will change in temperature as in Eq. (6.84). A well-mixed steam-jacketed kettle used for cooking foods is an example of a system where temperature change will follow Eq. (6.84). For a well-mixed steam-jacketed kettle, U, the overall heat transfer coefficient between the steam and the fluid inside the kettle will be used instead of the local heat transfer coefficient h used in deriving eq. 6.84.

**Example 6.17** A steam-jacketed kettle consists of a hemispherical bottom having a diameter of 69 cm and cylindrical side 30 cm high. The steam jacket of the kettle is over the hemispherical bottom only. The kettle is filled with a food product that has a density of 1008 kg/m³ to a point 10 cm from the rim of the kettle. If the overall heat transfer coefficient between steam and the food in the jacketed part of the kettle is 1000 W/(m² · K), and steam at 120 °C is used for heating in the jacket, calculate the time for the food product to heat from 20 °C to 98 °C. The specific heat of the food is 3100 J/(kg · K).

**Solution**
The surface area of a hemisphere is $(1/2)\pi d^2$; the volume is $0.5(1/6)\pi d^3$. Use Eq. (6.84) to solve for t. A = area of hemisphere with d = 0.69 m. $A = 0.5(\pi)(d^2) = 0.7479$ m²:

$$
\begin{aligned}
\text{Volume} &= 0.5(0.1666)(n)(.693) \\
&\quad + n[(0.69)(0.5)]2(0.30 - 0.10) \\
&= 0.08597 + 0.07478 = 0.16075 \text{ m}^3 \\
&= 0.16075 \text{ m}^3(1008 \text{ kg/m}^3) \\
&= 162.04 \text{ kg} T_m \\
&= 120; T_o = 20; U = 1000; C_p \\
&= 3100 \ln\left(\frac{120 - 98}{120 - 20}\right) = \\
&\quad -\frac{1000(0.7479)}{162.04(3100)} t = \\
-0.001489 \text{ tt} &= \frac{-\ln(0.22)}{0.001489} \\
&= 1016.9 \text{ s} \\
&= 16.95 \text{ min}
\end{aligned}
$$

## 6.6.2  Solids with Finite Thermal Conductivity

When k is finite, a temperature distribution exists within the solid. Heat transfer follows Fourier's second law, which was derived in the section "Heat Transfer by Conduction" (Eq. 6.19) for a rectangular parallelopiped. For other geometries, the following differential equations represent the heat balance (Carslaw and Jaeger 1959).

For cylinders with temperature symmetry in the circumferential direction:

$$
\frac{\delta T}{\delta t} = \alpha\left(\frac{\delta^2 T}{\delta r^2} + \frac{\delta T}{r\delta r} + \frac{\delta^2 T}{\delta Z^2}\right) \tag{6.85}
$$

For spheres with symmetrical temperature distribution:

$$
\frac{\delta T}{\delta t} = \alpha\left[\frac{\delta^2(rT)}{r\delta r^2} + \frac{1}{r^2 \sin\phi}\frac{\phi}{\delta\phi}\left(\sin\phi \frac{\delta T}{\delta\phi}\right)\right] \tag{6.86}
$$

Thermal diffusivity: $\alpha = k/(\rho C_p)$. An excel program for calculating $\alpha$ from values of k, $\rho$, and $C_p$ determined from food composition is given in Appendix 11.

These equations may be solved analytically by considering one-dimensional heat flow and obtaining a composite solution by a multiplicative superposition technique. Thus, the solution for a brick-shaped solid will be the product of the solutions for three infinite slabs, and that for a finite cylinder will be the product of the solution for an infinite cylinder and an infinite slab. Analytical solutions are not easily obtained for certain conditions in food processing, such as an interrupted process where a change in boundary conditions occurred at a midpoint in the process, or when the boundary temperature is an undetermined function of time. Equations (6.19), (6.85), and (6.86) may be solved using a finite difference technique if analytical solutions are not suitable for the conditions specified. Techniques for solving partial differential equations are discussed, and analytical solutions are given in Carslaw and Yaeger (1959).

### 6.6.3    The Semi-Infinite Slab with Constant Surface Temperature

A semi-infinite slab is defined as one with infinite width, length, and depth. Thus, heat is transferred in only one direction, from the surface toward the interior. This system is also referred to as a "thick solid," and under certain conditions, such as at very short times of heating, the surface of a solid and a point very close to the surface would have the temperature response of a semi-infinite solid to a sudden change in the surface conditions. Equation (6.19) may be solved using the boundary conditions: at time 0, the slab is initially at $T_o$ and the surface is suddenly raised to temperature $T_s$. The temperature T at any point x, measured from the surface, expressed as a dimensionless temperature ratio, $\theta$, is:

$$\theta = \mathrm{erf}\left[\frac{x}{(4\ \alpha\ t)^{0.5}}\right] \qquad (6.87)$$

**Table 6.4**  Values of the error function

| x | erf (x) | x | erf (x) | x | erf (x) |
|------|----------|------|----------|-----|----------|
| 0 | 0 | 0.70 | 0.677801 | 1.7 | 0.983790 |
| 0.05 | 0.056372 | 0.75 | 0.711156 | 1.8 | 0.989091 |
| 0.10 | 0.112463 | 0.80 | 0.742101 | 1.9 | 0.992790 |
| 0.15 | 0.157996 | 0.85 | 0.770668 | 2.0 | 0.995322 |
| 0.20 | 0.222703 | 0.90 | 0.796908 | 2.2 | 0.998137 |
| 0.25 | 0.276326 | 0.95 | 0.820891 | 2.4 | 0.999311 |
| 0.30 | 0.328627 | 1.00 | 0.842701 | 2.6 | 0.999764 |
| 0.35 | 0.37938 | 1.1 | 0.880205 | 2.8 | 0.999925 |
| 0.40 | 0.428392 | 1.2 | 0.910314 | 3.0 | 0.999978 |
| 0.45 | 0.47548 | 1.3 | 0.934003 | 3.2 | 0.999994 |
| 0.50 | 0.520500 | 1.4 | 0.952790 | 3.4 | 0.999998 |
| 0.55 | 0.563323 | 1.5 | 0.966105 | 3.6 | 1.00000 |
| 0.60 | 0.603856 | 1.6 | 0.976348 | 3.8 | 1.00000 |
| 0.65 | 0.64202 | | | 4.0 | 1.00000 |

*Source:* Kreyzig, E. 1963. Advanced engineering Mathematics, John Wiley & Sons, NY

$\theta = (T_s - T)/(T_s - T_O)$ and erf is the error function. The error function is a well-studied mathematical function, and values are found in standard mathematical tables. Values of the error function are given in Table 6.4.

The error function approaches 1.0 when the value of the argument is 3.6. In a solid, the point where erf = 1 is undisturbed by the heat applied at the surface. Thus, a penetration depth may be calculated, beyond which the thermal conditions are undisturbed. If this penetration depth is less than the half thickness of a finite slab, the temperature distribution near the surface may be approximated by Eq. (6.87). A finite body may exhibit a thick body response if the penetration depth is much less than the half thickness. The thick body response to a change in surface temperature is the temperature distribution expressed by Eq. (6.87) from the surface (x = 0) to a point in the interior s distant from the surface, when $\delta = 3.8(4\alpha t)^{0.5}$ is much less than the half thickness of a finite solid.

If the surface heat transfer coefficient is finite, the solution to Eq. (6.19) with the boundary condition, $T = T_o$ at time zero, heat is transferred at the surface from a fluid at temperature $T_m$, with a heat transfer coefficient h:

$$\theta = \ \mathrm{erf}\left[\frac{x}{\sqrt{(4\alpha t)}} + [e]^{hx/k+(h/k)^2\alpha t}\left[\mathrm{erfc}\left[\frac{x}{\sqrt{(4\alpha t)}} + h \ \ \mathrm{over} \ \ k\sqrt{(\alpha t)}\right]\right]\right]$$

$$(6.88)$$

where $\mathrm{erfc}[F(x)] = 1 - \mathrm{erf}[F(x)]$.

Equation (6.88) becomes Eq. (6.87) when h is infinite, as the complementary error function has a value of zero when the argument of the function is very large. Surface conductance plays a role along with the thermal conductivity in establishing if a body will exhibit a thick body response. Schneider (1973) defined a critical Fourier number when a body ceases to exhibit a thick body response:

$$\mathrm{Fo}_{critical} = 0.00756 \ \mathrm{Bi}^{-0.3} + 0.02 \quad \mathrm{for} \ \ 0.001 \le \mathrm{Bi} \le 1000$$

The Fourier number is: $\mathrm{Fo} = \alpha t/(L)^2$; $L = $ thickness/2; $\mathrm{Bi} = $ Biot number $= hL/k$.

**Example 6.18** A beef carcass at 38 °C is introduced into a cold room at 5 °C. The thickness of the carcass is 20 cm. Calculate the temperature at a point 2 cm from the surface after 20 min. The density is 1042 kg/m$^3$, the thermal conductivity is 0.44 W/(m · K), specific heat is 3558 J/(kg · K), and the surface heat transfer coefficient is 20 W/(m$^2$· K).

**Solution**

For the short exposure time, the material might exhibit a thick body response. This condition will be affirmed by calculating the Biot number and the Fourier number:

$$L = 20(.5) = 10 \ \mathrm{cm} \ = 0.10 \ \mathrm{m}$$
$$\mathrm{Bi} \ = \frac{hL}{K} = \frac{20(0.10)}{0.44} = 4.55$$
$$\mathrm{Fo} \ = \frac{\alpha t}{L^2} = \frac{0.44}{1042(3558)}\frac{20(60)}{(0.10)^2} = 0.0142$$
$$\mathrm{Fo}_{critical} = 0.00756(4.55)^{-0.3} + 0.02 = 0.024$$

The actual Fourier number is less than the critical value for the solid to cease exhibiting a thick body response; therefore, the error function solution can be used to calculate the temperature at the

designated point. Using eq. 88, $x = 0.02$ m from the surface:

$$\alpha \ = \frac{0.44}{(1042)(3558)} = 1.187 \times 10^{-7}\frac{x}{(2)\sqrt{(\alpha t)}}$$
$$= \frac{0.02}{(2)\left[(1.187 \times 10^{-7})(20)(60)\right]^{0.5}}$$
$$= 0.838 hx/k \ = 20(0.02)/0.44$$
$$= 0.909\left(\tfrac{h}{k}\right)^2\alpha t$$
$$= \left(\frac{20}{0.44}\right)(1.187 \times 10^{-7})(20)(60) = 0.294$$
$$= 0.542\left(\frac{h}{k}\right)\sqrt{(\alpha t)} = \left(\frac{20}{0.44}\right)$$
$$\times \left[(1.187 \times 10^{-7})(20)(60)\right]^{0.5}\theta$$
$$= \mathrm{erf}(0.838)$$
$$+[e]^{0.909+0.294}[\mathrm{erfc}(0.838 + 0.542)]$$
$$= 0.7663812 + (3.3301)(.059674) = 0.9335$$

For cooling, $\theta = (T - T_m)/(T_0 - T_m)$:

$$T = T_m + (T_o - T_m) = 5 + (0.9335)(38 - 5)$$
$$= 35.8°C$$

### 6.6.4    The Infinite Slab

This is a slab with thickness 2 L extending to infinity at both ends.

When h is infinite:

$$\theta \ = \ 2\sum_{n=0}^{\infty}\left[\frac{(-1)^n}{(n + 0.5)\pi}[e]^{-(n+0.5)^2\left(\pi^2\alpha t/L^2\right)}\right]$$
$$\times \left[\cos\left[\frac{(n + 0.5)\pi x}{L}\right]\right]$$

$$(6.89)$$

When h is finite:

$$\theta = 2\sum_{n=1}^{\infty} [e]^{-\delta_n^2 \alpha t/L^2} \left[\frac{\sin(\delta_n)\cos(\delta_n x/L)}{\delta_n + \sin(\delta_n)\cos(\delta_n)}\right]$$

(6.90)

$\delta_n$ are the positive roots of the transcendental equation:

$$\delta_n \, \tan(\delta_n) = \frac{hL}{k}$$

The center ($\theta_c$) and surface ($\theta_s$) temperature for an infinite solid with surface heat transfer is obtained by setting $x = 0$ and $x = L$ in Eq. (6.90):

$$\theta_c = 2\sum_{n=1}^{\infty} \frac{\sin(\delta_n)[e]^{-(\delta_n/L)^2 \alpha t}}{\delta_n + \sin(\delta_n)\cos(\delta_n)}$$

(6.91)

$$\theta_s = 2\sum_{n=1}^{\infty} \frac{\sin(\delta_n)\cos(\delta_n)[e]^{-(\delta_n)^2 \alpha t}}{\delta_n + \sin(\delta_n)\cos(\delta_n)}$$

(6.92)

### 6.6.5    Temperature Distribution for a Brick-Shaped Solid

The solution to the differential equation for a brick-shaped solid will be the product of the solution for infinite slabs of dimensions $L_1$, $L_2$, and $L_3$. The roots of the transcendental equation will be different for each dimension of the brick and will be designated $\delta_{n1}$, $\delta_{n2}$, and $\delta_{n3}$, respectively, for sides with half thickness $L_1$, $L_2$, and $L_3$.

Let $F_o$ = Fourier number, $\alpha t/L^2$:

$$F(x\delta_{ni}) = \frac{\sin(\delta_{ni})\cos(\delta_{ni}x/L)}{\delta_{ni} + \sin(\delta_{ni})\cos(\delta_{ni})}\theta$$
$$= 8\sum_{n=1}^{\infty} F(x\delta_{n1})F(x\delta_{n2})F(x\delta_{n3})[e]^{-\sum(\delta_{ni})^2 Fo_i}$$

(6.93)

Of interest to food scientists and engineers will be center temperature and surface temperature. Let:

$$F(c\delta_{ni}) = \frac{\sin(\delta_{ni})}{\delta_{ni} + \sin(\delta_{ni})\cos(\delta_{ni})}$$

$$F(s\delta_{ni}) = \frac{\sin(\delta_{ni})\cos(\delta_{ni})}{\delta_{ni} + \sin(\delta_{ni})\cos(\delta_{ni})}$$

The center or surface temperature may be calculated using $F(c\delta_{ni})$ or $F(s\delta_{ni})$ in place of $F(x\delta_{ni})$ in Eq. (6.93). Appendix 13 lists a computer program in Visual BASIC for calculating the surface and center temperature of a brick-shaped solid. An example of the output of the program is shown in Fig. 6.17. The surface temperature will be

**Fig. 6.17** Temperature at the surface and at the geometric center of two brick-shaped solids heated in an oven at 177 °C from an initial temperature of 4 °C. Solid A dimensions: 20.32 × 10.16 × 5.08 cm. Solid B dimensions: 30.48 × 15.24 × 5.08 cm. Parameters: $\rho = 1085$ kg/m$^3$; $C_p = 4100$ J/(kg · K); $k = 0.455$ W/(m·K); $h = 125$ W/(m$^2$ K)

affected by the thickness of the solid and will not assume the heating medium temperature immediately after the start of the heating process.

### 6.6.6    Use of Heisler and Gurney-Lurie Charts

Before the age of personal computers, calculations involving the transient temperature response of solids were a very laborious process. Solutions to the partial differential equations were plotted and arranged in a form that makes it easy to obtain solutions. Two of the transient temperature charts are the Gurney-Lurie chart shown in Fig. 6.18 and the Heisler chart shown in Figs. 6.19 and 6.20. These charts are for an infinite slab. Their use for a brick-shaped solid will involve the multiplicative superposition technique discussed earlier. Similar charts are available for cylinders and spheres.

Both charts are plots of the dimensionless temperature ratio against the Fourier number.

The Gurney-Lurie chart plots the dimensionless position ($x/L = n$; $n = 0$ at the center), and 1/Biot number $= m$, as parameters. The Heisler chart is good for low Fourier numbers, but a chart must be made available for each position under consideration. The Heisler chart shown in Fig. 6.19 is the temperature response at the surface, and Fig. 6.20 is the temperature response at the center of an infinite slab.

The use of these charts involves calculating the Biot number and the Fourier number. The appropriate curve is then selected which corresponds to the Biot number and the position under consideration. The charts are used to obtain a value for the dimensionless temperature ratio, $\theta$. If the solid is brick shaped, Fourier and Biot numbers are obtained for each direction, and the values of $\theta$ obtained for each direction are multiplied to obtain the net temperature response from all three directions.

**Example 6.19**  Calculate the temperature at the center of a piece of beef 3 cm thick, 12 cm wide,

**Fig. 6.18** Gurney-Lurie chart for the temperature response of an infinite slab. (Source: Adapted from McAdams, W. H. 1954. Heat transmission. 3rd. ed. McGraw-Hill, New York. Used with permission of McGraw-Hill, Inc.)

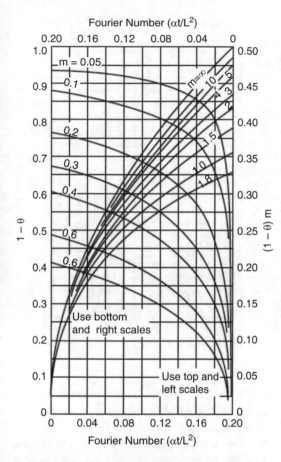

Fourier Number ($\alpha t/L^2$)

**Fig. 6.19** Heisler chart for the center temperature of a slab. (Adapted from Hsu, S.T. 1963. Engineering Heat Transfer. Van Nostrand Reinhold, New York)

and 20 cm long after 30 min of heating in an oven. The beef has a thermal conductivity of 0.45 W/ (K), a density of 1008 kg/m³, and a specific heat of 3225 J/(kg · K). Assume a surface heat transfer coefficient of 20 W/(m² · K). The meat was originally at a uniform temperature of 5 °C, and it was instantaneously placed in an oven maintained at 135 °C. Neglect radiant heat transfer and assume no surface evaporation.

**Solution**

Consider each side separately. Consider the x direction as the thickness (L = 0.5 (0.03) = 0.015 m), the y direction as the width (L = 0.5(0.12) = 0.06 m), and the z direction as the length (L = 0.5(0.2) = 0.1 m):

$$\alpha = \frac{k}{\rho C_p} = \frac{0.45}{(1008)(3225)} = 1.3843 \times 10^{-7} \, \text{m}^2/\text{s}$$

$$Fo_x = \left[1.3843 \times 10^{-7}(30)(60)\right]/(0.015)^2 = 1.107$$

$$Fo_y = \left[1.3843 \times 10^{-7}(30)(60)\right]/(0.06)^2$$
$$= 0.0692$$

$$Fo = \frac{\alpha t}{L^2}$$

$$Fo_z = \left[1.3843 \times 10^{-7}(30)(60)\right]/(0.1)^2$$
$$= 0.249$$

$$Bi = hL/k$$

$$Bi_x = 20(0.015)/0.45 = 0.667; m = 1.5$$
$$Bi_y = 20(0.06)/0.45 = 2.667; m = 0.374$$
$$Bi_z = 20(0.1)/0.45 = 4.444; m = 0.225$$

At the center, n = 0.

From Fig. 7.18, $\theta$ for m = 1.5 will be obtained by interpolation between values for m = 1 and m = 2. From Fig. 7.18, $Fo_x$ = 1.107, n = 0, m = 1, $\theta$ = 0.52, n = 0, m = 2, and $\theta$ = 0.7. For m = 1.5, $\theta$ = [0.52 + (0.70–0.52)(1.5–1)]/ 1 = 0.61.

The values of $\theta$ for $Fo_y$ = 0.0692 and $Fo_z$ appear to be almost 1.0 from Fig. 7.18. To verify, use Fig. 7.20 to see that at m = 0.4 and Fo = 0.07, 1 − $\theta$ = 0, and at m = 0.2 and Fo = 0.01, 1 − $\theta$ = 0. Thus, heating for this material occurs primarily from one dimension:

$$T = T_m - \theta(T_m - T_o)$$
$$= 135 - 0.61(135 - 5) = 55.7 °C.$$

## 6.7    Calculating Surface Heat Transfer Coefficients from Experimental Heating Curves

If heating proceeds for a long time, the series represented by Eq. (6.93) converges rapidly and the first term in the series is adequate. Equation (6.93) then becomes:

$$\theta_c = 2F(c\delta_1)F(c\delta_2)F(c\delta_3)[e]^{-Fo_1\delta_1{}^2-Fo_2\delta_2{}^2-Fo_3\delta_3{}^2}$$

Let $F(c\delta) = 2F(c\delta_1)F(c\delta_2)F(c\delta_3)$. Taking the logarithms of both sides of the equation:

**Fig. 6.20** Heisler chart for the surface temperature of a slab. (Adapted from Hsu, S. T. 1963. Engineering Heat Transfer. Van Nostrand Reinhold, New York)

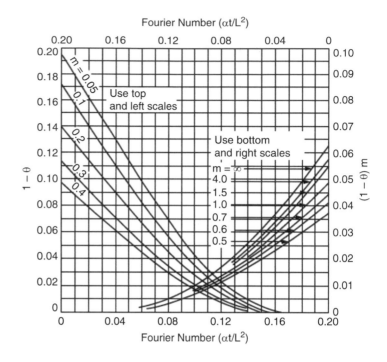

$$(\theta_c) = \log[F(C\delta)]$$

$$- [\alpha t \ \log(e)]\left[\frac{\delta_1^2}{L_1^2} + \frac{\delta_2^2}{L_2^2} + \frac{\delta_3^2}{L_3^2}\right] \quad (6.94)$$

Equation (6.94) shows that a plot of $\log(\theta_c)$ against t will be linear, and the surface heat transfer coefficient can be determined from the slope if $\alpha$ is known:

$$\text{Slope} = -[\alpha \ \log(e)]\left[\frac{\delta_1^2}{L_1^2} + \frac{\delta_2^2}{L_2^2} + \frac{\delta_3^2}{L_3^2}\right] \quad (6.95)$$

A computer program in BASIC shown in Appendix 14 can be used to determine the average heat transfer coefficient from the center temperature heating curve of a brick-shaped solid. Similar approaches may be used for cylindrical solids.

## 6.8 Freezing Rates

Temperature distribution in solids exposed to a heat exchange medium at temperatures below the solid's freezing point is complicated by the change in phase and the unique properties of the frozen and unfrozen zones in the solid. In addition, the ice front advances toward the interior from the surface, and at the interface between the two zones, a tremendous heat sink exists, in the form of the heat of fusion of water. A number of approaches have been used to mathematically model the freezing process, but the most successful in terms of simplicity and accuracy is the refinement by Cleland and Earle (1984) of the empirical equation originally attributed to Plank and cited in Pham Q. (2014).

Plank's original equation was:

$$t_f = \frac{\lambda}{T_f - T_a}\left[\frac{PD}{h} + \frac{RD^2}{k}\right] \quad (6.96)$$

where $t_f$ is freezing time for a solid with a freezing point $T_f$ and a thermal thickness D, which is the full thickness in the case of a slab or the diameter in the case of a sphere. P and R are shape constants (P = 1/6 for a sphere, 1/4 for a cylinder, and 1/2 for a slab; R = 1/24 for a sphere, 1/16 for a cylinder, and 1/8 for a slab). G is the latent heat of fusion per unit volume, h is heat transfer coefficient, and k is thermal conductivity of the frozen solid. Shape factors P and R have been developed for a brick-shaped solid and presented as a graph based on the dimensions of the brick. This graph can be seen in Charm (1971). Plank's equation has been refined by Cleland and Earle (1984) to account for the fact that not all water freezes at the freezing point and that the freezing process may proceed to specific final product temperature. The effect of an initial temperature different from the freezing point has also been included in the analysis. Cleland and Earle's (1984) equation is:

$$t_f = \frac{\Delta H_{10}}{(T_f - T_a)(\text{EHTD})} \left[ P\frac{D}{h} + R\frac{D^2}{k_s} \right]$$
$$\times \left[ 1 - \frac{1.65 \text{ STE}}{k_s} \ln\left( \frac{T_{fin} - T_a}{-10 - T_a} \right) \right]$$
$$\text{(6.97)}$$

$\Delta H_{10}$ = enthalpy change to go from $T_f$ to $-10^\circ$C in J/m$^3$.

$T_a$ = freezing medium temperature.

$T_{fin}$ = final temperature.

EHTD = equivalent heat transfer dimensionality, defined as the ratio of the time to freeze a slab of half thickness, D, to the time required to freeze the solid having the same D. EHTD = 1 for a slab with large width and length such that heat transfer is effectively only in one direction and 3 for a sphere. Bricks and cylinders will have EHTD between 1 and 3.

h = surface heat transfer coefficient.

$k_s$ = thermal conductivity of the frozen solid.

P, R = parameters that are functions of the Stefan number (STE) and the Plank number (PK):

$$\text{STE} = C_s\frac{(T_f - T_a)}{\Delta H_{10}}; \quad \text{PK} = C_1\frac{(T_i - T_f)}{\Delta H_{10}}$$

$C_s$ = volumetric heat capacity of frozen material, J/(m$^3$ · K).

$C_1$ = volumetric heat capacity of unfrozen material, J/(m$^3$ · K).

$T_i$ = initial temperature of solid.

P = 0.5 [1.026 + 0.5808 PK + STE(0.2296 PK + 0.105)].

R = 0.125 [1.202 + STE(3.41 PK + 0.7336)].

The major problem in using Equation (6.97) is the determination of EHTD. EHTD can be determined experimentally by freezing a slab and the solid of interest and taking the ratio of the freezing times. EHTD considers the total dimensionality of the material instead of just one dimension.

**Example 6.20** Calculate the freezing time for blueberries in a belt freezer where the cooling air is at $-35$ °C. The blueberries have a diameter of 0.8 cm and are to be frozen from an initial temperature of 15 °C to a final temperature of $-20$ °C. They contain 10% soluble solids, 1% insoluble solids, and 89% water. Use Choi and Okos' (1987) correlation for determining the thermal conductivity, Chang and Tao's (1981) correlation for determining the freezing point and enthalpy change below the freezing point, and Seibel's equation for determining the specific heat above and below freezing. The blueberries have a density of 1070 kg/m$^3$ unfrozen and 1050 kg/m$^3$ frozen. The heat transfer coefficient is 120 W/(m$^3$ · K).

**Solution**

For the specific heat above freezing, use Eq. (4.11), Chap. 4:

$$C_1' = 837.36(0.11) + 4186.8(0.89)$$
$$= 3818.3 \text{J}/(\text{kg.K})$$

Converting to J/(K):

$$C_1 = \frac{3818 \text{ J}}{\text{kg K}} \quad \frac{1070 \text{ kg}}{\text{m}^3}$$
$$= 4.0856 \times 10^6 \text{ J}/(\text{m}^3 \cdot \text{K})$$

For the specific heat below freezing, use Eq. 4.11, Chap. 4, but use the specific heat of ice (0.5×4186.8 = 2093.4) instead of the specific heat of liquid water:

$$C_s' = 837.36(0.11) + 2093.4(0.89)$$
$$= 1955.63 \text{ J}/(\text{kg} \cdot \text{K})$$

Converting to J/(m$^3$· K):

$$C_s = \frac{1955.63 \ \text{J}}{\text{kg} \cdot \text{K}} \frac{1050 \ \text{kg}}{\text{m}^3}$$

$$= 2.053 \times 10^6 \ \text{J}/(\text{m}^3 \ \text{K})$$

Use Chang and Tao's correlation for the enthalpy change between freezing point and $-10 \ °C$ (see Chap. 4, the section "Enthalpy Change with a Change in Phase"). From Eq. (4.20), Chap. 4:

$$T_f = 287.56 - 49.19(0.89) + 37.07(0.89)^2$$
$$= 273 \ \text{K}$$

Solving for the enthalpy at the freezing point, Eq. (4.22), Chap. 4: $H_f = 9792.46 + 405,096$ (0.89)

$H_f = 370327.9 \ \text{J/kg}$
$T = -10 \ °C = 263 \ \text{K}$
$T_r = (263-227.6)/(273-227.6) = 0.776$
$H = H_f [\alpha T_r + (1 - a)(T_r)^b]$
Solving for $\alpha$, Eq. (4.17), Chap. 4:

$$\alpha = 0.362 + 0.0498(0.89 - 0.73)$$
$$-3.465(0.89 - 0.73)^2 = 0.362$$
$$+0.007968 - 0.088704$$
$$= 0.281$$

Solving for b, Eq. (4.18), Chap. 4:

$$b = 27.2 - 129.04(0.89 - 0.23)$$
$$-481.46(0.89 - 0.23)^2$$
$$= 27.2 - 6.6151 - 1.265276$$
$$= 19.3196$$

Solving for H at $-10 \ °C$, $T_r = 0.776$; Eq. (4.23), Chap. 4:

$$H = H_f[0.281)(0.776) + (1 - 0.281)$$
$$(0.776)^{19.3196}] = 0.2234 H_f;$$
$$\Delta H_{10} = H_f(1 - 0.2234) = 0.7766(370327.9)$$

$$\Delta H_{10} = 2.876 \times 105 \ \text{J/kg}$$

The thermal conductivity will be determined using the spreadsheet program (Fig. 6.1) in the section "Estimation of Thermal Conductivity of Food Products." The amount of water unfrozen at the freezing point will be determined from the enthalpy at the freezing point using the base temperature for Chang and Tao's enthalpy

correlations, of 227.3 K as the reference, and assuming that all the water is in the form of ice. The heat of fusion of ice is 334,944 J/kg.

The specific heat below freezing calculated above was 1955.23 J/(kg · K):

$$H_f' = 0.89(334,944) + 1955.23(273 - 227.6)$$
$$= 386,867$$

Mass fraction unfrozen water at the freezing point:

$$= (386,867 - 370,327.9)/334,944 = 0.049$$

Using the BASIC program in Appendix 11, the following are entered with the prompts: $X_{water} = 0.049$; $X_{ice} = 0.89–0.049 = 0.8406$; $X_{carb} = 0.10$; $X_{fiber} = 0.01$; all other components are zero. $T = -10 \ °C$; the output is: $k_s = 2.067 \ \text{W}/(\text{m} \cdot \text{K})$. Solving for freezing time using Eq. (6.97):

$\Delta H_{10} = 2.876 \times 105 \ \text{J/kg} \ (1070 \ \text{kg/m}^3)$
$= 3.077 \times 108 \ \text{J/m}^3$
$T_a = -35°C; T_{fin} = -20°C$
$h = 120 \ \text{W}/(\text{m}^2 \cdot \text{K}); \text{EHDT} = 3$ for a sphere;
$D = 0.008 \ \text{m}$
$k_s = 2.067 \ \text{W}/(\text{m} \cdot \text{K}); T_i = 15 \ \text{EC};$
$C_s = 2.053 \times 10^6 \ \text{J/m}^3$
$C_1 = 4.0856 \text{J}/(\text{m}^3 \ \text{AK})$

$$\text{STE} = 2.053 \times 10^6 \left[\frac{0 - (-35)}{6.685 \times 10^7}\right] = 1.0749$$

$$\text{PK} = 4.0856 \times 10^6 \left[\frac{15 - 0}{6.685 \times 10^7}\right]$$

$= 0.9167P$
$= 0.5[1.026 + 0.5808(0.9167)$
$+1.0749[0.2296(0.9167) + 0.105]$
$= 0.9488R$
$= 0.125[1.202 + 1.0749[(3.41)(0.91674) + 0.7336]]$
$= 0.6688$

Substituting values in Eq. (6.97):

$$t_f = \frac{3.077 \times 10^8}{[0 - (-35)](3)} \left[0.9488\frac{0.008}{120} + 0.6688\frac{(0.008)^2}{2.067}\right]$$

$$= 0.029395 \times 10^8[0.00006325 + 0.00002071]$$
$$(1.4383) = 353.9\text{s}$$

$$\left[1 - \frac{1.65(1.0749)}{2.067} \ln\left[\frac{-20 - (-35)}{-10 - (-35)}\right]\right]$$

## Problems

6.1. How many inches of insulation would be required to insulate a ceiling such that the surface temperature of the ceiling facing the living area is within 2 °C of the room air temperature? Assume a heat transfer coefficient on both sides of the ceiling of 2.84 W/(m$^2$· K) and a thermal conductivity of 0.0346 W/(m · K) for the insulation. The ceiling is 1.27-cm-thick plasterboard with a thermal conductivity of 0.433 W/(m · K). Room temperature is 20 °C and attic temperature is 49 °C.

6.2. If a heat transfer coefficient of 2.84 W/(m$^2$ · K) exists on each of the two inside faces of 6.35-mm-thick glass separated by an air gap, calculate the gap that could be used such that the rate of heat transfer by conduction through the air gap would equal the rate of heat transfer by convection. What would be the rate of heat transfer if this gap is exceeded?

The thermal conductivity of air is 0.0242 W/(m · K). Solve for temperatures of 20 °C and −12 °C on the outside surfaces of the glass. Would the calculated gap change in value for different values of the surface temperatures? Would there be any advantage to increasing the gap beyond this calculated value?

6.3. A walk-in freezer 4 × 6 m and 3 m high is to be built. The walls and ceiling consist of 1.7-mm-thick stainless steel (k = 14.2 W/(m · K)), 10 cm thick of foam insulation (k = 0.34 W/(m · K)), a thickness of corkboard (k = 0.043 W/(m · K)), and 1.27 cm thick of wood siding (k = 0.43 W/(m · K)). The inside of the freezer is maintained at −40 °C. Ambient air outside the freezer is at 32 °C. The heat transfer coefficient is 5 W/(m$^2$ · K) on the wood side of the wall and 2 W/(m$^2$ · K) on the stainless steel side.

(a) If the outside air has a dew point of 29 °C, calculate the thickness of the corkboard insulation that would

prevent condensation of moisture on the outside wall of this freezer.

(b) Calculate the rate of heat transfer through the walls and ceiling of this freezer.

6.4. (a) Calculate the rate of heat loss to the surroundings and the quantity of steam that would condense per hour per meter of a 1.5-in. (nominal) schedule 40 steel pipe containing steam at 130 °C. The heat transfer coefficient on the steam side is 11,400 W/(m$^2$ · K) and on the outside of the pipe to air is 5.7 W/m$^2$ · K. Ambient air averages 15 °C in temperature for the year. The thermal conductivity of the steel pipe wall is 45 W/(m · K).

(b) How must energy would be saved in 1 year (365 days, 24 h/day) if the pipe is insulated with 5-cm-thick insulation having a thermal conductivity of 0.07 W/(m · K). The heat transfer coefficients on the steam and air sides are the same as in (a).

6.5. The rate of insolation (solar energy impinging on a surface) on a solar collector is 475 W/m$^2$. Assuming that 80% of the impinging radiation is absorbed (the rest is reflected), calculate the rate at which water can be heated from 15 °C to 50 °C in a solar hot water heater consisting of a spiral-wound, horizontal, 1.9-cm inside-diameter polyethylene pipe 30 m long. The pipe has a wall thickness of 1.59 mm. The projected area of a horizontal cylinder receiving radiation is the diameter multiplied by the length. Assume a heat transfer coefficient of 570 W/(m$^2$· K) on the water side and a heat transfer coefficient of 5 W/(m$^2$· K) on the air side, when calculating heat loss to the surroundings after the radiant energy from the sun is absorbed. The thermal conductivity of the pipe wall is 0.3 W/m · K. Ambient temperature is 7 °C.

6.6. A swept surface heat exchanger cools 3700 kg of tomato paste per hour from 93 °C to 32 °C. If the overall heat transfer coefficient based on the inside surface area is 855 W/m$^2$ · K, calculate the heating surface area required for concurrent flow and countercurrent flow. Cooling water enters at 21 °C and leaves at 27 °C. The specific heat of tomato paste is 3560 J/(kg · K).

6.7. Design the heating and cooling section of an aseptic canning system that processes 190 L per minute of an ice cream mix. The material has a density of 1040 kg/m$^3$ and a specific heat of 3684 J/(kg · K).

6.8. Calculate the number of units of swept surface heat exchangers required to heat the material from 39 °C to 132 °C. Each unit has an inside heat transfer surface area of 0.97 m$^2$. The heating medium is steam at 143 °C. Previous experience with a similar unit on this material was that an overall heat transfer coefficient of 1700 W/(m$^2$ · K) based on the inside surface area may be expected.

6.9. Calculate the number of units (0.9 m$^2$ inside surface area per unit) required for cooling the sterilized ice cream mix from 132 °C to 32 °C. The cooling jacket of the swept surface heat exchangers is cooled by Freon refrigerant from a refrigeration system at −7 °C. Under these conditions, a heat transfer coefficient of 855 W/(m$^2$ · K) based on the inside surface area may be expected.

6.10. A small swept surface heat exchanger having an inside heat transfer surface area of 0.11 m$^2$ is used to test the feasibility of cooking a slurry in a continuous system. When the slurry was passed through the heat exchanger at a rate of 168 kg/h, it was heated from 25 °C to 72°C. Steam at 110 °C was used. The slurry has a specific heat of 3700 J/(kg · K).

What is the overall heat transfer coefficient in this system?

If this same heat transfer coefficient is expected in a larger system, calculate the rate at which the slurry can be passed through a similar swept surface heat exchanger having a heat transfer surface area inside of 0.75 m$^2$, if the inlet and exit temperatures are 25 °C and 72 °C, respectively, and steam at 120 °C is used for heating.

6.11. A steam-jacketed kettle has an inside heat transfer surface area of 0.43 m$^2$ that is all completely covered by the product. The product needs to be heated from 10 °C to 99 °C. The product contains 80% water and 20% nonfat solids. Previous experience has established that an overall heat transfer coefficient based on the inside surface area of 900 W/(m$^2$ · K) may be expected. The kettle holds 50 kg of product. Condensing steam at 120 °C is in the heating jacket. The contents are well stirred continuously during the process.

Calculate the time required for the heating process to be completed.

Determine the nearest nominal size steel pipe that can be used to supply steam to this kettle if the rate of steam flow through the pipe is to average a velocity of 12 m/s.

6.12. A processing line for a food product is being designed. It is necessary to estimate the number of kettles that is required to provide a production capacity of 500 kg/h. The cooking process involving the kettles requires heating the batch from 27 °C to 99 °C, simmering at 99 °C for 30 minutes and filling the hot product into cans. Filling the kettles and emptying requires approximately 15 minutes. The specific heat of the product is 3350 J/(kg · K). The density is 992 kg/m$^3$.

Available for heating are cylindrical vessels with hemispherical bottoms with the hemisphere completely jacketed. The height of the cylindrical section is 25 cm. The diameter of the vessel is 0.656 m. Assume the vessels are filled to 85% of capacity each time. The overall heat transfer coefficient based on the inside surface area averages 600 W/(m$^2$ · K). How many kettles are required to provide the desired production capacity? Steam condensing at 120 °C is used in the jacket for heating.

6.13. A process for producing frozen egg granules is proposed where a refrigerated rotating drum that has a surface temperature maintained at −40 °C contacts a pool of liquid eggs at 5 °C. The eggs freeze on the drum surface and the frozen material is scraped off the drum surface at a point before the surface reenters the liquid eggs. The frozen material, if thin enough, will be collected as frozen flakes. In this process, the thickness of frozen egg that forms on the drum surface is determined by the dwell time of the drum within the pool of liquid. An analogy of the process, which may be solved using the principles discussed in the section on freezing water, 7.5 h is the freezing of a slab directly in contact with a cold surface (i.e., h is infinite). It is desired that the frozen material be 2 mm thick on the drum surface.

6.14. Calculate the dwell time of the drum surface within the liquid egg pool. Assume that on emerging from the liquid egg pool, the frozen material temperature will be 2 °C below the freezing point.

If the drum has a diameter of 50 cm and it travels 120E ($^1/_3$ of a full rotation) after emerging from the liquid egg pool before the frozen material is scraped off, calculate the rotational speed of the drum needed to satisfy the criterion stipulated in (a), and calculate the average temperature of the frozen material at the time it is scraped off the drum. The density of liquid eggs is 1012 kg/m$^3$ and frozen eggs 1009 kg/m$^3$. Calculate the thermophysical properties based on the following compositional data: 75% water, 12% protein, 12% fat, and 1% carbohydrates.

6.15. In an experiment for pasteurization of orange juice, a 0.25-in. outside-diameter tube with 1/32-in.-thick wall was made into a coil and immersed in a water bath maintained at 95 °C. The coil was 2 m long,

and when the juice was pumped at the rate of 0.2 L/min, the juice temperature changed from 25 °C to 85 °C. The juice contained 12% total solids. Calculate:

The overall heat transfer coefficient.

The inside local heat transfer coefficient if the ratio ho/hi is 0.8.

The inside local heat transfer coefficient, hi is directly proportional to the 0.8 power of the average velocity. If the rate at which the juice is pumped through the system is increased to 0.6 L/min, calculate the tube length needed to raise the temperature from 25 °C to 90 °C.

6.16. Calculate the surface area of a heat exchanger needed to pasteurize 100 kg/h of catsup by heating in a one-pass shell-and-tube heat exchanger from 40 °C to 95 °C. The catsup density is 1090 kg/m$^3$. The flow behavior index is 0.5, and the consistency index is 0.5 and 0.35 Pa ·s$^n$ at 25 °C and 50 °C, respectively. Estimate the thermal conductivity and specific heat using correlations discussed in Chaps. 5 and 7. The catsup contains 0.4% fiber, 33.8% carbohydrate, and 2.8% ash, and the balance is water. The catsup is to travel within the heat exchanger at a velocity of 0.3 m/s. Steam condensing at 135 °C is used for heating, and a heat transfer coefficient of 15,000 W/(m$^2$ · K) may be assumed for the steam side. The heat exchanger tubes are type 304 stainless steel with an inside diameter of 0.02291 m and an outside diameter of 0.0254 m.

6.17. A box of beef that was tightly packed was originally at 0 °C. It was inadvertently left on a loading dock during the summer when ambient conditions were 30 °C. Assuming that the heat transfer coefficient around the box averages 20 W/(m$^2$· K), calculate the surface temperature and the temperature at a point 2 cm deep from the surface after 2 h. The box consisted of 0.5-cm-thick fiberboard with a thermal conductivity of 0.2 W/(m · K) and had dimensions of 47 × 60 × 30 cm. Because the box material has very little heat capacity, it may be assumed to act as a surface resistance, and an equivalent heat transfer coefficient may be calculated such that the resistance to heat transfer will be the same as the combined conductive resistance of the cardboard and the convective resistance of the surface heat transfer coefficient. The meat has a density of 1042 kg/m$^3$, a thermal conductivity of 0.44 W/(m · K), and a specific heat of 3558 J/(kg · K).

6.18. In the operation of a microwave oven, reduced power application to the food is achieved by alternatively cutting on and off the power applied. To minimize excessive heating in some parts of the food being heated, the power application must be cycled such that the average temperature rise with each application of power does not exceed 5 °C followed by a 10-second pause between power applications. If 0.5 kg of food is being heated in a microwave oven having a power output of 600 W, calculate fraction of full power output that must be set on the oven controls such that the above temperature rise and pause cycles are satisfied. The food has a specific heat of 3500 J/(kg · K).

## Suggested Readings

ASHRAE. (1965). *ASHRAE guide and data book. Fundamentals and equipment for 1965 and 1966*. New York: American Society for Heating, Refrigerating and Air Conditioning Engineers.

Ball, C. O., & Olson, F. C. W. (1957). *Sterilization in food technology* (1st ed.). New York: McGraw-Hill Book Co..

Bennet, C. O., & Myers, J. E. (1962). *Momentum, heat, and mass transport*. New York: McGraw-Hill Book Co..

Carslaw, H. S., & Jaeger, J. C. (1959). *Conduction of heat in solids* (2nd ed.). London: Oxford University Press.

Charm, S. E. (1971). *Fundamentals of food engineering* (2nd ed.). Westport: AVI Publishing Co..

Chang, H. D., & Tao, L. C. (1981). Correlation of enthalpy of food system. *Journal of Food Science, 46*, 1493.

Choi, Y., & Okos, M. R. (1987). Effect of temperature and composition on thermal properties of foods. In M. Le Maguer & P. Jelen (Eds.), *Food engineering and process applications* (Vol. 1). New York: Elsevier.

Cleland, A. C., & Earle, R. L. (1984). Freezing time prediction for different final product temperature. *Journal of Food Science, 49*, 1230.

Geankoplis, C. J. (1993). *Transport processes and unit operations* (3rd ed.). Englewood Cliffs: Prentice Hall.

Heldman, D. R. (1973). *Food process engineering*. Westport: AVI Publishing Co..

Holman, J. P. (1963). *Heat transfer*. New York: McGraw-Hill Book Co..

Jacob, M., & Hawkins, G. A. (1957). *Heat transfer* (Vol. II). New York: John Wiley.

Leniger, H. A., & Beverloo, W. A. (1975). *Food process engineering*. Boston: D. Riedel Publishing Co..

McCabe, W. L., & Smith, J. C. (1967). *Unit operations of chemical engineering* (2nd ed.). New York: McGraw-Hill Book Co..

McCabe, W. L., Smith, J. C., & Harriott, P. (1985). *Unit operations in chemical engineering* (4th ed.). New York: McGraw-Hill.

McAdams, W. H. (1954). *Heat transmission* (3rd ed.). New York: McGraw-Hill.

Metzner, A. B., Vaughn, R. D., & Houghton, G. L. (1957). Heat transfer to non-Newtonian fluids. *AIChE Journal, 3*, 92.

Peters, M. S. (1954). *Elementary chemical engineering*. New York: McGraw-Hill Book Co..

Pham, Q. (2014). Analytical solution. In *Food freezing and theory calculations* (Springer Briefs in Food Health and Nutrition). New York: Springer.

Roshenow, W. M., & Hartnett, J. P. (1973). *Handbook of heat transfer*. New York: McGraw Hill Book Co..

Schneider, P. J. (1973). Conduction. In W. M. Roshenow & J. P. Hartnett (Eds.), *Handbook of heat transfer*. New York: McGraw-Hill.

Seider, E. N., & Tate, G. E. (1936). The viscosity correction factor for heat transfer to fluids. *Industrial and Engineering Chemistry, 28*, 1429 (cited by McCabe and Smith, 1967, op. cit.).

Sinnott, R. K. (1996). *Chemical engineering* (Vol. 6, 2nd ed.). Oxford: Butterworth-Heinemann.

Watson, E. L., & Harper, J. C. (1989). *Elements of food engineering*. Westport: AVI Publishing Co..

# Kinetics of Chemical Reactions in Foods

Chemical reactions occur in foods during processing and storage. Some reactions result in a quality loss and must be minimized, whereas others result in the formation of a desired flavor or color and must be optimized to obtain the best product quality. Kinetics is a science that involves the study of chemical reaction rates and mechanisms. An understanding of reaction mechanisms coupled with quantification of rate constants will facilitate the selection of the best conditions of a process or storage, in order that the desired characteristics will be present in the product.

## 7.1 Theory of Reaction Rates

Two theories have been advanced as a theoretical basis for reaction rates. The collision theory attributes chemical reactions to the collision between molecules that have high enough energy levels to overcome the natural repulsive forces among molecules. In gases, chemical reaction rates between two reactants have been successfully predicted using the equations derived for the kinetic energy of molecules and the statistical probability for collisions between certain molecules that possess an adequate energy level for the reaction to occur at a given temperature. The activation theory assumes that a molecule possesses a labile group within its structure. This labile group may be normally stabilized by

oscillating within the molecule or by steric hindrance by another group within the molecule. The energy level of the labile group may be raised by an increase in temperature, to a level that makes the group metastable. Finally, a chemical reaction results that releases the excess energy and reduces the energy level of the molecule to another stable state. The energy level that a molecule must achieve to initiate a chemical reaction is called the activation energy. Both theories for reaction rates will give a reaction rate constant, which is a function of the number of reacting molecules and the temperature.

Reactions may be reversible. Reversible reactions are characterized by an equilibrium constant, which establishes steady-state concentration of product and reactants.

## 7.2 Types of Reactions

### 7.2.1 Unimolecular Reactions

One type of chemical reaction that occurs during degradation of food components involves a single compound undergoing change. A part of the molecule may split off, or molecules may interact with each other to form a complex molecule, or internal rearrangement may occur to produce a new compound. These types of reactions are unimolecular and may be represented as:

© Springer International Publishing AG, part of Springer Nature 2018
R. T. Toledo et al., *Fundamentals of Food Process Engineering*,
https://doi.org/10.1007/978-3-319-90098-8_7

$$A \xrightarrow{k_1} products \qquad (7.1)$$

The reaction may occur in more than one step, and in some cases the intermediate product may also react with the original compound:

$$A \xrightarrow{k_1} B \xrightarrow{k_2} products \qquad (7.2)$$

$$A \xrightarrow{k_1} B + A \xrightarrow{k_3} products \qquad (7.3)$$

The reaction rate, r, may be considered the rate of disappearance of the reactant A or the rate of appearance of a reaction product. In reactions 7.1 and 7.2 dA/dt, the rate of disappearance of A, is proportional to a function of the concentration of A, while in reaction 7.3, dA/dt will be dependent on a function of the concentration of A and B. In reaction 7.1, the rate of formation of products will equal the rate of disappearance of the reactant, but in reactions 7.2 and 7.3, accumulation of intermediate products will result in a lag between product formation and disappearance of the original reactant. When intermediate reactions are involved, the rate of appearance of the product will depend on the rate constant $k_2$.

The rate constant, k, is the proportionality constant between the reaction rate and the function of the reactant concentration, F(A) or F(B). Thus, for reactions 7.1 and 7.2:

$$\frac{dA}{dt} = kF(A) \qquad (7.4)$$

For reaction 7.3:

$$\frac{dA}{dt} = k_1 F(A) + k_3 [F(A) + F(B)] \qquad (7.5)$$

In Eqs. (7.4) and (7.5), A and B represent concentrations of reactants A and B, and it becomes obvious that the reaction rate will increase with increasing reactant concentration. The concentration function, which is proportional to the reaction rate, depends on the reactant and could change with the conditions under which the reaction is carried out. When studying reaction rates, it is either the rate of appearance of a product or the rate of disappearance of reactants that will be of interest. On the other hand, if

intermediate reactions are involved in the process of transforming compound A into a final reaction product, and the rate constants and $k_2$ and $k_1$ are affected differently by conditions used in a process, it will be necessary to postulate rate mechanisms and evaluate an overall rate constant based on existing conditions in order to effectively optimize the process. Most of the reactions involving degradation of food nutrients are of the type shown in reaction 8.1.

## 7.2.2   Bimolecular Reactions

Another type of reaction involves more than one molecule. The second step of reaction 7.3 above is one example of a bimolecular reaction. In general, a bimolecular reaction is as follows:

$$aA + bB \xrightarrow{k_3} cC + dD \qquad (7.6)$$

In this type of reaction, the rate may be based on one of the compounds, either the reactant or the product, and the change in concentration of other compounds may be determined using the stoichiometric relationships in the reaction:

$$\begin{aligned} r_{3c} &= \frac{dC}{dt} = -\frac{cdB}{b\,dt} = -\frac{cdA}{a\,dt} = \frac{cdD}{d\,dt} \qquad (7.7) \\ &= kF(A)F(B) \end{aligned}$$

An example of the use of the relationship shown in Eq. (7.7) is the expression of productivity of fermentation systems as reduction of substrate concentration, increase in product concentration, or increase in mass of the microorganism involved in the fermentation.

## 7.2.3   Reversible Reactions

Some reactions are reversible:

$$nA \underset{k_2}{\overset{k_1}{\rightleftharpoons}} bB + cC \qquad (7.8)$$

Again, F(A), F(B), and F(C) are functions of the concentrations of A, B, and C.

The net reaction rate expressed as a net disappearance of A is:

$$r = -\frac{dA}{dt} = k_1 F(A) - k_2 [F(B) \cdot F(C)] \quad (7.9)$$

Expressing B and C in terms of A: B = (n/b) $(A_0 - A)$ and C = (n/c)$(A_0 - A)$, where $A_0$ = initial concentration of A. Let F(A) = A, F(B) = B, and F(C) = C (i.e., the reaction rate is directly proportional to the concentrations of the reactants):

$$r = -\frac{dA}{dT} = k_1 A - k_2 \left[ \left[\frac{n}{b}(A_0 - A)\right] \left[\frac{n}{c}(A_0 - A)\right] \right] \quad (7.10)$$

At equilibrium, $r_1 = r_2$ and:

$$k_1 A = k_2 \left[ \left[\frac{n}{b}(A_0 - A)\right] \left[\frac{n}{c}(A_0 - A)\right] \right]$$

Clearing fractions Eq. (7.11) becomes:

$$k_{eq} = \frac{k_1}{k_2} = \left[ \frac{\left[\frac{n}{b}(A_0 - A)\right] \left[\frac{n}{c}(A_0 - A)\right]}{A} \right] \quad (7.11)$$

Once a constant of equilibrium is known, it will be possible to determine the concentrations of A, B, and C from the stoichiometric relationships of the reaction. An example of this type of reaction is the dissociation of organic acids and their salts.

**Example 7.1** Cottage cheese whey containing initially 4.3% (w/w) lactose when treated with β-galactosidase showed maximum hydrolysis of 80% of the lactose. Calculate the equilibrium constant. In one experiment, an enzyme was added to the whey, and half of the lactose was hydrolyzed for 25 min. Calculate the rate constant for the forward and reverse reactions, and the time required to obtain 77% conversion of lactose, under the conditions given.

**Solution**
The reaction involved in the hydrolysis of lactose is:

$$C_{12}H_{22}O_{11} + H_2O \underset{k_2}{\overset{k_1}{\rightleftharpoons}} C_6H_{12}O_6 + C_6H_{12}O_6$$

Lactose           Glucose   Galactose

Water is not rate limiting in this reaction; therefore, its contribution to the reaction rate is ignored. Let L = the concentration of lactose, G = the concentration of glucose, and C = the concentration of galactose. The rate equations are:

$$r_1 = -\frac{dL}{dt} = k_1 L \; ; \; r_2 = \frac{dL}{dt} = k_2 G C$$

At equilibrium, $r_1 = r_2$, and $L = L_{eq}$; $G = G_{eq}$ and $C = C_{eq}$:

$$k_1 L_{eq} = k_2 G_{eq} C_{eq}$$
$$k_{eq} = \frac{k_1}{k_2} = \frac{G_{eq} C_{eq}}{L_{eq}}$$

Let f = fraction of lactose converted. From the stoichiometry of the reaction, $G_{eq} = C_{eq} = L_o - L_{eq} = L_o(f)$; then, $L_{eq} = L_o(1 - f)$:

Basis : 1 L of whey. $L_o = 4.3(10)/342 = 0.1257$ moles/L

$L_{eq} = (1 - 0.80)(0.1257) = 0.02514$

$G_{eq} = C_{eq} = 0.80(0.1257) = 0.10056$ moles/L

$k_{eq} = (0.10056)^2/0.02514 = 0.4022$

The rate constants for the reaction will be calculated by first setting up the rate equations and integrating.

The rate of disappearance of lactose is:

$$r = -dL/dt = k_1 L - k_2 GC$$
$$L = L_0(1 - f); dL/dt = -L_0(df/dt)$$
$$G = f L_0; C = f L_0; k_1 = k_2 k_{eq}$$
$$L_0 df/dt = k_2 k_{eq}(1 - f)L_0 - k_2 f^2 L_0^2$$
$$k_2 \, dt = \frac{df}{k_{eq}(1 - f) - f^2 L_0}$$

The equation is evaluated by graphical integration. Using the trapezoidal rule and using a BASIC program shown in Fig. 7.1, the integral has a value of 1.814:

$$k_2 = 1.814/25 = .0726$$
$$k_1 = k_2 k_{eq} = .0726(0.4022) = 0.1805$$

To obtain the time, the integral is evaluated using the limits 0 to 0.77. The same BASIC program will be used except that the second line of the program is changed to go from f = 0 to 0.77.

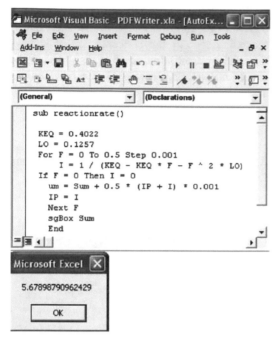

**Fig. 7.1** Visual BASIC program to solve kinetic parameters for Example 8.1

The value of the integral is 5.679:

$$t = 5.679/k2 = 5.679/(0.0726) = 78.2 \text{ minutes}$$

## 7.3    Enzyme Reactions

Enzymatic reactions encountered in food processing occur at a rate that is limited by the concentration of enzymes present. There is usually an abundance of the substrate (reactant) so that changes in substrate concentration do not affect the reaction rate. The activity of an enzyme is defined as the rate at which a specified quantity of the enzyme will convert a substrate to product. The reaction is followed by either measuring the loss of reactant or the appearance of product. The specific activity of an enzyme is expressed as activity/mass of protein. When using an enzyme in a process, the activity of added enzyme must be known in order that the desired rate of substrate conversion will be achieved.

Let $S$ = substrate concentration. If there is no product inhibition:

$$-\frac{dS}{dt} = a \qquad (7.12)$$

where $a$ is the enzyme activity. Thus, an enzymatic reaction without product inhibition suggests a linear change in substrate concentration, at the early stages of the reaction when substrate concentration is so high that enzyme concentration is rate limiting. Most enzymatic reactions however proceed in a curvilinear pattern which approaches a maximum value at infinite time. Consider an enzyme reaction where product inhibition exists. The reaction rate will be:

$$-\frac{dS}{dt} = a_0 - B$$

$B$ = enzyme bound to the product.

Assuming that $B$ is proportional to the amount of product formed, and using $ki$ as the constant representing the product inhibitory capacity:

$$-\frac{dS}{dt} = a_0 - k_i P$$

$P$ = product concentration.

Consider a reaction where $S$ is converted to $P$. If $f$ = the fraction of substrate converted to product: $S = S_0(1 - f)$ and $P = S_0 f$. Substituting into the rate equation, and letting $S_0$ = initial substrate level and $a_0$ = initial enzyme activity:

$$S_0\left[\frac{df}{dt}\right] = a_0 - k_i S_0 f \qquad (7.13)$$

$$\frac{df}{dt} = \frac{a_0}{S_0} - k_i f \qquad (7.14)$$

The differential equation can be easily integrated by separation of variables if the transformation $f = a_0/S_0 - k_i f$ is used:

$$\frac{df}{dt} = -\frac{1}{k_i}\frac{df'}{dt}; \ \frac{df'}{dt} = k_i f'$$

Integrating, substituting $f'$ for $f$, and using the boundary condition $f = 0$ at $t = 0$:

$$\ln\left(1 - \frac{k_i f S_0}{a_0}\right) = -k_i t; f = \frac{a_0}{k_i S_0}\left(1 - e^{-k_i t}\right) \quad (7.15)$$

A plot of the fraction of substrate converted against time will show a curvilinear plot that will level off at a certain value of f at infinite time. The maximum conversion will be $f_{max}$:

$$f_{max} = \frac{a_0}{k_i S_0}; \quad \ln\left(1 - \frac{f}{f_{max}}\right) = k_i t$$

Depending on the enzyme and substrate concentration, the conversion may not immediately follow the exponential expression. For example, if the initial enzyme activity is quite high relative to substrate concentration, the fraction converted may initially be a linear function of time. However, once product accumulates and substrate concentration drops, the influence of product inhibition becomes significant and substrate conversion occurs exponentially as derived above. With product inhibition, the product competes for active sites on the enzyme making these sites unavailable for making a complex with the substrate. Eventually, all enzyme active sites are occupied by the product and the reaction stops. Enzyme activities are determined as the slope of the substrate conversion vs. time curves at time zero (initial reaction velocity) to avoid the effect of product inhibition.

## 7.4   Reaction Order

Reaction order is the sum of the exponents of reactant concentration terms of the rate equation. Table 7.1 lists various heat-induced deteriorative reactions in foods and the kinetic constant based on a first-order reaction.

### 7.4.1   Zero-Order Reactions

$$r = \frac{dA}{dt} = k$$
$$A = A_0 + kt \quad (7.16)$$

A characteristic of a zero-order reaction is a linear relationship between the concentration of reactant or product and time of the reaction, t.

### 7.4.2   First-Order Reactions

$$r = -\frac{dA}{dt} = kA$$
$$\ln\left(\frac{A}{A_0}\right) = kt \quad (7.17)$$

$A_0$ is the concentration of A at time = 0. A first-order reaction is characterized by a logarithmic change in the concentration of a reactant with time. Most of the reactions involved in the processing of foods as shown in Table 7.1 are first-order reactions.

### 7.4.3   Second-Order Reactions

$$r = -\frac{dA}{dt} = kA^2$$
$$\frac{1}{A} - \frac{1}{A_0} = kt \quad (7.18)$$

Second-order unimolecular reaction is characterized by a hyperbolic relationship

**Table 7.1** Kinetic constants of reactions occurring in foods

| Factor | $D_o$ (min) | Z (°C) | References |
| --- | --- | --- | --- |
| Ascorbic acid (peas) | 246 | 50.5 | Rao et al. (1981) |
| Browning reaction (milk) | 12.5 | 26 | Burton (1954) |
| Carotene (beef liver paste) | 43.6 | 25.5 | Wilkinson et al. (1981) |
| Chlorophyll (peas) | 13.2 | 38.8 | Rao et al. (1981) |
| Overall sensory | 12.5 | 26 | Lund (1977) |
| Thiamin | 158 | 21 | Holdsworth (1985) |
| Pectin methyl esterase (citrus) | 0.053 | 14 | Williams et al. (1986) |

between concentration of the reactant or product and time. A linear plot will be obtained if $1/A$ is plotted against time. Second-order bimolecular reactions may also follow the following rate equation:

$$r = -\frac{dA}{dt} = kAB$$

where A and B are the reactants. The differential equation may be integrated by holding B constant to give:

$$\ln\left(\frac{A}{A_0}\right) = -k't \qquad (7.19)$$

$k'$ is a pseudo-first-order rate constant: $k' = kB$.

A second-order bimolecular reaction will yield a similar plot of the concentration of the reactant against time as a first-order unimolecular reaction, but the reaction rate constant will vary with different concentrations of the second reactant. An example of a second-order bimolecular reaction is the aerobic degradation of ascorbic acid. Oxygen is a reactant, and a family of pseudo-first-order plots will be obtained when ascorbic acid degradation is studied at different levels of oxygen availability.

### 7.4.4  *n*th-Order Reactions

$$r = -\frac{dA}{dt} = kA^n; \quad n > 1 \qquad (7.20)$$

The integrated Eq. (7.20) is:

$$A^{1-n} - A_0^{1-n} = -(1-n)kt$$

Evaluation of reaction order is a trial and error process that involves assuming various values for n and determining which value would result in the best fit with the nth-order equation above.

## 7.5  Reactions Where Product Concentration Is Rate Limiting

This type of reactions is usually followed not in terms of the concentration of the reactant but by some manifestation of the completion of the reaction in terms of a physical property change. Examples are protein gelation measured as an increase of the strength of the gel, nonenzymatic browning reaction in solid foods, textural changes during cooking, sensory flavor scores during storage, and so forth. The magnitude of the attribute measured usually levels off not because of depletion of the reactants, but because the measuring technique could no longer detect any further increase in intensity. Reactions of this type could be fitted to the following:

$$\ln\left(1 - \frac{C}{C^*}\right) = \pm kt \qquad (7.21)$$

where $C^*$ is the value of the measured attribute when it remained constant at long reaction times, and C is the value at any time during the transient stage of the process.

### Example 7.2

The following data shows the firmness of a protein gel as a function of time of heating. Derive an appropriate equation to fit this data and determine the rate constant for the reaction.

| Time (min) | 0 | 1 | 2 | 3 | 4 | 5 | 10 | 20 |
|---|---|---|---|---|---|---|---|---|
| Firmness (g) | 0 | 6.01 | 8.41 | 9.36 | 9.75 | 9.90 | 10.2 | 10.2 |

### Solution

The data shows firmness to reach a constant value at 10 minutes of heating. The data will be fitted to the equation:

$$\ln\left(1 - \frac{F}{F^*}\right) = -kt$$

$F$ = firmness value and $F^*$ = the final firmness value.

The following are the transformed data that will be analyzed by linear regression to obtain the slope, which will be the value of $k$.

| x (time) | y [ln (1 − F/F*)] |
|---|---|
| 0 | 0 |
| 1 | −0.889 |
| 2 | −1.74 |
| 3 | −2.497 |
| 4 | −3.121 |
| 5 | −3.526 |

Linear regression of $x$ and $y$ using Microsoft Excel gives a slope of $-1.3743$. $k = 1.3743$ min$^{-1}$.

A plot of the data will show a linear fit except for the last point, which deviated slightly from linearity.

## 7.6 The Reaction Rate Constant

The reaction rate constant defines the reaction rate. There are several ways in which the speed of a chemical reaction can be reported for first-order reactions which predominate in food systems.

*Rate Constant, k, for an Exponential Model of Concentration Change* This rate constant has units of reciprocal time and is the slope of a plot of ln(c) against time. This rate constant is defined for the various types of reactions in section "Reaction Order."

*The D Value* This method of representing the rate constant for a reaction had its origins in thermobacteriology, where the inactivation rate of microorganisms during heating is expressed as a decimal reduction time. This approach was later applied to chemical reactions, in order that the same computational scheme can be used for determining microbial inactivation and nutrient degradation during a thermal process for sterilization of foods.

The D value is defined as:

$$\log\frac{C}{C_0} = -\frac{t}{D} \qquad (7.22)$$

Thus, the D value is the negative reciprocal of the slope of a plot of log(C) against t. C in the above equation is the concentration of a reactant. D is based on common logarithms, in contrast with k, which is based on natural logarithms. D and k are related as follows:

$$\ln\left(\frac{C}{C_0}\right) = \ln(10)\log\left(\frac{C}{C_0}\right) = -kt$$

Thus:

$$\frac{1}{D} = \frac{k}{\ln(10)}; \quad D = \frac{\ln(10)}{k} \qquad (7.23)$$

*The Half-Life* This method of expressing the rate of a reaction is commonly used in radioisotope decay. It is easier to visualize the rate of the reaction when expressed as a half-life rather than a rate constant based on natural logarithms. The half-life is the time required for the reactant to lose half of its original concentration. The half-life is related to k and D as follows:

$$\ln(10.5) = -k(t_{0.5}); \quad t_{0.5} = -\frac{\ln(0.5)}{k}$$
$$\log(0.5) = -\frac{t_{0.5}}{D}; \quad t_{0.5} = -D\log(0.5) \qquad (7.24)$$

## 7.7 Temperature Dependence of Reaction Rates

### 7.7.1 The Arrhenius Equation

The activated complex theory for chemical reaction rates is the basis for the Arrhenius equation which relates reaction rate constants to the absolute temperature. The Arrhenius equation is:

$$k = A_0[e]^{-E_a/RT} \qquad (7.25)$$

$E_a$ is the activation energy, and $A_0$ is the rate constant as T approaches infinity. Another form of the Arrhenius equation involves the reaction rate constant at a reference temperature.

Let $T_0$ = the reference temperature at which $k = k_0$:

$$k_0 = A_0[e]^{-E_a/RT_0}; \quad k = A_0[e]^{-E_a/RT}$$

Taking the ratio of the two equations:

$$\frac{k}{k_0} = [e]^{(-E_a/R)(1/T - 1/T_0)} \qquad (7.26)$$

The negative sign is placed on the exponent of the Arrhenius equation in order that a positive activation energy will indicate an increasing reaction rate constant with increasing temperature. Using Eq. (7.26), the rate constant at any temperature can be determined from the activation energy and the rate constant ko at a reference temperature, $T_0$.

## 7.7.2  The $Q_{10}$ Value

The $Q_{10}$ value of a reaction is often used for reporting temperature dependence of biological reactions. It is defined as the number of times a reaction rate changes with a 10 °C change in temperature. If a reaction rate doubles with a 10 °C change in temperature, the $Q_{10} = 2$. For reactions such as enzymatically induced color or flavor change in foods, degradation of natural pigments, nonenzymatic browning, and microbial growth rate, the $Q_{10}$ is usually around 2. Thus the general rule of thumb in food storage is that a 10 °C reduction in storage temperature will increase shelf life by a factor of 2. The relationship between the $Q_{10}$ value and the activation energy is derived as follows:

Let $k_1$ = rate constant at $T_1$ and $k_2$ − rate constant at $T_2$.

From the definition of the $Q_{10}$:

$$k_2 = k_1[Q_{10}]^{(T_2 - T_1)/10} \qquad (7.27)$$

Taking the logarithm of Eq. (7.27):

$$\ln\left(\frac{k_2}{k_1}\right) = \frac{T_2 - T_1}{10} \ln Q_{10} \qquad (7.28)$$

Substituting $k_2$ for k, $k_1$ for $k_0$, $T_2$ for T, and $T_1$ for $T_0$ in Eq. 7.26:

$$\frac{k}{k_1} = [e]^{(-E_a/R)(1/T_2 - 1/T_1)} \qquad (7.29)$$

Taking the logarithm of Eq. (7.29):

$$\ln\left[\frac{k_2}{k_1}\right] = \frac{-E_a}{R}\left[\frac{1}{T_2} - \frac{1}{T_1}\right] \qquad (7.30)$$

$$\ln\left[\frac{k_2}{k_1}\right] = \frac{-E_a}{R}\left[\frac{T_1 - T_2}{T_2 T_1}\right] \qquad (7.31)$$

The negative sign on $E_a$ in Eq. (7.31) drops out when the signs on $T_1$ and $T_2$ in the numerator are reversed. Equating Eqs. (7.28) and 7.31) and solving for $E_a/R$:

$$\frac{E_a}{R} = \frac{\ln(Q_{10})}{10} T_2 T_1 \qquad (7.32)$$

$$Q_{10} = [e]^{(E_a/R)(10/T_2 T_1)} \qquad (7.33)$$

The $Q_{10}$ is temperature dependent and should not be used over a very wide range of temperature.

## 7.7.3  The z Value

The z value had its origins in thermobacteriology and was used to represent the temperature dependence of microbial inactivation rate. z was defined as the temperature change needed to change microbial inactivation rate by a factor of 10. The z value has also been used to express the temperature dependence of degradative reactions occurring in foods during processing and storage. The z value expressed in terms of the reaction rate constant is as follows:

$$k_2 = k_1[10]^{(T_2 - T_1)/z} \qquad (7.34)$$

Taking the logarithm:

$$\ln\left[\frac{k_2}{k_1}\right] = \frac{T_2 - T_1}{z}\ln(10) \qquad (7.35)$$

Equating the right-hand side of Eqs. (7.31) and (7.35):

$$\frac{\ln(10)}{z} = \left[\frac{E_a}{R}\right]\left[\frac{1}{T_2 T_1}\right] \qquad (7.36)$$

$$z = \frac{\ln(10)}{(E_a/R)} T_1 T_2$$

Solving for $E_a/R$ in Eqs. (7.32) and (7.36) and equating:

$$z = \frac{10\ln(10)}{\ln(Q_{10})} \qquad (7.37)$$

**Example 7.3** McCord and Kilara (J. Food Sci. 48:1479, 1983) reported the kinetics of inactivation of polyphenol oxidase in mushrooms to be first order, and the rate constants at 50 °C, 55 °C, and 60 °C were 0.019, 0.054, and 0.134 $\min^{-1}$, respectively. Calculate the activation energy, the z value, and $Q_{10}$ value for the inactivation of polyphenol oxidase in mushrooms.

**Solution**
The absolute temperatures corresponding to 50 °C, 55 °C, and 60 °C are 323, 328, and 333 K, respectively. A regression of ln(k) against 1/T gives a slope of $-21009.6$; thus, $-E_a/R = -21009.6$ $E_a/R = 21,009$ $K^{-1}$; $R = 1.987$ Cal/(gmole$K); and $Ea = 41.746$ kcal/gmole:

$$\ln(Q_{10}) = 10\left(\frac{E_a}{R}\right)\left(\frac{1}{T_1 T_2}\right) = \frac{10(21,009)}{323(333)} = 1.95$$

$$Q_{10} = 7.028$$

$$z = \frac{\ln(10)}{E_a/R}T_1 T_2 = \frac{323(333)\ln(10)}{21,009} = 11.8\,^\circ C$$

## 7.8  Determination of Reaction Kinetic Parameters

The reaction rate constant is usually determined at constant temperature by measuring changes in the concentration of reactant or concentration of reaction product with time. Because reaction rate can be affected by the presence of interfering compounds, pH, and water activity, kinetic parameters are determined with the reacting compound contained in a specific substrate. Model systems may be used to ensure that the substrate composition is constant during the determination of the kinetic parameters. However, literature data indicate variations in the value of the kinetic parameters for the same reaction in different food products.

The temperature dependence of the reaction rate constant is determined by conducting the kinetic studies at several constant temperatures and determining the z value or activation energy for the reaction.

A typical technique for determination of the kinetic parameters is to derive a linear form of the reaction rate equation and applying regression analysis on the transformed data. This approach however has been shown to have limitations because of the smoothing out effect of the transformation used to linearize the data. One method that can be used is nonlinear curve fitting. The use of statistical software packages for determining the reaction rate constant has been shown in the section on "Nonlinear Curve Fitting" in Chap. 1.

Another approach is the use of the Solver feature in Microsoft Excel. Although Solver only allows the manipulation of one variable to minimize the least square error, an iteration method may be employed to fit two-parameter equations to the data.

**Example 7.4** Data on softening of carrots at 90 °C from Paulus and Saguy (J. Food Sci. 45:239, 1980) given as (x, y) where y = rupture stress in $kg/cm^2$ and x = time in min are as follows: (30, 0.85), (40, 0.60), (50, 0.40), (60, 0.29), and (70, 0.24). Determine the rate constant k, defined as $\ln(y) = kt + b$.

**Solution**
Transformation of y to ln(y) and conducting a linear regression on Excel gives a slope of $-0.03256$ and an intercept of 0.7772. Thus, linear regression finds a k value of $-0.03256$.

To use the Solver feature in Microsoft Excel, enter the data in the spreadsheet. Designate two cells to contain values of k and b. For example, time values are in cells A2 to A6, and force values are in B2 to B6. Designate B7 to hold the value of k and B8 to hold the value of b. Calculate y in column C as y = exp. (kt + b). Then calculate the error square in column D as the difference between the calculated and experimental value squared. For example in column D2, enter (C2 − B2)^2. Calculate the sum of squares error, e.g., in D7 enter Sum(D2.D6).

Go to Tools and choose Solver on the menu. Assume a value for k and b, e.g., enter in B7 the value −0.03 and in B8 the value 0.7772.

Solver asks to set target cell (in this case the sum of squares, error in D7) to a minimum by changing the value of k in B7. Then click solve. The sum of squares and the value of k is displayed. Repeat the process this time designating the value of b in B8 as the designated cell to change. Then click solve and Solver will find a value of b that minimizes the sum of squares of error. Results of Solver gives k = −0.3248 and b = 0.7868. Plotting the calculated y and experimental y shows better agreement with the Solver solution compared to that obtained by linear regression after data transformation.

## 7.9    Use of Chemical Reaction Kinetic Data for Thermal Process Optimization

Data on temperature dependence of chemical degradation reactions in foods are valuable in determining the loss or gain in product quality that might result with an elevation in processing temperature. When microbial inactivation is a major objective of the heating process, the accompanying chemical reactions become an unwanted consequence of the heating process. An acceptable process must satisfy the microbial inactivation constraint, and processing temperature and time must be selected to minimize the extent of unwanted chemical reactions. The

application of kinetic data in optimizing quality factors during thermal processing is discussed further in the section "Quality Factor Degradation" in Chap. 8.

D and z values of some chemical reactions that degrade food quality and of enzyme inactivation are shown in Table 7.1. The z value of most chemical reactions associated with loss in food quality is at least two times higher than those for microbial inactivation (generally z = 10 °C). Thus, it is a well-known practice in the food industry to increase the processing temperature to shorten the processing time for microbial inactivation and minimize the extent of quality factor degradation.

When only relative quality indices are needed to optimize the temperature required for a process, the actual rate of reaction may not be needed in the calculations, and only the temperature dependence of that reaction will be required. Z values for different quality factor degradation reactions are given in Table 7.2.

## Problems

7.1. Nagy and Smoot (J. Agr. Food Chem 25:135, 1977) reported the degradation of ascorbic acid in canned orange juice to be first order, and the following first-order rate constants can be calculated from their data. At T = 29.4 °C, 37.8 °C, and 46.1 °C, k in day$^{-1}$ was 0.00112, 0.0026, 0.0087, respectively. Calculate the activation energy, $Q_{10}$, D value, and half-life at 30 °C.

7.2. The following data were collected for the sensory change in beef stored while exposed directly to air at −23 °C (From: Gokalp et al. J. Food Sci. 44:146, 1979). Sensory scores were 8.4, 6.2, 5.5, and 5.1 at 0, 3, 6, and 9 months in storage. Plot the data and determine an appropriate form of an equation to which the data can be fitted to obtain the reaction rate constant.

7.3. Accelerated shelf life testing is often done to predict how food products would behave in the retail network. If a food product is expected to maintain acceptable quality in

**Table 7.2** Z values for quality degrading chemical reactions in foods

| Texture | Z (Celsius) | Overall sensory | Z (Celsius) |
|---|---|---|---|
| Apples | 21 | Beans, green | 29 |
| Apples | 28 | Beets | 19 |
| Beans, black | 35 | Broccoli | 44 |
| Beans, navy | 37 | Carrots | 17 |
| Beans, soy | 42 | Corn kernels | 32 |
| Beef | 4 | Peas | 32 |
| Beets | 40 | Potatoes | 26 |
| Brussel sprouts | 21 | Squash | 26 |
| Carrots | 18 | | |
| Potatoes | 11 | **Color loss** | |
| Shrimp | 23 | Green pigment | 30 |
| | | Red pigment | 31 |
| | | Browning | 32 |
| **Nutrients** | | **Enzymes** | |
| Carotene | 25 | Peroxidase | 28 |
| Thiamine | 27 | Catalase | 8 |
| Pyridoxine | 29 | Lipoxygenase | 9 |
| Folic acid | 20 | Polyphenol oxidase | 8 |
| Ascorbic acid | 27 | Pectinesterase | 16 |

the retail network for 6 months at 30 °C, how long should this product be stored at 40 °C prior to testing in order that the results will be equivalent to 6 months at 30 °C? Assume that the temperature dependence of the sensory changes in the product is similar to that for the nonenzymatic browning reaction in Table 7.1.

7.4. Ascorbic acid degradation in sweet potatoes at a water activity of 0.11 is first order with a rate constant of $0.001500 \, h^{-1}$ at 25 °C. If the $Q_{10}$ for this reaction is 1.8, calculate the amount of ascorbic acid remaining in dried sweet potato stored at 30 °C after 3 months in storage if the initial ascorbic acid content was 33 mg/100 g.

7.5. In the example problem on B-galactosidase action on lactose in acid whey, calculate the lactose conversion that can be expected using the same level of enzyme addition as in the example, if the whey is preconcentrated prior to treatment to have a lactose content of 12.5%, after a treatment time of 60 min.

7.6. Pectin methyl esterase in orange juice has a D value at 85 °C of 8.3 min and the z value is

14 °C. Calculate the target juice temperature for pasteurization such that at least 99% of the enzyme will be inactivated after a 1-min hold time followed by immediate cooling.

# Suggested Reading

Burton, H. (1954). Color changes in heated and unheated milk. I. The browning of milk on heating. *The Journal of Dairy Research, 21*, 194.

Holdsworth, S. D. (1985). Optimization of thermal processing- A review. *Journal of Food Engineering, 4*, 89.

Holdsworth, S. D. (1990). Kinetic data, what is available and what is necessary. In P. Zeuthen, J. C. Cheftel, C. Eriksson, T. R. Gormley, P. Linko, & K. Paulus (Eds.), *Processing and quality of foods* (Vol. I, pp. 74–90). New York: Elsevier.

Lathrip, P. J., & Leung, H. K. (1980). Rate of ascorbic acid degradation during thermal processing of canned peas. *Journal of Food Science, 45*(1), 152.

Lund, D. B. (1977). Maximizing nutrient retention. *Food Technology, 31*(2), 71.

Perry, R. H., Chilton, C. H., & Kirkpatrick, S. D. (1963). *Chemical engineers handbook* (4th ed.). New York: McGraw-Hill.

Rao, M. A., Lee, C. Y., Katz, J., & Cooley, H. J. (1981). A kinetic study of the loss of vitamin C, color, and firmness during thermal processing of canned peas. *Journal of Food Science, 46*, 636.

Skinner, G. B. (1974). *Introduction to chemical kinetics.* New York: Academic.

Street, J. A., & Tong, C. H. (1996). Degradation kinetics of green color and chlorophylls in peas by calorimetry and HPLC. *Journal of Food Science, 61*(5), 924.

Wilkinson, S. A., Earl, M. D., & Cleland, A. C. (1981). Kinetics of vitamin A degradation in beef lived puree' on heat processing. *Journal of Food Science, 46*, 32.

Williams, D. C., Lim, M. H., Chen, A. O., Pangborn, R. M., & Whitaker, J. R. (1986). Blanching of vegetables for freezing – Which indicator enzyme to use. *Food Technology, 40*(6), 130.

# Thermal Process Calculations

Inactivation of microorganisms by heat is a fundamental operation in food preservation. The concepts learned in this chapter are not only applicable in canning but in any process where heat is used to inactivate microorganisms and induce chemical changes that affect quality. The term "sterilization" used in this chapter refers to the achievement of commercial sterility, defined as a condition where microorganisms that cause illness, and those capable of growing in the food under normal nonrefrigerated storage and distribution, are eliminated.

## 8.1 Processes and Systems for Stabilization of Foods for Shelf-Stable Storage: Systems Requirements

Different systems are available for treating foods to make them shelf stable. Suitability of a system depends on the type of food processed, production rates, availability of capital, and labor costs. Product quality and economics are the major factors to be considered in system selection. Because the major production costs are overhead and labor, plants with high production capacity are inclined to use systems that have high capitalization and low labor requirements. Products with superior quality will result from systems capable of high-temperature, short-time treatments.

### 8.1.1 In-Container Processing

This food sterilization technique is also referred to as "terminal sterilization." The simplest and oldest method of modern food preservation, this sterilization technique involves filling an unsterilized product into a container, sealing, and heating the sealed containers under pressure. Different types of pressure vessels or retorts are used.

#### 8.1.1.1 Stationary Retorts

These retorts are cylindrical vessels oriented vertically or horizontally. Crates are used to facilitate loading and unloading of cans. The cans are stacked vertically in the crates, and perforated metal dividers separate the layers of cans. In vertical retorts, the crates are lowered or raised using electric hoists. In horizontal retorts, the crates are of rectangular profile with dimensions to fit the cylindrical retort. The crates are mounted on a carrier with wheels, and tracks within the retort guide the wheels of the crate carrier during introduction and retrieval.

Stationary retorts for processing of canned foods must be equipped with an accurate temperature controller and recording device. In addition, steam must be uniformly distributed inside the retort. A steam bleeder continuously vents small amounts of steam and promotes steam flow within the retort. A fluid-in-glass thermometer is required to provide visual monitoring of the retort temperature by the operator. At the start of the

© Springer International Publishing AG, part of Springer Nature 2018
R. T. Toledo et al., *Fundamentals of Food Process Engineering*,
https://doi.org/10.1007/978-3-319-90098-8_8

**Fig. 8.1** Horizontal
stationary retort and crate

process, the retort is vented to remove air and
ensure that all cans are in contact with saturated
steam. Figure 8.1 shows a horizontal stationary
retort and crate.

Pressure-resistant hatches for the retorts are of
various design. Some are secured with hinged
bolt-like locks, but more recent designs facilitate
opening and closing of the retort. A wheel-type
lock advances or retracts locking bars that slip
into a retaining slot to secure the cover. Another
cover design consists of a locking ring that can be
engaged or disengaged with a turn of a lever. A
locking ring type of cover assembly is shown in
Fig. 8.1.

Typically, stationary retorts are operated by
loading the cans, venting the retort, and
processing for a specified time at a specified temperature. The time from introduction of steam to
attainment of processing temperature is called the
"retort come-up time." The process is "timed"
when the retort reaches the specified processing
temperature. A timed record of the retort temperature for each batch processed is required to be
maintained in a file. Cooling may be done inside
the retort. However, slow cooling cans may be
removed as soon as internal pressure has dropped
to just slightly above atmospheric, and cooling is
completed in canals, where circulating, cold
chlorinated water contacts the crates, which are
suspended and moved through the water by overhead conveyors.

### 8.1.1.2   Hydrostatic Cooker

A photograph of this type of retort is shown in
Fig. 8.2. It consists of two water legs that seal
steam pressure in the main processing section.
When processing at 121.1 °C, the absolute pressure of steam is 205,740 Pa; therefore, if the
atmospheric pressure is 101.325 Pa, a column of
water 10.7 m high must be used to counteract the
steam pressure. Thus, hydrostatic cookers are
large structures that are often in the open. With
non-agitating hydrostatic cookers, heat penetration parameters for thermal process calculations
are obtained using a stationary retort, and
specified processes are similar to those for a stationary retort. The specified process is set by
adjusting the speed of the conveyor, which carries
cans in and out of the retort such that the residence time in the steam chamber equals the
specified process time.

### 8.1.1.3   Continuous Agitating Retorts

One type of continuous agitating retort consists of
a cylindrical pressure vessel equipped with a
rotating reel that carries cans on its periphery.
When the reel rotates, cans alternately ride on
the reel or roll along the cylinder wall. Figure 8.3
is a photograph of a three shell continuous retort
viewed from the end that shows the drive system
for the reel. Also shown at the top of the retort is
the rotary valve, which receives the cans and
introduces them continuously into the retort

without losing steam from the retort, and the transfer valves, which transfer cans from one retort to the other. Figure 8.4 is a cutaway view showing the reel and the automatic can transfer valve. Agitation is induced by shifting of the headspace as the cans roll. Agitation is maximum in fluid products with small suspended particles, and no agitation exists with a semisolid product such as canned pumpkin. Agitation minimizes heat-induced changes in a product during thermal processing, when products are of low viscosity.

The speed of rotation of the reel determines the rate of heating and the residence time of the cans in the retort; therefore, the heat penetration parameters must be obtained at several reel speeds to match the residence time at a given reel speed to the processing time calculated using heat penetration parameters from the same reel speed. The processing time (t) in a continuous retort is determined as follows:

$$t = \frac{N_t}{N_p \Omega} \qquad (8.1)$$

where $N_t$ is total number of cans in the retort if completely full, $N_p$ is number of pockets around the periphery of the reel, and $\Omega$ is rotational speed of the reel.

A simulator, called the "Steritort," is used to determine heat penetration parameters at different rotational speeds of the reel.

Cans enter a continuous retort and are instantaneously at the processing temperature; therefore, the process time is exactly the residence time of the cans within the retort.

### 8.1.1.4   Crateless Retorts

The crateless retort is one that has a labor-saving feature over conventional stationary retorts, and it appeals to processors whose level of production cannot economically justify the high initial cost of a hydrostatic or continuous rotational agitating retort. Figure 8.5 shows a diagram of a crateless retort. A system of pumps and hydraulic operated

**Fig. 8.2** Hydrostatic cooker. (Courtesy of Food Machinery Corporation, Canning Machinery Division)

**Fig. 8.3** Multiple-shell, continuous rotary retort. (Courtesy of Food Machinery Corporation, Canning Machinery Division)

locks alternately opens the retort, fills it with water, receives the cans, which drop into the retort at random, seals the retort for pressure processing with steam, introduces cold water for cooling, and drops the cans and cooling water into a pool of cold chlorinated water for final cooling and retrieval by a conveyor. Energy is saved if hot water used to initially fill the retorts to receive the cans is stored and reused. Steam waste by venting is eliminated because steam displaces water at the initial phase of the process. The largest saving from this system is in labor and elimination of maintenance cost of the retort crates. Thermal process parameters are determined in the same manner as for stationary retorts.

**Fig. 8.4** Cross section of a continuous rotary retort showing can positioning on the reel and the can transfer valve, which continuously introduces the cans into the retort without releasing the pressure. (Courtesy of Food Machinery Corporation, Canning Machinery Division)

### 8.1.2  Processing Products Packaged in Flexible Plastic Containers

Containers made out of plastic do not have the strength to resist sudden changes in internal pressure during thermal processing. Thus, the heating and cooling steps must be carried out slowly, or air overpressure must be applied inside the retort all the time during the process to ensure that the pressure in the retort is always greater than the pressure inside the container. Processing with air overpressure, however, occurs under a nonsaturated steam atmosphere. Heat transfer is slower than in a saturated steam or saturated water medium. Another problem with steam-air processing medium is the possibility of large variations in the temperature within the retort and the difference in heat transfer coefficients with different concentrations of air in the medium. To solve these problems, processing may be done by complete immersion of the product in water, or water may be sprayed on the product throughout the process, or water may be cascading over the product during the process. One retort design (Fig. 8.6) has a blower that circulates the steam-air heating medium within the retort, but water may also be sprayed on the product to improve heat transfer and minimize temperature variations at different points in the retort.

Figure 8.7 is a picture of a retort that is capable of processing by complete immersion or spraying water on the product. In order to save energy and minimize the come-up time of the retort, a second shell is added to accommodate hot water under pressure that is inside the processing shell at the termination of the process. At the start of the

**Fig. 8.5** Schematic diagram of the operation of a crateless retort. (Courtesy of Food Machinery Corporation, Canning Machinery Division)

**Fig. 8.6** Retort system designed for steam-air mixtures as heating medium. (Courtesy of Societe' Lagarde)

**Fig. 8.7** Retort system designed for full water immersion or water spray as heating medium. (Courtesy of Stock America)

process, the hot water from the upper shell is pumped into the processing vessel along with steam. At the termination of the process, steam is cutoff, and while air overpressure is maintained in the processing vessel, the hot water is pumped into the upper shell. Then cold water is sprayed over the product for cooling.

### 8.1.3 Processing in Glass Containers

The inability of glass to withstand sudden temperature changes requires a gradual heating and cooling process. Products in glass containers are processed in stationary retorts by first filling the retort with water after the containers are loaded and heating the water by direct injection of live steam. A recirculating pump draws water from a point near the top of the water level in the retort and forces this back into the retort through the bottom. This procedure ensures uniform water temperature and uniform water velocity across all containers in the retort. Water is heated slowly thereby prolonging the come-up time and eliminating thermal shock to the glass. Cooling is accomplished by slowly introducing cold water at the termination of the scheduled process. Water temperature drops slowly eliminating thermal shock to the glass.

Evaluation of thermal precesses in hot water systems is best done using the general method for integrating process lethality. A minimum come-up time to the processing temperature, hold time at the specified temperature, and minimum cool down time must be part of the process specifications.

### 8.1.4    Flame Sterilization Systems

This relatively recent development in thermal processing systems is used primarily for canned mushrooms. The system consists of a conveyor that rotates the cans as they pass over an open flame. The cans themselves act as the pressure vessel, which sterilizes the contents. The fluid inside the cans must be of low viscosity, such as brine, water, or low sugar syrups because rapid heat exchange between the can walls and contents is needed to prevent scorching of product on the inner can surface. Internal vacuum at the time of filling must be at the maximum that can be achieved without paneling of the cans. This ensures that a saturated steam atmosphere will exist inside the can and internal pressure from expanding air and steam will not be too excessive during the high temperatures required for sterilization.

Thermal process determination requires a simulator that rotates the can over an open flame. Internal temperature must be monitored in the geometric center of the largest particle positioned in the geometric center of the can.

### 8.1.5    Continuous Flow Sterilization: Aseptic or Cold Fill

Fluids and small particle suspensions can be sterilized by heating in heat exchangers. Figure 8.8 is a schematic diagram of an aseptic processing system. The liquid phase reaches the processing temperature very rapidly; therefore the small sterilization value of the heating phase of the process is generally neglected.

More information on aseptic processing and aseptic packaging will be discussed in Chap. 9 "Aseptic Processing."

The specified process for sterilization in continuously flowing systems is a time of residence in a holding tube, an unheated section of the piping system that leads the fluid from the heat exchangers for heating to the heat exchangers for cooling. A back pressure valve or a positive displacement timing pump is positioned after the cooler to maintain the pressure within the system at a level needed to keep the product boiling temperature higher than the processing temperature. After cooling, the sterile product must be handled in a sterile atmosphere; therefore the process is called *aseptic processing*. The time of residence is set by the volume of the holding tube and the rate of fluid flow delivered by a positive displacement pump.

$$t_{avg} = \frac{A_c L}{Q} \qquad (8.2)$$

where $t_{avg.}$ is average fluid residence time, $A_c$ is cross-sectional area of the holding tube, L is length of the holding tube, and Q is volumetric rate of flow.

The average velocity ($V_{avg.} = Q/A_c$) may also be used to calculate the time in the holding tube.

$$t_{avg} = \frac{L}{V_{avg}} \qquad (8.3)$$

In most cases, however, the time of residence of the fastest flowing portion of the fluid is used as the required hold time in the thermal process calculations. This is because the highest probability of survivors from the thermal process is contributed by the section of fluid flowing close to the geometric center of the tube. The minimum time is:

$$t_{min} = \frac{L}{V_{max}} \qquad (8.4)$$

The maximum velocity for Newtonian fluids in laminar flow is:

**Fig. 8.8**  Schematic diagram of an aseptic processing system for product sterilization

$$V_{max} = 2V_{avg} \qquad (8.5)$$

For power flow fluids in laminar flow:

$$V_{max} = \frac{(3n + 1)}{(n + 1)} V_{avg} \qquad (8.6)$$

For Newtonian fluids in turbulent flow, the following equation was derived by Edgerton and Jones (1970) for $V_{max}$ as a function of the Reynolds number based on the average velocity:

$$V_{max} = \frac{V_{avg}}{0.00336 \log(Re) + 0.662} \qquad (8.7)$$

An equation similar to Eq. (8.7) can be derived by performing a regression analysis on data by Rothfus et al. (AIChE J. 3:208, 1957) for Reynolds number greater than $10^4$.

### 8.1.6    Steam-Air Mixtures for Thermal Processing

A recent development in thermal processing is the use of a mixture of steam and air instead of water or saturated steam for heating. This system has been touted as ideal in processing of products in retortable pouches and glass. The advantages are elimination of a need for exhausting, and no sudden pressure changes on heating or cooling preventing breakage of the fragile containers.

Heating rates on which the scheduled process is dependent are strongly dependent on the heat transfer coefficient when steam-air is used for heating. The heat transfer coefficient is a function of velocity and mass fraction of steam. Thus, a retort designed for steam-air heating must be equipped with a blower system to generate adequate flow within the retort to maintain uniform velocity and uniform temperature. Accurate and separate controllers must be used for pressure and temperature. The mass fraction steam ($X_s$) in a steam-air mixture operated at a total pressure P is given by:

$$X_s = \frac{P_s}{P}\left(\frac{18}{29}\right) \qquad (8.8)$$

$P_s$ is the saturation pressure of steam at the temperature used in the process.

## 8.2    Microbiological Inactivation Rates at Constant Temperature

### 8.2.1    Rate of Microbial Inactivation

When a suspension of microorganisms is heated at constant temperature, the decrease in number of viable organisms follows a first-order reaction. Let N = number of viable organisms.

$$-\frac{dN}{dt} = kN \qquad (8.9)$$

k is the first-order rate constant for microbial inactivation. Integrating Eq. (8.9) and using the initial condition, $N = N_0$ at $t = 0$:

$$\ln\left(\frac{N}{N_0}\right) = -kt \qquad (8.10)$$

Equation (8.10) suggests a linear semilogarithmic plot of N against t. Equation (8.10) expressed in common logarithms is:

$$2.303 \log\left(\frac{N}{N_0}\right) = -kt; \log\left(\frac{N}{N_0}\right) = \frac{-kt}{2.303}$$

or

$$\log\left(\frac{N}{N_0}\right) = \frac{-t}{D} \qquad (8.11)$$

Equation (8.11) defines D, the decimal reduction time, the time required to reduce the viable population by a factor of 10. D = 2.303/k. Thus, the decimal reduction time and the first-order kinetic rate constant can be easily converted for use in equations requiring the appropriate form of the kinetic parameter.

N, the number of survivors, is considered to be the probability of spoilage if the value is less than 1. Any value of $N \geq 1$ means certain spoilage (probability of spoilage = 1 or 100%).

### 8.2.2   Shape of Microbial Inactivation Curves

Microbial inactivation proceeds in a logarithmic function with time according to Eq. (8.11). However, although the most common inactivation curve is the linear semilogarithmic plot shown in Fig. 8.9a, several other shapes are encountered in practice. Figure 8.9b shows an initial rise in numbers followed by first-order inactivation. This has been observed with very heat-resistant spores and may be attributed to heat activation of some spores that otherwise would not germinate and form colonies, before the heat treatment reached

the severity needed to cause death to the organism. Figure 8.9c shows an inactivation curve that exhibits an initial lag or induction period. Very little change in numbers occurs during the lag phase. The curve represented by Fig. 8.9c can be expressed as:

$$\log \frac{N_0}{N} = 1 + \left(\frac{t - t_L}{D}\right); t > t_L \qquad (8.12)$$

where $t_L$ is the lag time, defined as time required to inactivate the first 90% of the population. In most cases, the curved section of the inactivation curve does not extend beyond the first log cycle of inactivation; therefore defining $t_L$ as in Eq. (8.12)

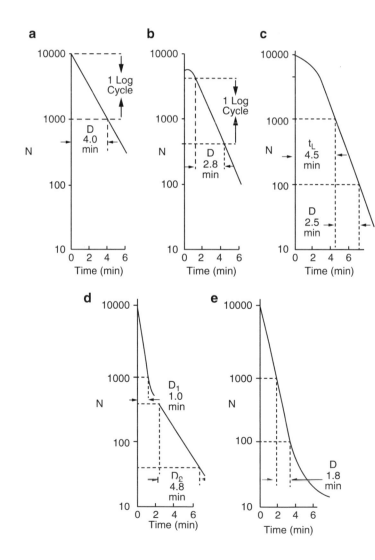

**Fig. 8.9** Microbial inactivation curves. (**a**) First-order inactivation rate. (**b**) Initial rise in numbers followed by first-order inactivation. (**c**) Initial lag in the inactivation curve. (**d**) Inactivation curve exhibited by a mixed culture. (**e**) Tailing of an inactivation curve

eliminates the arbitrary selection of the lag time from the point of tendency of the curved and the straight line portion of the inactivation curve. In general, $t_L$ approaches D as $N_0$ becomes smaller and as the temperature increases. When $t_L = D$, the first-order inactivation rate starts from the initiation of heating, and Eq. (8.12) reduces to Eq. (8.11). Equation (8.12) is not often used in thermal process calculations unless the dependence of $t_L$ on $N_0$ and T is quantified. Microbial inactivation during thermal processing is often evaluated using Eq. (8.11).

Figure 8.9d represents the inactivation curve for a mixed culture. The inactivation of each species is assumed to be independent of each other.

From Eq. (8.11), the number of species A and B having decimal reduction times of $D_A$ and $D_B$ at any time is:

$$N_A = N_{A0}(10)^{-(t/D_A)}; N_B = N_{B0}(10)^{-(t/D_B)}$$
$$N = N_{A0}(10)^{-(t/D_A);} + N_{B0}(10)^{-(t/D_B)}$$

$$(8.13)$$

If $D_A < D_B$, the second term will be relatively constant at small values of t, and the first term predominates as represented by the first line segment in Fig. 8.9d. At large values of t, the first term approaches zero, and microbial numbers will be represented by the second line segment in Fig. 8.9d.

The required heating time to obtain a specified probability of spoilage from a mixed species with known D values will be the longest heating time calculated using Eq. 8.11) for any of the species.

Figure 8.9e shows an inactivation curve that exhibits tailing. Tailing is often associated with very high $N_o$ values and with organisms which have a tendency to clump. As in the case of a lag in the inactivation curve, the effect of tailing is not considered in the thermal process calculation unless the curve is reproducible and the effect of initial number and temperature can be quantified.

**Example 8.1** Figure 8.10 shows data on inactivation of spores of FS 1518 reported by Berry

**Fig. 8.10** Inactivation curve for spores of FS 1518. (From Berry, M. R. et al., J. Food Sci. 50:815, 1985)

et al. (J. Food Sci. 50:815, 1985). When $6 \times 10^6$ spores were inoculated into a can containing 400 g of product and processed at 121.1 °C, the processed product contained 20 spores/g. Calculate the equivalent heating time at 121.1 °C to which the product was subjected.

**Solution**
Both lines in Fig. 8.10 are parallel, and the D value of 3.4 min is independent of initial number, as would be expected from either Eq. (8.11) or (8.12). However, both lines show a departure from linearity at the initial stage of heating. To determine if a lag time should be considered when establishing survivors from a heating process, data from both thermal inactivation curves will be fitted to Eq. (8.12) to determine if $t_L$ is consistent with different initial numbers. A point on each plot is arbitrarily picked to obtain a value for N and t. Choosing a value of $N = 10/g$, the time required to reduce the population from $N_0$ to N is 20 and 16.2 min, respectively for $N_0 = 6 \times 10^6$ and $4 \times 10^6$. Using Eq. (8.12): For $N_0 = 6 \times 10^6$:

$t_L = 20 - 3.4 \left[\log(6 \times 10^6/10) - 1\right]$

$= 3.75$ min

For $N_0 = 4 \times 10^6$ :

$t_L = 16.2 - 3.4 \left[\log(4 \times 10^6/10) - 1\right] = 0.56$ min

Thus, $t_L$ is not consistent at the two levels of $N_0$ tested. At $N_0 = 6 \times 10^6$, $t_L$ is almost equal to the D value of 3.4 min; therefore inactivation was first order at the very start of heating. However, $t_L$ was only 0.56 min when $N_0 = 4 \times 10^6$. Thus, the departure from linearity at the start of heating may be considered an anomaly, and for thermal process calculations, the inactivation curve may be considered as first order from the very start of heating. Reduction of viable numbers will be calculated using Eq. (8.11).

$t = -D\log(N/N_0)$

$N = 20 \dfrac{\text{spores}}{\text{g product}} \times 400\,\text{g product} = 8000$ spores

$N_0 = 6 \times 10^6; t = 3.4 \log \dfrac{6 \times 10^6}{8000} = 9.77$ min

Thus, the equivalent lethality of the process is 9.77 min at 121.1 °C.

**Example 8.2** A suspension containing $3 \times 10^5$ spores of organism A having a D value of 1.5 min at 121.1 °C and $8 \times 10^6$ spores of organism B having a D value of 0.8 min at 121.1 °C is heated at a uniform constant temperature of 121.1 °C. Calculate the heating time for this suspension at 121.1 °C needed to obtain a probability of spoilage of 1/1000.

**Solution**

Using Eq. (8.11):

For organism A : t $= 1.5\log(3 \times 10^5/0.001)$

$= 12.72$ minFor organism B : t

$= 0.8\ \log(8 \times 10^6/0.001)$

$= 7.92$ min

Thus, the required time is 12.72 min.

## 8.2.3 Sterilizing Value or Lethality of a Process

The basis for process lethality is Eq. (8.11), the destruction of biological entities in a heated material. The following may be used as means of expressing the sterilizing value of a process:

$S = $ number of decimal reduction $= \log N_0/N$

$F_T$ is process time at constant temperature T, which has the equivalent lethality of the given process. Usually F values are expressed at a reference temperature (121.1 °C for sterilization processes or 82.2 °C for a pasteurization process). The F and D value at 121.1 °C is $F_0$ and $D_0$, respectively. In a constant temperature process, S and F values can be easily converted between each other using Eq. (8.11), with F values substituted for t.

$$S = \frac{F_T}{D_T} \qquad (8.14)$$

However, if the suspension is heated under changing temperature conditions such as the interior of a can during thermal processing, Eq. (8.14) can be used only if the F value is calculated using the same z value as the biological entity represented by $D_T$. The z value is the parameter for temperature dependence of the inactivation rate and will be discussed in more detail in the section "Effect of Temperature on Thermal Inactivation of Microorganisms."

The use of S values to express process lethality is absolute (i.e., S is the expected effect of the thermal process). However, an S value represents a specific biological entity, and when several must be inactivated, the S value for each entity may be calculated from the F value using Eq. (8.14). The use of F values for expressing process lethality and its calculation under conditions of changing temperature during a process will be discussed in more detail in the section "Sterilizing Value of Processes Expressed as $F_0$."

### 8.2.4 Acceptable Sterilizing Value for Processes

A canned food is processed to achieve commercial sterility. Commercial sterility implies the inactivation of all microorganisms that endanger public health to a very low probability of survival. For canned foods, the critical organism is *Clostridium botulinum*. The 12D concept as a minimum process for inactivation of *C. botulinum* in canned foods is accepted in principle by regulatory agencies and the food industry. However, its interpretation has undergone a process of evolution, from a literal 12 decimal reduction, to what is now generally accepted as a probability of survival of $10^{-12}$.

The latter interpretation signifies a dependence of minimum processes according to the 12D concept on initial spore loads. Thus, packaging materials that have very low spore loads will not require as severe a process as products such as mushrooms which may have very high spore levels. Table 8.1 shows $N_0$ and $N$ values that may be used as a guide in selecting target $N_0/N$ values for thermal processing.

Spoilage from microorganisms that pose no danger to public health is called economic spoilage. Spoilage microorganisms often have higher heat resistance than *C. botulinum*, and their inactivation is the basis for the thermal process design. A high level of spoilage will be expected from the minimum process based on the 12D concept for *C. botulinum* inactivation.

**Table 8.1** N and $N_o$ values used to obtain target in ($N_o$/N) values for commercial sterility of canned foods

| Factor | N | $N_o$ | $D_{mo}$ |
|---|---|---|---|
| Public health | $10^{-9}$ | General 10 | 0.2 |
| | | Meats $10^2$ | |
| | | Mushrooms 10 4 | |
| | | Packaging $10^{-5}$ | |
| Mesophilic spoilage | $10^{-6}$ | General 10 | 0.5 |
| | | Meats $10^3$ 2 | |
| Thermophilic spoilage | $10^{-2}$ | General 10 | 1.5 |

*Source:* From Pflug, I. V., *J. Food Protect. 50:342, 50:347, 50:528, 1987.* Reprinted from Toledo, R. T., *Food Technol. 44(2):72, 1990*

**Example 8.3** The F value at 121.1 °C equivalent to 99.999% inactivation of a strain of *C. botulinum* is 1.2 min. Calculate the $D_0$ value of this organism.

**Solution**

A 99.999% inactivation is 5 decimal reductions (one survivor from 100,000). S = 5. Using Eq. (8.14):

$$D_0 = \frac{F0}{S} = \frac{1.2}{5} = 0.24\,min$$

**Example 8.4** Calculate $F_0$ based on the 12D concept using the $D_0$ value of *C. botulinum* in Example 8.3 and a most likely spore load in the product of 100.

**Solution**

$$S = \log 100 - \log(10 - 12) = 14$$
$$F_0 = 14\,(0.24) = 3.30\,min$$

**Example 8.5** The sterilizing value of a process has been calculated to be an $F_0$ of 2.88. If each can contained 10 spores of an organism having a $D_0$ of 1.5 min, calculate the probability of spoilage from this organism. Assume the $F_0$ value was calculated using the same z value as the organism.

**Solution**

Using Eq. (8.11) for a process time of 2.88 min:

$$\log\frac{N_0}{N} = \frac{2.88}{1.5};$$
$$N = N_0[10]^{-(2.88/1.5)}$$
$$N = 10\left(10^{-1.92}\right) = 0.12;$$
$$P_{spoilage} = 12\ in\ 100\ cans.$$

**Example 8.6** The most probable spore load in a canned food is 100, and the $D_0$ of the spore is 1.5 min. Calculate a target $F_0$ for a thermal process such that the probability of spoilage is 1 in 100,000. If under the same conditions *C. botulinum* type B has a $D_0$ of 0.2 min, would the target $F_0$ value satisfy the minimum 12D process for *C. botulinum*? Assume an initial spore load of 1 per can for *C. botulinum*.

**Solution**

For the organism, $S = \log(100/10^{-5}) = 7$. Using Eq. (8.14): $F_0 = 7(1.5) = 10.5$ min. For *C. botulinum*: $S = \log(1/10^{-12}) = 12$. Using Eq. (8.14): $F_0 = 12(0.2) = 2.4$ min. The $F_0$ for the spoilage organism satisfies the minimum process for 12D of *C. botulinum*.

### 8.2.5  Selection of Inoculation Levels in Inoculated Packs

In order to be reasonably sure of the safety of a process, products may be inoculated with an organism having known heat resistance, processed, and the extent of spoilage compared to the probability of spoilage designed into a process. The organism used must have a higher heat resistance than the background microflora in the product. The level of spoilage and the inoculation levels are set such that spoiled cans can be easily evaluated. The use of an organism that produces gas facilitates the detection of spoiled cans because spoilage will be manifested by swelled cans. If a flat sour organism is used, it will be necessary to open the cans after incubation to determine the number of spoiled cans. If the whole batch of inoculated cans is incubated, the fraction spoiled will be equivalent to the decimal equivalent of the number of surviving organisms. Inoculation levels can be calculated based on Eq. (8.11).

The two examples below represent procedures used for validation of thermal processes using inoculation tests. The first example represents an incubation test, and the second example represents a spore count reduction test.

**Example 8.7**  A process was calculated such that the probability of spoilage from an organism with a $D_0$ value of 1 min is 1 in 100,000 from an initial spore load of 100. To verify this process, an inoculated pack is made. Calculate the level of inoculum of an organism having a $D_0$ value of 1.5 min that must be used on 100 cans such that a spoilage rate of 5 cans will be equivalent in lethality to the calculated process.

**Solution**

The $F_0$ of the calculated process is determined using Eq. (8.11) with $F_0$ substituted for t.

$$F_0 = 1\left[\log\left(100/1 \times 10^{-5}\right)\right] = 7 \text{ min}$$

For the inoculum:

$$\log N_0 - \log(5/100) = F_0/D = 7/1.5 = 4.667$$
$$N_0 = 0.05(10)^{4.667} = 2323 \text{ spores}$$

**Example 8.8**  In an incidence of spoilage, the isolated spoilage organism was found to have a $D_0$ value of 1.35 min. It is desired that the probability of spoilage from this organism be 1 in 100,000. Initial spore loads were generally of the order 10/can. Calculate the required $F_0$ for this process to achieve the desired probability of spoilage. If an inoculated pack of FS 1518 is to be made, and an initial inoculation level of $5 \times 10^5$ spores is made into cans that contained 200 g of product, what will be the spore count in the processed product such that the lethality received by the can contents will be equivalent to the desired process for eliminating spoilage from isolated organism. The $D_0$ value of FS 1518 = 2.7 min.

**Solution**

The $F_0$ of the process can be calculated from the heat resistance and spore reduction needed for the spoilage organism.

$$F_0 = D(\log N_0 - \log N)$$
$$= 1.35\left[\log 10 - \log(1/100000)\right] = 8.1 \text{ min}$$

For FS 1518:

$$\log N - \log\left(5 \times 10^5\right) = -8.1/2.7 = -3.00$$

$$N = 10^{(-3.00+5.698)} = 500$$

Spore count after processing
$$= 500 \text{ spores}/200 \text{ g} = 2.5/\text{g}$$

### 8.2.6   Determination of D Values Using the Partial Sterilization Technique

This technique developed by Stumbo et al. (Food Technol. 4:321, 1950) and by Schmidt (J. Bacteriol. 59:433, 1950) allows the determination of D values using survivor data at two heating times. Appropriate selection of heating times to exceed the lag time for the heated suspension to reach the specified test temperature eliminates the need to correct for temperature changes occurring during the transient period of heating. When this procedure is used, the lag time to bring the suspension to the desired temperature must be established, and the two heating times used for D value determination must exceed the lag time. If $t_1$ and $t_2$ are the heating times and $N_1$ and $N_2$ are the respective number of survivors, the D value is determined by:

$$D = \frac{t_2 - t_1}{\log(N_1) - \log(N_2)} \qquad (8.15)$$

**Example 8.9** Sealed tubes containing equal numbers of spores of an isolate from a spoiled canned food were heated for 10 and 15 min at 115.5 °C. The survivors were, respectively, 4600 and 160. Calculate the D value. The lag time for heating the tubes to 115.5 °C was established in prior experiments to be 0.5 min.

**Solution**
Because the heating times are greater than the lag time for the tubes to attain the desired heating temperature, Eq. (8.15) can be used.

$$D = \frac{15 - 10}{\log(4600) - \log(160)} = \frac{5}{1.458} = 3.42 \, \text{min}$$

### 8.2.7   The Heat Resistance of Spoilage Microorganisms

The heat resistance of microorganisms is expressed in terms of a D value at a reference temperature, and the z value, the temperature

dependence of the thermal inactivation rate. The use of the z value in determining D values at different temperatures from D at the reference temperature is discussed in the section "Effect of Temperature on Thermal Inactivation of Microorganisms." Reference temperatures are 121.1 °C (250 °F) for heat-resistant spores to be inactivated in commercial sterilization processes and 82.2 °C (180 °F) for vegetative cells and organisms of low resistance, which are inactivated in pasteurization processes. D at 121.1 °C (250 °F) is $D_0$.

Tables 8.2 and 8.3 list the resistance of microorganisms involved in food spoilage.

The type of substrate surrounding the organisms during the heating process affects heat resistance. For a more detailed discussion of techniques for determination of microbial heat resistance and the effect of various factors on thermal inactivation rates, the reader is referred to Stumbo's (1973) book *Thermobacteriology in Food Processing* and NFPA's *Laboratory Manual for Canners and Food Processors*, which are listed in the "Suggested Reading" section at the end of this chapter.

### 8.2.8   F₀ Values Used in Commercial Sterilization of Canned Foods

$D_0$, $N_0$, and N values that can be used as a guide for determining $F_0$ values for food sterilization are listed in Table 8.1. $F_0$ must be based on microorganisms involved in economic spoilage, as these have higher heat resistance than *C. botulinum*. Table 8.4 lists $F_0$ values previously used commercially for different types of foods in various size containers. Data in Table 8.1 or Table 8.4 may be used as a base for selection of $F_0$ values needed to calculate thermal process schedules for sterilization.

### 8.2.9   Surface Sterilization

Packaging materials used in aseptic packaging systems and surfaces of equipment may be sterilized using moist heat, dry heat, hydrogen

**Table 8.2**  Heat resistance of spoilage microorganisms in low-acid canned foods

| Organism | Product | $D_0$ (min) | z (°F) | (°C) |
|---|---|---|---|---|
| *Clostridium botulinum* 213-B | Phosphate buffer (pH 7) | 0.16 | 18 | 10 |
| | Green beans | 0.22 | 22 | 12 |
| | Peas | 0.22 | 14 | 8 |
| *Clostridium botulinum* 62A | Phosphate buffer (pH 7) | 0.31 | 21 | 12 |
| | Green beans | 0.22 | 20 | 11 |
| | Corn | 0.3 | 18 | 10 |
| | Spinach | 0.25 | 19 | 11 |
| *Clostridium* spp. PA 3679 | Phosphate buffer (pH 7) | 1.45 | 21 | 12 |
| | Asparagus | 1.83 | 24 | 13 |
| | Green beans | 0.70 | 17 | 9 |
| | Corn | 1.20 | 18 | 10 |
| | Peas | 2.55 | 19 | 10 |
| | Shrimp | 1.68 | 21 | 12 |
| | Spinach | 2.33 | 23 | 13 |
| *Bacillus stearothermophilus* FS 1518 | Phosphate buffer (pH 7) | 3.28 | 17 | 9 |
| | Asparagus | 4.20 | 20 | 11 |
| | Green beans | 3.96 | 18 | 10 |
| | Corn | 4.32 | 21 | 12 |
| | Peas | 6.16 | 20 | 11 |
| | Pumpkin | 3.50 | 23 | 13 |
| | Shrimp | 3.90 | 16 | 9 |
| | Spinach | 4.94 | 21 | 12 |

*Source:* Reed, J. M., Bohrer, C. W. and Cameron, E. J., *Food Res. 16:338–408.*
*Reprinted from:* Toledo R. T. *1980.* Fundamentals of Food Engineering. AVI Pub. Co., Westport, CT

**Table 8.3**  Heat resistance of spoilage microorganisms in acid and in pasteurized foods

| Organism | Temperature °F | °C | D (min) | Z ° F | °C |
|---|---|---|---|---|---|
| *Bacillus coagulans* | 250 | 121.1 | 0.07 | 18 | 10 |
| *Bacillus polymyxa* | 212 | 100 | 0.50 | 16 | 9 |
| *Clostridium pasteurianum* | 212 | 100 | 0.50 | 16 | 9 |
| *Mycobacterium tuberculosis* | 180 | 82.2 | 0.0003 | 10 | 6 |
| *Salmonella* spp. | 180 | 82.2 | 0.0032 | 12 | 7 |
| *Staphylococcus* spp. | 180 | 82.2 | 0.0063 | 12 | 7 |
| *Lactobacillus* spp. | 180 | 82.2 | 0.0095 | 12 | 7 |
| *Yeasts and molds* | 180 | 82.2 | 0.0095 | 12 | 7 |
| *Clostridium botulinum*, Type E | 180 | 82.2 | 2.50 | 16 | 9 |

*Source:* (1) Anderson, E. E., Esselen Jr., W. B. and Fellers, C. R. *Food Res. 14:499–510, 1949.* (2) Crissley, F. D., Peeler, J. T., Angelotti, R. and Hall, H. E., *J. Food Sci. 33:133–137, 1968.* (3) Stumbo, C. R. *77termobacteriology in Food Processing,* Academic Press, New York, *1973.* (4) Townsend, C. T. *Food Res. 4:231–237, 1939.* (5) Townsend, C. T., and Collier, C. P. *Proc. Technical Session of the 48th Annual Convention of the National Canners Association (NCA). NCA information Newsl. No. 1526,* February 28, 1955 (6) Winter, A. R., Stewart, G. F., McFarlane, V. H. and Soloway, M. Am. J. Pub. Health 36:451–460, 1946. (7) Zuccharo. J. B., Powers, J. J., Morse, R. E. and Mills W. C. Food. Res. 16:3038, 1951
*Reprinted from:* Toledo, *1980.* Fundamentals of Food Process Engineering, lst. ed. AVI Pub. Co. Westport, Conn

peroxide, high-intensity ultraviolet, and ionizing radiation from either gamma rays or high-energy electron beams. The latter three methods have not been adopted in commercial food packaging, but various forms of heat and hydrogen peroxide combined with heat are commercially utilized.

Dry heat includes superheated steam and hot air. Resistance of microorganisms in these heating media is shown in Table 8.5. Inactivation occurs at a slower rate in dry heat compared with moist heat at the same temperature. The z value in dry heat is also higher than in moist heat. A

similar principle is utilized in evaluating microbial inactivation in dry heat as for moist heat.

The D values shown in Table 8.5, and the recommended N and $N_0$ values for commercial sterilization in Table 8.1, may be used to determine exposure times to the sterilant. In dry heat sterilization of surfaces, surface temperatures must be used as the basis for the process rather than the temperature of the medium. The surface heat transfer coefficient and the temperature on the opposite side of the surface sterilized determine the actual surface temperature.

Hydrogen peroxide is the only chemical sterilant allowed for use on food contact surfaces. This compound at 35% (w/w) concentration is applied to the surface by atomizing, by spraying, or by dipping, in the case of packaging materials in sheet form, followed by heating to vaporize the hydrogen peroxide and eliminate residue from the surface. A maximum tolerance of residual hydrogen peroxide in the package of 0.1 parts per million is required by federal regulations in the United States. Resistance of various microorganisms in hydrogen peroxide is summarized in Table 8.6. Inactivation rate is also temperature dependent.

**Table 8.4** Values of $F_o$ for some commercial canning processes

| Product | Can sizes | $F_o$(min) |
|---|---|---|
| Asparagus | All | 2–4 |
| Green beans, brine packed | No. 2 | 3.5 |
| Green beans, brine packed | No. 10 | 6 |
| Chicken, boned | All | 6–8 |
| Corn, whole kernel, brine packed | No. 2 | 9 |
| Corn, whole kernel, brine packed | No. 10 | 15 |
| Cream style corn | No. 2 | 5–6 |
| Cream style corn | No. 10 | 2.3 |
| Dog food | No. 2 | 12 |
| Dog food | No. 10 | 6 |
| Mackerel in brine | 301 × 411 | 2.9–3.6 |
| Meat loaf | No. 2 | 6 |
| Peas, brine packed | No. 2 | 7 |
| Peas, brine packed | No. 10 | 11 |
| Sausage, Vienna, in brine | Various | 5 |
| Chili con carne | Various | 6 |

*Source:* Alstrand, D. V., and Ecklund, O. F., *Food Technol. 6(5):185, 1952*

## 8.3    Effect of Temperature on Thermal Inactivation of Microorganisms

Microbial inactivation is a first-order chemical reaction and the temperature dependence of the rate constant can be expressed in terms of an

**Table 8.5** Resistance of microorganisms to microbicidal agents

| Organism | Heating medium | $D_{176.6 °c}$ (min) | $D_{121 °c}$ (min) | z (°C) |
|---|---|---|---|---|
| Bacillus subtilis | Superheated steam | 0.57 | 137 | 23.3 |
| Bacillus stearothermophilus (FS 1518) | Superheated steam | 0.14 | 982 | 14.4 |
| Bacillus polymyxa | Superheated steam | 0.13 | 484 | 15.6 |
| Clostridium sporogenes (PA 3679) | Air | 0.30 | 109 | 21.7 |
| Clostridium botulinum | Air | 0.21 | 9 | 33.9 |
| Bacillus subtilis | $N_z$ + He | 0.17 | 285 | 17.2 |
| Clostridium sporogenes (PA 3679) | He | 0.45 | 161 | 21.7 |
| Bacillus subtilis | A + $CO_z$ + Oz | 0.13 | 218 | 17.2 |

*Source:* Adapted from Miller, B. M., and Litskey, W. eds, 1976. *Industrial Microbiology,* McGraw-Hill. Used with permission of McGraw-Hill

**Table 8.6** Resistance of food spoilage microorganisms to inactivation in hot hydrogen peroxide

| Organism | $D_{80 °C}$ (min) | z (°C) |
|---|---|---|
| *Clostridium botulinum 169B* | *0.05* | *29* |
| *Bacillus subtilis ATCC 9372* | *0.063* | *41* |
| *Bacillus subtilis A* | *0.037* | *25.5* |
| *Bacillus subtilis* | *0.027* | *27* |
| *Bacillus stearothermophilus* | *0.07* | *22* |

*Source:* Toledo, R. T., *AIChE Symp. Ser. 78(218):81, 1982.* Reproduced by permission of the American Institute of Chemical Engineers, © *1982*, AIChE

activation energy or a z value. Eq. (7.21) of Chap. 7 expresses the rate constant for inactivation in terms of the activation energy. The activation energy is negative for reactions that increase in rate with increasing temperature.

$$\frac{k}{k_0} = e^{[(Ea/R)] \cdot [(1/T)-(1/T_0)]} \qquad (8.16)$$

Because $k = 2.303/D$ from Eq. 7.23, Sect. 7.6, the temperature dependence of the D value in terms of the activation energy will be:

$$\frac{D}{D_0} = e^{[(Ea/R)] \cdot [(1/T)-(1/T_0)]} \qquad (8.17)$$

Because $E_a$ is positive when reaction rates increase with temperature, Eq. (8.17) represents a decrease in D value with increasing temperatures.

Thermobacteriologists prefer to use the z value to express the temperature dependence of chemical reactions. The z value may be used on the target F value for microbial inactivation or on the D value to determine required heating times for inactivation at different temperatures. It may also be used on heating times at one temperature, to determine equivalence in lethality at a reference temperature. A semilogarithmic plot of heating time for inactivation against temperature is called the thermal death time plot. Equations based on this linear semilogarithmic plot are called the thermal death time model equations for microbial inactivation at different temperatures. Equation (8.18) is the equation of the thermal death time plot of log (F) against T. This linear plot has a slope of -1/z and passes through points $(T_0, \log(F_0))$ and $(T, \log(F_t))$.

$$\log\frac{F}{F_0} = \frac{T_0 - T}{z} \qquad (8.18)$$

Equations (8.19) and (8.20) are similar to Eq. (8.18) when the plot is made with log(D) vs. T (Eq. (8.19) or with log(t) vs. T (Eq (8.20)). When using Eq (8.20), $t_0$ = the equivalent heating time at $T_0$, and $t_t$ = heating time at T. When $t_T$ = 1 min, $t_0$ = equivalent min. of heating at $T_0$ which has the same lethal effect as 1 min. at T. This defines the lethality factor L, in Eq. (8.21).

$$\log\frac{D}{D_0} = \frac{T_0 - T}{z} \qquad (8.19)$$

$$\log\frac{t_0}{t_T} = -\frac{T_0 - T}{z} \qquad (8.20)$$

$$L = [10]^{T-T_0/Z} \qquad (8.21)$$

The inverse of L is the heating time at T equivalent to 1 min at 250 °F and is the parameter $F_i$ used in thermal process calculations.

$$F_i = [10]^{T_0-T/Z} \qquad (8.22)$$

The use of the z value is not recommended when extrapolating D values over a very large temperature range. Equation (8.17) shows that D will deviate from a linear semilogarithmic plot against temperature when the temperature range between T and $T_0$ is very large.

**Example 8.10** The $F_0$ for 99.999% inactivation of *C. botulinum* type B is 1.1 min. Calculate $F_0$ for 12D inactivation and the F value at 275 °F (135 °C) when z = 18 °F.

**Solution**
99.999% inactivation is equivalent to S = 5. For S = 12; $F_0$ = $SD_0$ = 12(0.22) = 2.64 min. Thus the $D_0$ value is 1.1/5 = 0.22 min. $F_{275}$ can be calculated using Eq. (8.18) or using Eq. (8.19) on the D value to obtain D at 275 °F.

Using Eq. (8.18): $(T_0 - T)/z = -25/18 = -1.389$

$$F_{275} = F_0(10^{-1.389}) = 2.64\ (0.0408)$$
$$= 0.1078\ \text{min}$$

Using Eq. (8.19): $D_{275} = 0.22(10^{-1.389}) = 0.00898$ min

$$F_{275} = S\,D = 12(0.00898) = 0.1077 \text{ min}$$

**Example 8.11** The $D_0$ for PA 3679 is 1.2 min and the z value is 10 °C. Calculate the process time for 8D inactivation of PA 3679 at 140.5 °C using the "thermal death time" model (Eq. 8.18) and the Arrhenius equation (Eq. 8.17). The z value was determined using data on D at 115.5 to 121.1 °C.

**Solution**

Using Eq. (8.18):

$$(T - T_0)/z = (140.5 - 121.1)/10 = -1.94$$

$$F_0 = 8(1.2) = 9.6 \text{ min}; F_{140.5} = 9.6(10^{-1.94})$$
$$= 9.6(0.01148) = 0.11 \text{ min}$$

From Eq. (7.36), (Chap. 7):

$$z = \frac{\ln(10)R}{E_a} T_1 T_2; \quad E_a = \frac{\ln(10)R T_2 T_1}{z}$$

$$T_2 = 121.1 + 273 = 394.1 \text{ K}; T_1$$
$$= 115.5 + 273 = 388.5 \text{ K};$$

$$R = 1.987 \text{ cal}/(\text{gmole} \cdot \text{K})$$

$$E_a = [\ln(10)](1.987)(388.5)(394.1)/10$$
$$= 70,050 \text{ cal/gmole};$$

$$E_a/R = 35,254$$

$$T = 140.5 + 273 = 413.5 \text{ K}; T_0 = 394.1 \text{ K}$$

$$(1/T - 1/T_0) = -0.000119$$

$$F_{140.5} = F_0[e]^{35254(-0.000119)} = 9.6(0.015067)$$
$$= 0.1446 \text{ min}$$

These calculations show that when extrapolating D values to high sterilization temperatures using the z value based on data obtained below 250 °F (121.1 °C), use of the Arrhenius equation will result in a safer process compared with the TDT (thermal death time) model.

## 8.4 Inactivation of Microorganisms and Enzymes in Continuously Flowing Fluids

The methods for calculating the extent of microbial inactivation that results from a heat treatment are the same regardless of the severity of the heat treatment. Mild heat treatments designed to inactivate vegetative cells of microorganisms are called *pasteurization*. Pasteurized foods will be shelf stable if they are high acid (pH ≤ 4.6), but they would require refrigeration and are perishable if the pH is >4.6. *Sterilization* is a high-temperature heat treatment designed to inactivate heat-resistant spores to produce a shelf-stable product when stored at ambient temperature. Pasteurization is used on food products to avoid public health hazards from pathogenic microorganisms. On perishable products, pasteurization reduces the number of viable spoilage microorganisms thereby increasing product shelf life. In a system for continuous pasteurization or sterilization of flowing fluids, the product is heated to a specified temperature and held at that temperature for a specified time.

The processing system used to heat treat fluids to destroy unwanted microorganisms or enzymes is discussed in the section "Continuous Flow Sterilization: Aseptic or Cold Fill." The process is essentially a constant temperature heating process with the residence time in the holding tube considered as the processing time and the temperature at the point of exit from the holding tube as the processing temperature.

### 8.4.1 Time and Temperature Used in the Pasteurization of Fluid Foods

The basis for pasteurization time and temperature for high-acid foods that will be refrigerated post-pasteurization is the resistance of pathogenic microorganisms. Liquid egg and egg product pasteurization and milk pasteurization are regulated by local and federal agencies. Data in Table 8.7 can be used to calculate minimum time/

**Table 8.7** Heat resistance of spoilage and pathogenic microorganisms in pasteurized food products

| Organism | Substrate | D (min) 60 °C | z °C | Reference |
|---|---|---|---|---|
| Low-acid foods (pH ∃ 4.6) | | | | |
| Aeromonas hydrophila | LWE | 0.04 | 5.6 | 1 |
| Mycobacterium tuberculosis | Milk | 14.1 | 4.4 | 2 |
| Listeria monocytogenes | Milk | 2.03 | 5.5 | 3 |
| Listeria monocytogenes | Egg yolk | 1.34 | 9.4 | 4 |
| | Egg white | 2.29 | 9.4 | 4 |
| Salmonella spp. | Egg white | 0.58 | 4.3 | 4 |
| | Egg yolk | 0.91 | 4.3 | 4 |
| E. Coli O157:H7 | Milk | 1.5 | 6.9 | 3 |
| Salmonella spp. | Milk | 0.52 | 6.9 | 3 |
| Listeria monocytogenes | LWE | 1.27 | 7.2 | 5 |
| Salmonella enteritidis | LWE | 0.25 | 6.9 | 5 |
| Staphylococcus aureus | Milk | 0.9 | 9.5 | 3 |
| Aeromonas hydrophila | Milk | 0.18 | 7.7 | 3 |
| Yersinia enterocolitica | Milk | 0.51 | 5.8 | 3 |
| High-acid foods (pH ≤ 4.6) | | | | |
| E. Coli 0157:H7 | Apple juice | 0.33 | 6.9 | 6 |
| Salmonella enteritidis | Model pH 4.4 | 0.24 | 6.9 | 3 |
| Leuconostoc | SSOJ | 0.19 | 7 | 7 |
| Lactobacilli | SSOJ | 1.30 | 7 | 7 |
| Yeast | SSOJ | 1.63 | 6 | 7 |
| Leuconostoc | 42BxOJ | 0.36 | 10 | 7 |
| Lactobacilli | 42BxOJ | 0.091 | 18 | 7 |
| Yeast | 42BxOJ | 1.02 | 9 | 7 |

LWE liquid whole eggs, SSOJ single-strength orange juice, 42BxOJ 42 brix orange juice concentrate
References: 1. Schuman JD, Sheldon BW, and Foegeding PM. (1997) J. Food Prot. 60:231; 2. Keswani J, and Frank JF (1998). J. Food Prot. 61:974; 3. ICMSF 1996 Microorganisms in Foods, Book 5. Blackie Publishers, London; 4. Schuman JD, and Sheldon BW (1997). J. Food Prot. 60:634; 5. Foegeding PM and Leasor SB (1990). J. Food Prot. 53:9; 6. Splitstoesser, DF. (1996) J. Food Prot. 59:226; 7. Murdock et al. (1953) Food Research 18:85

temperature combinations for safety of pasteurized low-acid products. The following data have been compiled from the literature and from discussions with processors of the different food products listed. In general, the main objective of pasteurization is the inactivation of microorganisms of public health significance. For high-acid foods, heat resistance of pathogenic and aciduric spoilage microorganisms is relatively low; therefore pasteurization processes can produce a commercially sterile product.

*Pasteurization processes for high-acid products (pH≤ 4.6) or acidified products*: The following time and temperature combinations are used as a guideline by the food industry for producing shelf-stable high-acid food products. Times and temperature are dependent on the product pH.

Acidified or naturally high-acid products: pH < 4.0 = 1 min at 87.8 °C (190 °F); pH 4.0 = 30 s at 96.1 °C (205 °F); pH 4.1 = 30 s at 100 °C (212 °F); pH 4.2 = 30 s at 102.2 °C (216 °F); pH > 4.2 to 4.5 = 30 s at 118.3 °C (245 °F). If sugar or starch is added to the product, the time/temperature for the next higher pH should be used. For example, if starch or sugar is a component of a product with a pH of 4.1, use a process of 30 s at 102.2 °C (process for pH 4.2 if no sugar or starch is added).

Tomato products: The pasteurization process is based on an equivalent F value at 200 °F (93.3 °C). The following $F_{200F}$ values are generally used: 1 min at pH 4.1, 3 min at pH 4.2, and 5 min at pH 4.3. If temperatures other than 93.3 °C are used, use a z value of 16 °F (8.9 °C) to obtain the equivalent process time from the

$F_{200F}$ values. For products with starch or sugar added at pH 4.3, use an $F_{250F}$ of 0.5 min (z = 16 °F) as a basis for the process.

Pineapple juice: The following $F_{200F}$ values are used: pH > 4.3, F = 10 min; pH between 4.0 and 4.3, F = 5 min. Use a z value of 15 °F (8.33 °C) for other temperatures. Example: Juice with a pH of 4.0 processed at 210 °F (98.9 °C) will need a process time of 1.06 min.

Other juices, flash pasteurization: Flash pasteurization refers to very rapid heating to the processing temperature followed by an appropriate hold, cooling, and aseptic filling. Hold time for peach juice, pH < 4.5 is 30 s at 110 °C; orange juice, 1 min at 90 °C or 15 s at 95 °C; and grapefruit juice, 16 s at 74 °C or 1 s at 85 °C. FDA regulations mandate a minimum pasteurization requirement of 5 log reduction of pathogenic microorganism capable of growing in the product (FDA 1998). For a typical flash pasteurization temperature of 80 °C, products with pH < 4.4 will require 0.2 s to obtain 5 log reduction of *E. coli* 0157:H7 and *Salmonella* spp. Thus, the time and temperature specified above are more than adequate to meet the minimum pasteurization requirement.

*Pasteurization requirements for milk*: (US Grade A Pasteurized milk ordinance—1978 Recommendations of the USPHS/FDA) specify the following temperature and times, 63 °C (145 °F) for 30 min, 72 °C (161 °F) for 15 s, 89 °C (192.2 °F) for 1 s, 90 °C (194 °F) for 0.1 s, and 96 °C (204 °F) for 0.05 s.

*Pasteurization requirements for liquid whole eggs*: the USDA specifies a pasteurization time and temperature of 3.5 min at 60 °C (140 °F).

Comparing the process times and temperature and the heat inactivation kinetics of pathogens in Table 8.7 gives the following log reduction of pathogens. For milk, a 15-s process at 72 °C = 58 log reduction of *M. tuberculosis*, = 18.7 log reduction of *L. monocytogenes*, > 100 log reduction of *E. coli* 0157:H7, and > 100 log reduction of *Salmonella* spp. For liquid whole eggs, a 3.5-min process at 60 °C = 2.75 log reduction of *L. monocytogenes*; = 14 log reduction of *Salmonella enteritidis*.

*Commercial pasteurization processes*: Commercial processors may use processing time and temperature higher than those specified by regulations to obtain long product shelf life. For example, shelf life (defined as time for CFU to reach $10^6$/mL) at 5 °C of milk processed at different time and temperature (from Kessler and Horak, Milchwissenschaft 39:451, 1984) are as follows: 21 days at 74 °C/40 s or 78 °C/15 s, 17 days at 74 °C/15 s or 71 °C/40 s, 16 days at 78 °C/14 s or 85 °C/15 s, and 12 days at 71 °C/15 s.

For liquid whole eggs, TetraPak recommends processing at 70 °C for 90 s to obtain a 3-month shelf life at 5 °C.

For acidified milk (flavored, yogurt-like), von Bockelman (1998) gave the heat treatment conditions to obtain a 3-month shelf life at ambient temperature storage as follows:

pH and temperature (°C) for a 40 s continuous flow process: pH 4.6, 140 °C; pH 4.5, 130 °C; pH 4.4, 120 °C; pH 4.3, 110 °C; pH 4.2, 100 °C; pH 4.1, 98 °C; pH 4.0, 94 °C; and pH 3.9, 90 °C.

pH and temperature (°C) for a 20 min batch process: pH 4.6, 115 °C; pH 4.5, 110 °C; pH 4.4, 105 °C; pH 4.3, 100 °C; pH 4.2, 95 °C; pH 4.1, 90 °C; pH 4.0, 85 °C; and pH 3.9, 75 °C.

### 8.4.2    Microbial Inactivation in Continuously Flowing Fluids

The time and temperature for pasteurization in the previous section are based on the residence time of the fluid in a holding tube following elevation of the product temperature to the designated processing temperature. The residence time is a function of the fluid velocity. Because there is a velocity distribution in fluids flowing through a pipe, residence time of the fluid will vary at different positions in the pipe. The integrated lethality of microorganisms in the fluid leaving a hold tube must be used as the basis for the process.

**Table 8.7a**  Heat treatment conditions for long-life pasteurized products

| pH | Temperature for a 40 s process | Temperature for a 20 min. process |
|----|--------------------------------|-----------------------------------|
| 4.6 | 140 °C | 115 °C |
| 4.5 | 130 | 110 |
| 4.4 | 120 | 105 |
| 4.3 | 110 | 100 |
| 4.2 | 100 | 95 |
| 4.1 | 98 | 90 |
| 4.0 | 94 | 85 |
| 3.9 | 90 | 75 |

The velocity distribution in a fluid flowing through a pipe has been derived for a power law fluid in laminar flow. Equation (5.16), Chap. 5 is:

$$V = \bar{V}\left(\frac{3n+1}{n+1}\right)\left[1 - \left[\frac{r}{R}\right]^{(n+1)n}\right] \quad (8.23)$$

Consider an area element of thickness dr within a tube of radius R: If $N_o$ is the total number of organisms entering the tube, N is the number leaving, $n_o$ is the organisms/unit volume entering, and n is organisms/unit volume leaving:

$$N_0 = n_O\left[2\pi\int_0^R Vrdr\right] \quad (8.24)$$

The expression in brackets in Eq. (8.24) is the volumetric rate of flow, $nR^2V$. Thus Eq. (8.24) becomes:

$$N_0 = nR^2n_0\bar{V} \quad (8.25)$$

The residence time of fluid within the area element is L/V, and the number of survivors, $N = N_o[10^{-L/(VD)}]$. Thus:

$$N = 2\pi n_o\int_0^R Vr[10]^{-L/(VD)}dr \quad (8.26)$$

In Eq. (8.23), let: A = (3n + 1)/(n + 1); B = (n + 1)/n; y = r/R. Equation (8.23) becomes:

$$V = A\left(1 - y^B\right)\bar{V} \quad (8.27)$$

The ratio $L/(\bar{V} \cdot D)$ is the lethality, $S_v$, based on the average velocity. Substituting Eq. (8.27) into Eq. (8.26), and $S_v = L/(\bar{V} \cdot D)$; r = yR; dr = Rdy. The integrated lethality, Si will be log $N_0/N$:

$$N = 2\pi n0\int_0^1 \bar{V}(A)\left(1 - y^B\right)$$
$$\times[10]^{-S_v/[A(1-y^B)]}R^2y \, dy \quad (8.28)$$

Dividing Eq. (8.25) by Eq. (8.28):

$$\frac{N_0}{N} = \frac{\pi R^2 n_0\bar{V}}{2\pi n_0\int_0^1 \bar{V}(A)\left(1 - y^B\right)[10]^{-Sv/[A(1-y^B)]}R^2ydy}$$

Canceling $\pi R^2n_0\bar{V}$, taking the logarithm of both sides, and making the logarithm of the denominator of the right-hand side negative to bring it to the numerator:

$$\log\left(\frac{N_0}{N}\right) = -\log\left[\int_0^1 \left[2A\left(1 - y^B\right)[10]^{-S_v/[A(1-y^B)]}y \, dy\right]\right] \quad (8.29)$$

log($N_0/N$) in Eq. (8.29) is the integrated lethality, $S_i$.

Equation (8.29) can be evaluated by graphical integration. The integrated sterility in the holding tube is independent of the size of the tube and the average velocity if the length and average velocity are expressed as the sterilization value based on the average velocity, $S_v$.

- Thus Eq. (8.29) can be solved, and the solution will yield a generalized table for the integrated lethality of heat in the holding tube when fluid

**Table 8.8** Integrated lethality in the holding tube of a continuous sterilization system for fluids in laminar flow

Integrated lethality, $S_i$ when fluid flow behavior index = n

| $S_v$ | 0.4 | 0.5 | 0.6 | 0.7 | 0.8 | 0.9 | 1.0 |
|---|---|---|---|---|---|---|---|
| 0.1 | 0.093 | 0.093 | 0.092 | 0.091 | 0.091 | 0.091 | 0.091 |
| 0.5 | 0.426 | 0.419 | 0.414 | 0.409 | 0.406 | 0.401 | 0.401 |
| 1 | 0.809 | 0.792 | 0.779 | 0.768 | 0.759 | 0.747 | 0.747 |
| 2 | 1.53 | 1.49 | 1.46 | 1.44 | 1.42 | 1.40 | 1.38 |
| 4 | 2.92 | 2.82 | 2.75 | 2.69 | 2.64 | 2.59 | 2.56 |
| 6 | 4.27 | 4.11 | 3.99 | 3.89 | 3.81 | 3.74 | 3.68 |
| 8 | 5.60 | 5.38 | 5.20 | 5.06 | 4.95 | 4.86 | 4.78 |
| 10 | 6.92 | 6.63 | 6.41 | 6.23 | 6.08 | 5.96 | 5.86 |
| 12 | 8.23 | 7.88 | 7.60 | 7.38 | 7.20 | 7.05 | 6.92 |
| 14 | 9.54 | 9.12 | 8.79 | 8.52 | 8.31 | 8.13 | 7.98 |
| 16 | 10.84 | 10.35 | 9.97 | 9.66 | 9.41 | 9.20 | 9.03 |
| 18 | 12.14 | 11.58 | 11.14 | 10.79 | 10.51 | 10.27 | 10.07 |
| 20 | 13.44 | 12.81 | 12.32 | 11.93 | 11.60 | 11.34 | 11.11 |
| 22 | 14.73 | 14.03 | 13.49 | 13.05 | 12.69 | 12.40 | 12.15 |
| 24 | 16.03 | 15.26 | 14.66 | 14.18 | 13.79 | 13.46 | 13.18 |
| 26 | 17.32 | 16.48 | 15.83 | 15.30 | 14.87 | 14.52 | 14.22 |

$S_v$ = sterilization value = decimal reductions based on the average velocity = $L/(D \cdot \overline{V})$ Where L = length of hold tube and D = the D value of microbiological inactivation

is flowing through that tube in laminar flow. Values of $S_i$ at different values of n and $S_v$ obtained by graphical integration of Eq. (8.29) are shown in Table 8.8. The data in Table 8.8 shows that $S_i/S_v$ is approximately $\overline{V}/V_{max}$. Thus, use of the maximum velocity to calculate lethality produces results much closer to the integrated lethality than if the average velocity is used.

**Example 8.12** A fluid food product with a viscosity of 5 cP and a density of 1009 kg/m$^3$ is to be pasteurized in a continuous system that heats the food to 85 °C followed by holding in a 1.5-in. sanitary pipe from which it leaves at 82.2 °C. The process should give 12 decimal reduction of *Staphylococcus aureus*, which has a $D_{82.2°C}$ of 0.0063 min. Calculate the length of the holding tube if the flow rate is 19 L/min.

**Solution**

From Chap. 5, Table 5.2, the inside diameter of a 1.5-in. sanitary pipe is 0.03561 m.

$$\overline{V} = \frac{19L}{min} \frac{1m^3}{1000\,L} \frac{1\,min}{60\,s} \frac{1}{\pi[0.5(0.03561)]^2 m^2}$$

$\overline{V}$ = 0.318 m/s. Re = (0.03561)(0.318)(1009)/5 (0.001) = 2285. Flow is in the transition zone from laminar to turbulent flow. For safety, assume flow is laminar, and $V_{max}$ = 2 $\overline{V}$ = (0.318) = 0.636 m/s.

Using Eq. (8.14): The F value is the required process time that must equal the residence time in the tube for the fastest flowing particle. F = $t_{min}$ = S · D = 12(0.0063) = 0.0756 min.

Using Eq. (8.4):

$$L = 0.0756\,min(60s/min)(0.636\,m/s)$$
$$= 2.88\,m$$

Length based on the integrated lethality for a Newtonian fluid: From Table 8.8, by interpolation, to obtain $S_i$ = 12, n = 1, $S_v$ = 22 − 0.17 (12.17 − 11.13)/2 = 21.91. Because $S_v$ = L/(V · D), L = 21.91(0.318)(0.0063)(60) = 2.63 m.

The integrated lethality gives a length much closer to that based on the maximum velocity compared to results using the average velocity.

At the Reynolds number in the range 3000 to 5000, the maximum velocity cannot be obtained using Eq. (8.7). Thus, the safest approach to determination of holding tube length in this range of Reynolds number will be to assume laminar flow.

**Example 8.13** An ice cream mix having a viscosity of 70 cP and a density of 1015 kg/m$^3$ is being canned aseptically in a system which uses a 100-ft-long, 1.0-in. sanitary pipe for a holding tube. Flow rate through the system is 5 gal/min. The fluid temperature at the exit from the holding tube is 285 °F (140.6 °C).

Calculate (a) the sterilizing value of the process for PA 3679 (D$_0$ = 1.83 min; z = 24 °F).

**Solution**

From Table 5.2, the dimensions of a 1.0 in. sanitary pipe are D = 0.02291 m, R = 0.01146 m.

Solving for the average velocity:

$$\bar{V} = \frac{5 \text{ gal}}{\text{min}} \frac{3.78541 \times 10^{-3} \text{m}^3}{\text{gal}}$$

$$\times \frac{1 \text{ min}}{60 \text{s}} \frac{1}{\pi(0.01146)^2 \text{ m}^2}$$

$\bar{V} = 0.765$ m/s. Re $= 0.02291(0.765)(1015)/70$ $(0.001) = 254$. Flow is laminar. L = 100 ft(1 m/ 3.281 ft) = 30.48 m.

(a)  Solving for V$_{max}$:

$$V_{max} = 2\bar{V} = 2(0.785) = 1.57 \text{ m/s}$$

The residence time of the fastest flowing particle is:

$$t = L/V_{max} = 19.41 \text{ s}.$$

The process occurs at 285 °F; therefore, the number of survivors from the process is calculated using the D value at 285 °F for D in Eq. (8.11). The D value at 285 °F is determined using Eq. (8.19):

$$\begin{aligned}
\log D &= [\log D_0 + (T_o - T)]/z \\
&= [\log[(1.83)(60)] + (250 - 285)]/24 \\
&= 2.0406 - 1.458 \\
D &= 10^{0.5826} \\
&= 3.8247 \text{ s at } 285°\text{F}
\end{aligned}$$

Using Eq. (8.11): S = 19.41/3.8247 = 5.075. The process will result in at least 5.075 decimal reduction of PA 3679.

The number of decimal reductions based on the integrated lethality is:

$$\begin{aligned}
S_v &= L/(\bar{V} \cdot D) = 30.48/[(0.765)(3.8247)] \\
&= 10.42
\end{aligned}$$

From Table 8.8, S$_v$ = 10.42, n = 1, by interpolation:

$$S_i = 5.86 + 0.42(6.93 - 5.86)/2 = 6.08$$

The integrated lethality is 6.08 decimal reductions of PA 3679.

## 8.4.3   Nutrient Degradation

Nutrient degradation can be calculated the same way as for microbial inactivation. Equations (8.24), (8.25), (8.26), (8.27), (8.28), and (8.29) are also applicable for loss of nutrients if the appropriate kinetic parameters of D and z values are used.

Nutrient retention with increasing processing temperature can be derived by simultaneously solving the rate equations for microbial inactivation and nutrient degradation. Using the subscript c and m to signify parameters for nutrient degradation and microbial inactivation, respectively, Eqs. (8.11) and (8.19) may be combined to determine the heating time at various temperatures for microbial inactivation.

$$t = D_{mo}\left[\log\left(\frac{N_0}{N}\right)\right][10]^{(T0-T)/z_m} \quad (8.30)$$

A similar(expression) can be formulated for nutrient degradation.

$$\log\left(\frac{C}{C_0}\right) = -\frac{t}{D_{co}[10]^{(T_0-T)/z_c}} \quad (8.31)$$

Substituting S$_v$ for [log (N$_0$/N)] and combining Eqs. (8.30) and (8.31):

$$\log\left(\frac{C}{C_0}\right) = -\left[\frac{D_{mo}}{D_{co}}S_v\right][10]^{(T_0-T)(1/z_m-1/z_c)}$$

$$(8.32)$$

Equation (8.32) is based on constant temperature processes and is applicable for estimating nutrient retention in holding tubes of aseptic processing systems. The derivation is based on plug flow of fluid at the average velocity through the tube, and adjustments will have to be made to obtain the integrated lethality and nutrient degradation. Table 8.8 can be used to determine $S_v$ needed to obtain the microbial lethality required expressed as $S_i$. Log($C/C_0$) will be the $S_v$ for nutrient degradation which can be converted to $S_i$ using Table 8.8.

**Example 8.14** The data for thiamin inactivation at 95 °C to 110 °C (from Morgan et al. J. Food Sci. 51:348, 1986) in milk show a $D_{100°C}$ of $3 \times 10^4$ s and a z value of 28.4 °C. Chocolate milk with a flow behavior index of 0.85 and a consistency index of 0.06 Pa s$^n$ is to be sterilized at 145 °C. The density is 1006 kg/m$^3$. If the rate of flow is 40 L/min, and the holding tube is 1.5-in. sanitary pipe, calculate the holding tube length necessary to give an integrated lethality of 7 decimal reductions of an organism having a D$_o$ value of 0.5 min and a z value of 10 °C. Assume the z value of the organism was determined in the temperature range that included 145 °C. Calculate the retention of thiamin after this process.

**Solution**
Because the z value for thiamin was determined at low temperatures, Eq. (8.17) will be used to extrapolate the D value to 145 °C. From Eq. (7.25), Chap. 7:

$$E_a = \frac{[\log(10)]R}{z}T_1T_2$$
$$= \frac{[\log(10)](1.987)(368)(383)}{28.4}$$

$$E_a = 22,706; E_a/R = 11,427;$$

$$T = 145 + 273 = 418\,K; Tr = 100 + 273 = 373\,K.$$

Using Eq. (8.17) and the subscript c to represent chemical degradation:

$$[1/T - 1/Tr] = (1/418 - 1/373)$$
$$= -0.000289D_c = 3 \times 10^4(e)^{11427(-0.000289)}$$
$$= 1109\,s$$

Equation (8.19) is used to determine $D_m$ at 145 °C.

$$[(T_0 - T)/z] = [(121.1 - 145)/10] =$$
$$-2.39D_m$$
$$= 0.5(10)^{-2.39} = 0.002037\,min$$
$$= 0.1222\,s$$

Equation (8.32) will be used, but because $D_m$ and $D_c$ are already calculated at 145 °C, T = $T_0$, the exponential term $10^{(T_0 - T)(1/z_m - 1/z_c)} = 1$ and:

$$\log\left(\frac{C}{C_0}\right) = -\frac{D_m}{D_c}S_{vm} \qquad (8.30a)$$

$S_{vm}$ is the sterilizing value for the microorganisms based on the average velocity.

The Reynolds number (Eq. 5.50, Chapter 5) is:

$$Re = \frac{8V^{2-n}R^n\rho}{K[3 + 1/n]^n}$$

From Table 5.2, R = 0.5(0.03561) = 0.017805 m

$$\bar{V} = \left[\frac{m^3}{s}\right] \cdot \frac{1}{Area\ (m^2)}$$

$$m^3/s = 40(0.001)/60 = (6.667) \cdot (10)^{-4}$$

$$Area = \pi \cdot (0.017805)^2 = (9.9594) \cdot (10)^{-4}\,m^2$$

$$\bar{V} = \frac{(6.667) \cdot (10)^{-4}}{(9.9594) \cdot (10)^{-4}} = 0.669\,m/s$$

$$\frac{(3) \cdot (n) + 1}{n} = \frac{2.55 + 1}{0.85} = 4.1764$$

Reynolds Number = Re

$$= \frac{8(0.669)^{1.15}(0.017805)^{0.85}(1006)}{0.06(4.1764)^{0.85}}$$

$$Re = 817$$

Flow is laminar, and the integrated lethality can be evaluated using Table 8.8. The integrated sterilizing value for the microorganism, $S_{mi}$, is the desired outcome of the process, and the value is 7.0. The sterilizing value based on the average velocity must be determined to obtain the length of the holding tube.

From Table 8.8, $S_{vm}$ is obtained for $S_i = 7$ and $n = 0.85$, by interpolation:

At $S_i = 7, n = 0.8, S_v = 10 + [2/1.11](0.91)$
$$= 11.64$$

$$n = 0.9, S_v = 10 + [2/1.1](1.04) = 11.89$$

Solving for $S_v$ at $n = 0.85$, $S_v = 11.64 + [0.25/0.1](0.05) = 11.77$

The sterilizing value based on the average velocity $= 11.77$ and the $D_m$ value at 145 C $= 0.1222$ s. Average retention time at 145 C $= 11.77(0.1222) = 1.438$ s. The length of the holding tube, $L = \bar{V} \cdot t = 0.669(1.438) = 0.962$ m.

The thiamine retention will be calculated using Eq. (8.30a): $D_c = 1109$ s, $D_m = 0.1222$ s, and $S_{vm} = (11.77.)$

$$\log\left(\frac{C}{C_0}\right) = -\frac{0.1222}{1109}11.77$$
$$= -0.001317$$

$$C/C_0 = (10)^{-0.001317}$$
$$= 0.997 \text{ or } 99.7\% \text{ retention.}$$

**Example 8.15** Tomato paste ($n = 0.5$, $k = 7.9$ Pa. s$^n$, $\rho = 1085$ kg/m$^3$) is sterilized at 95 °C using a holding tube with an inside diameter of 0.03561 m. The system operates at the rate of 50 L/min. Each package to be filled with the sterilized product contains 200 L, and the probability of spoilage to be expected from the process is 1 in 10,000 from spores of *Bacillus polymyxa*, which has $D_{80E}C = 0.5$ min and $z = 9$ °C

The unprocessed paste contains 4 spores/mL. Calculate the length of the holding tube necessary to achieve an integrated sterility, which is the desired spoilage probability. Calculate the extent of nonenzymatic browning that occurs during this process expressed as percentage increase over brown color before processing. Assume the $D_0$ value of nonenzymatic browning is 125 min at 80 °C and the z value is 16 °C.

**Solution**

Determine if flow is laminar or turbulent.

$$\bar{V} = \frac{50\,L}{min}\frac{1\,min}{60\,s}\frac{m^3}{1000\,L}\frac{1}{\pi[(0.5)(0.03561)]^2\,m^2}$$
$$= 0.837\,m/s$$

$$Re = \frac{8(\bar{V})^{2-n}R^n\rho}{K[3+1/n]^n}\left[\frac{(3n+1)}{n}\right]^n = \left[\frac{2.5}{0.5}\right]^{0.5} = 2.34$$

$$Re = \frac{8(0.837)^{1.5}(0.017805)^{0.5}(1085)}{7.9(2.24)} = 50$$

Flow is laminar. For each 200-L package, $N_0 = 80,000$. $N = 1/10,000$. $\text{Log}(N_0/N) = 9.9$. This is to be evaluated on the integrated sterility; therefore, $S_i = 9.9$. From Table 8.8, to obtain $S_i = 9$, for $n = 0.5$, $S_v$ is obtained by interpolation:

$$S_v = 14 + (2/1.24)(9.9 - 9.12) = 15.26$$
$$= L/(V - D)$$

$$D_{95} = D_{80}[10]^{(80-95)/9} = 0.0215(0.5)$$
$$= 0.01077 \text{ min}$$

Substituting and solving for L:

$$L = S_v$$
$$\bar{V}D = 15.26(0.837\,m/s)(0.01077\,min)(60\,s/min)$$
$$= 8.25\,m.$$

Equation (8.32) will be used for determining the increase in brown color that results from the process. Because the reaction involves the appearance of a brown color, the negative sign in Eq. (8.32) is changed to positive. The reference temperature, $T_0 = 80$ °C, $T = 95$ °C, $z_m = 9$, and $z_c = 16$.

$$\log\left(\frac{C}{C_0}\right) = \frac{0.5}{125}(15.26)[10]^{(80-95)(1/9-1/16)}$$
$$= 0.011$$

At very small values of S, $S_v = S_i$, therefore integrated $C/C_0$ at $S = 0.011 = 100.011 = 1.026$. An increase of 2.6% in the intensity of browning will be expected in the process. Calculations of integrated lethality for fluids in turbulent flow are not possible without an expression for velocity distribution within the tube. Currently available expressions for velocity distributions in turbulent flow are too unwieldy for a generalized treatment of the integrated sterility as was done with Eq. (8.27). A possible approach to determination of an integrated lethality will be to determine experimentally the fluid residence time distribution and express this as a distribution function for velocity within the tube relative to the average velocity. In the absence of velocity distribution functions, lethality in the holding tube of continuous sterilization systems in turbulent flow must be calculated using the maximum velocity (Eq. (8.7), Re > 5000). Examples shown above for nutrient degradation during high-temperature, short-time sterilization demonstrate very low values for log $(C/C_0)$ such that the integrated value approaches the value based on the average velocity. Thus, log$(C/C_0)$ can be based on the average velocity. On the other hand, holding tube length calculations must be based on the integrated lethality.

## 8.4.4   High-Pressure Pasteurization

The momentum toward adoption by the food industry of high pressure as a means of food preservation has increased because reliable equipment is now available for applying the process, and adequate data have been accumulated on inactivation of pathogenic and spoilage microorganisms to increase user's confidence in the process. High-pressure processes have the advantage of inactivating microorganisms with minimal exposure of the food product to heat. Thus, the potential for preserving the fresh-like quality of the preserved food is very good. High-pressure food preservation may be applied as a pasteurization or a sterilization process. High-pressure pasteurization of high-acid products may produce a product that is stable during ambient storage. The use of the process on acid products may extend refrigerated shelf-life post-treatment. High-pressure sterilization is considered by the US Food and Drug Administration as a nonconventional process; therefore careful scrutiny and proof of safety must be provided

before the process can be approved for commercial use. The units of pressure used in high-pressure processing are usually expressed in megapascals (MPa). Some reports express the pressure in Bars. A Bar is a technical atmosphere and is equivalent to 14.5 $lb_f/in^2$ (psi) or 0.1 MPa.

### 8.4.4.1   High-Pressure Systems

High-pressure systems may be classified as batch or continuous processes. The batch process is also known as "Isostatic High Pressure" or "High Hydrostatic Pressure" (HHP). The HHP system consists of a pressure vessel that holds the product to be treated; a pressure intensifier that raises the pressure of the pressurizing fluid to the target operating pressure and pumps this high-pressure fluid into the processing vessel, a hydraulic pump to operate the intensifier, and appropriate controls for pressurization, temperature control, and depressurization. Because time for loading and sealing the pressure vessel then unsealing and unloading postprocess are all part of the batch cycle, reducing these elements of the batch cycle time is just as important as reducing the high-pressure exposure time itself. HHP pressure vessels are equipped with seals that easily engage and disengage to shorten the time for loading and unloading the pressure vessel. For large systems, a product carrier is used to enable loading and unloading a full load of product rapidly. Product treated by HHP may be prepackaged and may contain large particulate material. Water is usually used as the pressurizing fluid, hence the term "hydrostatic." Because temperature is as important as pressure in effectively inactivating microorganisms, it is important that the system incorporate temperature monitors that can measure actual temperature inside a product within the pressure vessel.

A continuous high-pressure system can be used only on homogeneous liquids. The system consists of a feed pump, a pressure intensifier, a hydraulic system to operate the intensifier pistons, and a throttling device to reduce the pressure from high pressure to atmospheric pressure. Because the throttling device also generates extremely high shear rates on the fluid and because particle size reduction or homogenization also occurs

**Pressure Intensifier  a=low pressure fluid inlet; b=high pressure fluid exit; c=exit valve; d=inlet valve; e=piston; f = hydraulic fluid drive inlet and return**

**High hydrostatic pressure system.**

**Fig. 8.11** Pressure intensifier system for generating high pressure in high-pressure processing systems and photograph of a high hydrostatic pressure processing system.

High hydrostatic pressure system reprinted from Food Technology. Used with permission

during depressurization, the continuous high-pressure system may be considered a "high-pressure homogenization" system. High-pressure piston-type homogenizers have been used for this purpose in the past, but microbial reduction has been inadequate for successful pasteurization because of the pressure cycling inherent in piston-type pumps. Recent designs of continuous flow intensifiers incorporating two separate pistons operated with programmable logic controllers to time the cycling between the two pistons have greatly reduced the amplitude of pressure oscillation. Thus, a continuous flow high-pressure system can now be used for pasteurization of fluid food products.

Figure 8.11 shows a pressure intensifier system for continuous flow pasteurization of liquids and a HHP system. An important feature of a HHP system is ease of loading and unloading. In Fig. 8.11, the pressure chamber can be lowered and moved out of the cover to expose the chamber for loading and unloading. Pressure intensifiers have the same basic features whether used to directly pasteurize fluids or used to generate the pressurizing fluid for a HHP processing vessel. The high-pressure intensifier consists of in-line dual cylinders where a drive piston is directly

connected to the intensifier piston. The ratio of the area of the drive piston to that of the intensifier piston determines the pressure intensification. Standard hydraulic pumps that generate hydraulic fluid pressure of 20 MPa can be used to produce high-pressure fluid at 400 MPa if the drive and intensifier piston diameter ratio is 20:1. Each intensifier cylinder must have an inlet valve that opens to admit liquid into the intensifier and an exit valve that opens when the intensifier piston is discharging fluid at high pressure. The two valves alternately open and close, while the inlet valve is open, the exit valve is closed and vice versa. The opening and closing of the valves are timed to the position of the drive piston. When the drive piston is at the apex of its forward travel, the exit valve closes and the inlet valve opens. The hydraulic fluid in the drive piston is released so that the drive piston is pushed back to the farthermost backward position by the pressure of the fluid entering the intensifier. At this point, the inlet valve closes and the exit valve opens, and high-pressure hydraulic fluid is fed into the drive piston, thus allowing the intensifier piston to advance to deliver the high-pressure fluid. It is very important that the inlet and exit valves maintain positive closure when closed and that the timing of the opening and closing be properly set. A programmable logic controller may be used to control the timing of opening and closing of the valves, or a mechanical/pneumatic system may be used as a control system.

### 8.4.4.2 High-Pressure Pasteurization

Data in Table 8.9 can be used to determine the time and temperature needed for successful HHP pasteurization of food products. Microbial inactivation at high pressure has been shown to be first order, and inactivation rate can be expressed as a D value as with thermal processes. Processes at 50 °C will take less time than those at ambient temperature. Minimum pressure for inactivating microorganisms by HHP is 300 MPa. Increasing pressure to 700 MPa rapidly decreases the D value, but a point of diminishing returns occurs at pressures greater than 700 MPa. Microbial

inactivation should be at least 5D for pathogenic microorganisms and 8D for spoilage microorganisms. Using this criteria, HHP processing of meat products at 500 MPa and 50 °C will require 15 min to obtain 5D reduction of *Staphylococcus aureus*. *Staphylococcus aureus* in milk will require 12.5 min at 50 °C, and 500 MPa orange juice will require 7.8 min at 500 MPa and 37 °C, while apple juice will require 2.3 min at 40 °C and 450 MPa to obtain a 8 log reduction of *Saccharomyces cerevisiae*.

Microbial inactivation kinetics in continuous flow pasteurization does not follow the same trend as HHP. Hold time at high pressure prior to pressure reduction is only in the order of 0.5 s, yet substantial microbial inactivation results. Pressurizing to 242 MPa and releasing the pressure to 1 atm permit very rapid flow of fluid across the pressure reducing valve resulting in a

**Table 8.9** Resistance of microorganisms important in high-pressure pasteurization processes

| Organism | Substrate | D | P | T | References |
|---|---|---|---|---|---|
| *Solid medium* | | | | | |
| *Salmonella enteritidis* | Meat | 3 | 450 | N/A | 1 |
| *Salmonella typhimurium* | Meat | 1.48 | 414 | 25 | 2 |
| | Meat | 0.6 | 345 | 50 | 3 |
| *E. coli* | Meat | 2.5 | 400 | NA | 4 |
| *S. aureus* | Meat | 3 | 500 | 50 | 4 |
| *L. monocytogenes* | Meat | 2.17 | 414 | 25 | 2 |
| *L. monocytogenes* | Pork | 1.89–4.17 | 414 | 25 | 5 |
| *L. monocytogenes* | Pork | 0.37–0.63 | 414 | 50 | 3 |
| *C. botulinum* type E (beluga) | Crab meat | 3.38 | 758 | 35 | 6 |
| *C. botulinum* type E (Alaska) | Crab meat | 1.76 | 827 | 35 | 6 |
| *Liquid medium* | | | | | |
| *Salmonella typhimurium* | Milk | 3.0 | 350 | N/A | 1 |
| *E. coli* | Milk | 1.0 | 400 | 50 | 7 |
| *E. coli* 0157H7 | Milk | 3.0 | 400 | 50 | 4 |
| *S. aureus* | Milk | 2.5 | 500 | 50 | 4 |
| *L. monocytogenes* | Milk | 3.0 | 375 | N/A | 1 |
| *L. innocua* | Eggs | 3.0 | 450 | 20 | 8 |
| *C. botulinum* type E (Alaska) | Buffer | 2.64 | 827 | 35 | 6 |
| *Saccharomyces cerevisiae* | Orange juice | 0.97 | 400 | 37 | 9 |
| *Saccharomyces cerevisiae* | Apple juice | 0.28 | 450 | 40 | 9 |

*References:* 1. Patterson ME, Quinn M, Simpson R, and Gilmour A. (1955) J. Food Prot. 58:524; 2. Ananth V, Dickson JS, Olson DG, and Murano EA (1998) J. Food Prot. 61:1649; 3. Kalchayanand N, Sikes A, Dunne CP, and Ray B. (1998) J. Food Prot. 61:425; 4. Patterson MF and Kilpatrick DJ. (1998) J. Food Prot. 61:432; 5. Murano EA, Murano PS, Brennan RE, Shenoy K and Moriera R. (1999). J. Food Prot. 62:480; 6. Reddy NR, Solomon HM, Fingerhut G, Balasubramanian VM, and Rodelhamel EJ. (1999) NCFST, Sumit-Argo, IL. 7. Gervilla R, Capellas M, Farragut V and Guamis B. (1997) J. Food Prot. 60:33; 8. Ponce E, Pla R, Mor-Mur M, Gervilla R and Guamis B. (1998) J. Food prot. 61:119; 9. Zook CD, Parish ME, Braddock RJ and Balaban MO. (1999) J. Food Sci. 64:533

breakdown of microbial cell walls killing the microorganisms. Continuous flow high-pressure pasteurization is very similar to the action of homogenizers to break down microbial cells to release cellular proteins. However, the constant pressure and controlled flow of fluid through the throttling valve results in the exposure of all suspended cells to the same high shear rates thus resulting in effective pasteurization. Cells of *Saccharomyces cerevisiae* in orange juice are reduced 8 log at 242 MPa. Cells of *Listeria innocua* are reduced 5 log. Cells of *Lactobacillus sake* are reduced 5 log by this treatment. Orange juice processed by continuous flow pasteurization at 242 MPa retained the fresh-squeezed orange juice flavor and was stable both microbiologically and biochemically when stored for 90 days at 4 °C. Temperature rise during continuous flow pasteurization must be minimized. Temperature increases instantaneously on reduction of pressure due to the conversion of potential energy at the high pressure into heat. Temperature rise is about 24 °C/100 MPa pressure. To minimize exposure to high temperature, feed temperature should be kept at the lowest temperature above the freezing point, and liquid must be cooled rapidly through a heat exchanger after exiting the throttling valve. This process is mild enough to permit the inactivation of 6 log of *Listeria innocua* in liquid whole eggs without coagulating the protein.

### 8.4.4.3  High-Pressure Sterilization

Because high pressure alone is inadequate to inactivate spores, it is necessary to raise the temperature during the high-pressure treatment to above 121.1 °C. For HHP, the temperature rise that occurs with pressurization can be utilized to advantage in sterilization processes. Temperature rise on subjecting water to HHP ranges from 2.8°C to 4.4 °C/100 MPa pressure. The higher the initial temperature, the larger the temperature rise with increase in pressure. Foods with low density and specific heat such as oils will have higher temperature rise than water, while most fluid foods that are high in moisture content

such as fruit juices have similar temperature rise as water. A food product with a temperature rise of 3 °C/100 MPa pressure will exhibit a temperature rise of 21 °C going from one atmosphere to 700 MPa. Thus, if the product initial temperature prior to pressurization is 100 °C, temperature after pressurization will be 121 °C, a temperature that is lethal to spore-forming microorganisms. Of interest is inactivation of heat-tolerant spores of *Clostridium botulinum*. Type 62A in pH 7 buffer at 75°C and 689 MPa has a D value of 10.59 min. Because sterilization will require at least 12D of *C. botulinum*, a 132-min exposure time will be required. There are currently no data on high-pressure inactivation of spores at T > 108 °C. An advantage of HHP sterilization is that upon release in pressure, the temperature instantaneously drops to the temperature at the start of pressurization. The same principles may be used in continuous flow sterilization. If a fluid is heated to 80 °C while under pressure at 241 MPa, release of that pressure to atmospheric pressure will result in a temperature rise of about 55 °C resulting in an exit fluid temperature of 135 °C after pressure release. At this temperature, the D value for *C. botulinum* is about 0.7 s; therefore a 9-s hold will be adequate to achieve 12D inactivation of *C. botulinum*.

Sterilization by high pressure is still not an approved process for low-acid products by US regulatory agencies.

### 8.4.5  Sterilization of Fluids Containing Discreet Particulates by Heat

Discreet particulates within a flowing fluid will be heated by heat transfer from the suspending fluid. Thus, heat transfer coefficients between the particle and the fluid play a significant role in the rate of heating. Simplified equations for heat transfer will not be applicable because the fluid temperature is not constant as the mixture passes through the heater, and temperature of fluid in an unheated holding tube may not be constant because of heat exchange between the fluid and the suspended particles. Taking a conservative approach of ignoring the heat absorbed by the particles in the heaters can result in a significant overprocessing, particularly if the suspended particles have less than 0.5 cm as the thickness of the dimension with the largest area for heat transfer. Finite element or finite difference methods for solving the heat transfer equations with appropriate substitutions for changes in the boundary conditions when they occur are the only correct method to determine the lethal effect of heat in the holding tube. Residence time distribution of particles must also be considered, and as in the case of fluids in turbulent flow, the use of a probability distribution function for the residence time in the finite difference or finite element methods will allow calculation of an integrated sterility.

A discussion of the finite difference methods for evaluating heat transfer is beyond the scope of this textbook.

## 8.5   Sterilizing Value of Processes Expressed as $F_0$

The sterilizing value of a process expressed as the number of decimal reduction of a specific biological entity has been discussed in the section "Selection of Inoculation Levels in Inoculated Packs." When comparing various processes for their lethal effect, it is sometimes more convenient to express the lethality as an equivalent time of processing at a reference temperature. The term, the $F_0$, is a reference process lethality expressed as an equivalent time of processing at 121.1 °C calculated using a "z" value of 10 °C (18 °F). If the z value used in the determination of F is other than 10 °C, the z value is indicated as a superscript, $F^z_0$.

For constant temperature processes, the $F_0$ is obtained by calculating L in Eq. (8.21) and multiplying L by the heating time at T. Thus, $F_0 = L \cdot t$.

For processes where product is subjected to a changing temperature, such as the heating or

cooling stage in a canning process, a lethality is calculated over the length of the process. The process is separated into small time increments, $\Delta t$. The average temperature at each time increment is used to calculate L using Eq. (8.21). The $F_0$ will be $3L_T \Delta t$. The z value of E used in Eq. (8.21) to calculate L is 10 °C to determine $F_0$. If the $F_0$ value is to be used later to express lethality as the number of decimal reduction of a particular biological entity, the appropriate z value for that entity has to be used to calculate $F^z_0$.

## 8.6   Thermal Process Calculations for Canned Foods

When sterilizing foods contained in sealed containers, the internal temperature changes with time of heating. Lethality of the heating process may be calculated using the "general method," which is a graphical integration of the lethality-time curve, or by "formula methods," which utilize previously calculated tabular values of parameters in an equation for the required process time or process lethality. Problems in thermal process calculation can either be (I) the determination of process time and temperature to achieve a designed lethality or (II) the evaluation of a process time and temperature. These are referred to by some authors as Type I or Type II problems. Fundamental in the evaluation of thermal process schedules is heat transfer data, an equation or experimental data for temperature in the container as a function of time. Lethality may be expressed as the value achieved in a single point (i.e., the slowest heating point in the container), or it may be expressed as an integrated lethality. For microbial inactivation, where microbial numbers are nil at regions within the container nearest the wall, lethality at a single point is adequate and results in the safest process schedule or a most conservative estimate of the probability of spoilage. However, when evaluating quality factor degradation, an integrated lethality must be determined because a finite level of the factor in question exists at all points in the container.

### 8.6.1  The General Method

Process lethality is calculated by graphical integration of the lethality value (Eq. 8.21) using time-temperature data for the process.

$$F_0{}^z = \int\limits_0^t L_t dt$$

Equation (8.14) may also be used to determine process lethality. However, because D is not constant, the sterilizing value expressed as the number of decimal reductions of microorganisms will be integrated over the process time, using the value of D at various temperatures in the process.

$$S = \int\limits_0^t \frac{dt}{D_t}$$

If a process schedule is to be determined, the heating and cooling curves are used to determine the lethality curve, which is then graphically integrated to obtain either $F_0$ or S. The process lethality must equal the specified values for $F_0$ or S for the process to be adequate. If the lethality value differs from the specified, the heating time is scaled back, a cooling curve parallel to the original is drawn from the scaled back heating time, and the area is recalculated. The process is repeated until the specified and calculated values matches. Evaluation of the process lethality is done directly on the time-temperature data. Simpson's rule may be used for integration. "Graphical integration" following Simpson's may be used in thermal process determination by the general method as follows: Select time increments δt such that at the end of process time t, t/δt will be an even number. Using i as the increment index, with i = 0 at t = 0, i = 1 at t = δt, i = 2 at t = 2 δt, i = 3 at t = 3δt... and so forth.

$$A = \left(\frac{\delta t}{3}\right)[L_0 + 4L_1 + 2L_2 + +4L_3$$
$$+ 2L_4 + \cdots\cdots 2L_{i-2} + 4L_{i-1} + L_t]$$

The area under the cooling curve may be evaluated separately from that under the heating curve.

**Example 8.16** The following data represent the temperature at the slowest heating point in a canned food processed at a retort temperature of 250 °F. Calculate the $F_0$ value for this process. What will be the required process time to have a lethality equal to an $F_0$ of 9 min.

| Time (min) | Temp. (°F) | Time (min) | Temp. (°F) |
|---|---|---|---|
| 0 | 140 | 55 | 238 |
| 5 | 140 | 60 | 241 |
| 10 | 140 | 65 | 235 |
| 15 | 140 | 70 | 245 |
| 20 | 163 | 75 | 246.3 |
| 25 | 185 | 80 | 247.3 (cool) |
| 30 | 201 | 85 | 247.0 |
| 35 | 213 | 90 | 245.2 |
| 40 | 224 | 95 | 223.5 |
| 45 | 229.4 | 100 | 175 |
| 50 | 234.5 | 105 | 153 |

**Solution**

From t = 0 to t = 80 min, δt = 5 min will give 16 increments. L is calculated using Eq. (8.21). $T_0 = 250$ °F. The values of the lethality are as follows:

$$L_0 = 10^{-0.05555(110)} = 0 = L_1 = L_2 = L_3$$
$$L_4 = 10^{-0.05555(87)} = 1.5 \times 10^{-5}$$
$$L_5 = 10^{-0.05555(65)} = 2.45 \times 10^{-4}$$
$$L_6 = 10^{-0.05555(54)} = 0.001001$$
$$L_7 = 10^{-0.05555(32)} = 0.016688$$
$$L_8 = 10^{-0.05555(26)} = 0.035938$$
$$L_9 = 10^{-0.05555(20.6)} = 0.071725$$
$$L_{10} = 10^{-0.05555(15.5)} = 0.1377$$
$$L_{11} = 10^{-0.05555(12)} = 0.2155$$
$$L_{12} = 10^{-0.05555(9)} = 0.3163$$
$$L_{13} = 10^{-0.05555(6.5)} = 0.4354$$
$$L_{14} = 10^{-0.05555(5)} = 0.5275$$
$$L_{15} = 10^{-0.05555(3.7)} = 0.6229$$
$$L_{16} = 10^{-0.05555(2.7)} = 0.7079$$

The area under the heating curve will be:

$$4(L_1 + L_3 + L_5 + L_7 + L_9 + L_{11} + L_{13} + L_{15})$$
$$= 5.4498$$

$$2(L_2 + L_4 + L_6 + L_8 + L_{10} + L_{12} + L_{14})$$
$$= 2.0363$$

$$A = (5/3)(0 + 5.4498 + 2.0363 + 0.7079)$$
$$= 13.57$$

The area under the cooling curve will be:

$$L_0 = 10^{-0.05555(2.7)} = 0.70797$$
$$L_1 = 10^{-0.05555(3)} = 0.61829$$
$$L_2 = 10^{-0.05555(4.8)} = 0.54187$$
$$L_3 = 10^{-0.05555(26.5)} - 0.03371$$
$$L_4 = 10^{-0.05555(75)} = 0$$
$$L_5 = 0$$
$$4(L_1 + L_3 + L_5) = 2.8600$$
$$2(L_2 + L_4 + L_6) = 1.08374$$
$$A = (5/3) \times (0.70797 + 2.8600 + 1.0837 + 0)$$
$$= 7.753$$
Total area $= 13.57 + 7.753 = 21.32$

The cooling curve contributed about one-third of the total lethality in this example.

The calculated total lethality is much higher than the specified $F_0$ of 9 min. Thus, it will be necessary to reduce the heating time. Reduction of heating time will result in a reduction of the can temperature prior to cooling. Let the heating time be equal to 60 min. There are now only 12 area increments. The can temperature at 60 min of heating is 241 °F. The cooling curve will start at 241 °F. The cooling temperature will be parallel to the cooling curve of the original process (Fig. 8.12). Using Simpson's rule on the new heating and cooling curve, Table 8.10 may be constructed.

The area under the heating curve is
: $(1.17479 + 0.35019 + 0.316228)(5/3)$
$= 1.8420(5/3) = 3.07$.

The area under the cooling curve is
$(1.114881 + 0.430887 + 0.316228)\,(5/3)$
$= 1.96199(5/3) = 3.27$.

The total area $= F_o = 6.3$.

Ten more minutes of heating will add approximately 3 min to the $F_0$ because 1 min of heating at 241 °F is equivalent to 0.31 min at 250 °F. Thus

**Fig. 8.12** Graph showing retort temperature, lethality of the heat treatment, and the procedure for adjusting the processing time to obtain the specified process lethality

the heating time will be 70 min to give an $F_0$ of approximately 9 min.

This example underscores the importance of the contribution of the cooling part of the process to the total lethal value of the process. The relative contribution of the cooling curve to total lethality increases when product characteristics or processing conditions result in a slow rate of cooling.

### 8.6.2 Heat Transfer Equations and Time-Temperature Curves for Canned Foods

In the section "Heating of Solids Having Infinite Thermal Conductivity" Chap. 6 (Sect. 6.6.1), the transient temperature of a solid having an infinite thermal conductivity was derived. Equation (6.84) in Sect. 6.6.1 may be used to represent the temperature at a single point in a container. If the point considered is at the interior of the container, a time lag will exist from the start of heating to the time temperature at that point

**Table 8.10** Lethality of a process calculated using the general method and Simpson's rule for graphical integration

| Time (min) | Temp (°F) | L | 4L | 2L | L |
|---|---|---|---|---|---|
| 0 | 140 | 0 | – | – | 0 |
| 5 | 140 | 0 | 0 | – | – |
| 10 | 140 | 0 | – | 0 | – |
| 15 | 140 | 0 | 0 | – | – |
| 20 | 163 | 0 | – | 0 | – |
| 25 | 185 | 0.000245 | 0.000980 | – | – |
| 30 | 201 | 0.001896 | – | 0.003791 | – |
| 35 | 213 | 0.008799 | 0.035214 | – | – |
| 40 | 244 | 0.035938 | – | 0.071858 | – |
| 45 | 229.4 | 0.071706 | 0.286824 | – | – |
| 50 | 234.5 | 0.137686 | – | 0.275371 | – |
| 55 | 238 | 0.215443 | 0.861774 | – | – |
| 60 | 241 | 0.316228 | – | – | 0.316228 |
| Sum | | | 1.17479 | 0.351019 | 0.316228 |
| 60 | 241 | 0.316228 | – | – | 0.316228 |
| 65 | 240 | 0.278256 | 1.113024 | – | – |
| 70 | 238 | 0.215443 | – | 0.430887 | – |
| 75 | 190 | 0.000464 | 0.001857 | – | – |
| 80 | 149 | 0.052079 | – | 0 | – |
| 85 | 142 | 0 | 0 | – | – |
| 90 | 140 | 0 | – | – | 0 |
| Sum | | | 1.114881 | 0.430887 | 0.316228 |

actually changes. The following symbols are used for thermal process heat penetration parameters.

$I$ = initial temperature difference = $(T_r - T_0)$; $T_r$ = heating medium temperature = retort temperature; $T_0$ = can temperature at the start of the heating process.

$g$ = unaccomplished temperature difference at the end of a specified heating time = $(T_r - T)$;

$T$ = temperature at the point considered at any time, t, during the heating process.

$j$ = lag factor, also known as the intercept index for the linear semilogarithmic temperature versus time plot of the heating curve. jh refers to the heating curve, and jc refers to the cooling curve.

$f$ = the slope index of the linear semilogarithmic temperature versus time plot of the heating curve. If the heating curve consists of n line segments, $f_i (i = 1$ to $n)$ is used to represent the slope index of each line segment with 1 representing the first line segment from the start of the heating process. fc refers to the cooling curve.

Expressing Eq. (6.84), Chap. 6, in terms of the above parameters:

$$\log\left(\frac{g}{jI}\right) = -\frac{t}{f_h} \tag{8.33}$$

$$\log\left(\frac{(T_r - T)}{jI}\right) = -\frac{t}{f_h} \tag{8.34}$$

$$\log(T_r - T) = \log(jI) - \frac{t}{f_h} \tag{8.35}$$

$$T = T_r - jI \cdot 10^{\frac{-t}{f_h}} \tag{8.36}$$

Equation (8.35) shows that a semilog plot of the unaccomplished temperature difference $(T_r - T)$ against time will have a slope of $-1/f_h$ and an intercept of log (jI). The latter is the reason that j is called the intercept index. j is also called the lag factor because the higher the value of j, the longer it will take for the temperature at the point being monitored, to respond to a sudden change in the heating medium temperature.

The cooling curve expressed in a form similar to Eq. (8.35) with $T_c$ as the cooling water temperature is:

$$\log(T - T_c) = \log(j_c I_c) - \frac{t_c}{f_c} \qquad (8.37)$$

where $I_c = (T_g - T_c)$; $T_g$ = temperature at the end of the heating process = $(Tr - g)$.

Equation (8.37) shows that a semilogarithmic plot of $\log(T - T_c)$ versus time $t_c$ (with $t_c = 0$ at the start of cooling) will be linear, and the slope will be $-1/f_c$. The temperature at any time during the cooling process is:

$$T = T_c + j_c I_c [10]^{-t_c/f_c} \qquad (8.38)$$

Equation (8.38) represents only part of the cooling curve and is not the critical part that contributes significantly to the total lethality. The initial segment just after the introduction of cooling water is nonlinear and accounts for most of the lethality contributed by the cooling curve. Thus, a mathematical expression which correctly fits the curved segment of the temperature change · on cooling will be essential to accurate prediction of the total process lethality.

The initial curved segment of cooling curves has been represented as hyperbolic, circular, and trigonometric functions. A key parameter in any case is the intersection of the curved and linear segments of the cooling curve. The linear segment represented by Eq. (8.38) can be easily constructed from $f_c$ and $j_c$, and temperature at any time within this segment can be calculated easily using Eq. (8.38) from the point of intersection of the curved and linear segments. Hayakawa (Food Technol. 24:1407, 1970) discussed the construction of the curved segment of the cooling curve using a trigonometric function. The equations which are valid for $1 \le j_c \le 3$ are:

$$T = T_c + \left[T_g - T_c\right]^{\cos(Bt_c)} \qquad (8.39)$$

$$B = \frac{1}{t_L}\left[\arccos\left[\frac{\log(j_c I_c) - t_L/f_c}{\log(I_c)}\right]\right] \qquad (8.40)$$

The cosine function in Eq. (8.39) uses the value of the angle in radians as the function argument.

The arccos function in Eq. (8.40) returns the value of the angle in radians.

$t_L$ is the time when the curved and linear segments of the cooling curve intersect. $t_L$ may be derived from the intersection of a horizontal line drawn from the temperature at the initiation of cooling and the linear segment of the cooling curve represented by Eq. (8.38). At the intersection, $(T - T_c) = I_c$, and $t_c = t_L$. Substituting in Eq. (8.38), solving for $t_L$ and introducing a factor k to compensate for the curvature in the cooling curve:

$$t_L = f_c \log\left(\frac{j_c}{k}\right) \qquad (8.41)$$

The factor k in Eq. (8.41) may be determined from the actual cooling curves, when plotting heat penetration data. $k = 0.95$ has been observed to be common in experimental cooling curves for canned foods. Equation (8.41) represents cooling data for canned foods better than the equation for $t_L$ originally given by Hayakawa (1970).

### 8.6.3    Plotting Heat Penetration Data

Raw time-temperature data may be plotted directly on semilog graphing paper to produce the linear plot needed to determine $f_h$ and j, by rotating the paper 180 degrees. The numbers on the graph that mark the logarithmic scale are marked as $(Tr - T)$, and the can temperature is marked on the opposite side of the graphing paper. Figures 8.13 and 8.14 show how the can temperature is marked on three-cycle semilog graphing paper for retort temperature of 250 °F and 240 °F, respectively.

#### 8.6.3.1    Determination of $f_h$ and j

Can temperature is plotted on the modified graphing paper, and a straight line is drawn

**Fig. 8.13** Diagram showing how the axis of semilogarithmic graphing paper is marked for plotting heat penetration data. Retort temperature = 250 °F

**Fig. 8.14** Diagram showing how the axis of semilogarithmic graphing paper is marked for plotting heat penetration data. Retort temperature = 220 °F

connecting as much of the experimental data points as possible. There will be an initial curvature in the curve, but the straight line is drawn all the way to t = 0.

In any simulator used for heat penetration data collection, the retort temperature does not immediately reach the designated processing temperature. The time from introduction of steam to when processing temperature (Tr) is reached is the

retort come-up time, $t_{come-up}$. Sixty percent of the retort come-up time is assumed to have no heating value; therefore heating starts from a pseudo-initial time $t_{pi}$ which is $0.6t_{come-up}$. The pseudo-initial temperature, $T_{pi}$, is the intersection of the line drawn through the points and the line representing $t = t_{pi}.(Tr - T_{pi}) = jI$. The intercept index, j, is:

$$j = \frac{T_r - T_{pi}}{T_r - T_0} = \frac{jI}{I} \tag{8.42}$$

The slope index, $f_h$, is the time for the linear section of the heating curve to traverse one log cycle on the graph.

### 8.6.3.2 Determination of $f_c$ and $j_c$

Using 180-degree rotated semilog graphing paper, the cooling curve is plotted on the paper with the marked side labeled $(T - T_c)$, and the opposite side labeled the can temperature, T. The abscissa is the cooling time, $t_c$. At $t_c = 0$, steam is shut off, and cooling water is introduced. The retort is assumed to reach cooling water temperature immediately; therefore the intercept of the cooling curve is evaluated at $t_c = 0$. $j_cI_c$ is the intercept of the line drawn between the data points, and $t_c = 0$. $f_c$ is the slope index of the cooling curve and is the time for the linear section of the curve to traverse one log cycle. If the cooling data does not complete one log cycle within the graph, $f_c$ may be evaluated as the negative reciprocal of the slope of the line. Let $(T_1 - T_c)$ and $(T_2 - T_c)$ represent the unaccomplished temperature difference at $t_{c1}$ and $t_{c2}$, respectively:

$$f_c = -\frac{t_{c1} - t_{c2}}{\log(T_1 - T_c) - \log(T_2 - T_c)} \tag{8.43}$$

**Example 8.17** Determine the heat penetration parameters, j, $f_h$, $j_c$, and $f_c$, for a canned food that exhibited the following heating data when processed in a retort at 250 °F. It took 3 min from introduction of steam to the time retort reached 250 °F. Cooling water temperature is 60 °F.

| Time (min) | Temp. (°F) | Time (min) | Temp. (°F) |
|------------|-----------|------------|-----------|
| 0          | 180       | 30         | 245       |
| 5          | 190       | 30 (cool)  | 245       |
| 10         | 210       | 35         | 235       |
| 15         | 225       | 40         | 175       |
| 20         | 235       | 45         | 130       |
| 25         | 241       | 50         | 101       |

Calculate the temperature at various times during heating and cooling of this product processed at $Tr = 251 °F$ if $T_0 = 160 °F$, and $t = 35$ min from steam introduction. $T_c = 70 °F$. Retort come-up time $= 3$ min. Calculate the $F_o$ of this process using the general method and Simpson's rule for graphical integration of the lethality.

**Solution**

The heating and cooling curves are plotted in Figs. 8.15 and 8.16. The value for $f_h = 22$ min and how it is determined is shown in Fig. 8.14. $T_{pi} = 152$ min is read from the intersection of the line through the data points, and $t_{pi} = 0.6t_{come-up} = 1.8$ min. jI can be read by projecting the intersection to the axis labeled $(Tr - T)$. $jI = 98 °F$

or by subtracting $T_{pi}$ from $Tr$. $I = (Tr - T_0) = 250 - 180 = 70 °F$. $j = 98/70 = 1.40$.

The cooling curve is shown in Fig. 8.16. The intercept of the linear portion of the curve with $t_c = 0$ is projected to the side marked $(T - T_c)$, and $j_c I_c$ is read to be 333 °F. The initial temperature difference for cooling, $I_c = 245 - 60 = 185 °F$. Thus $j_c = j_c I_c / I_c = 333/185 = 1.8$. The slope index for the cooling curve is calculated from the points $(t_c = 0; (T - T_c) = 333)$ and $(t_c = 20; (T - T_c) = 41)$.

$$f_c = -\frac{0 - 20}{\log(333) - \log(41)} = 22 \text{min}$$

The curved section of the cooling curve intersects the linear section at $t_c = 6$ min. Thus, $t_L = 6$ min, and for this cooling curve, k in Eq. (8.41) is:

$$k = \frac{j_c}{[10]^{t_L/f_c}} = \frac{1.8}{[10]^{6/22}} = 0.96$$

The temperature during heating and cooling at the same point within a similar-sized container can be

**Fig. 8.15**  A plot of the heating curve showing how the heating curve parameters $f_h$ and $j$ are determined

**Fig. 8.16** A plot of the cooling curve showing how the cooling curve parameters $f_c$ and $j_c$ are obtained

**Table 8.11** Time and temperature during heating for Example 8.17

| Time (min) | A | $T_r - jIa$ | Temp (F) |
|---|---|---|---|
| 0 | 1.207 | (97) | 160 |
| 2 | 0.979 | (126) | 160 |
| 4 | 0.794 | (150) | 160 |
| 6 | 0.644 | 168 | 168 |
| 8 | 0.522 | 184 | 184 |
| 10 | 0.423 | 197 | 197 |
| 12 | 0.344 | 207 | 207 |
| 14 | 0.279 | 215.5 | 215.5 |
| 16 | 0.226 | 222 | 222 |
| 18 | 0.184 | 227.6 | 227.6 |
| 20 | 0.149 | 232 | 232 |
| 22 | 0.121 | 235.6 | 235.6 |
| 24 | 0.098 | 238.5 | 238.5 |
| 26 | 0.079 | 241.6 | 241.6 |
| 28 | 0.064 | 242.8 | 242.8 |
| 30 | 0.052 | 244.3 | 244.3 |

determined for any retort temperature, initial can temperature, or cooling water temperature once the heating and cooling curve parameters are known. Equation (8.36) is used to determine the temperature during heating. The initial heating period is assumed to be constant at $T_0$ until calculated values for T exceeds $T_0$. This assumption does not introduce any errors in the calculation of the lethality of heat received, since at this low temperature, lethality is negligible. Exceptions are rare and apply to cases where a very high initial temperature exists.

As previously discussed, when heating is carried out under conditions where a come-up time exists, the first 60% of the come-up time is assumed to have no heating value; therefore, the time variable in Eq. (8.36) should be zero when $t = t_{come-up}$. If t used in Eq. (8.36) is based on the time after "steam on," then 60% of $t_{come-up}$ must be subtracted from it when used in Eq. (8.36).

Let the exponential term in Eq. (8.36) = A.

$$A = 10^{-\{[t-0.6(3)]/fh\}} = 10^{-[(t-1.8)/22]}$$

$$T = Tr - jIA = 251 - 1.4(251 - 160)A$$

Calculated temperatures are shown in Table 8.11.

During cooling, the curved portion of the cooling curve is constructed with the temperatures calculated using Eqs. (8.39), (8.40), and (8.41). The time when the linear and curved portions of the cooling curve intersect is calculated using Eq. (8.39) and the previously calculated value of k of 0.96.

$$t_L = 22 \log\left(\frac{1.8}{0.96}\right) = 6.1\,\text{min}$$

Solving for parameter B in Eq. (8.38):

$$B = \frac{1}{6.1}\left[\arccos\left[\frac{\ln(1.8)(174.3) - (6.1/22)}{\ln(174.3)}\right]\right]$$

$$= \frac{1}{6.1}\arccos(0.99018)$$

$$= \frac{1}{6.1}\left[8.031°\left[\frac{2\pi\,\text{rad}}{360°}\right]\right]$$

$$= 0.02337$$

**Table 8.12** Time and temperature during cooling for Example 8.17

| Time ($t_c$) | E | A | (Temp °F) $70 + 174.3^E$ | $70 + (313.74)^A$ |
|---|---|---|---|---|
| 0 | 1 | 1 | 244.3 | (383.7) |
| 2 | 0.998 | 0.811 | 243.3 | (324.5) |
| 4 | 0.993 | 0.658 | 240.4 | (276.4) |
| 6 | 0.984 | 0.534 | 235.7 | (237.3) |
| 8 | | 0.433 | | 199.4 |
| 10 | | 0.351 | | 180.2 |
| 12 | | 0.285 | | 159.4 |
| 14 | | 0.231 | | 142.5 |
| 16 | | 0.187 | | 128.8 |

The temperature is then calculated using Eq. (8.39):

$$T = 70 + (174.3)^{\cos[(0.02337)(tc)]}$$

Let $E = \cos[(0.0296)(t_c)]$;

$$T = 70 + (174.3)^E$$

For the linear portion of the cooling curve, Eq. (8.38) is used. The temperature calculated using Eq. (8.30) will represent the actual can temperature only when $t_c > t_L$ because $t_L$ is the intersection of the curved and linear portions of the cooling curve. Equation (8.38): $T = 70 + 1.8$ $(174.3)(10)^{(-tc/22)}$

Let $A = 10^{(-tc/22)}$; $T = 70 + (313.74)^A$

The calculated temperatures for cooling, and the lethality and area elements for area calculation using Simpson's rule, are shown in Tables 8.12 and 8.13, respectively.

$$\text{Area} = F_0 = (2/3)(1.7904 + 5.5823 + 0.9698)$$
$$= 5.6 \text{ min}$$

The lethality of the process expressed as an equivalent heating time at 250 °F is 5.3 min. The complete procedure used in the evaluation of the lethality of a heating process in this example is the general method. The general method may be used on data obtained experimentally or on time-temperature data reconstructed from values of heating and cooling curve parameters calculated from experimental data.

The use of a spreadsheet will greatly facilitate thermal process calculations using the general method. Time increments used in the lethality calculations can be made very small to increase the accuracy of calculated process lethality.

### 8.6.4   Formula Methods for Thermal Process Evaluation

Formula methods are based on tabulated values for lethality expressed as the parameter $f_h/U$, which have been previously calculated for various conditions of heating and cooling when unaccomplished temperature difference is expressed as the parameter "g." Two formula methods are available: Stumbo's (1973) and Hayakawa's (1970). The differences in the two methods are as follows:

Stumbo's $f_h/U$ versus g tables combine lethalities of both heating and cooling. A major assumption used in the calculation of lethality is that $f_h = f_c$. When actual conditions do not meet this assumption, Stumbo recommends using the general method. Hayakawa (1970) calculated the lethality of the heating and cooling stages in the process and presented separate tables to use in calculating cooling and heating process lethality. Thus, the method may be used even when rates for heating and cooling are different. Hayakawa's tabular values allow substitutions for different values of z, simplifying calculation of specific $F_0^z$ values for different z.

**Table 8.13** Time, temperature, and Simpson's rule factors for area calculations for Example 8.17

| Time (min) | Temp (°F) | L | 2L | 4L | L |
|---|---|---|---|---|---|
| Heating | | | | | |
| 0 | 160 | 0 | | | |
| 2 | 160 | 0 | | | 0 |
| 4 | 160 | 0 | | 0 | |
| 6 | 169 | 0 | 0 | | |
| 8 | 184 | 0.0002 | | 0.0008 | |
| 10 | 197 | 0.0011 | 0.0022 | | |
| 12 | 207 | 0.0042 | | 0.0168 | |
| 14 | 215.5 | 0.0121 | 0.0242 | | |
| 16 | 222.2 | 0.0284 | | 0.1136 | |
| 18 | 227.6 | 0.0571 | 0.114 | | |
| 20 | 232 | 0.1005 | | 0.402 | |
| 22 | 235.6 | 0.1588 | 0.3177 | | |
| 24 | 238.5 | 0.2304 | | 0.9215 | |
| 26 | 241.8 | 0.3414 | 0.6829 | | |
| 28 | 242.8 | 0.3977 | | 1.5908 | |
| 30 | 244.3 | 0.4899 | | | 0.4849 |
| Cooling | | | | | |
| 0 | 244.3 | 0.4849 | | | 0.4849 |
| 2 | 243.3 | 0.4244 | | 1.6976 | |
| 4 | 240.4 | 0.2929 | 0.5858 | | |
| 6 | 235.7 | 0.1605 | | 0.6421 | |
| 8 | 199 | 0.0015 | 0.0030 | | |
| 10 | 180 | 0.0001 | | 0.0004 | |
| 12 | 159 | 0 | 0 | | |
| 14 | 142 | 0 | | 0 | |
| 16 | 129 | 0 | | | 0 |
| Sums | | | 1.7297 | 5.3866 | 0.9698 |

In this section, process calculations for products that exhibit simple heating curves will be discussed. Calculations for broken heating curves will be discussed in the section "Broken Heating Curves." Both formula methods are based on the equation for the heating curve (Eq. 8.33). Let $g$ = unaccomplished temperature difference, $(T_r - T)$ at the termination of the heating period; and $B_b$ = the heating time at that point. $B_b$ is the scheduled sterilization process. For products with simple heating curve, Eq. (8.33) becomes:

$$B_b = f_h[\log(jI) - \log(g)] \qquad (8.44)$$

$B_b = t - 0.6t_{come-up}$. $t$ is time evaluated from steam introduction into the retort. In practice, $B_b$ is timed from the point where the retort temperature reaches the processing temperature to avoid the probability of errors arising from the operator having to correct for the come-up time in setting the timing for the process.

$g$ is obtained from the tables, using a specified $F_0$ and $z$ value. Stumbo's tables are simpler to use for thermal process determinations when $f_c = f_h$. Hayakawa's tables will require an iterative procedure involving an assumption of the value of $g$, calculating the $F_0$, and calculations are repeated until the calculated matches the specified $F_0$. Thus, most thermal processing specialists use Stumbo's procedure.

Because Hayakawa's procedure is seldom used, the procedure will not be discussed in this edition. However, a detailed discussion of the procedure is available in the 3rd. edition of this textbook.

The following parameters are used in the formula method: let $F_i$ = heating time at temperature T equivalent to 1 min of heating at 250 °F.

$$F_i = 10^{(250-T)/z} \qquad (8.45)$$

$U = F_0 F_i$
= time of processing at Tr equivalent to $F_0$.

### 8.6.4.1 Stumbo's Procedure

Stumbo's tabulated $f_h/U$ versus g with $j_c$ as a parameter. $j_c$ strongly influences the contribution of the cooling part of the process to the total lethality as discussed in the section "Sterilizing Value of Processes Expressed as $F_0$." In general, $j_c$ values are higher than $j_h$. In the absence of $j_c$, $j_h$ may be used, and the error will be toward a longer process time or the safe side relative to spoilage. Condensed $f_h/U$ versus g tables for z values from 14 to 22 and for z from 30 to 45 are shown in Tables 8.14 and 8.15. Table 8.14 is used for microbial inactivation, and Table 8.15 is used for nutrient degradation. It is possible to interpolate between values in the table for other z values. Thermal process determinations can easily be made from specified $F_0$ values and product heat penetration parameters by solving for $f_h/U$ determining the corresponding value for g and solving for $B_b$ using Eq. (8.44).

### 8.6.5 Evaluation of Probability of Spoilage from a Given Process

This procedure is used to determine if a process that deviated from specifications will give a safe product. The procedures discussed in this section will also be useful in cases of spoilage outbreaks where a spoilage organism is isolated, its heat resistance determined, and it is desired to determine if process schedule adjustment is necessary to prevent future occurrences of spoilage. Another useful application of these procedures is the conversion of standard $F_0$ values to $F^z_0$ values for specific microorganisms.

### 8.6.5.1 Constant Temperature Processes

A process time at a constant retort temperature and an initial temperature is given. The procedure is similar to Example 8.18. A g value is calculated using Eq. (8.33). Table 8.14 is then used to determine U at a specified z value, from which $F^z_0$ is calculated. The probability of spoilage is then calculated by substituting $F^z_0$ for t in Eq. (8.11).

**Example 8.18** The following data represents the heating characteristics of a canned product. $f_h = f_c = 22.5$ min; $j = j_c = 1.4$. If this product is processed for 25 min at 252 °F from an initial temperature of 100 °F, calculate (a) the $F_0$ and (b) the probability of spoilage if an organism with a $D_0$ value of 0.5 min and a z value of 14 at an initial spore load of 10/can.

**Solution**

g is determined from the process time, using Eq. (8.33):

$$g = [10]^{\log(jI)-t/f_h}$$
$$= [10]^{\log[(1.4)(252-100)]-25/22.5}$$
$$= [10]^{1.2168} = 16.5°F$$

(a) Because $f_h = f_c$, Stumbo's procedure can be used. The $F_0$ is determined using Table 8.14 for z = 18 °F and $j_c = 1.4$. Inspection of Table 8.14 reveals that to obtain g = 16.5 when j = 1, f $_h$/U has to be between 40 and 45. However, the interpolation factor $\Delta g/\Delta j$ is about 6; therefore, because $j_c = 1.4$, g in the table will increase by 0.4(6) or 2.4. Thus, the entry for $f_h/U = 30$ will be considered, and after calculating g at $j_c = 1.4$, the other entry to be used in the interpolation will be selected.

$$\frac{f_h}{U} = 30; \quad g_{j=1.4} = 14.60 + 0.4(5.3) = 16.72$$

**Table 8.14**  $f_h/U$ vs. g table used for thermal process calculations by Stumbo's procedure

| $f_h/U$ | z = 14 | $\Delta g/\Delta j$ | z = 18 | $\Delta g/\Delta j$ | z = 22 | $\Delta g/\Delta j$ |
|---|---|---|---|---|---|---|
| 0.2 | 0.000091 | 0.0000118 | 0.0000509 | 0.0000168 | 0.0000616 | 0.0000226 |
| 0.3 | 0.00175 | 0.00059 | 0.0024 | 0.00066 | 0.00282 | 0.00106 |
| 0.4 | 0.0122 | 0.0038 | 0.0162 | 0.0047 | 0.020 | 0.0067 |
| 0.5 | 0.0396 | 0.0111 | 0.0506 | 0.0159 | 0.065 | 0.0197 |
| 0.6 | 0.0876 | 0.0224 | 0.109 | 0.036 | 0.143 | 0.040 |
| 0.7 | 0.155 | 0.036 | 0.189 | 0.066 | 0.25 | 0.069 |
| 0.8 | 0.238 | 0.053 | 0.287 | 0.103 | 0.38 | 0.105 |
| 0.9 | 0.334 | 0.07 | 0.400 | 0.145 | 0.527 | 0.147 |
| 1.0 | 0.438 | 0.009 | 0.523 | 0.192 | 0.685 | 0.196 |
| 2.0 | 1.56 | 0.37 | 1.93 | 0.68 | 2.41 | 0.83 |
| 3.0 | 2.53 | 0.70 | 3.26 | 1.05 | 3.98 | 1.44 |
| 4.0 | 3.33 | 1.03 | 4.41 | 1.34 | 5.33 | 1.97 |
| 5.0 | 4.02 | 1.32 | 5.40 | 1.59 | 6.51 | 2.39 |
| 6.0 | 4.63 | 1.56 | 6.25 | 1.82 | 7.53 | 2.75 |
| 7.0 | 5.17 | 1.77 | 7.00 | 2.05 | 8.44 | 3.06 |
| 8.0 | 5.67 | 1.95 | 7.66 | 2.27 | 9.26 | 3.32 |
| 9.0 | 6.13 | 2.09 | 8.25 | 2.48 | 10.00 | 3.55 |
| 10 | 6.55 | 2.22 | 8.78 | 2.69 | 10.67 | 3.77 |
| 15 | 8.29 | 2.68 | 10.88 | 3.57 | 13.40 | 4.60 |
| 20 | 9.63 | 2.96 | 12.40 | 4.28 | 15.30 | 5.50 |
| 25 | 10.7 | 3.18 | 13.60 | 4.80 | 16.9 | 6.10 |
| 30 | 11.6 | 3.37 | 14.60 | 5.30 | 18.2 | 6.70 |
| 35 | 12.4 | 3.50 | 15.50 | 5.70 | 19.3 | 7.20 |
| 40 | 13.1 | 3.70 | 16.30 | 6.00 | 20.3 | 7.60 |
| 45 | 13.7 | 3.80 | 17.00 | 6.20 | 21.1 | 8.0 |
| 50 | 14.2 | 4.00 | 17.7 | 6.40 | 21.9 | 8.3 |
| 60 | 15.1 | 4.3 | 18.9 | 6.80 | 23.2 | 9.0 |
| 70 | 15.9 | 4.5 | 19.9 | 7.10 | 24.3 | 9.5 |
| 80 | 16.5 | 4.8 | 20.8 | 7.30 | 25.3 | 9.8 |
| 90 | 17.1 | 5.0 | 21.6 | 7.60 | 26.2 | 10.1 |
| 100 | 17.6 | 5.2 | 22.3 | 7.80 | 27.0 | 10.4 |
| 150 | 19.5 | 6.1 | 25.2 | 8.40 | 30.3 | 11.4 |
| 200 | 20.8 | 6.7 | 27.1 | 9.10 | 32.7 | 12.1 |

*Source:* Based on $f_h/U$ vs. g tables in Stumbo, C. R. 1973. *Thermobacteriology in Food Processing,* 2nd ed. Academic Press, New York

Notes on using Table 8.14: To use for values of *j* other than l, solve for $g_j$ as follows:

$$g_j = g_{j-1} + (j-1)\left[\frac{\Delta g}{\Delta j}\right] \quad (8.46)$$

Example: g for$(f_n/U) = 20$ and j = 1.4 and z = 18: $g_{j\,=\,1.4} = 12.4 + (0.4)(4.28) = 14.11$.

Reprinted from: Toledo, R. T. 1980. Fundamentals of Food Process Engineering, ist ed. AVI Pub. Co. Westport, CT

The value of g is greater than 16.5; therefore, the next lower value of $f_h/U$ in the tables will be used to obtain the other value of g to use in the interpolation.

$$\frac{f_h}{U} = 25; \quad g_{j=1.4} = 13.6 + 0.4(4.8) = 15.52$$

Interpolating between g = 16.72 and g — 15.52 to obtain $f_h/U$ corresponding to g = 16.5:

**Table 8.15**  $f_h/U$ vs. $g$ table used for thermal process calculation by Stumbo's procedure

| $f_h/U$ | $z = 60$ | | $z = 70$ | | $z = 80$ | | $z = 90$ | |
|---|---|---|---|---|---|---|---|---|
| | $g_{j=1}$ | $\frac{\Delta g}{\Delta j}$ | $g_{j=1}$ | $\frac{\Delta g}{\Delta j}$ | $g_{j=1}$ | $\frac{\Delta g}{\Delta j}$ | $g_{j=1}$ | $\frac{\Delta g}{\Delta j}$ |
| 0.2 | 0.00018 | 0.00015 | 0.000218 | 0.000134 | 0.000253 | 0.00017 | 0.000289 | 0.000208 |
| 0.3 | 0.0085 | 0.000475 | 0.0101 | 0.0062 | 0.000253 | 0.00017 | 0.0134 | 0.0097 |
| 0.4 | 0.0583 | 0.032 | 0.0689 | 0.0421 | 0.0118 | 0.00775 | 0.0919 | 0.0661 |
| 0.5 | 0.185 | 0.1025 | 0.0219 | 0.0134 | 0.0802 | 0.0545 | 0.292 | 0.208 |
| 0.6 | 0.401 | 0.2225 | 0.474 | 0.292 | 0.255 | 0.17 | 0.632 | 0.452 |
| 0.7 | 0.699 | 0.3875 | 0.828 | 0.510 | 0.552 | 0.3675 | 0.101 | 0.791 |
| 0.8 | 0.064 | 0.595 | 0.263 | 0.777 | 0.963 | 0.6425 | 0.678 | 1.205 |
| 0.9 | 1.482 | 0.8325 | 1.76 | 1.08 | 1.469 | 0.9775 | 2.34 | 1.68 |
| 1.0 | 1.94 | 1.075 | 2.30 | 1.42 | 2.05 | 1.45 | 3.06 | 2.19 |
| 2.0 | 7.04 | 4.025 | 8.35 | 5.19 | 2.68 | 1.775 | 11.03 | 7.88 |
| 3.0 | 11.63 | 6.65 | 13.73 | 8.58 | 9.68 | 6.475 | 18.0 | 12.8 |
| 4.0 | 15.40 | 9.00 | 18.2 | 11.4 | 12.92 | 8.65 | 23.6 | 16.7 |
| 5.0 | 18.70 | 10.75 | 21.9 | 13.7 | 15.85 | 10.65 | 28.2 | 19.7 |
| 6.0 | 21.40 | 12.50 | 25.1 | 15.6 | 18.5 | 12.5 | | |
| 7.0 | 23.80 | 13.75 | 27.9 | 17.2 | 20.9 | 14.0 | | |
| 8.0 | 26.00 | 15.00 | 30.3 | 18.6 | 23.1 | 15.5 | | |
| 9.0 | 27.90 | 16.00 | 32.5 | 19.8 | 25.1 | 16.75 | | |

*Source:* Based on $f_h/U$ vs. $g$ tables in Stumbo, C.R. 1973. *Thermobacteriology in Food Processing*, 2nd ed. Academic Press, New York

$$\frac{f_h}{U} = \frac{25 + 5(16.5 - 15.52)}{(16.72 - 15.52)} = 29.1$$

$$U = \frac{f_h}{\left(\frac{f_h}{U}\right)_{g=16.5}} = \frac{22.5}{29.1} = 0.7736 = F_0 F_i$$

$$F_i = (10)^{-2/18} = 0.774$$

$$F_0 = \frac{0.773}{0.774} = 0.999 \text{ min}$$

(b) To evaluate lethality to the organism with $z = 14\ °F$, $F^{14}{}_0$ must be determined. Using Table 8.14 for $z = 14\ °F$:

$$\frac{f_h}{U} = 50; \quad g_{j=1.4} = 14.2 + 0.4(4.00) = 15.8$$

For $g = 16.5$:

$$\frac{f_h}{U} = 60; g_{j=1.4} = 15.1 + 0.4(4.3) = 16.82$$

$$\frac{f_h}{U} = \frac{50 + 10(15.5 - 15.8)}{(16.82 - 15.8)} = 56.9$$

$$U = \frac{f_h}{(f_h/U)_{g=16.5}} = \frac{22.5}{56.9} = 0.395 = F_0 F_i$$

$$f_i = (10)^{-2/14} = 0.7197$$

$$F_0^{14} = \frac{U}{F_i} = \frac{0.395}{0.7197} = 0.549 \text{ min}$$

The number of survivors will be:

$$N = 10[10] - F_0/D_0 = 10(10)^{0.549/0.5} = 0.798$$

The probability of spoilage is 79.80%.

### 8.6.5.2 Process Temperature Change

When the process temperature changes, errors in the formula method are magnified because evaluation of $f_h$ and $j$ is based on an original uniform initial temperature, while the starting temperature distribution with in-process temperature deviations is no longer uniform. The most accurate method for evaluating the effect of process temperature changes is by using finite difference methods for evaluation of temperature at the critical point and using the general method for determining process lethality. If process deviation occurs before the temperature at the critical point exceeds 200 °F, and the deviation simply involves a step change in processing temperature at $t = t_1$ from $T_{r1}$ to $T_{r2}$ and remains constant for the rest of the process, lethality may be approximated by the formula methods. The part of the process before the step temperature change is considered to have negligible lethality (if the temperature at the critical point did not exceed 200 °F), and the temperature at $t_1$ is considered the initial temperature for a process at $T_{r2}$. Procedures for evaluation will be the same as in Example 8.18.

### 8.7 Broken Heating Curves

Broken heating curves are those that exhibit a break in continuity of the heating rate at some point in the heating process. Thus, two or more line segments will be formed when the heat penetration data are plotted on semilogarithmic graphing paper. This type of heating behavior will occur when the product inside the can undergoes a physical change that changes the heat transfer characteristics. A typical broken heating curve is shown in Fig. 8.17. The slope indices of the curve are designated as $f_{h1}$ for the first line segment and $f_{h2}$ for the second line segment. The retort-can temperature difference at the point of intersection of the first and second line segments is designated $g_{bh}$. The rest of the parameters of the heating curve is the same as for a simple heating curve.

The equation of the first line segment of the broken heating curve is:

$$\log\left(\frac{jI}{g_{bh}}\right) = \frac{t_{bh}}{f_{h1}} \qquad (8.47)$$

The equation of the second line segment is:

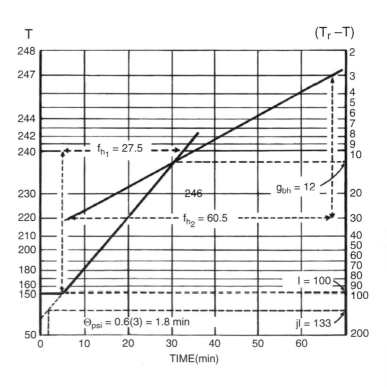

**Fig. 8.17** Diagram of a broken heating curve showing the heating curve parameters

$$\log\left(\frac{g_{bh}}{g}\right) = \frac{t - t_{bh}}{f_{h2}} \qquad (8.48)$$

Combining Eqs. (8.46) and (8.47):

$$t = f_{h1}\log\left(\frac{jI}{g_{bh}}\right) + f_{h2}\log\left(\frac{g_{bh}}{g}\right) \qquad (8.49)$$

Equation (8.49) is used to calculate a process time to obtain g. An expression for g can be obtained by rearranging Eq. 8.49).

$$g = [10]^{1/f_{h2}[f_{h1} \log(jI) - (f_{h1} - f_{h2}) \log(g_{bh}) - t]} \qquad (8.50)$$

Stumbo's procedure for evaluating thermal processes for products exhibiting broken heating curve involves evaluation of lethality of individual segments of the heating curve. Because the $f_h/U$ versus g tables were developed to include the lethality of the cooling part of the process, a correction needs to be made for the lethality of the non-existent cooling attributable to the first line segment of the heating curve. The "r" parameter was used to express the fraction of the total process lethality attributed to the heating part of the process.

For the first line segment which ends when $(T_r - T) = g_{bh}$:

$$U_1 = r\frac{f_{h1}}{[f_h/U]_{g_{bh}}} \qquad (8.51)$$

The second line segment begins when $(T_r - T) = g_{bh}$ and ends when $(T_r - T) = g$. The lethality from the $f_h/U$ tables considers heating process with the same $f_h$ value starting from time zero; therefore the effective lethality up to $(T_r - T) = g_{bh}$ must be subtracted from the total.

$$U_2 = \frac{f_{h2}}{(f_h/U)_g} - r\frac{f_{h2}}{(f_h/U)_{g_{bh}}} \qquad (8.52)$$

Thus, the total U for the process is:

$$U = \frac{f_{h2}}{(f_h/U)_g} + \frac{r(f_{h1} - f_{h2})}{(f_h/U)_{g_{bh}}} \qquad (8.53)$$

The denominator, $(f_h/U)_g$ or $(f_h/U)_{g_{bh}}$ in Eqs. (8.51), (8.52), and (8.53), represents tabular

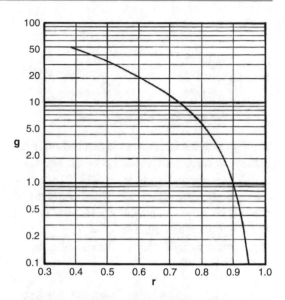

**Fig. 8.18** Values of the parameter r corresponding to the value of g. (Source: Anonymous. 1952. Calculation of process for canned food. American Can Company, Technical Services Division, Maywood, IL)

values for $f_h/U$ corresponding to g or $g_{bh}$. The parameter "r" is a function of g. Figure 8.18 can be used to obtain r corresponding to g.

**Example 8.19** For the product that exhibited the heating curve shown in Fig. 8.17, assume $f_c = f_{h2}$ and $j_c = j$. This product is processed for 50 min at 248 °F, from an initial temperature of 140 °F. Calculate $F_0$ and the probability of spoilage from an organism having a $D_0$ value of 1.5 min and a z value of 16 °F in cans given this process. The initial spore load is 100/can. The cooling water temperature is 60 °F.

**Solution**
Note that the conditions under which the product is processed are different from those under which the heat penetration parameters were derived. Figure 8.17 is simply used to determine the heat penetration parameters, and these parameters are utilized in the specific process to evaluate the process lethality. From Fig. 8.17: $f_{h1} = 27.5$ min; $f_{h2} = 60.5$ min; $g_{bh} = 12$ °F; I= 248 − 140 = 108 °F; j = $j_c$ = 1.33. Solving for g using Eq. (8.50):

The exponent is:

$$g = [10]^{\left(\frac{1}{f_{h1}}\log(jI) - (f_{h1} - f_{h2})\log(g_{bh}) - t\right)}$$

$$\text{Exponent} = \frac{27.5\log(1.33)(108) - (27.5 - 60.5)\log(12) - 50}{60.5}$$

$$\text{Exponent} = \frac{[27.5(2.157) + 33(1.0792) - 50]}{60.5} = 0.743$$

$$g = (10)^{0.743} = 5.53\,°F$$

The $F_0$ value for the process is determined using Table 8.14 for z = 18 °F and $j_c$ = 1.33. The value of $f_h/U$ corresponding to g = 5.53 °F and that corresponding to $g_{bh}$ = 12 °F will have to be determined by interpolation. Inspection of Table 8.12 shows that for $j_c$ = 1, g = 5.40 corresponds to $f_h/U$ and the interpolating factor $\Delta g/\Delta j$ = 1.59. Thus:

$$\frac{f_h}{U} = 5; g_{j=1.33} = 5.40 + 1.59(0.33) = 5.925$$

This value is higher than 5.53; therefore, a lower value of $f_h/U$ is picked for the second set of values to use in the interpolation. Thus:

$$\frac{f_h}{U} = 4; g_{j=1.33} = 4.41 + 1.34(0.33) = 4.852$$

The value of $f_h/U$ corresponding to g = 5.53 is calculated by interpolating between the above two tabular values of g:

$$\left(\frac{f_h}{U}\right)_{g=5.53} = 4 + \frac{(5 - 4)(5.53 - 4.852)}{(5.925 - 4.852)}$$

$$= 4.632$$

Next, the value of $f_h/U$ corresponding to $g_{bh}$ needs to be evaluated. $g_{bh}$ is entered into Table 8.14 as a value of g, and the corresponding $f_h/U$ is determined.

Inspection of Table 8.14 shows that for $j_c$ = 1, z = 18, and $f_h/U$ = 15 corresponds to g = 10.88 with the interpolating factor $\Delta g/\Delta j$ = 3.57. Thus:

$$\frac{f_h}{U} = 15; g_{j=1.33} = 10.88 + 0.33(3.57) = 12.058$$

The value of g is greater than 12; therefore, the next lower tabular entry is selected as the other set of values used in the interpolation.

$$\frac{f_h}{U} = 10; g_{j=1.33} = 8.78 + 0.33(2.69) = 9.668$$

Interpolating between these two values to obtain $f_h/U$ corresponding to g = 12:

$$\left(\frac{f_h}{U}\right)_{g_{bh}} = \left(\frac{f_h}{U}\right)_{g=12}$$

$$= 10 + 5(12 - 9.668)/(12.058 - 9.668)$$

$$= 14.88$$

Equation (8.53) is used to determine U from values of $(f_h/U)_g$, and $(f_h/U)_{g\,bh}$. r needs to be evaluated from Fig. 8.18 to correspond to $g_{bh}$ = 12. From Fig. 8.18, r = 0.71. Using Eq. (8.53):

$$U = \frac{60.5}{4.622} + \frac{0.71(27.5 - 60.5)}{14.88}$$

$$= 13.08 - 1.574 = 11.51$$

Using Eq. (8.45):

$$F_i = (10)^{2/18} = 1.291$$

$$F_0 = U/F_i = 11.51/1.291 = 8.91 \text{ min}$$

The probability of spoilage: Because the organism has a z value of 16 °F, $F_0$ cannot be used in Eq. (8.11) to determine the probability of spoilage. It is necessary to calculate $F_0^{16}$ for the conditions used in the process. Because the process is the same, g calculated in the first part of this problem is the same; g = 5.53. Evaluating $f_h/U$ corresponding to this value of g, however, should be done using z = 16 °F in Table 8.14. A double interpolation needs to be done. Values for $f_h/U$ corresponding to g = 5.53 are determined for z = 14 and z = 18; and the two values are interpolated to obtain $f_h/U$ at z = 14.

$$f_h/U = 6; g_{j=1.33, z=14} = 4.63 + 0.33(1.56) = 5.145$$
$$g_{j=1.33, z=18} = 6.5 + 0.33(1.82) = 6.850$$
$$f_h/U = 6; g_{j=1.33, z=16} = 5.145$$
$$+ [(18 - 16)(6.850 - 5.145)]/(18 - 14)$$
$$= 5.997$$
$$f_h/U = 5; g_{j=1.33, z=14} = 4.02 + 0.33(1.32) = 4.455$$
$$f_h/U = 5; g_{j=1.33, z=18} = 5.40 + 0.33(1.59) = 0.925$$
$$f_h/U = 5; g_{j=1.33, z=16} = 4.455 + (18 - 16)$$
$$(5.925 - 4.455)/(18 - 14) = 5.190$$

Interpolating between the two values of $f_h/U$ at z = 16, which straddles g = 5.53:

$(f_h/U)_{g=5.53, z=16} = 5 + (6 - 5)(5.53 - 5.190)$
$\times / (5.997 - 5.190) = 5.42$

The same procedure is used to determine $f_h/U$ corresponding to $g_{bh} = 12$

$f_h/U = 20; g_{j=1.33, z=14} = 9.63 + 0.33(2.96) = 10.61$
$f_h/U = 20; g_{j=1.33, z=18} = 12.4 + 0.33(4.28) = 13.81$
$f_h/U = 20; g_{j=1.33, z=16} = 10.61 + (16 - 14)$
$\qquad (13.81 - 10.61)/(18 - 14) = 12.21$
$f_h/U = 15; g_{j=1.33, z=18} = 10.88 + 0.33(3.57) = 12.06$
$f_h/U = 15; g_{j=1.33, z=14} = 8.29 + 0.33(2.68) = 9.17$
$f_h/U = 15; g_{j=1.33, z=16} = 9.7 + (16 - 14)$
$\qquad (12.06 - 9.17)/(18 - 14) = 11.14$

Interpolating between the two values of $f_h/U$ at $z = 16$, which correspond to values of g, which straddle $g_{bh} = 12$:

$(fh/U)_{g_{bh}} = 15 + (12 - 11.14)(20 - 15)$
$\times / (12.21 - 11.14) = 19.02$

Because r is dependent only on $g_{bh}$, the same value as before is obtained from Fig. 8.18. For $g_{bh} = 12$, $r = 0.71$. Using Eq. (8.53):

$U = \dfrac{60.5}{5.41} + \dfrac{0.71(27.5 - 60.5)}{19.02}$
$\quad = 11.182 - 1.232 = 9.95 F_i = (10)^{2/16}$
$\quad = 1.333 F_0^{16} = U/F_i = 9.95/1.333$
$\quad = 7.465$

The number of survivors can now be calculated by substituting $F_0^{16}$ for t in Eq. (8.11). Using Eq. (8.11):

$\log(N/100) = -F_0/D_0 = -7.465/1.5$
$\qquad\qquad = -4.977$
$N = 10^{-4.977} = 0.0011$

The probability of spoilage is 11 in 10,000.

## 8.8   Quality Factor Degradation

Quality factor degradation has to be evaluated on the basis of integrated lethality throughout the container. Unlike microbial inactivation, which leaves practically zero survivors at regions in the can near the surface, there is substantial nutrient retention in the same regions. Quality factor

degradation can be determined by separating the container into incremental cylindrical shells, calculating the temperature at each shell at designated time increments, determining the extent of degradation, and summing the extent of degradation at each incremental shell throughout the process. At the termination of the process, the residual concentration is calculated by integrating the residual concentration at each incremental cylindrical shell throughout the container. The procedure is relatively easy to perform using a computer, but the calculations can be onerous if done by hand.

For cylindrical containers, Stumbo (1973) derived an equation for the integrated residual nutrient based on the following observations on the temperature profiles for conduction heat transfer in cylinders.

(a)   In a container, an isotherm exists where the j value at that point, designated $j_v$, is 0.5 j. The g at that point at any time, designated $g_v$, is 0.5 g, and the volume enclosed by that isotherm is 19% of the total volume.

(b)   If v is the volume enclosed by the isotherm, and if the F value at the isotherm and at the critical point are $F_v$ and F, respectively, then the difference, $(F_v - F)$, is proportional to $\ln(1 - v)$.

The following expression was then derived:

$$\bar{F} = F + D\log\left(\frac{D + 10.92(F_v - F)}{D}\right) \quad (8.54)$$

Equation (8.54) is an expression for the integrated lethal effect of the heating process (F) on nutrients based on the lethality at a the critical point, F, and the lethality $F_v$ evaluated at a point where $g_v = 0.5$ g and $j_v = 0.5$ j. Equation (8.54) has been found to be adequate for estimating nutrient retention in cylindrical containers containing foods that heat by conduction.

**Example 8.20**   A food product has a j value of 1.2, a $j_c$ value of 1.4, and $f_h = f_c = 35$ min. This product is processed at 255 °F from an initial temperature of 130 °F to an $F_0$ of 5.5. Calculate the residual ascorbic acid remaining in this product after the process if the initial concentration was 22 mg/g. The $D_0$ value for ascorbic acid in the product is 248 min, and the z value is 91 °F.

**Solution**

Stumbo's procedure will be used. The $F_0$ value is used to determine a value of g using the $f_h/U$ tables for $z = 18$ and $j_c = 1.4$. From this value of g, a new $f_h/U$ is determined for a z value of 91 °F. The $F^{91}_0$ obtained is the value of F in Eq. (8.54). $F_v$ is determined from $f_h/U$, which corresponds to $g_v = 0.5$ g and $j_v = 0.5 j_c$.

$$U = F_0 F_i = 5.5(10)^{-5/18} = 2.901 \text{ min}$$
$$f_h/U = 35/2.901 = 12.06$$

From Table 8.14, for $z = 18$:

$$\frac{f_h}{U} = 10; g_{jc=1.4} = 8.78 + 2.69(0.4) = 9.956$$
$$\frac{f_h}{U} = 15; g_{jc=1.4} = 10.88 + 3.57(0.4) = 12.308$$

For $f_h/U = 12.06$:

$$g = 9.936$$
$$+ (12.308 - 9.956)(12.06 - 10)/(15 - 10)$$
$$= 10.925$$

For $g = 10.925$, using Table 8.15, $z = 90$ °F:

$$\frac{f_h}{U} = 1; g_{jc=1.4} = 3.06$$

$$+0.4(2.19) = 3.936 f_h/U \text{ for } g = 10.925 : \frac{f_h}{U} = 2;$$
$$g_{jc=1.4} = 11.03 + 0.4(7.88) = 14.182 = 1$$
$$+\frac{(2 - 1)(10.925 - 3.936)}{(14.182 - 3.936)} = 1.682$$

U at the geometric center $= 35/1.682 = 20.81$ min.

$$F_i = 10^{-5/90} = 0.8799.$$

F at the geometric center $= U/F_i = 23.65$ min.

At a point where $g_v = 0.5$ g and $j_{vc} = 0.5 j_c$, $g = 0.5(10.925) = 5.463$; $j_{vc} = 0.5(1.4) = 0.7$. Using Table 8.15, $z = 90$ °F:

$$\frac{f_h}{U} = 1; g_{jc=0.7} = 3.06 - 0.3(2.19) = 2.403$$

$f_h/U$ for $g_v = 5.463$:

$$\frac{f_h}{U} = 2; g_{jc=0.7} = 11.03 - 0.3(7.88) = 8.666 = 1$$
$$+\frac{(2 - 1)(5.463 - 2.403)}{(8.666 - 2.403)} = 1.488$$

U at point where $g_v = 0.5$ g $= 35/1.488 = 23.52$ min.

$$F_v = U/F_I = 23.52/0.8799 = 26.73 \text{ min.}$$

Substituting in Eq. (8.54):

$$\bar{F} = 23.65 + 248$$
$$\cdot log \left[ \frac{248 + 10.92(26.73 - 23.65)}{248} \right]$$

$$\bar{F} = 23.65 + 248(0.055233) = 37.348$$

$$log \left( \frac{C}{C_o} \right) = -\frac{F}{D} = \frac{-37.348}{248} = -0.1056$$

$$\frac{C}{C_o} = 10^{-.1056} = 0.706$$

The percent retention of ascorbic acid is 70.6%.

The residual ascorbic acid content is 0.706 (22) = 15.55 mg/g.

## Problems

8.1. Calculate the D value of an organism that shows 30 survivors from an initial inoculum of $5 \times 10^6$ spores after 10 min at 250 °F.

8.2. What level of inoculation of PA 3679 ($D_0 = 1.2$ min) is required such that a probability of spoilage of 1 in 100 attributed to PA 3679 would be equivalent to 12D inactivation of *C. botulinum*? Assume the same temperature process and the same z values for both organisms. The $D_0$ value of *C. botulinum* is 0.22 min.

8.3. Calculate the length of a holding tube in high-temperature processing in an aseptic packaging system that would be necessary to provide a 5D reduction of spores of PA 3679 ($D_{250} = 1.2$ min) at 280 °F. Use a z value of 20 °F. The rate of flow is 30 gal/min, density is 65 lb/ft$^3$, and viscosity is 10 cp. The tube has 1.5-in. outside diameter and has a wall thickness of 0.064 in.

8.4. If the same system were used on another fluid having a density of 65 lb/ft$^3$ and a viscosity of 100 cp, calculate the probability of spoilage when the process is carried out at 280 °F (z = 20). The initial inoculum is 100 spores/can ($D_{250} = 1.2$ min). The rate of flow is 30 gal/min on a 1.5-in. outside diameter tube (wall thickness 0.064 in.).

8.5. If an initial inoculum of 10 spores/g of produce ($D_{250} = 1.2$ min) and a spoilage rate of 1 can in 100,000 is desired, calculate an F value for the process that would give the desired level of inactivation. Calculate the $F_{280}$ for a z value of 18 °F.

8.6. If an organism has a D value of 1.5 at 250 °F and a z value of 15 °F, calculate the $F_{240}$ for a probability of spoilage of 1 in 10,000 from an initial inoculum of 100 spores/can.

8.7. Figure 8.19 shows an air sterilization system that supplies sterile air to a process. Calculate the length of the holding tube necessary to sterilize the air. The most heat-resistant organism that must be avoided requires 60 min of heating at 150 °C sterilization and has a z value of 70 °C. The inside diameter of the holding tube is 0.695 in. Assume plug flow ($V_{max} = V_{avg}$).

8.8. (a) A food product in a 303 × 407 can has an $f_h = 5$ and a $j = j_c = 0.8$. For an initial temperature of 80 °F and a retort temperature of 250 °F, calculate the process time $B_b$. Use an $F_0 = 4$ min and z = 18 °F.

(b) The product in part (a) is processed in a stationary retort, and it takes 4 min for the retort to reach 250 °F from the time the steam was turned on. How many minutes after turning the steam on should the steam be turned off?

(c) In one of the retorts where the cans were processed, there was a mis-process, and the record on the retort temperature chart showed the following:

| Time (min) | Retort temperature °F |
| --- | --- |
| 0 | 70 |
| 3 | 210 |
| 10 | 210 |
| Sudden jump from 210°FE to 250 °F at 10 min | |
| 15 | 250 |
| 16 | Steam off, cooling water on |

What is the $F_0$ of this process. The can temperature at time 0 was 80 °F.

**Fig. 8.19** Diagram of an air sterilization system by heat (for Problem 7)

Compressed Air

2.25 atm, 30°C
15 m$^3$/hr

T=450°C

Heater

Holding Tube

L

T=400°C

Sterile Air
40°C, 1 atm

Cooler

8.9. In a given product, PA 3679 has a D value of 3 min at 250 °F and a z value of 20 °F. If the process was calculated at 280 °F for a z value of 18 °F, how many minutes of heating is required for a 5D process? What would be the actual probability of spoilage of PA 3679 if $N_0$ is 100 spores/can?

8.10. The following heat penetration data were obtained on chili con carne processed at 250 °F in a retort having a come-up time of 3 min. Assume $j = j_c$.

(a) Calculate the $f_h$ and j values and processes at:
250 °F, z = 18, $F_0$ = 8 (initial temperature = 120 °F)
240 °F, z = 18, $F_0$ = 8 (initial temperature = 120 °F)
260 °F, z = 18, $F_0$ = 8 (initial temperature = 120 °F)

(b) Calculate the probability if spoilage from FS 1518 that might occur from the process calculated at 240 °F, z = 18, $F_0$ = 8 if FS 1518 has a D value at 250 of 4 min and a z value of 22 °F for an initial spore load of 50/can.

Heat penetration data

| Time (min) | Temp. (°F) | Time (min) | Temp. (°F) |
|---|---|---|---|
| 0 | 170 | 35 | 223 |
| 5 | 170 | 40 | 228 |
| 10 | 180 | 45 | 235 |
| 15 | 187 | 50 | 236 |
| 20 | 200 | | |
| 25 | 209 | | |
| 30 | 216 | | |

8.11. A canned food having an $f_h$ of 30 and a $j = j_c$ of 1.07 contains a spore load of 56 organisms per can, and this organism had a $D_{250}$ value of 1.2 min. A process with an $_0$ of 6 min was calculated for this product at a retort temperature of 250 °F and an initial temperature of 150 °F. Subsequent analysis revealed that the spores actually have a z value of 14 °F instead of 18 °F. If the same time as the above process was used at a retort

temperature of 248 °F, calculate the probability of spoilage.

8.12. A process for a pack of sliced mushrooms in 303 × 404 cans on file with the FDA specifies a processing time at 252 °F for 26 min from an initial temperature of 110 °F. A spoiled can from one pack was analyzed microbiologically and was found to contain spore-forming organisms. Data on file for similar products show j values ranging from 0.98 to 1.15 and $f_h$ values ranging from 14 to 18 min.

(a) Would the filed process be adequate to provide at least a 12D reduction in spores of *C. botulinum* (D = 0.25 min; z = 14 °F)?

(b) If the spores of the spoilage organism have $D_0$ of 1.1 min and z of 16 °F, what would have been the initial number of organisms in the can to result in a probability of spoilage of 1 in 10,000 after the process?

(c) If you were evaluating the process, would you recommend a recall of the pack for inadequate processing? Would you be calling for additional technical data on the process before you made a recommendation? Explain your action and provide as much detail to convince a nontechnical person (i.e., lawyers and judges) that your action is the correct way to proceed.

8.13. The following data were collected in a heat penetration test on a canned food for thermal process determination.

| Time (min) | Temp. (°F) | Time (min) | Temp. (°F) |
|---|---|---|---|
| 0 | 128 | 35 (cool) | 245 |
| 3 | 128 | 40 | 243 |
| 5 | 139 | 45 | 240 |
| 10 | 188 | 50 | 235 |
| 15 | 209 | 55 | 185 |
| 20 | 229 | 60 | 145 |
| 25 | 238 | 65 | 120 |
| 30 | 242 | 70 | 104 |

The processing temperature was 250 °F and the retort come-up time was 2 min. Cooling water temperature was 60 °F.

Calculate:

(a) The values of $f_h$, $f_c$, $j_h$, and $j_c$.
(b) If this product is processed from an initial temperature of 150 °F at 248 °F, how long after retort temperature reaches 248 °F must the process be carried out before the cooling water is turned on if the final can temperature at the time of cooling must reach to within 2 degrees of the retort temperature?
(c) If the process is to be carried out at 252 °F from an initial temperature of 120 °F, calculate a process time such that an organism with a $D_0$ value of 1.2 min, and a z of 18 °F will have a probability of spoilage of 1 in 10,000 from an initial spore load of 100/can.

8.14. Beef stew is being formulated for canning. The marketing department of the company wants large chunks of meat in the can, and they stipulate that the meat should be 5-cm cubes.

The current product utilizes 2-cm cubes of all vegetables (carrots and potatoes) and meat, and the process time used is 50 min at 250 °F from an initial temperature of 150 °F. Marketing thinks that the change can be made without major alteration of the current process.

Heat penetration data for the current product obtained from the files did not specify if the thermocouple was embedded in a particle during the heating process. The $f_h$ value was reported to be 35 min and j of heating was 1.55. There was no data available on j of cooling. There was, however, an inoculated pack done where an inoculum of 1000 spores of an organism having a $D_0$ value of 1.2 min and a z value of 18 °F injected into a single meat particle in each can resulted in a spoilage rate of three cans in 1000.

(a) Calculate the $F_0$ of the process used on the current product based on the heat penetration data available.
(b) Calculate the $F_0$ of the process based on the survivors from the inoculated microorganisms. Does the inoculated pack data justify the assumption of a safe process on the current product?
(c) Is it likely that the heat penetration data was obtained with the thermocouple inside a particle, or was it simply located in the fluid inside the can? Explain your answer.
(d) Calculate the most likely value for the $f_h$ if the thermocouple was located in the center of a particle, based on the inoculated pack data, assuming that the j value would be the theoretical j for a cube of 2.02.
(e) If the $f_h$ varies in direct proportion to the square of the cube size, and j remains the same at 2.02, estimate the process time for the 5-cm size cube such that the $F_0$ value for the process will be similar to that based on the inoculated pack data on the present product.

8.15. A biological indicator unit (BIU), which consists of a vial containing a spore suspension and installed at the geometric center of a can, was installed to check the validity of a thermal process given a canned food. The canned food has an $f_h$ value of 30 min and a j value of 1.8. Of the 1000 spores originally in the BIU, an analysis after the process showed a survivor of 12 spores. The spores in the BIU has a $D_0$ of 2.3 min and a z value of 16 °F. The process was carried out at 248 °F from an initial temperature of 140 °F. Calculate:

(a) The $F_0$ value received by the geometric center of the can
(b) The process time
(c) The sterilizing value of the process expressed as a number of decimal reductions [(log $N_0/N$) of C. botulinum having a $D_0$ value of 0.21 min and a z value of 18 °F]

8.16. A canned food with an $f_h$ of 30 min and j of 1.2 is to be given a process at 250 °F with an $F_0$ of 8 min and a z of 18 °F. In order to verify the adequacy of the process, an inoculated pack is to be performed using an organism with a $D_0$ of 1.5 min and a z of 22 °F. If the process to be used on the inoculated pack is at 250 °F from an initial temperature of 130 °F, calculate the number of spores that must be inoculated per can such that a spoilage rate of 10 in 100 will be equivalent in lethality to the process $F_0$ desired. Assume j of heating and cooling are the same.

## Suggested Reading

Aiba, S., Humphrey, A. E., & Millis, N. F. (1965). *Biochemical Engineering*. New York: Academic.

Anon. (1952). *Calculation of process for canned food*. Maywood: American Can, Technical Services Division.

Ball, C. O., & Olson, F. C. W. (1957). *Sterilization in food technology* (1st ed.). New York: McGraw-Hill Book.

Charm, S. E. (1971). *Fundamentals of food engineering* (2nd ed.). Westport: AVI Publishing.

Cleland, A. C., & Robertson, G. L. (1986). Determination of thermal process to ensure commercial sterility of food in cans. In S. Thorn (Ed.), *Developments in food preservation—3*. New York: Elsevier.

Edgerton, E. R., & Jones, V. A. (1970). Effect of process variables on the holding time in an ultra high temperature steam injection system. *Journal of Dairy Science, 53*, 1353–1357.

Hayakawa, K. (1970). Experimental formulas for accurate estimation of transient temperature of food and their application to thermal process evaluation. *Food Technology, 24*, 1407.

Institute of Food Technologists. (2001). Kinetics of microbial inactivation for alternative food processing technologies. *Journal of Food Science, 66*(Suppl), 46–64.

Leniger, H. A., & Beverloo, W. A. (1975). *Food process engineering*. Boston: D. Riedel Publishing.

Lewis, M., & Heppell, W. (2000). *Continuous thermal processing of food*. Gaithersburg: Aspen Publishers.

Richardson, P. (2004). *Improving the thermal processing of foods*. Boca Raton: CRC Press.

Ruthfus, R. R., Archer, D. H., Klimas, I. C., & Sikchi, K. G. (1957). Simplified flow calculations for tubes and parallel plates. *AICHE Journal, 3*, 208.

Saravacos, G. D., & Kostaropoulos, A. E. (2002). *Handbook of food processing equipment*. New York: Kluwer Acdemic/Plenum Publishers.

Stumbo, C. R. (1973). *Thermobacteriology in food processing* (2nd ed.). New York: Academic.

Toledo, R. T. (2007). *Fundamentals of food process engineering* (3rd ed.). New York: Springer.

Zeuthen, P, Chefter, J. C., Eriksson, C., Gormley, T. T., Linko, P., and Paulus, K., Eds. 1990. Processing and quality of foods I. Elsevier, New York.

# Aseptic Processing

9

Aseptic processing and packaging has come a long way since C. Olin Ball, one of the greats in the field of food science and technology first proposed the idea in the 1920s of improving the quality of preserved foods by sterilizing the product outside of the container and filling the product into sterile containers in an aseptic environment. The initial attempts were on fruit juices which were heated to temperatures that would be sufficient to inactivate bacterial vegetative cells, yeast, and molds, followed by filling the hot juice into glass or metal containers and applying a hermetic seal. The hot juice adequately sterilized the containers before the filled containers were cooled. Later, it was recognized that the desired quality may not be achieved because of slow cooling of the hot-filled containers. This is particularly the case when a viscous product is filled into large containers. Thus, the concept of "cold-fill" was introduced and modern Aseptic Processing and Packaging progressed to enable successful applications in food preservation for a range of products filled into metal, glass, and even plastic containers. The term "aseptic" refers to a "germ-free" condition and is now referred to as "commercial sterility," a condition where pathogenic and spoilage microorganisms that can grow in the product during storage and distribution are eliminated during product sterilization, packaging

material sterilization, and filling and packaging machine sterilization.

The product suitable for aseptic processing has to be pumpable. Only liquid foods or liquids with particulates that can flow through pipes in the heating, cooling, and filling stages of an aseptic processing system are suitable for use in aseptic processing and packaging technology. The product characteristics important for successful operation and design of the system relate to fluid flow behavior and heating/cooling characteristics of the product. Several products, such as puddings, pastes, sauces, and beverages using aseptic processing, have been in the marketplace for several decades. The technology involves separate sterilization of a pumpable food, containers, and closures followed by the cooling of the food and filling and sealing in containers under a sterile environment (Clark, P.J. 2004, *Food Technology 58(11):80–81, 89*).

## 9.1 The System and Its Elements

The aseptic processing system consists of several key elements: (1) the product, (2) flow control (pumps), (3) deaerator (optional), (4) product heating, (5) hold tube, (6) product cooling, (7) surge tank, and (8) filler for packaging (Fig. 9.1).

**Fig. 9.1** Schematic of an aseptic processing system showing different components. *PD* positive displacement, *A* and *B* are two surge tanks

## 9.2    Characteristics of Specific Elements

Each element in the aseptic processing system is designed to function in concert with the whole system. Thus, the performance of the component may affect the performance of the entire system. When considering the resulting quality of the finished product, the processor should not view the equipment as individual components to be linked together but as an entire system. Before an aseptic processing system is put together, a team of experts should look at some key design elements. These include the type of process, goal, and criteria. The process used is heating of product for a certain length of time to extend its shelf life. The goal of process is to determine proper time-temperature combination to produce a safe, extended shelf life and a nutritious product.

### 9.2.1    Flow Control

The process relies on a continuous flow of product and a continuous flow of packages. Properly selected pumps achieve the continuous flow of product. FDA regulations state, "A metering

pump shall be located upstream from the holding tube and shall be operated to maintain the required rate of product flow. A means of preventing unauthorized speed changes shall be provided." The metering pump (sometimes referred to as the timing pump) is a positive displacement type because this type is less sensitive to pressure drop than are centrifugal pumps. Centrifugal pumps are used in aseptic processing systems especially for initial sterilization of the system with hot water and for cleaning in place (CIP).

Selection of the type of positive displacement pump is dependent on product characteristics such as viscosity and presence of particulates, pressure drop in the system, and cost. When the pressure drop in the system is less than 150 psig and the product is homogeneous or contains small particles (e.g., less than one-eighth inch), a rotary positive displacement pump is usually the most economical choice (Fig. 9.2). The pump can be either fixed rate or variable speed. If the pump is variable speed, a means must be provided to prevent unauthorized speed changes so that the maximum product feed rate is not exceeded. The type of pumps most commonly used for this purpose are reciprocating piston type for low to

**Fig. 9.2** (**a**) Face view of a lobe-type and (**b**) a piston-type (Marlen) positive displacement pump

medium viscosity products or interlocking lobe type as shown in Fig. 9.2. At higher pressure drop with homogeneous products, a high-pressure piston-type pump would be required, and at pressures over 2350 psig, this type of pump is the only one capable of performing. For products with particulates up to three inches in size, the reciprocating piston pump is the metering pump of choice. Centrifugal pumps should not be used as timing pumps because they are not positive displacement pumps and flow rate will be affected by system pressure.

Some means must also be developed to verify the specified product flow rate. One method is to correlate pumping rate of water through a cold system under a no-load (no back pressure) recording tachometer to provide an indirect record of the product flow rate. The presence or absence of back pressure in the system should be noted. Instruments in the form of various flow measuring devices, which give a direct indication of product flow, have been developed.

### 9.2.2  Heat Transfer/Cooling

Increasing the temperature of the product to the holding temperature and subsequently cooling the product prior to packaging is performed continuously using heat exchangers. Various methods are used, but essentially the systems can be divided into direct and indirect heat exchange methods.

#### 9.2.2.1  Direct Heating/Cooling

The product is heated by direct contact with steam, which condenses giving up its latent heat to increase the temperature of the product. Steam can contact the product in one of two ways: (1) steam injection or (2) steam infusion. Both systems are used commercially for aseptic processing of food fluids. In steam injection (Fig. 9.3), steam is dispersed into the product stream by an injector and the mixture of product with steam reaches desired temperature. Steam infusion, on the other hand, relies on dispersing the product into a steam atmosphere (Fig. 9.4). The product can be sprayed into the steam or distributed as free falling sheets, depending on equipment manufacturer.

As a rule of thumb, approximately one pound of steam will condense for every 10 pounds of product heated by 100 °F. Thus, the dilution rate of the product is 10%. Added water must be accounted for in product formulation or it must be removed. Usually the water is removed by flash cooling the product. The hot product is dispersed into a vacuum vessel operating at a pressure such that the amount of water that was added during steam heating will flash evaporate. The latent heat for flash evaporation comes from the sensible heat of the product, resulting in cooling the product. The "rule of thumb" in operation of a direct steam infusion heater and vacuum flash cooling system is that the "product temperature leaving the flash cooler should be

**Fig. 9.3** Direct steam injector (http:// primetechejectors.com/ steam_injection.htm)

**Fig. 9.4** Direct steam infusion (the PDX steam cooker)

equal to the product temperature entering the steam infusion heater" to avoid dilution of the product by the infused steam. The desired temperature is achieved by maintaining a vacuum in the "flash cooler" where the absolute pressure on the product would give a product boiling point at the desired exit temperature.

Since product volume increases in direct steam heating due to added water, it is necessary to account for this situation in sizing the hold tube to achieve the scheduled process.

Direct heating has several advantages, including (1) relatively low initial cost, (2) most rapid method of heating and cooling (nearly instantaneous), (3) minimum burn-on (fouling), (4) minimum floor space, (5) no moving parts, and (6) deaeration of the product upon flash cooling.

The disadvantages associated with direct heating are as follows: (1) water from steam condensate dilutes the product; (2) culinary steam, free of non-condensible gases, is required; (3) the product may be destabilized due to high shear during steam condensation and flashing from the product; (4) controls are required to insure no boiler chemicals carryover; (5) volatile compounds may be stripped from the product resulting in flavor changes; and (6) it may be difficult to control final temperature and solid content in the product.

### 9.2.2.2    Indirect Heating/Cooling

For indirect heating, the product does not come into contact with the heating or cooling medium, but, rather, heat is transferred through a surface. For heating/cooling liquids, there are three major types of indirect heating units: plate, tubular, and scraped-surface heat exchangers.

Plate heat exchangers consist of a number of corrugated, gasketed, thin stainless steel plates ordered together in a frame such that the product flows on one side of the plate and the heating or cooling medium flows on the other side (Fig. 9.5). The plate-type heat exchanger is commonly used for nonviscous products, such as dairy and juice products. There are some newer plate heat exchangers with wide gaps that can handle fibers/small seeds or even moderate viscosity products. The heating medium is usually hot water or steam. Hot water is used when the product is particularly prone to fouling.

Frequently product-to-product heat exchange is used in plate heat exchangers to reduce the cost of heating and cooling. This practice, called regeneration, cools the hot product while preheating the cold product. Since the hot sterile

**Fig. 9.5** Diagram of a typical plate heat exchanger

product is used indirectly to preheat the cold unsterile product, it is absolutely necessary to operate the regenerative section **so that the pressure on the sterilized product side is at least 1 psig greater than the pressure on the nonsterilized product side.** There must be an accurate differential pressure recorder-controller installed on the regenerator. One pressure sensor must be installed at the sterilized product outlet and the other at the nonsterilized product inlet.

Some advantages of plate heat exchangers are (1) low initial cost with high efficiency for heat transfer, (2) small hold up volume in the heat exchanger, (3) usually easy inspection of the unit by disassembling, and (4) flexibility of the system since it can be readily expanded or contracted to perform multiple duties. The main disadvantages are (1) the plate exchanger is generally limited to low viscosity products; (2) re-gasketing costs can be high; and (3) fouling tendencies exist for some products.

Tubular heat exchangers (Fig. 9.6) use concentric stainless steel tubes to separate the product from the heating/cooling medium or consist of a coiled tube in a shell. In the double tube and coiled tube-in-a-shell types, the product flows in the annular area. In each type, the heating or cooling medium flows countercurrent to the product to maximize heat exchange efficiency.

The main advantages of the tubular unit are (1) there are few gaskets, thereby minimizing maintenance costs; (2) the units are compact; and (3) there are no moving parts. The disadvantages include (1) excessive pressure drops; (2) inability to open and inspect the surfaces; (3) maximum regeneration of 70–75%, which is lower than that for plate exchangers; (4) fouling tendencies because of low shear; and (5) restriction of products by viscosity and particle size.

Highly viscous foods and those containing large particulates are sterilized as well as cooled in scraped-surface heat exchangers (SSHE) (Fig. 9.7). Sometimes a product may require one type of heat exchanger for heating and another type for cooling. For example, cheese sauce can be sterilized in steam injection system but cooled by a scraped-surface heat exchanger. It consists of

**Fig. 9.6** (**a**) Tube in shell and (**b**) concentric tubular heat exchangers

**Fig. 9.7** Scraped-surface heat exchanger

a mutator shaft with scraper blades concentrically located within a jacketed, insulated heat exchange tube with clearances of one-fourth to two inches. The blades rotating at 40–400 rpm continuously remove the product from the wall, increasing the heat transfer and reducing burn-on. The heating or cooling medium flows on the outside of the product tube and, in some designs, can also flow through the mutator tube. The units are generally mounted vertically and the product is introduced at the bottom to eliminate air. Advantages of the

SSHE include (1) the ability to process any pumpable product, including those containing particulate matter; (2) flexibility to expand the system by adding surface area; (3) low pressure drops; and (4) ability to process at very high temperatures without burn-on.

Significant disadvantages are (1) high cost per units area, (2) high operating costs, (3) large floor space requirement compared to other types, and (4) disintegration of fragile particles. Heat transfer from the heating medium (in the case of heating) to the foodstuff in a SSHE is done by a series of mechanisms. Firstly, the heat is transferred by conduction through the barrel wall from heating medium to inside wall of the exchanger. Considering the case of heating of a product containing particulates, a small portion of heat transfer is then achieved by conduction to the liquid phase. However, the majority of the heat is transferred by convection caused by the blades paring away material (Hayes, 1988, *AIChE Symp. Ser. 84: 251*).

There also exists convection heat transfer between bulk liquid and the surface of the particle and finally conduction heat transfer within the particle where it is thermally the most difficult part to reach (Heppell, 1985, In Proc. IUFOST Symp. Aseptic Processing and Packaging of Foods, Tylosand, Sweden, pp. 108–114).

Each of the heat exchangers described in this section is used in continuous processing systems, which require a pumpable product. The heating profile in a direct heat exchanger is square-shaped due to faster heating and quick flash cooling. A plate heat exchanger operating under high velocity regime will give similar temperature profile, but the profiles in tubular and scraped-surface heaters show some lag.

If particles are present, they must be relatively small and resistant to shear. For particulate products, two systems have been developed. Acid foods such as fruits are subjected to steam heating in a continuous steamer. The operating system consists of a device to evenly feed the product to a chamber maintained slightly above atmospheric pressure by steam injection. The product is gently conveyed in contact with steam-heated surfaces and in hot compatible liquids and juices. The product is held for the required time and is discharged through an aseptic pressure lock to the cooling vessel. The cooling vessel is pressurized with sterile nitrogen, and the product is cooled with sterile cooling media such as syrup, water, juice, or brine.

For low-acid piece-form foods, several commercial systems have been designed and built. These systems have generally consisted of a batch-type unit for heating and cooling the particulates, coupled to a continuous system for processing the sauce, water, or other liquids. In one system (Hersom and Shore 1981), the sterilized liquid is pumped into the vessel with the sterilized particles and, from there, the mixture is transferred to the filling machine. The major advantages of this system are the ability to process large particulates and to heat and cool the particulates without sauce or liquid, thereby eliminating overprocessing of the liquid. The major disadvantages include that it is a batch system, that it is very expensive, and that there are a number of transfers, which must be made aseptically.

### 9.2.3   Hold Tube

For continuous systems, the hold tube is designed to accomplish the lethality for the product. Thus, the temperature must be maintained at a specified minimum in the hold tube, and the minimum residence time must be established so that, at the hold temperature and residence time combination, product sterility is assured. The time-temperature combination is calculated according to principles discussed earlier (Chap. 8). Since the hold tube is the heart of the process to insure commercial sterility, certain precautions must be followed. It must be sloped upwards at least 0.25 inch per foot to eliminate air pockets and ensure self-draining. It must be constructed with good sanitary design criteria, such as a smooth interior without protrusions, easily disassembled for inspection and a fail-safe system for reassembly so the hold tube is not shortened or changed in diameter (this would change the minimum residence time). The tube should not be exposed to condensate drip or draft of cold air, which could affect product temperature in the tube. Heat cannot be applied at any point along the tube. Finally, the pressure in the hold tube must be maintained sufficiently above the vapor pressure of the product at the process temperature to prevent flashing or boiling. A back pressure valve usually accomplishes this.

The temperature of the product in the hold tube is monitored at both the inlet and outlet of the hold tube (i.e., at the exit from the final heater and at the end of the hold tube prior to the cooler, respectively). A temperature recorder-controller is used at the inlet so that a deviation of temperature from set point can initiate an appropriate action in the heater to return the temperature to set point. At the outlet of the hold tube, between the exit of the tube and the entrance to the cooler, an automatic recording thermometer must be used with chart graduations not exceeding 2 °F (1 °C) within a range of 10 °F (6 °C) of the desired product sterilization temperature. The product temperature measured at the exit of the holding tube is used to determine the adequacy of the process for commercial sterilization.

### 9.2.4   Deaerators

Most products which are aseptically processed must be deaerated prior to packaging. The air is removed to prevent undesirable oxidative reactions which occur as the product temperature is increased during the process. The deaerator generally consists of a vessel in which the product is exposed to a vacuum on a continuous flow. The location of the deaerator is dictated by the product. At higher temperature the noncondensable gases are less soluble but the flavor constituents are also more volatile. Thus, more flavor will be lost as the temperature of deaerating is increased. However, it is more economical to operate the deaerator at higher temperatures if the product will tolerate it. For aseptic processing, the residual oxygen in the product becomes even more important since many of the packaging materials exhibit an oxygen permeability that, even though small and not affecting asepsis, can lead to product quality deterioration over storage time.

### 9.2.5   Aseptic Surge Tank

Aseptic surge tanks are used in aseptic processing systems to allow the accumulation of a sterile product prior to packaging. Ranging in capacity from about 100 gallons to several thousand gallons, the tanks provide flexibility, especially for systems in which the flow rate of sterile

product is not compatible with packaging capacity. Valving that connects the surge tank between the end of the cooling section and the packaging system allows the processor to carry out the processing and packaging functions more or less independently. The product is pumped into the surge tank and is removed by maintaining a positive pressure in the tank with sterile air or other sterile gas. The positive pressure must be monitored and controlled to protect the tank from contamination.

### 9.3   Thermal Process for the Product

The product characteristics important for successful operation and design of the system relate to flow and heating/cooling characteristics of the product. During processing, the product travels through a series of equipment or unit operations where specific treatments are given. Properties of the product dictate the possible options in designing an aseptic processing system. Product considerations include (1) liquid or liquid plus particulates, (2) size of particulates, (3) product viscosity, and (4) thermal sensitivity of product color and flavor.

Thermal process causes change in many quality-related factors in addition to the destruction of microorganisms. Such quality changes include nutrient loss (destruction), flavor and color change, and coagulation of proteins. Quantitatively, destruction rate of microbial spores is different from the destruction rate of vitamins. Therefore, a process can be designed to cause high destruction of spores and less destruction of vitamins. In those cases where destruction of vitamins, flavor, or other key components is of major concern, those components can be aseptically added after cooling the product prior to packaging.

### 9.3.1   Influence of Product Characteristics

One of the most important product considerations is its pH. Generally, low-acid foods (pH > 4.6) require more severe heat treatment in order to attain commercial sterility than do the acid or acidified food products (pH < 4.6). The fact that the processing requirements are different for high-acid versus low-acid products may simplify equipment needs, because the demand for heat exchange capacity will be less for acid/acidified foods than that for low-acid foods. The temperature requirement for sterilization of low-acid foods (milk, cheese sauce, puddings, etc.) must exceed 121.1°C, whereas the acid/acidified products (fruit juices and drinks) could be sterilized at temperatures below 100°C for a few seconds (Table 8.8). The thermal conditions needed to produce commercial sterility in the product depend on (a) nature of the food (e.g., pH and water activity), (b) storage conditions following the thermal process (refrigerated versus room temperature), (c) heat resistance of the vegetative microorganisms or spores, (d) heat transfer to the food, and (e) the initial load of microorganisms.

A product's flow characteristics dictate the type of heat exchanger required. For example, relatively viscous products, such as puddings or sauces, may best be processed through a scraped-surface heat exchanger, while products such as fruit juices may best be processed through equipment which allows maximum retention of volatile flavor compounds, such as plate or tubular heat exchangers. In SSHE, the heat exchanger tube is lined with specific material to minimize corrosion of the tube. This material is usually specific for a given food group. The use of this tube for different food groups may damage the heat exchanger surface. Special coatings or linings may also decrease fouling of the heat exchanger surface (Sandu and Singh 1991, Food Technol. 45 (12):84–91). Some products like cheese sauce, milk, and egg proteins are very susceptible to heat treatment, and they can cause fouling in heat exchangers (Li et al. 2004, Lebensm. Wiss. Technol. 37:565–572). The same researchers found that the process conditions with high mean velocity, regardless of temperature, did not produce detectable fouling deposit. Maximum fouling was obtained in the holding section after the heating section of the SSHE when processing at 121.1°C and for 7 h due to inverse solubility salts and high-temperature exposure. Formation of deposit was extremely sensitive to temperature, and no fouling was found in the cooling sections (Li et al. 2004). Nonetheless, the fouled surface in the heat exchanger may cause product burn and thus causing quality defects.

## 9.3.2   Thermal Process Calculations

Thermal process calculations rely on the destruction of microorganisms under a certain time and temperature profile. This simply means that if the food is subjected to a constant temperature for longer time, there will be more destruction of microorganism than if it was subjected to the same temperature for shorter time. However, subjecting the food at a constant temperature for excessively long times would decrease the probability of surviving microorganisms. Furthermore, a higher temperature process will provide more microbial killing than a relatively lower temperature process for the same time period exposure of food to the heat. In case of fruit juices, the process is intended to inactivate native enzymes (1–2 log reduction) which also sufficient for inactivation of microorganisms.

The microbial destruction is a logarithmic decay process with respect to the processing time. The time required to reduce the microbial population by tenfold at a constant temperature is called the D value of the process (Chap. 8, Sect. 8.1.1). The D value provides a quantitative index for heat resistance of microbial cells or spores. The process given to a food material is quantified in terms of multiples of D value at a reference temperature (i.e., a 6D process or 12D process, etc.), which is also referred as F value at reference condition ($F_0$). The accepted reference temperature for sterilization is 121.1 °C and for pasteurization, it is 82.2 °C. The D values of each microbial species also depend on process temperatures. Higher temperatures are more lethal than lower ones, meaning that higher temperature would give lower D values for the same microorganism. The temperature change required to cause a tenfold change in D value is called z value and is used for comparing processes at various temperatures. An example of some D and z values is given in Tables 8.2 and 8.3. The

**Table 9.1** Approximations of heat processes for destruction of *C. botulinum* and commercial sterility

| Process | Storage | Heat/hold | Sterilizing value[a] |
|---|---|---|---|
| Ultrapasteurization | Refrigerated | 138 °C, 2 s | $F_0$ 1.5 min |
| Minimum to destroy *C. botulinum* spores | Nonrefrigerated | 138 °C, 4 s | $F_0$ 3.0 min[b] |
| Commercial sterility | Nonrefrigerated | 138 °C, 8 s | $F_0$ 6.0 min |
| European UHT range | Nonrefrigerated | 135 °C, 3 s | $F_0$ 1.22 min |
|  |  | 140 °C, 3 s | $F_0$ 3.87 min |

*Source:* McGarrahan, E.T. 1982, *J. Dairy Sci. 65*:2023–2034
[a]Sterilizing values were calculated for $Z = 10$ °C
[b]Minimum for public health safety varies based on product and process

lower D and z values for vegetative cells imply that they can be destroyed faster than spores. Enzyme destruction takes more time than the destruction of microorganisms. Quality factors and nutrients are relatively stable under high-temperature short-time thermal process.

The target for sterilizing value or $F_0$ is set based on the product type and its intended use (refrigerated or shelf-stable). The target for setting-up the thermal process could be based on microbial kinetics or enzyme. In case of fruit juices and several high-acid beverages, the target is enzyme inactivation (Kim et al. 1999a. J. Food Proc. Eng. 22(5):395–403). Approximations of heat process for destruction of *C. botulinum* and commercial sterility are shown in Table 9.1 (McGarrahan 1982). The reported sterilization values are for actual processes in-practice but similar values can be calculated for other processes with different targets as a multiple of D value.

**Example 9.1** *Clostridium botulinum* spores suspended in phosphate buffer had $D_{121°C} = 0.37$ min. How long would it take to reduce a population of *C. botulinum* spores in phosphate buffer from $10^{12}$ spores to $10^0$ spores (1 spore) at 121 °C?

**Solution**
(See details in Example 8.6)

Since $10^{12}$ to $10^0$ is 12 log reductions, then the time required is 12 × 0.37 min = 4.44 min.

**Example 9.2** Pectinesterase (PE) enzyme in orange juice pulp is a heat-resistant enzyme, and it is inactivated in the aseptic processing of orange juice with pulp. Its D and z values are available from the literature as, at 90 °C, $D_{90} = 0.53$ min

(31.8 s), $z = 6.5$ °C (Wicker & Temelli 1987. *J. of Food Sci.*). Calculate at 100 °C.

**Solution**
(Use Eq. 8.11)

$$D_{100} = 31.8 * 10^{(90-100)/6.5} = 0.92 \ s$$

What is D value for PE at 96.5 °C? $D_{96.5} = 0.053$ min (3.18 s).

Note that a rise of one z value reduces the D value by tenfold.

**Example 9.3** A fluid milk processing plant is designed to sterilize the milk for aseptic packaging. The various unit operations involved and the temperature profile is shown in Fig. 9.8. Using the thermal processing calculations from Chap. 8, calculate the total lethality ($F_0$) for this process. For simplicity, use the following assumptions: (1) uniform product temperature in each section and (2) logarithmic bacteria kill over the process temperatures.

Heating profile of process:

| 0–1 Preheater | 50–71 °C | | |
|---|---|---|---|
| 1–2 Homogenizer | 71–74 °C | 2.5 s | |
| 2–3 Regenerative heating | 74–95 °C | 4.5 s | |
| 3–4 Hold | 95 °C | 2 s | |
| 4–5 Steam heater (HE) | 95–139 °C | 17.5 s | $T = 142.3$ [$1 - 0.3324$ $e^{-0.1524t}$] |
| 5–6 Hold tube (HO) | 139 °C | 2.4 s | |
| 6–7 Regenerative cooling (CO) | 139–50 °C | 21.6 s | $T = 139 - 4.12 \ t$ |

**Fig. 9.8** Heating curve for milk sterilization in a plate heat exchanger with preheating

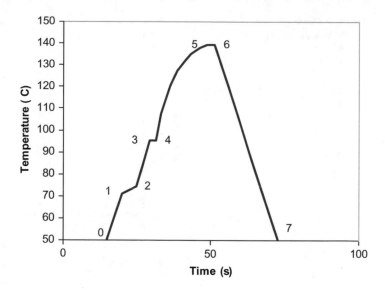

**Table 9.2** Lethality calculation for $L_{HE}$ by numerical integration

| Time (s) | Temperature (°C) | L.R. | Area under curve (s) |
|---|---|---|---|
| 0 | 95 | 0.0025 | 0.0 |
| 2.92 | 112 | 0.126 | 0.19 |
| 5.84 | 123 | 1.54 | 2.43 |
| 8.76 | 130 | 7.68 | 13.46 |
| 11.68 | 134 | 21.5 | 42.60 |
| 14.6 | 137 | 41.58 | 92.10 |
| 17.5 | 139 | 63.31 | 152.09 |

Add the area column and, thus, total area = 302.87 s = 5.05 min

**Solution**

Total Lethality = $F_0$

$$= L_{HE} + L_{HO} + L_{CO} \quad (9.1)$$

Lethal rate (L.R.) = $10^{(T-121)/z}$ where 121 °C is $T_{ref}$ and z = 10 °C.

Then  $L_X = \int L.R.dt = \int 10^{(T(t)-121)/z}dt \quad (9.2)$

For heating section: 17.5 s, where T(t) = 142.3 $[1-0.3324e^{-.1524t}]$.

$$L_{HE} = \int 10^{[142.3[1-0.3324e^{-0.1524t}]-121]/10}dt \quad (9.3)$$

$$L_{HE} = \int_0^{17.5} 10^{2.13}\, 10^{-4.73}\, 10^{e^{-0.01524t}} dt$$

$$= \int L.R.dt \quad (9.4)$$

Calculate L.R. for each time-temperature segment (Table 9.2) and graph it against the time (s) as shown in Fig. 9.9.

For hold tube: 2.4 s at 139 °C.

$$L_{HO} = L.R. \cdot \Delta t_{HO} = 10^{(139-121)/10} \cdot 2.4s \cdot \frac{1min}{60s}$$

$L_{HO}$ = 2.52 min

For cooling section: 21.6 s, where T(t) = 139–4.12 t.

**Fig. 9.9** Graph of
calculated L.R. versus
time (s)

$$L_{CO} = \int L.R.dt = \int 10^{[(139-4.12t)-121]/10}dt$$
$$= \int 10^{1.8-.412t}dt = 10^{1.8} \int 10^{-.412t}dt$$

$$L_{CO} = 10^{1.8} \cdot \frac{10^{-0.412t}\big|_0^{21.6}}{-0.412} = 153.14\,\text{seconds}$$
$$= 2.55\,\text{minutes}$$

$F = L_{HE} + L_{HO} + L_{CO}$
$F = 5.05 + 2.52 + 2.55 = 10.12$ min
Heating and cooling sections contribute ~3 ×
hold tube lethality = 7.6 min.

The sterility values for a heating and cooling
phase are calculated from the time-temperature
[T(t)] profile of the product as:

$$F_0 = \int 10^{[T(t)-TR]/z}dt \qquad (9.5)$$

where T (t) = product temperature (°C) as a
function of time t and TR = reference temperature
(°C). The limit for integration time is the time for
that particular segment of process where sterilization value needs to be estimated. Equation (9.5) is
useful to find out overprocessing of product or it
can be used for high-acid foods. In case of low-
acid foods, the measured temperature at the end of
hold tube is normally used for estimating the $F_0$
for the process but that value is an overestimate. If
process equipment has staged heating, the inlet
temperature value to a heat exchanger module can
be used for calculation of the sterilization value.

The process sterilization value must be equal or
higher than the target $F_0$ value; otherwise, the
product will be under-processed. If the product
has particulates mixed with carrier fluid, the temperature of fluid can be used for estimation of
product temperature and thus the sterility at the
center of particles (Lee and Singh 1990, *Chem.
Eng. Comm.* 87:21–51; Lee et al. 1990a, *J. Food
Eng.* 11:67–92; Bhamidipati and Singh 1994,
*Biotechnol. Progr.* 10:230–236). Verification of
sterilizing value in these situations has been done
by the microbial, enzymatic, or chemical
indexing methods (Bhamidipati and Singh 1996,
*J. Food Sci.* 61:1–5).

## 9.4    Flow Characteristics of Product

As mentioned earlier, aseptic processing is a continuous process and thus the product must be
pumped through a set of unit operations
connected with pipes. A number of parameters
are important for the flow of fluid foods. These
include fluid viscosity, fluid density, fluid velocity size and shape of the pipe, and surface roughness of equipment. The flow may be broadly
characterized as either laminar flow or turbulent
flow. Laminar flow is characterized by the fluid
particle  moving  smoothly  along  parallel
streamlines, whereas the turbulent flow is
characterized by fluid particles moving in a

**Fig. 9.10** Velocity profile
in a pipe flow

Laminar velocity profile (parallel
streamlines)

Wall of pipe

Center element

Flatter profile

Turbulent velocity profile (mixing of
stream lines)

random, tumbling, churning motion (Fig. 9.10).
The velocity of fluid flowing in a pipe is greatest
at the center and decreases toward the pipe wall
with the velocity of the fluid at the wall being
zero.

Temperature could easily distort the flow profile in a pipe flow. A constant temperature across
the section of flow will not cause distortion, but a
heated wall (heating section of heat exchanger)
will make the flow profile flatter by causing more
wall slippage (Chakrabandhu and Singh 2005,
*J. Food Eng.* 70:73–81), and cooling will cause
a much faster center velocity than the isothermal
condition.

Viscosity is a measure of resistance of fluid to
shear. Fluids are described as Newtonian or non-
Newtonian fluids based on the behavior of viscosity. The viscosity of a Newtonian fluid is not
influenced by shear rate (i.e., through homoge-
nizer, pumping, etc.), whereas viscosity of a
non-Newtonian fluid can be influenced by the
shear rate. Viscosity also depends on temperature
of the product. Most fluids display lower viscosity with increasing temperatures unless they are
heated beyond their phase transition (starch gelation or protein denaturation/aggregation). In case
of phase transition, the trend may reverse and the
increasing temperature could cause an increase in
viscosity.

The distinction between laminar and turbulent
flows is based on Reynolds number (Re) which is

calculated from fluid density, fluid velocity, fluid
viscosity, and the diameter of the pipe (Eq. 5.42).

$$Re = \frac{D\bar{V}\rho}{\mu}$$

If the flow is laminar (Re < 2300), the fastest
moving element moves twice as fast as the average moving element. If the flow is turbulent
(Re > 4000), then the fastest moving element
moves only 1.22 times faster than the average.
Maximum velocity of fluid element is used for
calculating the minimum hold time as:

$$t_{min} = L/V_{max} \tag{9.6}$$

where $t$ = time (s), $V_{max}$ = velocity of fastest
moving element (m/s), and $L$ = the length of the
hold tube (m). The velocity of the fastest moving
particle depends on the velocity distribution
within the hold tube.

For laminar flow of Newtonian fluids, the
maximum velocity is twice the average velocity,
that is, $V_{max}/V_{avg} = 2$, where $V_{avg}$ (the average
velocity of fluid) can be calculated by dividing the
flow rate by the cross-sectional area. In case of
turbulent flow condition, the velocity ratio ranges
from 1.22 to 2.

**Example 9.4** A processor uses the same equipment to process products flowing at a rate of
10.4 gal/min (39.4 L/min) with different

**Table 9.3** Flow behavior of soybean milk as influenced by the temperature and solids content

|          | 9% solids |        | 13% solids |        | 16% solids |        |
|----------|-----------|--------|------------|--------|------------|--------|
|          | 25 °C     | 65 °C  | 25 °C      | 65 °C  | 25 °C      | 65 °C  |
| n        | 0.77      | 0.77   | 0.67       | 0.68   | 0.58       | 0.60   |
| K (pa s$^n$) | 0.02  | 0.008  | 0.14       | 0.04   | 0.70       | 0.18   |

*Source:* Son and Singh 1998. *Int. J. Food Prop.* 1(1):57–70
n = power law index
K = consistency coefficient

viscosities. How will the flow profile change with changed viscosity product? If the processor changed the pipe (tube) diameter, how will the flow profile change?

**Solution**

Assume that the viscosity of cream was 10 cP and then the processor processed milk, viscosity 2 cP. Use Eq. 5.42 for pipe diameter of 2 inch.

Re for 10 cP = 1580 (laminar flow)
Change diameter to 1 inch, Rc = 3160 (transitional flow)
Viscosity change to 2 cP for 2 inch pipe, Re = 7900 (turbulent flow)

Flow properties of several food products have been characterized and used for prediction of flow in aseptic processing systems (Bhamidipati and Singh 1990, *J. Food Process Eng.* 12:275–293; Son and Singh 1998, *Int. J. Food Prop.* 1(1):57–70; Ditchfield et al. 2004, *Int. J. Food Prop.* 7(3):571–584, and Chakrabandhu and Singh 2005, *J. Food Eng.* 66:117–128). Most food fluids show non-Newtonian pseudoplastic behavior and can be modeled by the power law model:

$$\tau = K (\gamma)^n \qquad (9.7)$$

where $\tau$ = shear stress (Pa), K = consistency coefficient (Pa s$^n$), $\gamma$ = shear rate (s$^{-1}$), and n = power law index (dimensionless). If the value of n is less than 1, the fluid has a pseudoplastic or shear thinning behavior; if n is equal to 1, it is Newtonian; and if n is more than 1, then it is dilatant. The values of K and n are temperature dependent with K having a higher temperature dependency than n. Bhamidipati and Singh (1990) measured the flow behavior of

tomato sauce at 7° and 14° Brix with and without particulates. The values of n and K for 7° Brix sauce ranged from 0.30–0.86 and 0.04–0.34 (Pa s$^n$), respectively. For the 14° Brix sauce, these values were 0.27–0.53 and 0.20–0.60 (Pa s$^n$), respectively. The values of n decreased and K increased with increase in concentration of solids and particulates. Son and Singh (1998) modeled flow behavior of soy milk at various temperatures and solid concentrations and observed similar trend (Table 9.3).

Rheological properties of banana puree were measured at high temperatures and modeled as Herschel-Bulkley model which is extension of the power law model and accounts for the yield stress (Ditchfield et al. 2004). The puree presented a complex rheological behavior, which changed drastically with temperature (Table 9.4). In general, the higher temperature lowers the viscosity but the puree showed increased viscosity for temperature range of 50–60°C and 110–120°C. Such unexpected changes were attributed to the presence of starch in banana and the interactions of polysaccharides. Therefore, it is very important to have experimental data on rheological behavior of products to be processed aseptically.

### 9.4.1 Residence Time Distribution (RTD)

In a complete system, the velocity distribution leads to a residence time distribution (Singh and Morgan 2010, in *Nelson, P.E. (ed.) Principles of aseptic processing & packaging*). Experimental determination of residence time distribution is carried out by measuring the response of an inert tracer injected into the product stream—the so-called stimulus-response technique. The

**Table 9.4** Parameters obtained from experimental data fitting for banana puree flow curves to the Bingham, power law, and Herschel-Bulkley models for the increasing shear rate (up) and decreasing shear rate (down) at various temperatures

| Temperature (°C) | | Bingham model | | | Power law model | | | Herschel-Bulkley model | | | |
|---|---|---|---|---|---|---|---|---|---|---|---|
| | | $\sigma_0$ (pa) | K (pa.s$^n$) | $R^2$ | K (pa.s$^n$) | n | $R^2$ | $\sigma_0$ (pa) | K (pa.s$^n$) | n | $R^2$ |
| 30 | Up | 101.1 | 0.103 | 0.90 | 76.7 | 0.104 | 0.89 | 81.0 | 4.67 | 0.442 | 0.99 |
| | Down | 91.1 | 0.129 | 0.92 | 44.4 | 0.198 | 0.98 | 31.9 | 21.4 | 0.281 | 0.99 |
| 40 | Up | 100.4 | 0.081 | 0.92 | 64.9 | 0.116 | 0.83 | 82.2 | 3.15 | 0.457 | 0.96 |
| | Down | 78.5 | 0.110 | 0.89 | 74.1 | 0.071 | 0.59 | 50.0 | 9.12 | 0.354 | 0.98 |
| 50 | Up | 86.5 | 0.054 | 0.89 | 43.4 | 0.163 | 0.87 | 74.8 | 1.06 | 0.614 | 0.99 |
| | Down | 68.9 | 0.056 | 0.90 | 79.1 | 0.055 | 0.58 | 58.5 | 1.32 | 0.594 | 0.95 |
| 60 | Up | 83.6 | 0.119 | 0.84 | 81.7 | 0.054 | 0.72 | 69.0 | 4.92 | 0.412 | 0.96 |
| | Down | 77.5 | 0.085 | 0.95 | 82.4 | 0.041 | 0.53 | 68.0 | 1.70 | 0.552 | 1.0 |
| 70 | Up | 72.4 | 0.071 | 0.86 | 55.0 | 0.108 | 0.75 | 56.0 | 2.02 | 0.542 | 1.0 |
| | Down | 69.4 | 0.046 | 0.83 | 73.8 | 0.054 | 0.61 | 55.4 | 1.79 | 0.556 | 1.0 |
| 80 | Up | 65.2 | 0.063 | 0.90 | 71.7 | 0.046 | 0.75 | 56.4 | 1.71 | 0.523 | 0.97 |
| | Down | 57.1 | 0.055 | 0.93 | 24.6 | 0.216 | 0.87 | 52.4 | 0.13 | 0.889 | 0.95 |
| 90 | Up | 60.4 | 0.035 | 0.85 | 69.3 | 0.031 | 0.60 | 54.9 | 0.32 | 0.701 | 0.96 |
| | Down | 46.9 | 0.045 | 0.79 | 45.2 | 0.080 | 0.39 | 44.6 | 0.30 | 0.787 | 0.99 |
| 100 | Up | 59.1 | 0.040 | 0.87 | 71.6 | 0.035 | 0.71 | 50.9 | 0.17 | 0.809 | 0.93 |
| | Down | 42.7 | 0.056 | 0.87 | 45.8 | 0.073 | 0.58 | 35.2 | 0.43 | 0.709 | 1.0 |
| 110 | Up | 27.8 | 0.031 | 0.89 | 35.9 | 0.053 | 0.75 | 30.9 | 0.06 | 0.871 | 0.98 |
| | Down | 18.2 | 0.035 | 0.93 | 16.8 | 0.149 | 0.83 | 12.2 | 1.13 | 0.539 | 0.99 |
| 120 | Up | 36.2 | 0.031 | 0.80 | 47.3 | 0.054 | 0.70 | 40.9 | 0.03 | 0.966 | 0.95 |
| | Down | 27.9 | 0.033 | 0.85 | 28.5 | 0.092 | 0.80 | 28.8 | 0.97 | 0.516 | 0.97 |

$\sigma_0$ = yield stress, K = consistency coefficient, n = power law coefficient
*Source:* Ditchfield et al. 2004., *Int. J. Food Prop.* 7(3):571–584

tracers are usually a soluble dye, salt, or acid that can be detected at the exit of the test zone. The techniques are all based on the detection of changes at the output stream of a test zone (or entire system) based on known changes to the input stream which include (1) a step input, in which the input concentration (of a tracer or other measureable component) is changed from one steady level to another; (2) a pulse input, in which a relatively small amount of tracer is injected into the feed stream in the shortest possible time; and (3) a sinusoidal input, in which the frequency of the sinusoidal variation is changed, thus generating a frequency-response diagram for the system. In principle, any one of the three techniques can be transformed into the other two. However, it is experimentally easier to monitor a tracer response that approximates a step change or a pulse input because the measurements associated with sinusoidal variations are much more time consuming and require special equipment.

The RTD is measured by a salt (tracer) injection test for liquids and a slug of particles via a side loop for slurries. Quantification of RTD provides information on minimum and maximum times spent by food elements in the system. It also provides insight into channeling or dead spaces in the system. Experimental determination of RTD is required by the US Food and Drug Administration during the commissioning trials for a filed process. The regulatory requirement is only for the hold tube if only the hold tube sterilization value had to be accounted for in the process filing. The measurements should be done for actual product and the actual conditions.

In laminar flow, using coiled tubes instead of straight tubes has been suggested as a means to achieve radial mixing of the fluids; otherwise, the spot injection of tracer material will exit in the same streamline flow and will not provide the realistic distribution of flow (Levenspiel 1972). This is due to the secondary flow (Dean's effect) which occurs as a result of the imbalanced centrifugal force of the flow in the curved tube. The secondary flow pattern can be described as a double vortex circulation in the perpendicular plane to the main flow, superimposed on the velocity profile in the axial direction. In addition to the narrower RTD attained in the curved tube, it has also been suggested that the existence of the Dean effect helps enhance the heat transfer in tube flow (Kao, 1987, *J. Fluid Mechanics 184:335–356*). In order to create radial mixing of fluid near injection port and near the monitoring port, a small inline mixing section is used. However, after the commissioning trail, the inline mixing section is replaced with a same sized pipe to avoid entrapment of fibers or particles.

### 9.4.1.1    Mean Residence Time (t̄)

The mean residence time of a system can be calculated if the product's volumetric flow rate is known, and it is assumed that there is no velocity distribution in the flow stream (i.e., ideal flow conditions with all the particles traveling at the same speed). In such a situation, the mean residence time is simply the volume of test zone (components between the inlet and outlet where the tracer is injected and measured, respectively) divided by the product flow rate.

$$Volume\ of\ pipe\,(V) = \pi \cdot r^2 \cdot L \qquad (9.8)$$

where r = internal radius of pipe and L = length of pipe section.

$$Then\ \bar{t} = \frac{V}{Q} \qquad (9.10)$$

where Q = volume flow rate (gal/min).

**Example 9.5** Orange juice flows through a straight pipe (2″ i.d.) 21 feet long, at a flow rate of 3 gal/min. What is the mean residence time of orange juice in the pipe?

**Solution**

Volume of orange juice in the pipe $(V)$
$$= \pi \left(\tfrac{1}{12}ft\right)^2 21ft = 0.458\,ft^3$$

Flow rate of orange juice (Q)

$$= 3\,\frac{gal}{min} \cdot \frac{1ft^3}{7.48\,gal} = 0.401\,\frac{ft^3}{min}.$$

$$Mean\ residence\ time\ (\bar{t}) = \frac{0.458\,ft^3}{0.401\,\frac{ft^3}{min}} = 1.14\,min$$

### 9.4.1.2    RTD Using Step Change Tracer Method

The product stream at the inlet of the test zone (i.e., hold tube or other device) is subject to a step input change of a tracer. The tracer is usually a salt, acid, or dye which can be detected at the outlet using a conductivity, pH, or color sensor, respectively. The probability that any element of volume entering the system at time t = 0 has left the test zone within the time t is equal to F(t) or fraction eluting as a function of time. The probability that the element is still in the system and will leave at a time later than t is [1 F(t)].

It will always take a finite time for a fluid element to traverse the system, so F(t) = 0 at t = 0. Likewise, none of the material can remain in the flow system indefinitely, so F(t) = 1.0 at t = ∞. Let us examine these limiting values of F (t) curves for various types of ideal flow.

#### 9.4.1.2.1    Laminar Flow

As shown in Fig. 9.10, the laminar flow is characterized by its parabolic velocity profile. Product elements in the center of the stream move fastest and will thus leave the system before the elements closer to the walls. The response of a step tracer will be an F(t) curve for this type of flow as shown in Fig. 9.11.

Note that the F(t) curve for laminar flow is a smooth sigmoid shape. In the ideal situation, 50% of the product would exit the system before, and after, the mean residence time (t̄).

#### 9.4.1.2.2    Plug Flow

Ideal plug flow is characterized by its distinctive flat velocity profile, as shown in Fig. 9.12a. In this case, all of the fluid elements across a section of the flow system move at the same speed, and thus they all exit at a time equal to the mean residence time.

Figure 9.12b shows an F(t) curve for product in plug flow. Notice the "stair-step" nature of the F(t) curve for plug flow. This shows that 100% of the product entering the test zone at time = 0 exists at exactly the mean residence time, t̄.

### 9.4.1.2.3   Channeling

Channeling is encountered when a portion of tracer emerges from a vessel in an unusually short time. This portion of tracer that exits the

system before the bulk of tracer exits has found a "shortcut" through the vessel. This happens when the inlet and outlet of a system are close to each other. In such situations, a portion of input tracer immediately flows out of the vessel before it can be well mixed.

Figure 9.13 shows a typical F(t) curve for a vessel with channeling. The large initial rise in the F(t) curve is due to the tracer that short-circuits and immediately exits the system.

### 9.4.1.2.4   Dead Space

Dead space leads to the situation of a nonideal flow. In this case, part of the fluid entering the zone is caught in the dead space, thus reducing its flow velocity through the pipe. Dead space occurs in corners, baffles, or regions with obstruction to flow. Figure 9.14a shows dead space in a pipe. The F(t) curve for plug flow in a pipe with dead space is shown in Fig. 9.14b.

Notice that instead of the usual stair-step associated with plug flow, a step with a curved top is obtained. The curve is due to the portion of fluid which is caught in the dead space. The size

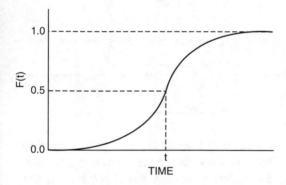

**Fig. 9.11**   F(t) curve for laminar flow

**Fig. 9.12a**   Ideal plug flow profile

**Fig. 9.12b**   F(t) curve for a plug flow

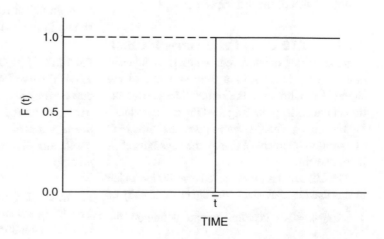

**Fig. 9.13** F(t) curve for a
system with channeling

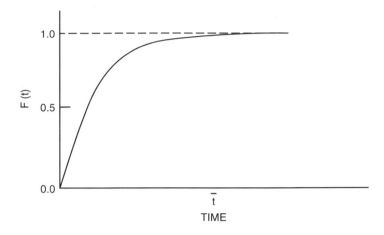

**Fig. 9.14a** Dead space
flow in a pipe

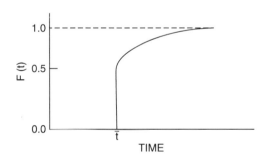

**Fig. 9.14b** F(t) curve for a dead space flow

and shape of the curve depend on the
characteristics of the dead space.

### 9.4.1.3  RTD Using Pulse Tracer Method

A pulse tracer of sufficient amount is injected
quickly into the inlet of a flow stream and the
signal is monitored at the outlet. The concentration at the outlet plotted against the time produces
the E(t) curve. The E(t) curve gives the "fraction"
of volume element exiting the system at a
particular time.

The E(t) curve is a normalized distribution which
means that the total area under the curve is unity or
$\int_0^\infty E(t)\,dt = 1$. According to this definition, the

fraction of volume elements in the exit stream that
have spent less time in the vessel than $t_i$ is equal to
the area under the curve from t = 0 to t = $t_i$. Or:

$$\int_0^{t_i} E(t)dt \tag{9.11}$$

Likewise, the fraction of volume elements in
the exit stream that have spent more time in the
vessel than $t_i$ is:

$$\int_{t_i}^{\infty} E(t)dt = 1 - \int_0^{t_i} E(t)dt \tag{9.12}$$

A graphical representation of an E(t) curve is
shown in Fig. 9.15a.

#### 9.4.1.3.1  E(t) Curves for Laminar Flow

The E(t) curve for a laminar flow is a bell-shaped
curve symmetrical about $\bar{t}$ (Fig. 9.15a). For laminar flow through a tube or pipe, the E(t) curve
displays a peak at $\bar{t}/2$ which corresponds to the
maximum velocity equal to twice the average
velocity.

#### 9.4.1.3.2  E(t) Curve for Plug Flow

The E(t) curves for plug flow under ideal and
nonideal situations are shown in Figs. 9.15b and

**Fig. 9.15a**  (Left) E(t) curve for laminar flow

**Fig. 9.15b**  (Right) E(t) curve for an ideal plug flow

**Fig. 9.15c**  E(t) curve for nonideal plug flow

9.15c, respectively. For ideal plug flow, all of the injected tracer exits at the same time which corresponds to the mean residence time, t̄.

#### 9.4.1.3.3    E(t) Curve for Channeling
The E(t) curve for a channeling has the same shape as the one without channeling, but it has a sharper initial drop. Short-circuiting fluid that leaves the vessel quickly accounts for the large initial drop.

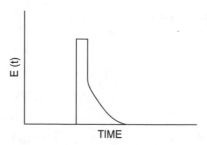

**Fig. 9.15d**  E(t) curve for dead space

#### 9.4.1.3.4    E(t) Curve for Dead Space
An E(t) curve for plug flow with dead space is shown in Fig. 9.15d. The tail following the pulse is caused by the volume element that was slowed down due to the dead space.

#### 9.4.1.4    Reduced Time (θ)
Time axis (i.e., *x-axis*) of both F(t) and E(t) curves can be made dimensionless by dividing time by t̄. This reduced time (θ) will equal 1.0 at the mean residence time (t̄) and allows systems of different sizes to be compared on the same scale.

**Example 9.6**  A pipe is 10 ft. long with a 2″ inner diameter. A chocolate milk beverage ($\mu = 4$ cP @ 25 °C) flows through the pipe at a rate of 3 gal/min. The E(t) curve for the chocolate milk is given in Fig. 9.16a. Draw the dimensionless E(t) curve.

**Solution**
**Step 1.** Calculate mean residence time ($\bar{t}$) as shown in Example 9.5:

$$\bar{t} = \frac{\pi \cdot \left(\frac{1}{12} ft^2\right) \cdot 10 ft \cdot 60 \frac{s}{min}}{0.401 \frac{ft^3}{min}} = 32.6\, s$$

**Step 2.** Divide time (t) on the x-axis by the mean residence time (32.6 s) to get E(θ) on the y-axis vs. reduced time (θ) on x-axis.

In Example 9.6, the Reynolds number is ~ 1183, which indicates laminar flow in the pipe

**Fig. 9.16a**  E(t) curve for
laminar flow in example 9.6

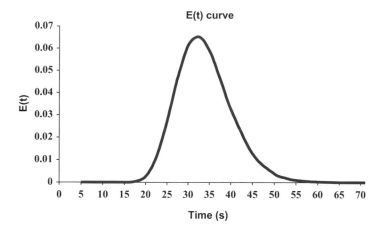

**Fig. 9.16b**  Dimensionless
E(t) curve [E(θ) curve) for
Example 2

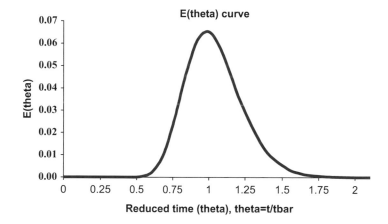

filled with the chocolate milk flowing at 3 gal/ min. The maximum velocity of the product for a laminar flow will be approximately twice the average velocity. Or, this can be stated as: the shortest residence time is approximately half of the mean residence time. In Figs. 9.16a or 9.16b, the shortest residence time can be identified by the time that the tracer first starts to show up in outlet stream, i.e., when the E(t) curve first becomes nonzero. In Fig. 9.16a, this occurs between 15 and 20 s. One-half of the mean residence time, in this case, is 32.6/2 s or 16.3 s (clearly in the expected range).

It should be pointed out that for pipe flow, the fastest moving fluid under any flow conditions for either Newtonian or Pseudoplastic fluids will never exceed twice the mean velocity (Pinheiro-

Torres and Oliveira, 1998, *J. Food Eng. 36:1–30*). Therefore, no portion of the product should exit the pipe in less than half the mean residence time.

### 9.4.1.5  Comparison Between E(t) and F(t) Curves

A simple example will be used to compare the two types of curves E(t) and F(t). Suppose that water is running through a vessel and, at some instant $t_0$, a 1 M solution of HCl replaces the water (a step change input). A pH meter is used to measure the concentration of acid in the outlet stream. At any time $t_i$ greater than $t_0$, only acid solution in the exit stream could have been in the vessel for a time shorter than $t_i-t_0$. The fraction of acid solution in the outlet stream = Fraction of outlet stream that has been in the vessel for a time less than $t_i$-$t_0$. The equation becomes:

$$F(t) = t_0 \int^{t_i} E(t)dt$$

if $t_0$ is set at 0

then : $\quad F(t) = \int_0^{t_i} E(t)dt \qquad (9.13)$

or, $\dfrac{dF(t)}{dt} = E(t)$

Thus, the derivative of the F(t) curve with respect to time gives the E(t) curve. Because of this relationship, an E(t) curve can easily be generated if an F(t) curve is known. Simply take the slope of the F(t) curve for various times and plot the slope vs. time (Figs. 9.17a and 9.17b).

Likewise the relationship

$$F(t) = \int_0^{t_i} E(t)dt \qquad (9.14)$$

can be used to generate the F(t) curve for a given E(t) curve.

Integrating an E(t) curve gives the corresponding F(t) curve. The integration can be accomplished by either graphical or numerical techniques. In order to use the graphical technique, choose a time $t_i$, and determine the area under the curve from time zero to $t_i$ which generates the value of F for that particular $t_i$. Thus by choosing a set of $t_i$, an F(t) curve can be generated (Figs. 9.18a and 9.18b).

**Example 9.7** A pulse input tracer of 0.1 M HCl was injected into the product stream at the beginning of a test zone. The pH of the output stream was recorded at 2 s intervals after the tracer first began to exit the section of tube. Make E(t) and F(t) curves from the following recorded data:

**Fig. 9.17a** Graphical differentiation of F(t) curve

**Fig. 9.17b** E(t) curve generated from the F(t) curve in Fig. 9.17a

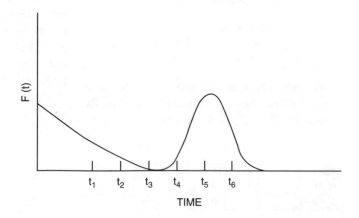

**Fig. 9.18a** Graphical
integration of an E(t) curve

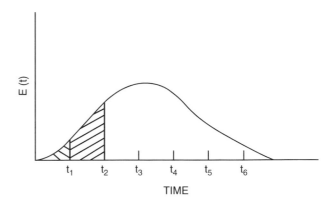

**Fig. 9.18b** F(t) curve
generated from E(t) curve
in Fig. 9.18a

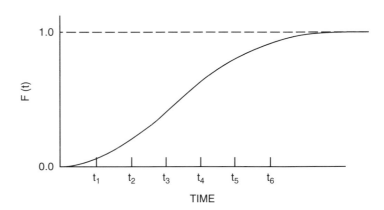

| Time (s) | 0 | 8 | 10 | 12 | 14 | 16 | 18 | 20 | 22 | 24 |
|---|---|---|---|---|---|---|---|---|---|---|
| pH | 11 | 11 | 8 | 7 | 6 | 6 | 7 | 9 | 10 | 11 |
| ΔpH | 0 | 0 | 3 | 4 | 5 | 5 | 4 | 2 | 1 | 0 |

**Solution**

An E(t) curve can be generated from this set of
data. The total area under the pH curve is:

$$A = \sum \Delta pH \times \Delta t$$
$$= (3 + 4 + 5 + 5 + 4 + 2 + 1) \times 2\,s$$
$$= 48\,pH \times s$$

The area under an E(t) curve must be unity.
This result can be accomplished by plotting the
change in pH, or ΔpH/48, vs. time.

The E(t) curve is plotted in Fig. 9.19a. Now
use the E(t) curve to determine the F(t) curve for
this pulse input injection.

Notice that there are ~100 squares under the
curve. (Each square is 0.004 high and 2.5 wide for
an area of 0.01 and the total area under the curve
is 1.0.) Pick a time and count how many squares
are under the curve from time zero to the chosen
time. Multiply the counted squares with the
proper scale factor (i.e., 0.01 in this case) to get
an F(t) curve value associated with the chosen
time. By repeating this over the entire time
range, an F(t) curve for the process is generated.

Figure 9.19b demonstrates the area under the E
(t) curves for times of 5 and 15.

| Time (s) | 8 | 10 | 12 | 14 | 16 | 18 | 20 | 22 | 24 |
|---|---|---|---|---|---|---|---|---|---|
| **ΔpH** | 0 | 3 | 4 | 5 | 5 | 4 | 2 | 1 | 0 |
| **ΔpH/48** | 0 | 0.063 | 0.083 | 0.14 | 0.104 | 0.083 | 0.042 | 0.021 | 0 |

**Fig. 9.19a**  E(t) curve for Example 9.7

**Fig. 9.19b**  Graphical determination of area under the E(t) curve in Fig. 9.19a. Areas shown for 0 to 10 s and 0 to 14 s

**Fig. 9.19c**  F(t) curve for Example 9.7

By calculating area data at 2 s intervals for Example 9.7 (Fig. 9.19c):

**Example 9.8**  A step change of 0.1 M HCl solution was introduced into the product stream at the beginning of a test zone. The pH was monitored at the end of the test zone and was recorded by a strip chart. The data and a best fit curve are shown in Fig. 9.20a. Obtain the F(t) and E(t) curves.

**Solution**

To obtain an F(t) curve, all one has to do is adjust the y-axis. When the pH reached 4.25, all of the acid had left the system. Therefore, 4.25 is given a value of unity. Conversely, a pH of 8 is given a value of 0. The F(t) curve value of any other pH and the curve is shown in Fig. 9.20b.

| Time (s) | 0 | 8 | 10 | 12 | 14 | 16 | 18 | 20 | 22 | 24 |
|---|---|---|---|---|---|---|---|---|---|---|
| Area | 0 | 0 | 6 | 21 | 40 | 60 | 79 | 92 | 98 | 100 |
| F(t) | 0.0 | 0.0 | 0.06 | 0.21 | 0.40 | 0.60 | 0.79 | 0.92 | 0.98 | 1.0 |

Fig. 9.20a  Stimulus response (pH) versus time curve for Example 9.8

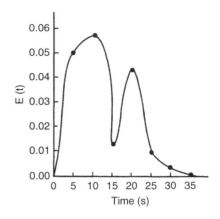

Fig. 9.20d  E(t) curve for Example 9.8

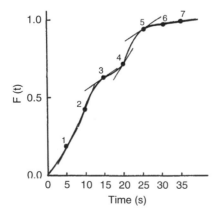

Fig. 9.20b  F(t) curve for Example 9.8

**Table 9.5** Slope of tangents (E value) drawn in Fig. 9.20c

| Time | Line # | Slope (E value) |
|------|--------|-----------------|
| 0    | 0      | 0               |
| 5    | 1      | .050            |
| 10   | 2      | .057            |
| 15   | 3      | .013            |
| 20   | 4      | .044            |
| 25   | 5      | .010            |
| 30   | 6      | .003            |
| 35   | 7      | 0               |

$$F(t) = \frac{8 - pH}{8 - 4.25} = \frac{8 - pH}{3.75} \qquad (9.15)$$

The F(t) curve can now be used to generate an E(t) curve. Choose a time and find its corresponding point on the F(t) curve. Draw a line tangent to the F(t) curve at this point. The slope of the line is the E(t) value for the selected time. By repeating these steps over the entire time range, one can generate an E(t) curve. This technique is illustrated in Fig. 9.20c. The E(t) curve data generated are listed in Table 9.5 and plotted in Fig. 9.20d.

### 9.4.1.6  Statistical Considerations

Sometimes it is required to characterize a distribution with just two or three numbers. The most important measure is the mean value of the distribution. The mean value is simply the median of

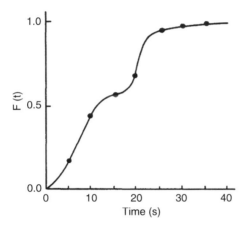

Fig. 9.20c  Tangent to F(t) curve for Example 9.8

**Fig. 9.21** E(t) curve for
Example 9.9

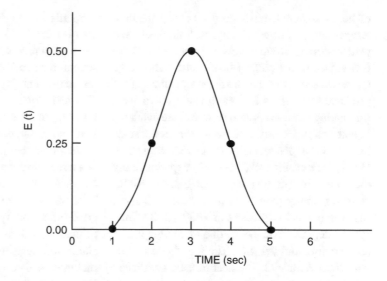

the distribution. For an E(t) curve, the mean is
given by:

$$\bar{t} = \int_0^\infty t\ E(t)dt = \sum t_i E(t_i) \Delta t \qquad (9.16)$$

**Example 9.9** Given the following E(t) curve
(Fig. 9.21), calculate the time using statistical
method at which the mean occurs.

**Solution**
Discretize points on the curve in Fig. 9.21 and
read their x and corresponding y values. Use
those values to create a table with time increment
($\Delta t$) of 1 s.

| Time (s) | 0 | 1 | 2 | 3 | 4 | 5 |
|----------|---|---|------|------|------|---|
| E value  | 0 | 0 | 0.25 | 0.50 | 0.25 | 0 |

$$\bar{t} = \sum t_i E(t_i)\ \Delta t = (0 \times 0 \times 1\ \ \text{sec})$$
$$+\ (1 \times 0 \times 1\ \ \text{sec}) + (2 \times .25 \times 1\ \ \text{sec})$$
$$+ (3 \times .5 \times 1\ \ \text{sec}) + (4 \times .25 \times 1\ \ \text{sec})$$
$$+\ (5 \times 0 \times 1\ \ \text{sec})$$
$$\text{Mean Time} = 3\ \ \text{sec}$$

Another important quantity is the variance ($\sigma^2$
). The variance gives a value to the spread of the
distribution and is expressed as units of (time)$^2$. It
is used mainly for comparing experimental curves
to theoretical curves. For an E(t) curve, the variance is given by:

$$\sigma^2 = \int_0^\infty (t - \bar{t})^2 E(t)dt$$
$$= \int_0^\infty t^2 E(t)dt - (\bar{t})^2 \qquad (9.17)$$

$$\frac{\sum t_i^2 \sum (t_i)}{\sum E(t_i)} - (\bar{t})^2 = \sum t_i^2 E(t_i)\ \Delta t$$
$$- (\bar{t})^2 \qquad (9.18)$$

**Example 9.10** Given the E(t) curve from
Example 9.9, calculate the variance of the curve.
Assume $\Delta t = 1$ s and $t = 3$ s.

$$\sigma^2 = \sum t_i^2 E(t_i)\ \Delta t - (\bar{t})^2$$
$$= (0^2 \times 0 \times 1\ \ \text{sec}) + (1^2 \times 0 \times 1\ \ \text{sec})$$
$$+ (2^2 \times .25 \times 1\ \ \text{sec})$$
$$+ (3^2 \times .5 \times 1\ \ \text{sec})$$
$$+ (4^2 \times .25 \times 1\ \ \text{sec})$$
$$+ (5 \times 0 \times 1\ \ \text{sec}) - (3^2\ \ \text{sec})$$
$$= 9.5\ \ \text{sec}^2 - 9\ \ \text{sec}^2$$
$$\sigma^2 = 0.5\ \ \text{sec}^2$$

## 9.5    Heat Transfer to Product

Indirect heat exchangers use a heating medium to
heat the wall, which contacts the product. The rate

of heat transfer depends on the heating medium temperature, its flow velocity, wall thickness and wall material, product flow velocity, etc. These resistances in the path of heat transfer should be known in order to calculate the temperature of the product. The values for heat transfer coefficients for heating medium (steam or water) are available in text books. However, those for the product side have to be experimentally measured. Kim et al. (1999, J. Food Eng. 42(2):79–84) experimentally measured the product side heat transfer coefficient for orange juice in a plate heat exchanger. The value was in the range of 2000–6300 W/m²-°C and depended on the heating or cooling mode, temperature, and velocity. The grand mean value was 3800 W/m²-°C. The heat transfer coefficient was higher in the cooling section than in heating section, and the values increased with increasing velocity of fluid.

If the product has suspended particles, then the fluid food will heat the particulates. For proper processing, the center of fastest moving particle should be sterile. The extent of heating at the center of particle will not only depend on the

properties of particles and the carrier fluid but also on the RTD of particles and their rotation. In order to predict the particle center temperature, several researchers have measured the heat transfer coefficient ($h_{fp}$) at the fluid-particle interface. The techniques to estimate $h_{fp}$ have aspects of both the inverse heat transfer (conduction) problem and parameter estimation. A survey of the literature shown in Table 9.6 indicates that there is a large variation in the values of $h_{fp}$ determined by different researchers (18–7870 W/m² °C). This is due to variations in experimental conditions and type of products. The different products used for these studies included water, starch solutions, and gum solutions. The different methods of measurement included calculation of temperature based on direct measurement of particle temperature with stationary particles and moving fluid as well as moving particle in stationary or moving fluids, or calculations based on the destruction of imbedded bacterial spores (Heppel 1985) or enzymes. Chakrabandhu and Singh (1998) measured the $h_{fp}$ values for rotating and stationary particles to simulate the actual

**Table 9.6** Heat transfer coefficient values between liquid and particles ($h_{fp}$) reported in literature

| Source | Range of $h_{fp}$ (W/m² °C) |
|---|---|
| Alhamdan et al.,1990 | 652–850 |
| Astrom and Bark, 1993 | 50–513 |
| Balasubramaniam and Sastry, 1994 | 986–2270 |
| Bhamidipati and Singh, 1995 | 108–195 |
| Chakrabandhu and Singh, 1998 | 120–2400 stationary particle 360–4300 rotating particle |
| Chakrabandhu and Singh, 2002 | 538–1450 coiled tube 204–1220 straight tube |
| Chandarana and Gavin, 1989 | 18.0–65.2 |
| Chang and Toledo,1989 | 146–303 |
| Heppel, 1985 | 930–7870 |
| Mwangi et al., 1993 | 58.3–1301.3 |
| Sastry et al., 1989 | 2039–2507 |
| Sastry et al., 1990 | 638–3005 |
| Weng et al., 1992 | 122.6–233.7 |

Sources: Alhamdan et al.,1990, *Trans. ASAE* 33(2):620–624; Astrom and Bark, 1993, *J. Food Eng. 21:97–125;* Balasubramaniam and Sastry, 1994, *J. Food Sci. 59(3):675–681;* Bhamidipati and Singh, 1995,*Trans. ASAE 38 (3):857–862;* Chakrabandhu and Singh, 1998, *J. Food Process Eng.* 21:327–350; Chakrabandhu and Singh, 2002, *Lebensm.Wiss. u – Technol.* 35:420–438; Chandarana and Gavin, 1989, *J. Food Sci. 54(1): 198–207;* Chang and Toledo,1989, *J. Food Sci. 54(4):1017–1023,* 1030; Heppel, 1985, In Proc. IUFOST Symp. Aseptic Processing and Packaging of Foods,Tylosand, Sweden, pp. 108–114; Mwangi et al., 1993, *J. Food Eng. 19:55–74;* Sastry et al., 1989, *Food Technol. 43(3): 132;* Sastry et al., 1990, *J. Food Process Eng. 13:239–253;* Weng et al., 1992*, J. Food Eng. 16:197–214.*

**Fig. 9.22** Lethality in liquid to that in particulates affected by particle size and holding time. This is on a semi-log scale, and y-axis indicates ratio of lethality. Source: deRuyter and Brunet (1973), *Food Technol. 27(7):44–51*

conditions in a tube flow. It is obvious that the rotating particles had higher $h_{fp}$ values than the stationary ones. In a subsequent study the same authors determined $h_{fp}$ for multiple particles in straight and coiled tube flow situations (Chakrabandhu and Singh 2002). Significantly higher $h_{fp}$ value was observed particles in coiled tube as compared to that in straight tube (Table 9.6).

Sterility or microbial lethality in liquid and particles are different because of the resistance to the heat transfer. The smaller particulates (<3 mm) get heated with the same rate as the liquid (de Ruyter and Brunet 1973, *Food Technol. 27(7):44–51*). However, larger particles may require significantly longer time to reach the intended temperature and thus causing over-processing of the liquid (Fig. 9.22). In the figure, one can notice that a 25.4-mm cube particle for a 2-min hold time would require 100-fold higher lethality in liquid to receive the intended lethality at the center of the particle. This overprocessing has hindered the adaptation of aseptic processing for low-acid particulates.

Once a minimum residence time is established, a process temperature can then be calculated for the liquid foods using the following:

$$T = TR + z \log(F/t) \qquad (9.19)$$

where T = process temperature (°C) measured at the end of the hold tube, TR = reference temperature (°C), z = temperature (°C) change

necessary for the D value to change by a factor of 10 or for one log or 90% reduction, t = hold time (min or s) calculated from flow rate and tube diameter, and F = sterilizing value needed to achieve commercial sterility for the product. However, the product temperature in the center of particulate will be lower than the liquid temperature, and it is impractical to measure it during the process. Therefore, numerical and analytical models have been developed to predict the least lethal particle temperature and use the predicted value for process verification.

In many cases, one has to combine the fluid flow and heat transfer characteristics to gain knowledge of the complex system. The complexity comes from the non-Newtonian nature of fluid, large particulates, and turbulent flow. Son and Singh (2002, *J. Food Eng. 52:177–184*) developed models (computer program) for simulation of turbulent flow and successfully verified with known analytical solutions of non-Newtonian laminar flow and by comparing between the bulk temperatures from the analytical calculation and the simulated average temperatures along the length of heat exchanger. For the purpose of verifying simulated velocity profiles, the RTD of soybean milk were measured in a helical type tubular heat exchanger. The experimentally determined velocity profiles were flatter than those simulated under laminar and turbulent flow conditions due to the secondary flow by a helically coiled tube. To compensate for the curvature effect, the original simulation was modified by a correction factor based on the curvature ratio (heat exchanger internal diameter/diameter of coil) and Reynolds number. Temperature profiles in a heating tube were also measured by inserting thermocouples at the end of heat exchanger. The measured temperatures of 9% soybean milk were consistent with the simulated temperatures. However, in the case of 16% soy milk, the measured temperatures were slightly higher than those of simulation. Through the successful simulation of velocity and temperature distributions under turbulent flow conditions, one can find the desirable processing conditions to maximize the quality of aseptically processed soybean milk without numerous experiments.

## 9.6    Filling and Packaging

The containers for aseptic packaging have ranged from consumer size retail packs of a few grams (60–180 g) to bulk storage containers up to more than about 4–6 m$^3$ (1.0–1.7 million gallons). The largest bulk containers used today carry single-strength orange juice from Brazil to different countries in tanker ships. The bulk storage of orange juice is refrigerated to control the chemical reactions in the juice and not for microbiological reasons. Other uses of bulk storage has been in tomato paste storage and shipment (190 L drums of tomato paste and banana puree, rail car, and truck containers) and 378,000 L (100,000 gal) tanks for storage of soy sauce and other liquids. Particulates such as diced tomatoes, pineapple cubes, strawberry toppings, etc. have been aseptically packed in 1140 L bags, which are kept in rigid boxes for support. The recent interest is to be able to package nutraceutical or functional foods, especially beverages in aseptic plastic containers. The heat-sensitive nature of some components in functional foods requires that the heat treatment be limited to relatively lower temperatures for shorter times. There are also opportunities for packaging semi-processed foods for institutional use. The products for the institutional market could be in 4–20 L plastic packages.

The containers for aseptic products are sterilized by heat or chemicals. Steam is used for sterilizing metal cans and drums, whereas heat of thermoforming sterilizes the thermoformed plastics. A combination of 30% hydrogen peroxide and heat is used in sterilizing preformed aseptic containers and closures (Clark 2004). The bulk bags are sterilized by irradiation whereas chemicals are used for sterilizing the bulk rigid containers. The bulk containers for storage of orange juice are sterilized by flooding with 25 μg/g (25 ppm) iodophor. The calculation of sterility for packaging material is done in the same manner as that for the product.

The filling environment is sterilized by heat or chemicals. After sterilization, the environment is maintained sterile using filtered air or inert gas (Clark 2004). The filters must be sterilized and validated to assure that there is no recontamination of the sterile environment.

## 9.7    Monitors and Controls

In order to verify that all critical factors are met or exceeded and that equipment is functioning as intended, adequate instrumentation, recorders, and controls must be in place. Particular attention should be given to the location of sensors and to the controlling logic. As previously noted, the temperature of every particle of the product must be at or above a specific temperature for a specified time. Achieving the proper hold time depends on proper sizing of the hold tube and the flow rate of product. For homogeneous products, temperature can be documented by indicating and recording devices properly located in the hold tube. The temperature at the inlet of the tube is monitored with a temperature recorder-controller located at the final heater outlet. An acceptable temperature-indicating device such as an accurate thermocouple recorder or a mercury-in-glass thermometer (MIG) must be installed between the hold tube outlet and the inlet to the first cooler. In addition, there must also be an automatic recording device located in the product stream at the hold tube outlet. The product temperature is measured at the hold tube. The temperature-recording device chart graduations must not exceed 1 °C (2 °F) within a range of 6 °C (10 °F) of the desired product sterilization temperature.

The processor should keep in mind that the temperature-indicating device (thermocouple or MIG) is considered the reference temperature instrument by the regulatory agencies. The temperature recorder must be adjusted to agree as nearly as possible with but in no event higher than the known accurate temperature-indicating device. The temperature-indicating device must be tested for accuracy against a known accurate standard thermometer upon installation and at least once a year thereafter or more frequently, if necessary, to ensure its accuracy.

If regeneration is used, the pressure on the sterile product must always be higher than the pressure of the raw product. One pressure sensor is typically located at the sterile product outlet and the other at the raw and sterile product must be recorded continuously. Other requirements and recommendations regarding temperature monitoring and recording devices are contained in the regulations. Those who will be in charge of these functions should become familiar with the appropriate regulatory requirements.

Temperature monitoring equipment, as well as pressure sensors, timers, etc., will typically be interfaced with various control systems. Control systems available for both processing and packaging equipment can be adapted for complete automation. However, this adaptation may not necessarily mean that these installations will meet regulatory requirements or that the operation is being properly controlled. The logic controlling such systems must be evaluated to verify that adequate controls, interlocks, and other safety features have been incorporated and are functioning as intended. The control software and system performance should be verified at installation and routinely thereafter. The interlocks (alarms) that monitor critical functions must be tested periodically to prove that they are functioning properly and the results of these challenges recorded.

Changes to the control software should only be made by authorized personnel.

When automatic control systems are implemented, appropriate gauges and recording devices should also be included so that operators can observe results and record the information in the daily production log. This precaution will also serve as a check against the automatic system. Further recommendation regarding automatic control systems can be found in NFPA's bulletin 43-L "Automatic Control Guidelines for Aseptic Systems Manufacturers and Companies Using Aseptic Processing and Packaging for Preserving Foods."

## 9.8 Processing System Sterilization

As required by regulatory agencies, all product contact surfaces downstream from the final product heater must be brought to a condition of commercial sterility prior to production. Most systems use saturated steam or pressurized hot water to sterilize the processing system. The system is sterilized by maintaining the temperature within all parts of the system at or above a specified temperature by continuously circulating the sterilizing medium for a specified period of time. If steam is used as a sterilizing medium, adequate provisions must be made to remove condensate from the system. Inadequate sterilization can result from condensate collecting in low spots within the system due to excessively low temperature.

Aseptic systems may include unique items such as aseptic pumps, flash tanks, and surge tanks in addition to heat exchangers, piping, and valves. Due to the large capacity of various surge tanks, these units are usually sterilized with saturated steam rather than hot water. Surge tank sterilization may proceed separately from the sterilization cycle for the rest of the processing system but should occur simultaneously with this cycle. The surge tank may require special attention, such as documentation of uniform heat distribution during sterilization. Microbiological challenges may also be needed to confirm sterility of inaccessible areas associated with the surge tank where temperatures cannot be directly measured.

The preproduction equipment sterilization cycle is usually monitored by temperature sensors located at the coldest point(s) within the system. The sterilization time should incorporate only that portion of the cycle where all temperatures are at or above the established minimum temperature. If the temperature drops below the minimum sterilization temperature, the timer should be restarted to zero, and timing should not restart until the proper temperature is reestablished. A permanent and continuous record of temperatures must be made during the sterilization period.

### 9.8.1    Maintenance of Sterility

After the system sterilization cycle is completed, a transition phase is initiated during which the system beyond the end of the hold tube is cooled and readied for the introduction of product. From the end of sterilization through the end of the production period, the system must be maintained in a sterile condition. One of the most common ways to prevent contaminants from entering the system is to maintain the product under constant positive pressure. As stated earlier, a back pressure device located after the product coolers and typically before the filler is used to maintain the product at a pressure which is usually 69–103 kPa (10–15 psig) over the pressure exerted by the product at its maximum temperature. This excess pressure not only prevents flashing of the product but also the recontamination after sterilization.

There are some areas where rotating shafts (aseptic) or reciprocating shafts (valve stems) may allow entrance of bacteria in the sterile product area (zone). An effective barrier against the entry of microorganisms needs to be provided at all potential contamination sites. Steam seals are commonly used for this purpose. Steam seals consist of an area where steam in a groove or trace in the housing forms a ring around a pump shaft or covers the total stroke of a valve stem.

Steam must be continually supplied to the seal area, and the proper operation of the steam seal must be verified by the operator. Operator verification usually consists of visually checking for steam discharge from the individual seal or an indication of temperature at seal discharge points.

An automatic flow diversion device may also be utilized in an aseptic processing system to prevent a potentially non-sterile product from reaching the packaging equipment. The flow diversion device, usually located after the back pressure valve, must be designed so that it can be adequately sterilized and operated in a reliable manner. It is recommended that the flow diversion device be activated by the control system which monitors temperature at the end of the hold tube and other devices which monitor critical factors such as differential pressures in a regenerator or positive pressure in a surge tank. Should there be a deviation in any of these critical factors, the flow diversion device should divert product flow away from the filler and prevent product from being packaged.

### 9.8.2    Process Confirmation

Once the hold tube length, process temperature, and product flow rate have been calculated and the system has successfully completed trial runs, inoculated packs of product or simulated product should be conducted for confirmation of proper system operation. These inoculated pack studies are conducted by batch inoculating the product with an appropriate test organism followed by processing at the maximum flow rate (minimum residence time) specified in the scheduled thermal process. Temperatures should be varied during the inoculated pack study to yield a product processed to pre-selected lethality values. Generally, five temperatures should be used and the pack is run continuously, starting with the highest temperature and then adjusting the temperature downward to the next target temperature.

At least 100 packages containing inoculated product should be collected from each temperature interval. These packages should be incubated and monitored for spoilage. Results of the inoculated pack should correlate with calculated lethality delivered by the system at each temperature and should confirm operation of the equipment.

The automatic controls and safety devices built into both the processing and packaging systems should also be challenged to verify proper function. For example, the flow diversion device should be challenged to verify that no product is packed following a temperature drop or other fault. Other interlocks, such as system shutdown in the event of a loss of sterile air pressure, or loss of proper pressure differentials, should also be verified. Agreement between sensing devices, indicators, and recorders should be checked, and the programming of automatic control devices should be verified. Calibration of all sensing and measuring devices must be completed before system start-up and at subsequent intervals as prescribed by regulations.

It is suggested that at least four small-scale production runs of uninoculated product be packed and incubated followed by 100% examination for evidence of spoilage. The packer must retain records of these commissioning trials, as well as other test results.

## Problems

9.1. In an aseptic processing system, sweet potato puree was flowing at a flow rate of 5.0 gal/min (measured) through the hold tube (1.87 inches i.d.); calculate the mean flow velocity.

   (a) Since the hold tube pipe was measured to be 51.4 ft. long, calculate the mean velocity of the fluid through the pipe using the residence time distribution (RTD) data, which suggests that the minimum RTD was 60 s, maximum 115 s, and mean around 87 s.

   (b) Is this flow closer to laminar or turbulent flow based on the Reynolds number? [There are publications in journals for sweet potato puree which will give you necessary data for calculation of Reynolds number.]

   (c) What would be the most conservative estimate of $F_0$ value of the product that is aseptically processed using the hold tube above? The product enters this hold tube at 220 °F and exits at 215 °F.

9.2. Aseptic processing is used for commercial sterilization of foods. What are the key components of an aseptic processing system? Calculate the length of a holding tube to meet the following requirement for an aseptic low-acid food:

Mass flow rate = 2500 kg/h, density of product = 0.98 g/mL.

The fluid is non-Newtonian with a consistency coefficient (K) = 0.10 g/cm-s.

Internal diameter of holding tube = 5 cm.

Hold temperature = 135 °C.

The $F_0$ needed for the process was 7.0 min, assume z value of 10 °C.

The ratio average to maximum velocity under turbulent flow is about 0.82, which can be used for the tube length calculation.

9.3. If the flow rate of chicken broth was 3.0 gal/min (measured) through the hold tube (30 ft. long and 1.37 inches i.d.), calculate the mean flow velocity. Assume viscosity of 1 cp. Explain whether it is a laminar or turbulent flow. What would be the Fo value of the chicken broth that is aseptically processed using the hold tube mentioned in "1" above? The broth enters this hold tube at 255 F and exits at 245 F.

9.4. The following data is available for residence time distribution in a holding tube. Plot the data for E(t) curve and also shoe F9 t) curve. Calculate the statistical mean residence time.

| Time (s) | 0 | 40 | 42 | 44 | 45 | 46 | 47 | 48 | 53 | 58 | 68 |
|---|---|---|---|---|---|---|---|---|---|---|---|
| Conductivity | 0.35 | 0.35 | 1.18 | 1.58 | 3.22 | 3.23 | 2.89 | 2.87 | 1.0 | 0.50 | 0.35 |

9.5. Using the Residence Time Distribution Curves Plotted in 9.4, Answer the Following Questions.

(a) Calculate the mean flow velocity based on the measured flow rate (gal/min) in the data file name above.

(b) Determine the mean velocity of the fluid thru the pipe using the RTD data you collected (hold tube was 37 feet long). Compare this value with that which you calculated 1 and 2.

(c) Explain whether this is laminar or turbulent flow. Show calculations when CMC has an effective viscosity of 200 cps.

(d) From the RTD data, determine the *maximum* velocity of the fluid.

(e) How does the experimental maximum velocity compare to the theoretical maximum for the type of flow?

(f) Based on the results of this RTD study, what would be the Fo value if the product enters and exits the hold tube at 230 F?

(g) Assuming this data was for sweet potato puree, calculate the integrated lethality, Fs, using your E(t) data and the equation on page 44 in the text. Recall that the term exp.$(-kt)$ is the same as $10^{(-t/D)}$ (you will need to determine/select an appropriate D value) (hint: numerically integrate).

(h) Explain whether this process and Fo is appropriate for aseptically processing a sweet potato puree in:
 (i) 16 oz. containers
 (ii) 300 gallon Scholle bags

Assume the # spores/ml is the same in each. If not adequate, what suggestions would you suggest?

## Suggested Reading

Clark, P. J. (2004). Aseptic processing: New and old. *Food Technology, 58*(11), 80–81 89.

David, J. R. D., Graves, R. H., & Szemplenski, T. (2013). *Handbook of aseptic processing and packaging* (2nd ed.). Boca Raton: CRC Press/Taylor & Francis Group 372 pp.

Hersom & Shore (1981). *Food Technology, 35*(5):53

Levenspiel, O. (1972). *Chemical reaction engineering* (2nd ed.). New York: Wiley.

Lewis, M., & Heppell, N. (2000). *Continuous thermal processing of foods*. Gaithersburg: Aspen Publishers 447 pp.

Li, et al. (2004) *Lebensm. Wiss. Technol.* 37:565–572

Nelson, P. E. (Ed.). (2010). *Principles of aseptic processing and packaging* (3rd ed.). West Lafayette: Purdue University Press 161 pp.

Reuter, H. (1993). *Aseptic processing of foods* (p. 313). Lancaster: Technomic Publ. Co.

Sastry, S. K., & Cornelius, B. D. (2002). *Aseptic processing of foods containing solid particulates*. New York: Wiley 250 pp.

Singh, R. K. (1995). *Food process design and evaluation*. Lancaster: Technomic Publications 257 pp.

Singh, R. K., & Nelson, P. E. (Eds.). (1992). *Advances in aseptic processing technologies*. London: Elsevier .

# Refrigeration

Cooling is a fundamental operation in food processing and preservation. Removal of heat could involve either the transfer of heat from one fluid to another or from a solid to a fluid, or it could be accomplished by adiabatic vaporization from liquid water. Knowledge of the principles of heat transfer is an essential prerequisite to the understanding of the design and operation of refrigeration systems.

Maintaining temperatures lower than ambient inside a system requires both the removal of heat and prevention of incursion of heat through the system's boundaries. The rate of heat removal from a system necessary to maintain the temperature is the refrigeration load. Refrigeration systems must be sized to adequately handle the refrigeration load. When heat has to be removed from a system continuously, at temperatures below ambient and for prolonged periods, a mechanical refrigeration system acts as a pump that extracts heat at low temperatures and transfers this heat to another part of the system where it is eventually dissipated to the surroundings at a higher temperature. The operation requires energy, and a well-designed system will allow the maximum removal of heat at minimum energy cost.

## 10.1 Mechanical Refrigeration System

### 10.1.1 Principle of Operation: The Heat Pump

The second law of thermodynamics mandates that heat will flow only in the direction of decreasing temperature. In a system that must be maintained at a temperature below ambient, heat must be made to flow in the opposite direction. A refrigeration system may be considered as a pump that conveys heat from a region of low temperature to another region that is at a high temperature.

The low-temperature side of a refrigeration system is maintained at a lower temperature than the system it is cooling to allow spontaneous heat flow into the refrigeration system. The high-temperature side must have a temperature higher than ambient to allow dissipation of the absorbed heat to the surroundings. In some instances, this absorbed heat is utilized as a heat source for use in heating processes.

Maintaining a high and a low temperature in a refrigeration system is made possible by the use of a refrigerant fluid that is continuously recirculated through the system. A liquid's boiling or condensation temperature is a function of

R. T. Toledo et al., *Fundamentals of Food Process Engineering*,
https://doi.org/10.1007/978-3-319-90098-8_10

the absolute pressure. By reducing the pressure, a low boiling temperature is made possible, allowing for absorption of heat in the form of the heat of the refrigerant's vaporization as it is vaporized at the low pressure and temperature. The vapors, when compressed to a high pressure, will condense at the high temperature, and the absorbed heat will be released from the refrigerant as it condenses back into liquid at the high temperature and pressure. Figure 10.1 shows the vapor pressure versus temperature of commonly used refrigerants. The diagram also illustrates how this pressure and temperature relationship is utilized for cooling and heating.

### 10.1.2   Refrigerants

Atmospheric contamination of refrigerants that contain chlorine and fluorine in the molecule (chlorofluorocarbon, CFC) has been shown to contribute to global warming and cause ozone depletion in the upper atmosphere. Thus, by international agreement, manufacturing of CFCs was stopped after 1996. CFCs have the highest ozone depletion potential (ODP) among the refrigerants and also has high global warming potential (GWP). Manufacturing of hydrochlorofluorocarbon (HCFC) refrigerants is to be phased out in 2030. HCFCs have lower ODP but also contribute to global warming. Because no new CFCs are being manufactured, replacement in existing refrigeration systems must come from recovered CFCs, or the refrigeration unit must be recharged with a completely new refrigerant. Non-CFC refrigerants must be used in new refrigeration systems. Existing refrigeration systems may be upgraded by utilizing a drop-in refrigerant (i.e., the new refrigerant is added after the original CFC refrigerant has been removed).

Refrigerants now used in the food industry include R502 for transport refrigeration, R502 and R22 for retail display cases and retail central storage, R502 for cold storage, R22 for refrigerated storage and refrigerated vending machines, and R717 for large freezers, frozen storage warehouses, and large refrigerated warehouses. R502 contains R12, which is no

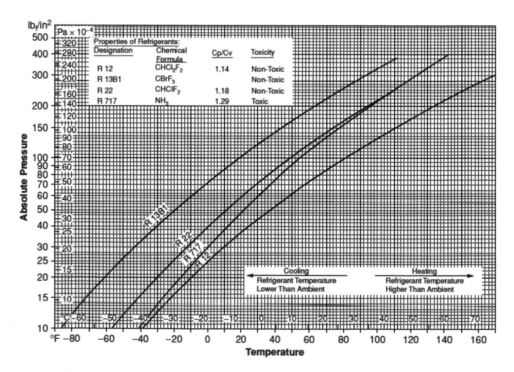

**Fig. 10.1** Vapor pressure of commonly used refrigerants as a function of temperature

**Table 10.1**  Refrigerants used in the food industry

| Refrigerant | Composition | Boiling point at 1 °C (°F) |
|---|---|---|
| R12 | $CCl_2F_2$ | −30 (−22) |
| R22 | $CHClF_Z$ | −41 (−41) |
| R32 | $C11_Z1^7_Z$ | −52 (−62) |
| R115 | $CClF_ZCF_3$ | −39 (−38) |
| R124 | $CHClFCF_3$ | −12 (10) |
| R125 | $CHF_ZCF_3$ | −49 (−56) |
| RI34A | $CH_ZFCH_3$ | −26 (−15) |
| R142B | $CH_3CCl_ZF$ | −10 (14) |
| R143A | $CH_3CF_3$ | −25 (−13) |
| R152A | $CH_3CHF_2$ | −25 (−13) |
| R290 | $CH_3CH_ZCH_3$ | −42 (−44) |
| C318 | CA (cyclic) | −6 (23) |
| R401A | R22, R152A, R124(53:13:34) | |
| R401B | R22, R152A, R124(61:11:28) | |
| R402A | R125, R290, R22 (60:2:38) | |
| R402B | R125, R290, R22 (38:2:60) | |
| R404A | R125, R143A, R134A (44:52:4) | |
| R405A | R22, R152A, R142B, C318 (45:7:5.5:42.5) | |
| R407A | R32, R125, R134A (20:40:40) | |
| R407B | R32, R125, R134A (10:70:20) | |
| R407C | R32, R125.R134A (23:25:52) | |
| R410A | R32, R125 (50:50) | |
| R502 | R22, R115 (48.8:51.2) | −45 (−49) |
| R507A | R125, R143A (50:50) | −46.7 (−52.1) |

R 502 and R507A are azeotropes, while the R400 series are zeotropes

*Source*: ASHRAE Standards 2000. Standards for Designation and Safety of Refrigerants

longer manufactured. R22 can still be used but will be phased out later. Thus, alternatives to these refrigerants must be selected for retrofits to existing installations or for new installations. For long-term retrofit, alternatives are R407A, R407B, and R507, R407C and R410A, and R134A may be used for R502, R22, and R12, respectively.

Table 10.1 lists the refrigerants used in the food industry and their boiling points. Also listed are alternative non-CFC refrigerants. Most alternatives are mixtures that are either azeotropes or zeotropes. Azeotropes are mixtures with a constant boiling point and where the vapor and liquid composition are the same. Zeotropes are mixtures where vapor and liquid compositions are different at different temperatures, and there is a large gap between the dew point (the temperature where liquid first starts to form) and the bubble point (the temperature where the mixture is completely

in the liquid phase). The use of alternative refrigerants may require different compressor lubricant, so this has to be taken into consideration in retrofits.

### 10.1.3   The Refrigeration Cycle

Figure 10.2 shows a schematic diagram of a mechanical refrigeration system. The heart of the system is the *compressor*.

When the compressor is operating, refrigerant gas is drawn into the compressor continuously. Low pressure is maintained at the suction side, and because of the low pressure, the refrigerant can vaporize at a low temperature. In the compressor, the refrigerant gas is compressed increasing in both pressure and temperature during the process. The hot refrigerant gas then flows into a heat exchange coil called the *condenser* where

**Fig. 10.2** Schematic
diagram of a refrigeration
system

heat is released in the process of condensation at
constant pressure and temperature. From the condenser, the liquid refrigerant flows into a liquid
refrigerant holding tank. In small systems, there
may be no holding tank, and the refrigerant just
continuously cycles through the system.

When there is a demand for cooling, the liquid
refrigerant flows from the holding tank to the
low-pressure side of the refrigeration system
through an *expansion valve*. The drop in pressure
that occurs as the refrigerant passes through the
expansion valve does not change the heat content
of the refrigerant. However, the temperature
drops to the boiling temperature of the liquid at
the low pressure. The cold liquid refrigerant then
flows to another heat exchange coil called the
*evaporator* where the system performs its cooling
function, and heat is absorbed by the refrigerant
in the process of vaporization at constant temperature and pressure. From the evaporator, the cold
refrigerant gas is drawn into the suction side of
the compressor thus completing the cycle.

A refrigeration system is usually equipped
with low-pressure and high-pressure cutoff
switches that interrupt power to the compressor
when either the high-pressure set point is
exceeded (this occurs if cooling capacity of the
condenser is inadequate) or when the pressure
drops below the low-pressure set point (this
occurs when the compressor is running, but the

cooling demand is much less than the capacity of
the refrigeration system). The low-pressure set
point can be used to control the evaporator temperature. In some systems, liquid refrigerant flow
through the expansion valve is thermostatically
controlled, interrupting flow when the evaporator
temperature is lower than the set point. In addition, refrigeration systems are equipped with
thermostats that interrupt power to the compressor when the temperature of the refrigerated room
reaches a set temperature.

A refrigeration system may also be used for
heating. A system that alternates heating and
cooling duty is called a *heat pump*. These units
are used extensively for domestic heating or
cooling in areas where winter temperatures are
not too severe. In these units, either a
low-pressure or a high-pressure refrigerant could
go through the heat exchange coils that constitute
the evaporator and condenser; thus, either could
act as a heating or cooling coil depending upon
the duty expected of the system. The ability of
heat pumps to deliver heat with low power consumption is due to the fact that power is used only
to pump energy from a low temperature to a
higher temperature. The heat is not derived
completely from the power supplied, but rather
it is extracted from cooler air from the surroundings. The efficiency of heat pumps in
transforming applied power to heat varies

inversely with the temperature differential between the temperature of the medium from which heat is extracted and the temperature of the system that is being heated. Heat pump systems can be used as a means of recovering energy from low-temperature heat sources for use in low-temperature heat applications such as dehydration.

### 10.1.4   The Refrigeration Cycle as a Series of Thermodynamic Processes

Starting from the compressor in Fig. 10.2, low-pressure gas is compressed adiabatically to high pressure, which should allow condensation at ambient temperature. Work is required to carry out this process, and this energy is supplied in the form of electrical energy to drive the compressor motor. The gas also gains in enthalpy during this compression process. At the condenser, the gas condenses and transfers the latent heat of condensation to the surroundings. There is a loss of enthalpy in a constant pressure process. At the expansion valve, the liquid expands to low pressure at constant enthalpy, while at the evaporator, the liquid evaporates at constant pressure and

gains in enthalpy. The two processes crucial to the efficiency of the refrigeration system are the adiabatic compression process where energy is applied and the isobaric expansion process where energy is extracted by the refrigerant from the system. Work and enthalpy change in an adiabatic compression process has been derived in Chap. 3 in the section "Work and Enthalpy Change on Isothermal Expansion or Compression of an Ideal Gas." The ratio of enthalpy change to work in adiabatic compression of an ideal gas is equal to the specific heat ratio.

$$\frac{\Delta H}{W} = -\gamma; \quad W = \frac{-\Delta H}{\gamma} \qquad (10.1)$$

The negative sign indicates that energy is being used on the system.

### 10.1.5   The Refrigeration Cycle on the Pressure/Enthalpy Diagram for a Given Refrigerant

The thermodynamic properties of refrigerants when plotted on a pressure-enthalpy diagram result in a plot that is useful in determining capacities and power requirement for a refrigeration system. Figure 10.3 is a schematic diagram of

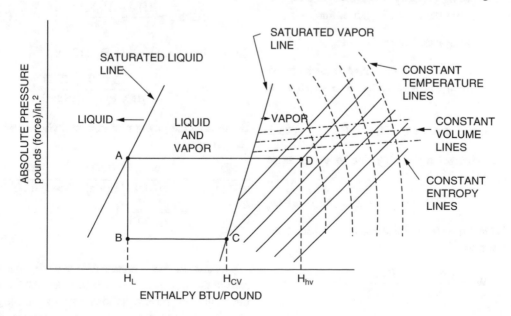

**Fig. 10.3** The refrigeration cycle on a pressure-enthalpy diagram

a refrigeration cycle on a pressure-enthalpy diagram. The diagram consists of lines representing the vapor and liquid pressure-enthalpy relationship, lines representing change in enthalpy with pressure during adiabatic compression (constant entropy), and in some charts, lines representing specific volumes at various pressures and enthalpies.

Point A in Fig. 10.3 represents the liquid refrigerant at high pressure entering the expansion valve. The refrigerant expands at constant enthalpy ($H_L$) as it goes through the expansion valve and leaves the valve as a mixture of liquid and vapor at a lower pressure, represented by point B. As the refrigerant absorbs heat in the evaporator, it gains in enthalpy represented by the line BC. The refrigerant leaves the evaporator as saturated vapor (represented by point C) having the enthalpy $H_{cv}$. The compressor raises the pressure, and the change is represented by the line CD that parallels the lines of constant entropy. As the compressed refrigerant gas leaves the compressor at point D, it will have an enthalpy represented by $H_{hv}$. At the condenser, heat is dissipated resulting in a drop in enthalpy represented by line AD. The liquid refrigerant leaves the condenser with the pressure and enthalpy represented by point A.

The cooling capacity of the refrigeration system is represented by the length of line BC:

$$\text{Cooling capacity} = H_{cv} - H_L \qquad (10.2)$$

The condenser heat exchange requirement or condenser duty is represented by the length of line AD:

$$\text{Condenser duty} = H_{hv} - H_L \qquad (10.3)$$

The change in enthalpy due to compression $\Delta H_c$ is:

$$\Delta H_c = H_{hv} - H_{cv} \qquad (10.4)$$

From Eq. (10.1), the work required for compression is:

$$W = \frac{-\Delta H_c}{\gamma} = \frac{-(H_{hv} - H_{cv})}{\gamma} \qquad (10.5)$$

If M is the mass of refrigerant recirculated through the system per unit time, the power requirement can be calculated as follows:

$$P = \frac{W}{\text{time}} = M \frac{-(H_{hv} - H_{cv})}{\gamma} \qquad (10.6)$$

The negative sign on the work and power indicates that work is being added to the system.

The efficiency of a refrigeration system is also expressed in terms of a *coefficient of performance* (COP), which is a ratio of the cooling capacity over the gain in enthalpy due to compression:

$$\text{COP} = \frac{H_{cv} - H_L}{H_{hv} - H_{cv}} \qquad (10.7)$$

The power requirement (P) can be expressed in terms of the coefficient of performance using Eqs. (10.6) and (10.7):

$$P = \frac{H_{cv} - H_L}{\gamma(\text{COP})} \cdot M \qquad (10.8)$$

The refrigeration capacity is expressed in tons of refrigeration, the rate of heat removal sufficient to freeze 1 ton (2000 lb) of water in 24 h. Because the heat of fusion of water is 144 BTU/lb., this rate of heat removal is equivalent to 12,000 BTU/h. The refrigeration capacity in tons is:

$$(\text{tons})_r = \frac{(H_{cv} - H_L)(M)}{12,000} \qquad (10.9)$$

Substituting Eq. (10.9) in Eq. (10.8):

$$P = \frac{(\text{tons})_r(12,000)}{\gamma(\text{COP})} \left[ \frac{\text{BTU}}{\text{h}(\text{ton})_r} \right]$$

Expressing the power requirement in horsepower (HP)/(ton)$_r$:

$$\frac{\text{HP}}{(\text{tons})_r} = \frac{12,000}{\gamma(\text{COP})} \frac{\text{BTU}}{\text{h}(\text{ton})_r} \frac{1\text{HP}}{2545\ \text{BTU/h}}$$
$$= \frac{4.715}{\gamma(\text{COP})}$$
$$\qquad (10.10)$$

Because there is a certain degree of slippage of refrigerant past the clearance between the cylinder and the piston, particularly at the high

Scale ──▶|◀── Change

Enthalpy – BTU/lb Above Saturated Liquid at –40°F

**Fig. 10.4** Pressure-enthalpy diagram for refrigerant 717 (NH₃). (Adapted from ASHRAE Guide and Data Book, Fundamentals and Equipment, 1965)

pressures and also because some frictional resistance occurs between the piston and the cylinder, the actual work expended would be higher than that determined using Eq. (10.10). The ratio between the theoretical horsepower as calculated from Eq. (10.10) and the actual horsepower expended is the efficiency of compression. The efficiency of compression depends upon the ratio of high side to low side pressure across the compressor and will be discussed later in the text.

Pressure-enthalpy diagrams for ammonia, Freon 12 and Freon 22 are shown in Figs. 10.4, 10.5, and 10.6.

### 10.1.5.1   Example Problems on the Use of Refrigerant Charts

**Example 10.1** A refrigerated room is to be maintained at 40 °F (4.44 °C) and 80% relative humidity. The contents of the room vaporize moisture requiring removal of moisture from the air to maintain the desired relative humidity. Assuming that airflow over the evaporator coil is such that the bulk mean air temperature reaches to within 2 °F (1.11 °C) of the coil temperature, determine the low side pressure for an ammonia refrigeration system such that the desired humidity will be maintained.

**Solution**

From a psychrometric chart, the dew point of air at 40 °F and 80% RH is 33 °F (0.56 °C). When room air is cooled to 33 °F passing over the evaporator coils, any moisture present in excess of the saturation humidity at this temperature will condense. If airflow over the coil is sufficient to allow enough moisture removal through condensation to equal the rate of moisture vaporization

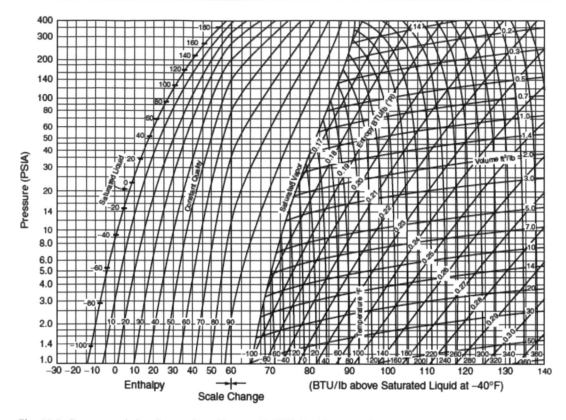

**Fig. 10.5** Pressure-enthalpy diagram for refrigerant 12 ($CCl_2f_2$). (Courtesy of E. I. Du Pont de Nemoirs Inc.)

into the air, then the desired relative humidity will be maintained. The coil temperature should be 31 °F (−0.56 °C). From Fig. 10.1 for ammonia, the pressure corresponding to a temperature of 31 °F is 61 psia (420 kPa absolute).

**Example 10.2** A heat pump is proposed as a means of heating a cabinet drier. Air is continuously recycled through the system passing through the condenser coil at the air inlet to the drying chamber where the air is heated. Prior to recycling, the hot moist air leaving the drying chamber is passed through the evaporator coil where moisture is removed by condensation. Inlet air to the drying chamber is at 140 °F (60 °C) and 4% relative humidity. Assuming that the evaporator and condenser coil design and airflow are such that air temperature approaches 5 °F (2.78 °C) of the evaporator coil temperature and log mean ΔT between condenser coil and air is 20°F (5.56 °C), determine the high

and low side pressure for the refrigeration system using refrigerant 22 as the refrigerant.

**Solution**
From a psychrometric chart, the dew point of air at 140 °F (60 °C) and 4% RH is 38 °F (3.33 °C). Thus, the evaporator coil temperature should be at 33 °F (0.56 °C). The condenser coil temperature and pressure will be calculated from the specified log mean ΔT.

From Fig. 10.1, for refrigerant 22, the low side pressure corresponding to 33 °F (0.56 °C) is 77 psia (530 kPa absolute).

At the condenser, air enters at 38 °F and leaves at 140 °F. If the system is set up in countercurrent flow, the log mean ΔT can be calculated as follows:

$$\overline{\Delta T} = \frac{(T_1 - 38) - (T_g - 140)}{\ln (T_1 - 38)/(T_g - 140)}$$

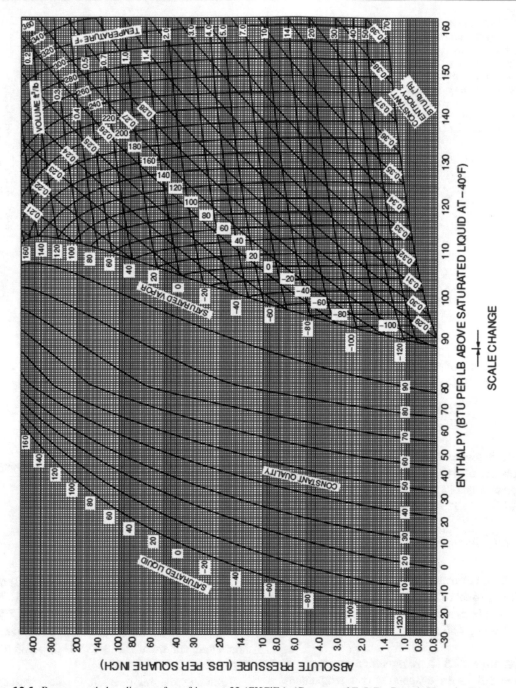

**Fig. 10.6**   Pressure-enthalpy diagram for refrigerant 22 (CHClF$_2$). (Courtesy of E. I. Du Pont de Nemoirs Inc.)

where $T_1$ and $T_g$ refer to the temperature of refrigerant liquid and gas, respectively.

Figure 10.7 shows the pressure-enthalpy diagram for the system under the conditions given.

The high pressure will be determined by trial and error from Fig. 10.6. Following the constant entropy line from a pressure of 77 psia, a high pressure of 250 psia will give a hot refrigerant gas

**Fig. 10.7** Pressure-enthalpy diagram of the heat pump system in Example 10.2. *Schematically drawn from pressure corresponding to refrigerant temperature in Fig. 10.1 and data given in Example 10.2*

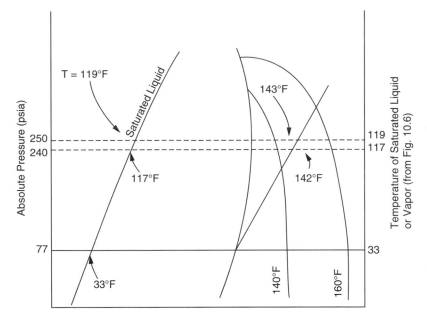

temperature of 143 °F and a liquid temperature of 119 °F. The log mean $\Delta T$ is:

$$\overline{\Delta T_1} = \frac{(119 - 38) - (143 - 140)}{\ln 92/3} = 23.6$$

This is higher than the 20 °F $\Delta T$ specified. At a pressure of 240 psia, the hot refrigerant gas temperature is 142 °F, and the liquid temperature is 117 °F:

$$\overline{\Delta T_1} = \frac{(117 - 38) - (142 - 140)}{\ln 79/2} = 20.9$$

This calculated value for the log mean $\Delta T$ is almost equal to the specified 20 °F; therefore, the high side pressure for the system should be at 240 psia (1650 kPa).

**Example 10.3** A refrigeration system is to be operated at an evaporator coil temperature of −30 °F (−34.4 °C) and a condenser temperature of 100 °F (37.8 °C) for the liquid refrigerant. For (a) refrigerant 12 and (b) refrigerant 717, determine:
1. The high side pressure

2. The low side pressure
3. Refrigeration capacity per unit weight of refrigerant
4. The coefficient of performance
5. The theoretical horsepower of compressor per ton of refrigeration
6. The quantity of refrigerant circulated through the system per ton of refrigeration

**Solution**

(a)  For refrigerant 12:

The high and low side pressures (1 and 2) are determined from Fig. 10.1. At −30 °F (−34.4 °C), pressure is 12.3 psia (85 kPa). The high side pressure corresponding to a refrigerant liquid temperature of 100 °F (37.8 °C) is 133 psia (910 kPa).

Knowing the high and low side pressures, Fig. 10.5 is used to construct the pressure-enthalpy diagram for the refrigeration cycle. This diagram is shown in Fig. 10.8.

The refrigeration capacity (Question 3) is calculated using Eq. (10.2):

**Fig. 10.8** Pressure-enthalpy diagram of refrigerant 12 for the problem in Example 10.3. *Schematically drawn from refrigerant enthalpy corresponding to temperature of R22 in Fig. 10.5 and data given in Example 10.3*

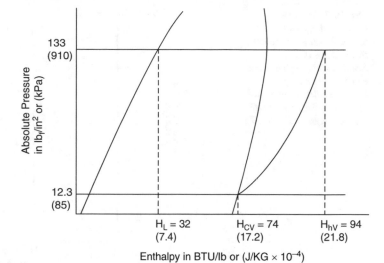

Refrigeration capacity $= (74 - 32)$
$$= 42 \text{ BTU/lb}$$

$$= (17.2 - 7.4) \times 10^4$$
$$= 98,000 \text{ J/kg}$$

The coefficient of performance (4) is calculated using Eq. (10.7):

$$\text{COP} = \frac{74 - 32}{94 - 74} = 2.1$$

The theoretical horsepower per ton of refrigeration (5) is calculated using Eq. (10.10):

$$\frac{C_p}{C_v} \text{ for Freon 12} = \gamma = 1.14$$

$$\frac{\text{HP}}{(\text{ton})_r} = \frac{4.715}{(\gamma)(\text{COP})}$$

$$= \frac{4.715}{(1.14)(2.1)} = 1.97$$

One ton of refrigeration (6) is equivalent to 12,000 BTU/h or 3517 W. Weight of refrigerant circulated/h is:

Weight
$$= \frac{\text{coolong capacity/ton of refrigeration}}{\text{coolong capacity/unit weight iof refrigerant}}$$
$$= \frac{12,000 \text{ BTU/h}}{42 \text{ BTU/lb}} = 286 \text{ lb refrigerant per hour}$$
$$= \frac{3517 \text{ J/s}}{98,000 \text{ J/kg}} = 0.0359 \text{ kg/s}$$
$$= 129 \text{ kg/h}$$

(b)  For refrigerant 717

The high and low side pressures (1 and 2) corresponding to $-30$ °F ($-34.4$ °C) and 100 °F (37.8 °C) are determined from Fig. 10.1. The low side pressure is 13.7 psia (94 kPa). The high side pressure is 207 psia (1440 kPa).

The pressure-enthalpy diagram (Question 3) is then constructed knowing the high and low side pressures using Fig. 10.4. Figure 10.9 shows this diagram.

Refrigeration capacity $= 600 - 155$
$$= 445 \text{ BTU/lb}$$
$$= (139 - 36) \times 10^4$$
$$= 1,030,000 \text{ J/kg}$$

The coefficient of performance (4):

$$\text{COP} = \frac{600 - 155}{784 - 600} = 2.42$$

The theoretical horsepower per ton of refrigeration (5) is calculated using a $C_p/C_v$ ratio of 1.29:

$$\frac{\text{HP}}{(\text{ton})_r} = \frac{4.715}{1.29(2.42)} = 1.51$$

Weight of refrigerant circulated per hour $(\text{ton})_r$ (6) is:

**Fig. 10.9** Presure-enthalpy diagram of ammonia for the problem in Example 10.3. *Schematically drawn from refrigerant enthalpy corresponding to pressure of R717 in Fig. 10.4 and data given in Example 10.3*

$$\text{Weight} = \frac{12,000 \text{ BTU/(h)(ton)}_r}{445 \text{ BTU/lb}}$$

$$= 26.97 \text{ lb/h(ton)}_r$$

$$= \frac{1317 \text{ J/s(ton)}_r}{1,030,000 \text{ J/kg}}$$

$$= 0.00341 \text{ kg/s(ton)}_r$$

$$= 12.3 \text{ kg/h(ton)}_r$$

### 10.1.6 The Condenser and Evaporator

Most refrigeration systems transfer heat between the refrigerant and air. Because heat transfer coefficients to air are usually very low, the air film resistance controls the rate of heat transfer. Very large heat transfer areas would be required to achieve the necessary heat transfer rate. To accomplish the necessary heat transfer rate and still have equipment that is reasonably sized, the heat transfer surface area of the tubes that comprise the evaporator or condenser coil is increased by the use of fins.

Finned heat exchange units are sized in terms of an effective heat transfer surface area which is the sum of the area of the bare tube $A_t$ and the effective area of the fin, which is the product of the fin surface area $A_s$ and the fin efficiency $\eta$.

$$A_{\text{eff}} = \eta A_f + A_t$$

The fin efficiency, $\eta$, decreases as the base area of the fin decreases and as the height of the fin increases. Furthermore, as the ratio of the heat transfer coefficient to air over the thermal conductivity of the metal that constitutes the fin ($h_o/k$) approaches zero, 6 approaches unity. Figure 10.10 shows fin efficiency for the most common systems. The fins in these units often consist of a stack of very thin sheets of metal pierced by a bundle of tubes. The equivalent fin surface area, $A_f$, is calculated as follows:

$$A_f = 2\pi \left( r_f^2 - r^2 \right) (n_t)(n_f)$$

where $n_t$ is number of tubes, and $n_f$ is number of fins along the length of each tube.

The capacity of a condensing or evaporating unit depends upon the surface area available for heat transfer and the mean temperature difference between the surfaces of the unit and the air going through the unit.

#### 10.1.6.1 Problems with Heat Exchange in Systems Using Zeotropic Refrigerants

The following examples are based on systems using a single compound or an azeotrope as the

**Fig. 10.10** Fin efficiency of fins consisting of a stack of sheet metal each sheet with thickness l, pierced by tubes of radius r. The tubes are arranged with centers on an equilateral triangle having distance b between centers

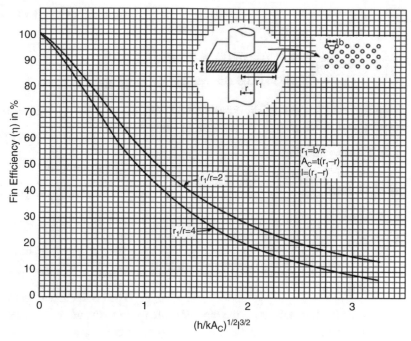

refrigerant. When an azeotrope is used as refrigerant, provision must be given to the fact that local heat transfer coefficient will be reduced on the refrigerant side of the heat exchanger because the composition of liquid refrigerant on the heat exchange surface will not be the same at different points in the heat exchanger. The thermal properties of the condensate will vary at different points in the heat exchanger, and the vapor phase of the other refrigerant components may also interfere with the heat transfer. Thus, the usual method of calculating the log mean $\Delta T$ in the heat transfer calculations is no longer correct. Depending upon the refrigerant, errors could go as high as 11% in the determined heat transfer surface area compared to using the standard heat transfer coefficient from condensing liquids and conventional log mean $\Delta T$ calculations. This error could result in undersized evaporator or condenser units.

**Example 10.4**  A condensing unit of a refrigeration system consists of 13 rows of 38.1-cm-long, 1.27-cm outside diameter copper tubes with 1.651-mm wall stacked four deep on 2.54-cm centers. The fin consists of 0.0254-cm thick aluminum sheets spaced 10 per 2.54 cm. If the rate of airflow through this condenser is such that

the outside heat transfer coefficient, h, is 9 BTU/(h $\cong$ ft$^2$ $\cong$ °F) or 51.1 W/(m$^2$ $\cong$ K), calculate the effective heat transfer surface area for this unit.

**Solution**
Refer to Fig. 10.10. Because the tubes are spaced on 2.54-cm centers, b = 2.54 cm. The thermal conductivity of aluminum (from Appendix Table A.9) is 206 W/(m $\cong$ K). Figure 10.10 can now be used to determine the fin efficiency η.

The tube radius, r = 0.00635 m

$$\text{Equivalent fin radius}, r_f = \frac{b}{\sqrt{\pi}} = \frac{0.0254}{1.772}$$
$$= 0.014334 \text{ m}$$

Fin cross-sectional area $A_c = t\,(r_f - r)$

$$= 0.000254\,(0.014334 - 0.00635)$$

$$= 2.0279 \times 10^6$$

The height of the fin, $P = (r_f - r)$

$$= (0.014334 - 0.00635) = 0.007984 - m$$

Values for the abscissa in Fig. 10.10 can now be calculated:

$$\left[\frac{h}{kA_c}\right]^{0.5}(\ell)^{1.5} = \left[\frac{51.1}{206(2.0279\times10^{-6})}\right]^{0.5}(0.007984)^{1.5}$$

$$= (349.7)(0.000713) = 0.2495$$

$$r_f/r = 0.014334/0.00635 = 2.26$$

From Fig. 10.10, $\eta = 0.9$. The equivalent fin area is calculated as follows:

$$n_t = 13(4) = 52 \text{ tubes}$$

$$n_f = 38.1 \text{ cm}\left[\frac{10 \text{ fins}}{2.54 \text{ cm}}\right]$$

$$= 150$$

$$A_f = 2\pi(r_f^2 - r^2)(n_t/n_f)$$

$$= 2\pi\left[(0.014334)^2 - (0.00635)^2\right](52)(150)$$

$$= 8.094 \text{ m}^2$$

The area of the bare tube $A_t$ is the total tube area minus the area covered by the fins. Let $P_t$ = length of the tube:

$$A_t = (2\pi r l_t)(n_t) - (2\pi r)(t)(n_t)(n_f)$$

$t$ = fin thickness = 0.000254 m:

$$A_t = 2\pi(0.00635)(0.381)(52)$$

$$- 2\pi(0.00635)(0.000254)(52)(150)$$

$$= 0.790 - 0.0790 = 0.711 \text{ m}^2$$

Effective area $A_{eff} = 0.91(8.094) + 0.711 = 8.0765 \text{ m}^2$.

**Example 10.5** If the conditions are such that a mean temperature difference $\Delta T$ of 45 °F (25 °C) exists in the condensing system and the refrigeration unit has a coefficient of performance of 2, how many tons of refrigeration can be supplied by the refrigeration system fitted with the condensing unit in Example 10.4? Assume the outside coefficient of heat transfer controls the heat transfer rate, $U = h_0 = 51.1 \text{ W/(m}^2 \cong \text{K)}$.

**Solution**

The rate of heat transferred through the unit is:

$$q = U A_{eff} \Delta T$$

$$= 51.1(8.0765)(25)$$

$$= 10,318 \text{ W}$$

From Eq. (10.7), let $R_r$ = refrigeration capacity:

$$COP = \frac{R_r}{H_{hv} - H_{cv}}$$

Where:

$$R_r = H_{cv} - H_L$$

$$H_L = H_{cv} - R_r$$

Condenser load:

$$C = H_{hv} - H_L = H_{hv} - H_{cv} + R_r.$$

$$(H_{hv} - H_{cv}) = C - R_r.$$

$$(\text{tons})_r = \frac{6878 \text{ W}}{3515.88 \text{ W/ton}} = 1.956 \text{ tons}$$

The coefficient of performance can be expressed in terms of the condenser load C and the refrigeration capacity $R_r$:

$$COP = \frac{R_r}{C - R_r}$$

Solving for $R_r$:

$$R_r = \frac{C(COP)}{1 + COP}$$

Substituting values for COP and C:

$$R_r = \frac{10,318(2)}{3} = 6878 \text{ W}$$

A similar procedure may be used to evaluate evaporators. The examples illustrate the dependence of the capacity of a refrigeration system on the heat exchange capacity of the condensing or evaporator unit in the system.

### 10.1.7  The Compressor

The compressor must be adequate to compress the required amount of refrigerant per unit time between the high side and low side pressures, to provide the necessary refrigeration capacity. Compressor capacities are determined by the displacement and the volumetric efficiency of the unit. The displacement is the volume displaced by the piston or rotary unit per unit time. For a reciprocating compressor, the displacement is calculated as follows:

Displacement
= (area of bore) (stroke) (cycles/min)

Theoretically, the quantity of gas passing through the unit per unit time should be independent of the pressure and should be a function only of the displacement of the unit. However, at high pressures, rotary units allow a certain amount of slip, and reciprocating units leave a certain amount of gas at the space between the cylinder head and the piston at the end of each stroke. Thus, the actual amount of gas passed through the unit per cycle is less than the displacement. The volumetric efficiency of a unit is the ratio of the actual volume delivered to the displacement. The volumetric efficiency decreases with increasing ratios of high side to low side pressures.

Large units that operate at very low pressures at the evaporator (e.g., units used for freezing) often are of a multistage type, where the refrigerant is compressed to an intermediate pressure, then partially cooled before it enters a second stage where it is compressed to the next higher pressure, and so forth, until the desired high side pressure is reached. The staging is designed such that equal work is accomplished in each stage and the volumetric efficiency remains high because of the low ratio of high side to low side pressure in each of the stages.

Figure 10.11 shows a comparison between the power requirement per ton of refrigeration between one-stage and two-stage units operating at various suction side pressures. For an ammonia compressor operating at a suction side pressure equivalent to a temperature of $-30\,^\circ$F, the motor horsepower per ton of refrigeration is 3.2 for a one-stage unit and is only 2.5 for a two-stage unit. In Example 10.3, the theoretical horsepower per ton of refrigeration under these conditions is only 1.51. Thus, the efficiency of the compressors represented by the graph in Fig. 10.11 is only 47% for a one-stage unit and 60% for a two-stage unit.

The presence of air in the refrigerant line reduces the efficiency of the compressor if expressed in terms of horsepower per ton of refrigeration as the same amount of power will

**Fig. 10.11** Motor horsepower and kW/Ton of refrigeration for one- and two-stage compressors. (From Food Eng. 41 (11):91, 1969)

be required to compress the gases to the required pressure. Yet, less refrigerant is passed through because part of the gas compressed is air.

## 10.2    Refrigeration Load

The amount of heat that must be removed by a refrigeration system per unit of time is the refrigeration load. The load may be subdivided into two categories: the *unsteady-state* load, which is the rate of heat removal necessary to reduce the temperature of the material being refrigerated to the storage temperature within a specific period of time, and the *steady-state* load, which is the amount of heat removal necessary to maintain the storage temperature. For food products, the temperature has to be reduced to the storage temperature in the shortest time possible to prevent microbiological spoilage and quality deterioration. Therefore, if large quantities need to be introduced into a refrigerated storage room intermittently, these materials are usually precooled to the storage temperature in a smaller precool or chill room or by other means prior to introduction into the large storage warehouse. This practice prevents the necessity of having to install oversized refrigeration units on the large storage warehouse that would operate at full capacity only during the short periods that materials are being introduced into the warehouse.

The unsteady-state load includes the sensible heat of the product, the heat of respiration if the product is fresh produce, and the heat of fusion of the water in the product if the product is to be frozen. The steady-state load includes heat incursion through the walls of the enclosure, through cracks and crevices and through doors, latent heat of condensation of moisture infiltrating into the room, and heat generated inside the room.

### 10.2.1    Heat Incursion Through Enclosures

Heat transfer rates through composite solids and techniques for determination of heat transfer coefficients were discussed thoroughly in Chap. 6. The ASHRAE guide and data book for 1965 recommends the design heat transfer coefficients to be used for various surfaces as shown in Table 10.2.

### 10.2.2    Heat Incursion Through Cracks and Crevices

A majority of heat transferred is due to fluctuations in pressure caused by cycling of temperature in the room. At the high-temperature point in the cycle, cold air will be expelled from the room. When the pressure drops at the low-temperature point in the cycle, warm air will be drawn into the room. The amount of air admitted into the room because of temperature cycling can be calculated using the principles discussed in Chap. 3.

**Table 10.2**  Heat transfer coefficient to air from various surfaces under various conditions

|  | Heat transfer coefficients | |
| --- | --- | --- |
|  | BTU | W |
| Surface | $h(ft^2)(°F)$ | $m^2 (K)$ |
| Inside wall (still air) | 1.5 | 8.5 |
| Outside wall or roof | 5.9 | 33.5 |
|   15 miles/hr or 24 km/h wind | | |
| Outside wall or roof | 4.0 | 22.7 |
|   7.5 miles/hr or 12 km/h wind | | |
| Horizontal surface, still air, upward heat **flow** | 1.7 | 9.7 |
| Horizontal surface, still air, downward heat **flow** | 1.1 | 6.25 |

*Source*: Adapted from ASHRAE, 1965. Design heat transmission coefficient. *ASHRAE Guide and Data Book. Fundamentals and Equipment for 1985 and 1986.* American Society of Heating, Refrigeration, and Air Conditioning Engineers, Atlanta, GA

### 10.2.3   Heat Incursion Through Open Doors

Opening doors allow entry of warm outside air and expulsion of cold air. The rate of heat incursion is dependent upon the size of the door and the temperature differential between the inside and outside. Data on rate of heat transfer through doors of refrigerated rooms have been determined empirically. The equation for rate of heat loss calculated from data published in *Food Engineering Magazine* [Food Eng. 41(11):91, 1969] for values of $\Delta T$ between 40 °F and 120 °F (22.2 °C and 66.7 °C) is:

$$q = 2126 \; W[e]^{0.0484(\Delta T)}(h)^{1.71} \qquad (10.11)$$

where q is rate of heat incursion into the room in watts, W is the width of the door in m, $\Delta T$ is the temperature difference in °C inside and outside the room, and h is the height of the door in m.

**Example 10.6**   The door to a refrigerated room is 3.048 m high and 1.83 m wide. It is opened and closed at least five times each hour with the door remaining open for at least 1 min at each opening. Calculate the refrigeration load due to the door opening if the room is maintained at 0 °C and ambient temperature is 29.4 °C.

**Solution**
h = 3.048 m, W = 1.83 m, and $\Delta T$ = 29.4 °C. Substituting into Eq. (10.11):

$$q = 2126(1.83)(e)^{0.0484(29.4)}(3.048)^{1.71}$$
$$= 108.6 \; KW$$

The total time the door was opened in 1 h is 300 s. Refrigeration load:
= 108.6 KW · (300 s) = 35.58 MJ.

### 10.2.4   Heat Generation

Motors inside a refrigerated room generate heat at the rate of 1025.5 W/hp. This rate drops to 732.48 W/hp if only the motor is inside and the

load it drives is outside the refrigerated room. Workers inside a room generate approximately 293 W per person. The rate of heat generation by personnel increases with decreasing room temperatures. Heat dissipated by light bulbs is the same as the wattage of the lamp.

Fruits and vegetables respire and the heat of respiration adds to the refrigeration load. The heat of respiration of fruits and vegetables is a function of temperature and can be calculated using the expression:

$$q = a(e)^{bT} \qquad (10.12)$$

q is the rate of heat generation (mW/kg in SI). Values for a and b for various fruits and vegetables are shown in Table 10.3.

During cooling, heat of respiration will decrease as temperature decreases. An average heat of respiration may be calculated between two temperatures $T_1$ and $T_2$ knowing the time $\Delta t$ for the temperature to change between $T_1$ and $T_2$ and assuming that the temperature change during this period is linear.

$$q = \frac{1}{\Delta t} \int_0^{\Delta t} a(e)^{bT} \; dt \qquad (10.13)$$

Assuming a linear temperature change:

$$T_1 - T = \frac{T_1 - T_2}{\Delta t}t \qquad (10.14)$$

Substituting Eq. (10.14) for T in Eq. (10.13):

$$q = \frac{a}{\Delta t} \int_0^{\Delta t} [e]^{bT_1 - \frac{T_1 - T_2}{\Delta t}bt} \; dt$$

Integrating and substituting limits:

$$q = \frac{ae^{bT_1}}{b(T_1 - T_2)}\left[1 - (e)^{-b(T_1 - T_2)}\right] \qquad (10.15)$$

Equation (10.15) is the average rate of heat generation over the time $\Delta t$. The units will be the same as the units of a, which in SI, is mW/kg as tabulated in Table 10.3. Multiplying q in Eq. (10.15) by $\Delta t$ will give the total heat generated for the time period under consideration.

**Table 10.3** Heat of respiration of fruits and vegetables in air. Values of the constants a and b in the equation $q = a(e)^{bT}$

| Product | T in °F | | T in °C | |
| | $q^1$ in $\frac{BTU}{(ton) \cdot (24\ hr)}$ | | $q^1$ in $\frac{mW}{kg}$ | |
| | a | b | a | b |
|---|---|---|---|---|
| Apples | 213 | 0.06 | 19.4 | 0.108 |
| Asparagus | 2779 | 0.048 | 173.0 | 0.086 |
| Beans (green or snap) | 829 | 0.064 | 86.1 | 0.115 |
| Beans (lima) | 376 | 0.071 | 48.9 | 0.128 |
| Beets (topped) | 1054 | 0.031 | 38.1 | 0.056 |
| Broccoli | 854 | 0.067 | 97.7 | 0.121 |
| Brussels sprouts | 1845 | 0.045 | 104.0 | 0.081 |
| Cabbage | 337 | 0.041 | 16.8 | 0.074 |
| Cantaloupes | 128 | 0.07 | 16.1 | 0.126 |
| Carrots (topped) | 498 | 0.046 | 29.1 | 0.083 |
| Celery | 237 | 0.58 | 20.3 | 0.104 |
| Corn (sweet) | 2465 | 0.043 | 131.0 | 0.077 |
| Grapefruit | 171 | 0.051 | 11.7 | 0.092 |
| Lettuce (head) | 416 | 0.049 | 26.7 | 0.088 |
| Lettuce (leaf) | 1188 | 0.041 | 59.1 | 0.074 |
| Onions | 89 | 0.055 | 6.92 | 0.099 |
| Oranges | 151 | 0.059 | 13.4 | 0.106 |
| Peaches | 104 | 0.074 | 14.8 | 0.133 |
| Pears | 42 | 0.096 | 12.1 | 0.173 |
| Peas (green) | 1249 | 0.059 | 111.0 | 0.106 |
| Peppers (sweet) | 796 | 0.040 | 33.4 | 0.072 |
| Spinach | 473 | 0.073 | 65.6 | 0.131 |
| Strawberries | 568 | 0.059 | 50.1 | 0.106 |
| Sweet potatoes | 796 | 0.034 | 31.7 | 0.061 |
| Tomatoes | 159 | 0.057 | 13.2 | 0.103 |
| Turnips | 590 | 0.037 | 25.8 | 0.067 |

*Source*: ASHRAE, 1974. Approximate rates of evolution of heat by certain fruits and vegetables when stored at temperatures indicated. In: *ASHRAE Handbook and Product Directory—1974 Applications*. American Society of Heating, Refrgerating, and Air Conditioning Engineers, Atlanta, GA

[1]*q* calculated using the constants *a* and *b* are maximum values in the range reported. Minimum values average 67% of the maximum

Although the temperature at any given point in a material during the process of cooling usually changes exponentially with time, materials that are cooled in bulk (e.g., head lettuce or boxed products) would have a temperature gradient with the interior parts at a higher temperature than the external sections exposed to the cooling medium. Thus, if the bulk mean temperature is considered, the deviation from the linear temperature change assumption would be minimal.

**Example 10.7** It is desired to cool cabbage from 32.2 °C to 4.44 °C in 4 h. Calculate the heat of respiration during this cooling period. Assume a linear temperature change.

**Solution**

$T_1 = 32.2$ °C; $T_2 = 4.44$ °C; and $\Delta t = 4$ h = 14,400 s. From Table 10.3 for cabbage, a = 16.8 and b = 0.074 in SI units. Examination of Eq. (10.12) will reveal that a will have the same units as q and b will have units of 1/EC.

$$q = \frac{16.8\ e^{0.074(32.2)}}{(0.074)(32.2 - 4.44)} \left[ 1 - (e)^{-0.074(32.2 - 4.44)} \right]$$

$$= 77.253\ mW/kg$$

Heat generated over the time period of 14,400 s is Q:

$$Q = (77.253 \times 10^{-3} \text{ W/kg})(14,400 \text{ s}) = 1112 \text{ J/kg}$$

**Example 10.8** Calculate the refrigeration load due to the heat of respiration of spinach at a constant temperature of 3.33 °C.

**Solution:**
T = 3.33 °C. From Table 10.3, a = 65.6 mW/kg and b = 0.131 °C$^{-1}$.

$$q = 65.6e^{0.131(3.35)} = 101.74 \text{ mW/kg}$$

### 10.2.5  The Unsteady-State Refrigeration Load

Procedures for calculating the heat capacity and sensible heat gain or loss are discussed in Chap. 5. When a change in phase is involved, the latent heat of fusion of the water must be considered. The heat of fusion of ice is 144 BTU/lb, 80 cal/g, or 0.334860 MJ/kg.

**Example 10.9** Calculate the refrigeration load when 100 kg/h of peas needs to be frozen from 30 °C to −40 °C. The peas have a moisture content of 74%. The freezing point is −0.6 °C.

**Solution**
The refrigeration load above freezing will be calculated using Siebel's equation. From Eq. (4.9), Chap. 4, the specific heat above freezing for a material that contains 74% water and 26% solids nonfat is:

$$C_p = 837.36(0.26) + 4186.8(0.74) = 3315.9 \text{ J/(kg K)}.$$

The refrigeration load required for cooling 1 kg peas from 30 °C to −0.6 °C = $Q_1$

$$Q_1 = 1(3315.9)(30.6) = 101,467 \text{ J}$$

The refrigeration load below freezing is calculated using the concepts and equations discussed in the section "Enthalpy Changes in Food Freezing" in Chap. 4. Molality of solutes in unfrozen water at different temperatures below freezing will be used as the basis for the calculations.

$$T_f = -0.6 \degree C;$$
$$T_2 = -40 \degree C;$$
$$K_f = 1.86;$$
$$\Delta H_f = 334,860 \text{ J}$$

The specific heat of ice = 2093.4 J/(kg K)
The specific heat of water is 4186.8 J/(kg K)
The specific heat of solids nonfat = 837.36 J/(kg K)
The initial moisture before freezing = $W_o$ = 0.740 kg
The liquid water at −40 °C = W = $W_o(-T_f / -T_2) = 740(0.6/40) = 0.0111$ kg
Ice at −40 °C = 0.740−0.0111 = 0.7289 kg
Change in sensible heat for ice from −0.6 °C to −40 °C
(From Eq. 4.22, Chap. 4)
$Q_{ice} = I \ C_{pi}W_o[(T_f - T_2) - (-T_f \ln(-T_2/-T_f)] = 0.0111 \ (2093.4) [39.4 - 0.6 \ln(40/0.6)] = 856.9$ JChange in sensible heat of water from −0.6 °C to −40 °C
Change in sensible heat of water from −0.6 °C to −40 °C
(From Eq. 5.22, Chap. 5)

$$Q_{water} = C_{pw}W_o(-T_f) \ln(-T_2/-T_f)$$
$$= 4186.8(0.7289)[0.6 \ln(40/0.6)]$$
$$= 7689.9 \text{ J}$$

$$Q_{snf} = 0.26 (837.36)[-0.6 - (-40)]$$
$$= 8577.9 \text{ J}$$

The amount of ice that is formed = 0.7289 kg

$$\Delta H_f = 0.7289 (334,860) = 244,079 \text{ J}$$

Total refrigeration load/kg:
$$= 101,467 + 856.9 + 8577.9 + 7689.9 + 244,079$$
$$= 362,670 \text{ J/kg}$$
For 100 kg, the refrigeration load will be 36.267 MJ.

**Table 10.4** Recommended storage conditions for fruits and vegetables

| (1) Storage temperatures 30°–32 °F (−1.11°–0 °C). Highest Freezing point 28 °F (−2.22 °C). 85–90% RH. | | | | |
|---|---|---|---|---|
| Apricots | Pears | | | |
| Cherries | Peaches | | | |
| Grapes | Plums | | | |
| (2) Storage temperatures 32 °F (0 °C) | | | | |
| 75%RH | 90% RH | 95% RH | 95% RH | 95% RH |
| Garlic | | | | |
| Onions | Mushrooms, oranges, tangerines | Artichokes, asparagus, lima, beans, beets, broccoli, brussels sprouts, cabbage, cauliflower, celery[a] | Sweet corn endives[b], escarole[b], leafy greens parsley, green peas, radishes, rhubarb, spinach[b] | Carrots, lettuce[b], parsnips, rutabagas, turnips |
| (3) Storage temperature 36 °F (2.22 °C). 95% RH. Apples | | | | |
| (4) Storage temperature 45 °F (7.22 °C). 90% RH. Subject to chill injury at temperatures below 45 °F | | | | |
| Green or snap beans | | | | |
| Ripe tomatoes | | | | |
| (5) Storage temperature 50 °F (10 °C). Subject to chill injury at temperatures below 50 °F | | | | |
| 85% RH | 90–95% RH | | | |
| Melons | | | | |
| Potatoes | Cucumbers | | | |
| Pumpkin | Eggplants | | | |
| Squash | Sweet peppers | | | |
| Green tomatoes | Okra | | | |
| (6) Storage temperatures 58°–60 °F (14.4–15.6 °C). 85–90% RH | | | | |
| Bananas | | | | |
| Grapefruit | | | | |
| Lemons | | | | |

*Source*: ASHRAE. 1974 ASHRAE Handbook and Product Directory—1974. Applications. American Society of Heating, Refrigeration, and Air Conditioning Engineers, New York
All others have freezing points below 31 °F(−0.56 °C).
[a]Highest freezing point 31.1 °F(−0.5 °C)
[b]Highest freezing point 31.9 °F (−40.06 °C)

## 10.3  Commodity Storage Requirements

Most food products would benefit from a reduction in the storage temperature provided that no freezing occurs. Reduced temperature reduces the rate of chemical reactions that deteriorate the product and also reduce microbiological activity. Freezing damages the cellular structure of fruits and vegetables and severely affects acceptability. Meat pigments darken irreversibly upon freezing. Acceleration of the development of a strong fishy flavor has been observed when fish were frozen at or near the freezing point and thawed. Thus, for perishable foods not specifically prepared for frozen storage, the freezing temperature should be the lowest acceptable limit for storage. Some fruits and vegetables are susceptible to chill injury at temperatures above the freezing point. The lowest safe temperatures have been defined for these commodities. Table 10.4 shows the temperature and humidity recommended for storage of various fruits and vegetables.

## 10.4  Controlled Atmosphere Storage

### 10.4.1  Respiration

Fruits and vegetables continue their metabolic activity after harvest. Maintenance of this metabolic activity is essential in the preservation of quality. Metabolic activity is manifested by respiration, a process where the material consumes oxygen and evolves carbon dioxide. A major nutrient metabolized during respiration is carbohydrate. Because respiration depletes nutrients in a product after harvest, the key to prolonging shelf life of fruits and vegetables is in reducing the rate of respiration. Early studies on the relationship between respiration and quality have established that storage life is inversely related to the rate of respiration and that for two products respiring at different rates, when the total quantity of $CO_2$ evolved by each product is equal, the products would have reached a comparable stage in their storage life.

Reduction of temperature is an effective means of reducing the rate of respiration. However, for some products subject to chill injury, respiration rate may still be quite high even at the lowest safe storage temperature. Controlled atmosphere storage has been developed as a supplement to refrigeration in prolonging the storage life of actively respiring fruits and vegetables.

The reaction involved in respiration is primarily the oxidation of carbohydrates:

$$C_6H_{12}O_6 + 6O_2 = 6CO_2 + 6H_2O$$

The respiratory quotient (RQ) defined as the ratio of $CO_2$ produced to $O_2$ consumed is 1 for the reaction. Most products show respiratory quotients close to 1. The heat evolved during respiration has also been shown to be related to the quantity of $CO_2$ produced in the same manner as that released during the combustion of glucose (10.7 J/mg $CO_2$). The rate of oxygen consumption and $CO_2$ evolution can be calculated from the heat of respiration and vice versa.

Controlled atmosphere (CA) storage is based on the premise that increasing the $CO_2$ level and decreasing the $O_2$ level in the storage atmosphere will result in a reduction of the rate of respiration. Indeed, calorimetric determination of heat of respiration of products in continuously flowing CA of the optimum composition for the given product has shown that CA could reduce the respiration rate to approximately 1/3 of the respiration rate in air at the same temperature.

**Example 10.10** One pound (0.454 kg) of head lettuce is packaged in an airtight container with a volume of 4 L. The product occupies 80% of the volume, the rest being air. If the product is at a constant temperature of 4 °C, calculate how long it will take for the oxygen content in the package to drop to 2.5%. Assume a RQ of 1.

**Solution**

From Table 10.3, the constants a and b for the heat of respiration of head lettuce (in mW/kg) are 26.7 and 0.088, respectively.

$$q = (26.7) \times (e)^{[(0.088) \times (4)]} = 38 \text{ mW/kg}$$
$$= 0.038 \text{ J/(s} \cdot \text{kg)}$$

$$\frac{\text{mg } CO_2}{h} = (0.038) \cdot \frac{\text{mg } CO_2}{10.7 \text{ J}} \cdot \frac{3600 \text{ s}}{h} (0.454 \text{ kg})$$

$$= 5.8$$

$$\frac{\text{gmoles } CO_2}{h} = \frac{0.0058 \text{ g } CO_2}{h} \cdot \frac{\text{g mole}}{44 \text{ g}}$$

$$= (1.318) \cdot 10^{-4}$$

Since RQ $= 1$, g moles $O_2$/h $= (1.3818) \cdot 10^{-4}$ g moles/h

Air is approximately 21% $O_2$ and 79% $N_2$. The total number of moles of air originally in the container is:

$$n_{air} = \frac{PV}{RT} \text{ where: } P = 1 \text{ atm}, V = 4 \text{ L}, T = (273 + 4)$$

$$R = 0.08206 \frac{L \cdot atm}{g \text{ mole} \cdot K} \text{ and}$$
$$n_{o2} = 0.2 \text{ moles } O_2/\text{mole air}$$

$$n_{O_2} = \frac{1 \cdot (4) \cdot (0.2)}{(0.08206) \cdot (277)} = 0.0352$$

Because the RQ is 1, there will be no net change in the total number of moles of gases inside the container. The number of moles of oxygen when the concentration is 2.5% is:

$n_{O_2 final} = 0.025(0.0352) = 0.00088$

The original number of moles of $O_2$ is:

$n_{O_2 initial} = 0.21(0.0352) = 0.007392$ moles

The number of moles of $O_2$ that must be depleted by respiration is:

$n_{O_2 depleted} = 0.007392 - 0.00088 = 0.006512$

The time required to deplete $O_2$ to the desired level is:

$$\text{Time} = \frac{0.006512 \text{ mole}}{0.0001318 \text{ mole/h}} = 49.4 \text{ h}$$

### 10.4.2 CA Gas Composition

CA storage has been used successfully with apples and pears. Experimental results have been very encouraging on CA storage of cabbage, head lettuce, broccoli, and brussels sprouts. Atmospheric modification of product containers during transit of vegetables is also gaining wider acceptance. CA is recommended when transit time is 5 days or more.

Reduction of oxygen concentration in the storage atmosphere to 3% or below has been shown to be most effective in reducing respiration rate, with or without $CO_2$. Too low an oxygen concentration, however, leads to anaerobic respiration resulting in the development of off-flavors in the product. The $O_2$ concentration for onset of anaerobic respiration ranges from 0.8% for spinach to 2.3% for asparagus. The optimum CA composition varies from one product to another and between varieties of a given product. CA compositions found to be effective for some products are shown in Table 10.5.

**Table 10.5** CA storage conditions suitable for use with various products

| Product | % $CO_2$ | % $O_2$ |
|---|---|---|
| Apples | 2–5 | 3 |
| Asparagus | 5–10 | 2.9 |
| Brussels sprouts | 2.5–5 | 2.5–5 |
| Beans (green or snap) | 5 | 2 |
| Broccoli | 10 | 2.5 |
| Cabbage | 2.5–5 | 2.5–5 |
| Lettuce (head or leaf) | 5–10 | 2 |
| Pears | 5 | 1 |
| Spinach | 11 | 1 |
| Tomatoes (green) | 0 | 3 |

Percentages are volume or mole percent. Balance is nitrogen. Reprinted from Toledo, R. T. 1980. *Fundamentals of Food Process Engineering*. AVI Pub. Co., Westport, Conn

To be effective, CA storage rooms must be reasonably airtight. To test for airtightness, the room should be pressurized to a positive pressure of 1 in. water gauge (wg) (249 Pa gauge). A room is considered airtight if at the end of 1 h, the pressure does not drop below 0.2 in. wg (49.8 Pa). This requirement is equivalent to an air incursion rate of 0.2% of the gas volume in the room per hour at a constant pressure differential between inside and outside of 0.5 in. wg (124 Pa). Current standards for refrigerated tractor trailers call for a maximum air incursion rate of 2 ft$^3$/min (3.397 m$^3$/h) at a pressure differential of 0.5 in. water (124 Pa) between the inside and outside of the trailer. This rate is approximately 15 times what would meet the specification for airtightness.

CA in storage warehouses are maintained at the proper composition by introduction of $CO_2$ and/or $N_2$ if product respiration is insufficient to raise $CO_2$ levels or lower $O_2$ levels. In a full, tightly sealed warehouse, ventilation with fresh air and scrubbing of $CO_2$ from the storage atmosphere may be necessary to maintain the correct CA composition and prevent anaerobic respiration. Successful operation of a CA storage facility necessitates regular monitoring and analysis of gas composition and making appropriate corrections when necessary. Figures 10.12 and 10.13 are schematic diagrams of CA storage systems for warehouses and for in-transit systems.

**Example 10.11** The rate of flow of fluids through narrow openings is directly proportional to the square root of the pressure differential across the opening. If the pressure inside a chamber of volume V changes exponentially with time from 1 in. wg (249 Pa) at time 0 to 0.2 in. wg (49.8 Pa) at 1 h, calculate the mean volumetric rate of flow of gases out of the chamber and the mean pressure at which this rate of flow would be expected if the pressure is held constant.

**Solution**

The expression for the volumetric rate of flow Q is:

$$Q = k(\Delta P)^{1/2}$$

**Fig. 10.12** Schematic diagram of a controlled atmosphere storage system for warehouses

CONTROLLED ATMOSPHERE STORAGE SYSTEM
(STATIONARY)

H$_2$O, DILUTE NaOH SOLUTION, OR MONOETHANOL AMINE SOLUTION

BLOWER

CO$_2$ ABSORPTION CHAMBER TRAPS EXCESS CO$_2$ AND OTHER GASES

WELL SEALED CHAMBER

MONOETHANOL AMINE REACTION:

ABSORPTION
$$2\ NH_2CH_2CH_2OH + CO_2 \longrightarrow (NH_2CH_2CH_2)_2CO_3 + H_2O$$

REGENERATION
$$(NH_2CH_2CH_2)_2CO_3 + H_2O \xrightarrow{\text{HEAT}} 2NH_2CH_2CH_2OH + CO_2 \uparrow$$

**Fig. 10.13** Schematic diagram of a transportable system for generating and maintaining controlled atmospheres

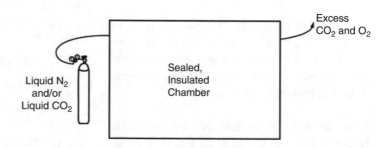

Excess CO$_2$ and O$_2$

Liquid N$_2$ and/or Liquid CO$_2$

Sealed, Insulated Chamber

The volume dQ expelled over a period of time dt is:

$$dQ = k(\Delta P)^{1/2}\ dt$$

Because P is exponential with time, substituting the boundary conditions, $\Delta P = 1$ at $t = 0$ and $\Delta P = 0.2$ at $t = 1$, will give the equation for $\Delta P$ as a function of t as follows:

$$\Delta P = (e)^{-(1.6094)\cdot t}$$

Substituting P in the differential equation and integrating:

$$Q = \int_0^1 \left[ k(e)^{0.5(-1.6094)t} \right]\ dt$$

The mean rate of flow in 1 h, $\bar{q} = Q/t$

$$\bar{q} = \frac{1}{1}\left[ \left( \frac{k}{-0.8047} \right)(e)^{-0.8047t} \right]_0^1$$
$$= 0.687\ k$$

Because $\bar{q} = k(\overline{\Delta P})^{1/2}$, the mean pressure $\overline{\Delta P}$ that would give the equivalent rate of flow as $\bar{q}$ is $(0.687)^2$ or 0.472.

One atmosphere is 406.668 in. of water. Using the ideal gas equation, the number of moles gas in the chamber at pressure $P_1$ and $P_2$ is:

$$n_1 = \frac{P_1 V}{RT} \text{ and } n_2 = \frac{P_2 V}{RT}$$

The number of moles lost with a drop in pressure from $P_1$ to $P_2$ would be:

$$n_1 = \frac{V}{RT}(P_1 - P_2)$$

Expressing the quantity expelled as volume at standard atmospheric pressure:

$$V_1 = n_1 \frac{RT}{P} = \frac{(P_1 - P_2)V}{P}$$

Substituting $P_1 = 1$ in.; $P_2 = 0.2$ in.; and $P = 406.668$:

$$V_1 = \frac{0.8V}{406.668} = 0.00196 \text{ V}$$

The specified conditions of a drop in pressure from 1 in. wg to 0.2 in. wg in 1 h is equivalent to a rate of loss of approximately 0.2% of the chamber volume per hour at an average pressure drop of approximately 0.5 in. wg.

**Example 10.12** A refrigerated tractor-trailer is loaded with head lettuce such that 20% of its total volume is air space. The trailer has a total volume of 80 m$^3$. Assuming that the trailer meets the specifications for the maximum air incursion rate of 2 ft$^3$/min (3.397 m$^3$/h) at a pressure differential of 0.5 in. wg (124 Pa), what would be the rate of addition of $N_2$ and $CO_2$ necessary to maintain the atmosphere inside the truck at 2% $O_2$ and 5% $CO_2$ when the truck is traveling at the rate of 50 miles/h (80 km/h). The average temperature of the product is 1.5 °C, and the trailer contains 18,000 kg of lettuce.

**Solution**

If the atmosphere inside the truck has been pre-equilibrated at the point of origin to the desired CA composition, the amount of nitrogen and $CO_2$ that must be added during transit would

be that necessary to displace the air infiltrating into the trailer. Let x = rate of $N_2$ addition, and y = rate of $CO_2$ addition. If $n_{ai}$ = the rate of air infiltration, $n_{rc}$ = the rate of $CO_2$ generation by the product, and $n_{ro}$ = the rate of oxygen consumption by the product during respiration, the following material balance may be set up for $CO_2$ and $O_2$ assuming that the expelled gases are of the same composition as those in the interior of the trailer.

$O_2$ balance (air is 21% oxygen):

$$n_{ai}(0.21) = (n_{ai} + x + y)(0.02) + n_{ro}$$

$CO_2$ balance:

$$y + n_{rc} = (n_{ai} + x + y)(0.05)$$

The heat of respiration of head lettuce in mW/Kg from values of a = 26.7 and b = 0.088 from Table 10.3 and for $_t = 1.5$ °C:

$$q = 26.7(e)^{1.5(0.088)} = 30.47 \text{ mW/kg}$$

$$\frac{\text{mg } CO_2}{\text{kg} \cdot \text{h}} = \frac{30.47 \times 10^{-3} \text{J}}{\text{kg} \cdot \text{s}} \cdot \frac{\text{mg } CO_2}{10.7 \text{ J}} \cdot \frac{3600 \text{ s}}{\text{h}}$$

$$= 10.25$$

$$\frac{\text{g moles } CO_2}{\text{kg} \cdot \text{h}} = \frac{\text{g moles } O_2}{\text{kg} \cdot \text{h}} = \frac{10.25 \times 10^{-3}}{44}$$

$$= 0.233 \times 10^{-3}$$

$$n_{rc} = (18,000 \text{ kg})\frac{0.233 \times 10^{-3} \text{ g moles}}{\text{kg} \cdot \text{h}}$$

$$= 4.194 \frac{\text{g moles}}{\text{h}}$$

$$n_{ro} = 4.194 \text{ g moles/h}$$

The rate of air infiltration must now be calculated at a velocity of 80 km/h. The velocity head of a flowing fluid, $\Delta P/\rho$, is $V^2/2$. $\rho$ is the fluid density:

$$\Delta P = \frac{(80,000)^2 \text{m}^2}{\text{h}^2}(0.5)\frac{\rho \text{kg}}{\text{m}^3}\frac{1 \text{ h}^2}{(3600)^2 \text{ s}^2}$$

$$= 247 \text{ } \rho \text{ Pa}$$

The specified infiltration rate of 3.397 m$^3$/h at a pressure of 124 Pa can be converted to the rate at the higher pressure by using the relationship that rate of flow is directly proportional to the square root of the pressure.

$$Q = 3.397 \left[\frac{247}{124}\right]^{\frac{1}{2}} = 4.79 \ \text{m}^3/\text{h}$$

Using the ideal gas equation, $n_{ai} = PV/RT$:

$$n_{ai} = \frac{(4.79 \ \text{m}^3/\text{h})(1000 \ \text{L/m}^3)(1 \ \text{atm})}{0.08206(\text{L})(\text{atm})/(\text{gmole} \ \text{K})(274.5 \ \text{K})}$$
$$= 212.7 \ \text{g moles/h}$$

Substituting values for $n_{ai}$, $n_{ro}$, and $n_{rc}$ in the $O_2$ and $CO_2$ balance equations:

$$212(0.21) = (212 + x + y)(0.02) + 4.176$$
$$x + y = \frac{44.52 - 4.176 - 4.24}{0.02} = 1805$$
$$x = 1805 - y$$
$$y + 4.176 = (212 + x + y)(0.05)$$
$$0.95y = 0.05x + 6.424$$
$$0.95y - 0.05(1805 - y) = 6.424$$
$$1.0y = 6.424 + 90.25$$
$$y = 96.67 \ \text{g moles} \ CO_2/\text{h}$$
$$x = 1708.3 \ \text{g moles} \ N_2/\text{h}$$

Note that because of air infiltration, it would not be possible for product respiration to maintain the necessary CA composition inside the trailer. Furthermore, the respiration rate for this particular product is so low that flushing of the trailer with $N_2$ and $CO_2$ would have to be done initially to bring the CA to the desired composition within a short period of time.

---

## 10.5  Modified Atmosphere Packaging

Modified atmosphere (MA) and CA are similar and are sometimes used interchangeably. A feature of CA is some type of control that maintains constant conditions within the storage atmosphere. MA packaging (MAP) involves changing the gaseous atmosphere surrounding the product at the time of packaging and using barrier packaging materials to prevent gaseous exchange with the environment surrounding the package. The use of gas permeable packaging materials that is tailored to promote gas exchange to maintain relatively constant conditions inside the package may be considered a form of CA packaging. Reduction of respiration rates of live organisms by lowering oxygen and elevating $CO_2$ has been discussed in the previous section. MA packaging, as is now practiced, is primarily designed to control the growth of microorganisms and in some cases, enzyme activity, in packaged uncooked or cooked food. Modified atmosphere packaging of washed packaged ready to eat salad greens is now widely practiced in the vegetable industry. Fresh ready to cook meats in modified atmosphere packages are also gaining consumer acceptability. The success of these types of packaging can be attributed to their prolonged shelf life and convenience to the consumer. For fruits and vegetables, MAP is used primarily for salad greens and precut fruits.

### 10.5.1  Modified Atmosphere Packaging of Fruit, Vegetables, Bakery Products, and Nuts

The storage atmospheres shown in Table 10.5 are a good guide in selection of the appropriate composition to use in MAP of produce. The atmosphere used, if the product itself is non-respiring, is usually 75% $CO_2$, 15% $N_2$, and 10% $O_2$. Bakery products, which are generally of lower moisture content than most foods, could be preserved longer in 100% $CO_2$. Dissolution of $CO_2$ and $O_2$ in the aqueous phase of the product is responsible for microbial inhibition. High $CO_2$ is inhibitory to most microorganisms. Although $CO_2$ lowers the pH, the inhibitory effect is more than from the pH reduction alone. The presence of $O_2$ has slight inhibitory activity against anaerobes resulting in slower growth compared with a 100% $CO_2$ atmosphere. Incidence of illnesses attributed to consumption of MAP packaged salad greens has led to the adoption of prepackaging steps to eliminate pathogens in the product. Large-scale corporate MAP operations start with antimicrobial treatments at the time of harvest where the product is trimmed of inedible parts and sprayed with chlorinated water before it is transported to the processing plant. At the plant, more trimming is done and leafy parts are separated so that each part can contact an antimicrobial agent, usually chlorinated water containing up to 15% active chlorine. This is achieved by transporting the product in a flume, elevated to drain as much of the entrained water, followed by a vibrating screen to remove as much of the entrained water as possible. A low-speed centrifuge is then used to produce a semidry product which is then filled into pouches using a form fill and seal packaging machine. The MA gas is introduced into the package before the package is sealed.

The type of packaging material used for pouches of MAP produce will depend on the product inside the package. Typically, a low oxygen atmosphere and a high CO2 atmosphere are used. However, fresh vegetables are live respiring products; therefore provision must be made to permit slow entry of oxygen to replace what was consumed by product respiration and to permit removal of $CO_2$ generated by respiration. Typically, packaging film with moderately low oxygen permeability and moderately high $CO_2$ permeability is used. Another factor to be considered is a moderately low moisture-vapor transmission to permit moisture produced by respiration to exit the package but not overly high rate of transmission, otherwise the product will lose turgidity and reduced sensory crunchiness.

### 10.5.2  MAP of Fresh Ready to Cook Meats

Successful food shelf-life extension of fresh meats by MAP is not just dependent on the atmosphere in the package. Successful shelf-life extension will depend on the following:

1. The initial microbial load. A low microbial count will extend the lag phase of growth of microorganisms. Proper plant sanitation, and a pretreatment with an antimicrobial before the packaging operation, ensures low microbial counts in the packaged product at the time of packaging.

2. Lowest possible product temperature at the time of packaging and during storage and distribution. There will be a twofold decrease in the generation time of microorganisms with every 10 C increase in temperature.

3. The presence of microbial inhibitors. For fresh meats, the use of microbial inhibitors is widely used to achieve an appreciable shelf-life extension. The US Dept. of Agriculture Food Safety and Inspection Service lists the following "Safe and Suitable Ingredients used in the production of meat, poultry and egg": buffered vinegar, lactate, diacetate, lauric arginate (lauramide arginine ethyl ester), octanoic acid, citric acid, and sodium diacetate.

4. Uniform dispersion of antimicrobial agent on product surfaces. The antimicrobial ingredients may be applied as a marinade added to the meat by injecting, tumbling, spraying, or dipping. The process should ensure consistent concentration of antimicrobial compound coating the product surfaces.

5. Accurate delivery of modified atmosphere gas concentration into the package. Typically, a vacuum is drawn on the package, and the modified atmosphere gas mixture containing the desired concentration of component gases is introduced into the package typically to achieve a pressure inside the package just slightly below atmospheric pressure to hold the pieces inside the package in place to avoid cross-contamination between pieces if they are allowed to rub against one another in the package. Excessive vacuum is to be avoided since this would create stresses as the packaging material stretches over the surfaces in contact with the product. Any air left inside the package before introducing the modified atmosphere gases will be a diluent of the gas composition, and the extent of dilution will depend on the vacuum drawn and the void spaces occupied by air in the package at the time of packaging.

The final package atmosphere after drawing a vacuum and replacing the package atmosphere with the appropriate gas mixture will depend on the void volume in the package and the vacuum drawn on the package before back flushing and sealing. Typically, 30% $CO_2$ and 70% $N_2$ are targeted for fresh poultry and fish, while package atmosphere for red meats is usually 30% $O_2$, 30% $CO_2$, and 40% $N_2$. MAP packaging material for fresh poultry and seafood would be a low oxygen permeability and low $CO_2$ permeability, while for red meats, a high $O_2$ permeability and low $CO_2$ permeability would be ideal.

Other MAP gas combinations and the products on which they have been used are 80% $CO_2$ and 20% $N_2$ for luncheon meats and 100% $CO_2$ for cheese, nuts, and bakery products.

It must be pointed out that packaging material suppliers have the most information on the type of packaging material and gas permeability of these materials; therefore it is highly recommended that these suppliers be contacted for their recommendations on the appropriate packaging material to use for an application.

## Problems

10.1. An ammonia refrigeration unit is used to cool milk from 30 °C to1 °C (86 °F–33.8 °F) by direct expansion of refrigerant in the jacket of a shell and tube heat exchanger. The heat exchanger has a total outside heat transfer surface area of 14.58 m² (157 ft²). To prevent freezing, the temperature of the refrigerant in the heat exchanger jacket is maintained at −1 °C (31.44 °F).

   (a) If the average overall heat transfer coefficient in the heat exchanger is 1136 W/m² $\cong$ K (200 BTU/h $\cong$ ft² $\cong$ °F) based on the outside area, calculate the rate at which milk with a specific heat of 3893 J/kg K(0.93 BTU/lb $\cong$ °F) can be processed in this unit.

   (b) Determine the tons of refrigeration required for the refrigeration system.

   (c) The high-pressure side of the refrigeration system is at 1.72 MPa (250 psia). Calculate the horsepower of the compressor required for the refrigeration system assuming a volumetric efficiency of 60%.

10.2. A single-stage compressor in a Freon 12 refrigeration system has a volumetric efficiency of 90% at a high side pressure of 150 psia (1.03 MPa) and a low side pressure of 50 psia (0.34 MPa). Calculate the volumetric efficiency of this unit if it is operated at the same high side pressure but the low side pressure is dropped to 10 psia (68.9 kPa). Assume R12 is an ideal gas.

10.3. Calculate the tons of refrigeration for a unit that will be installed in a cooler maintained at 0 °C (32 °F) given the following information on its construction and operation:

The cooler is inside a building:
Dimensions: $4 \times 4 \times 3.5$ m (13.1 $\times$ 13.3 $\times$ 11.5 ft)
Wall and ceiling construction:
3.175 mm (1/8 in.) thick polyvinyl chloride sheet inside (k of PVC = 0.173 W/m $\cong$ K or 0.1 BTU/h $\cong$ ft $\cong$ °F)
15.24 cm (6 in.) fiberglass insulation
5.08 cm (2 in.) corkboard
3.17 mm (1/2 in.) PVC outside
Floor construction:
3.175 mm (1/8 in.) thick floor tile (k = 0.36 W/(m $\cong$ k) or 0.208 BTU/(h $\cong$ ft $\cong$ °F)
10.16 cm (4 in.) concrete slab
20.32 cm (8 in.) air space
Concrete surface facing the ground at a constant temperature of 15°C (59°F)
Door:
1 m wide $\times$ 2.43 m high (3.28 $\times$ 8 ft)
Design for door openings that average four per hour at 1 min per opening
Air infiltration rate:
1 m³/h (35.3 ft³/h) at atmospheric pressure and ambient temperature
Ambient conditions:
32 °C (89.6 °F)
Product cooling load:
Design for a capability to cool 900 kg of product ($C_P$ = 0.76 BTU/lb $\cong$ °F or 3181 J/kg $\cong$ K) from 32 °C to 0 °C (89.6 °F to 32 °F) in 5 h. The freezing point of the product is −1.5 °C (29.3 °F).

10.4. The "stack effect" due to a difference in temperature between the inside and outside of a cooling room is often cited as the major reason for air infiltration. In this context, $\Delta P$ is positive at the lowest section of a cooler and is negative at the highest section, with a zone, called the neutral zone, at approximately the center of the room where the $\Delta P$ is zero. If the area of the openings at the lowest sections where $\Delta P$ is positive equals the area of the openings in the highest sections where $\Delta P$ is negative, air will enter at the top and escape at the openings in the bottom at the same volumetric rate of flow (assuming no pressure change inside the room). If the room allows air leakage at the rate of 2% of the room volume per minute at a $\Delta P$ of 0.5 in. wg (124 Pa), determine the rate of air infiltration that can be expected in a room that is 2 m (6.56 ft) high to the neutral zone if the interior of the room is at $-20\,^{\circ}C$ ($-4\,^{\circ}F$) and ambient temperature is $30\,^{\circ}C$ ($86\,^{\circ}F$). The rate of gas flow through the cracks is proportional to the square root of $\Delta P$. Assume air is an ideal gas. $\Delta P$ due to a column of air of height h at different temperatures $= g(\rho_1 - \rho_2)h$, where $\rho_1$ and $\rho_2$ are the densities of the columns of air.

10.5. For a 1-ton refrigeration unit (80% volumetric efficiency) using refrigerant 12 at a high side pressure of 150 psia (1.03 MPa) and a low side pressure of 45 psia (0.31 Mpa), operating at an ambient temperature of $30\,^{\circ}C$ ($86\,^{\circ}F$), determine the effect of the following on refrigeration capacity and on HP/(ton)$_r$. Assume the same compressor displacement in each case.

(a) Reducing the evaporator temperature to $-30\,^{\circ}C$ ($-22\,^{\circ}F$). High side pressure remains at 150 psia (150 MPa).

(b) Increasing ambient temperature to $35\,^{\circ}C$ ($95\,^{\circ}F$). (Low side pressure remains at 45 psia [0.31 Mpa]). The high side pressure is to change such that $\Delta T$ between the hot refrigerant gas and ambient air remains the same as in the original set of conditions.

(c) Air in the line such that the vapor phase of refrigerant always contain 10% air and 90% refrigerant by volume. Assume condensation temperature of hot refrigerant gas and temperature of cold refrigerant gas are the same as in the original set of conditions (partial pressure of refrigerant gas at the low and high side pressures are the same as in the original set of conditions, 45 and 150 psia or 0.31 and 1.03 MPa). Use R = 1.987 BTU/(lbmole $\cong\,^{\circ}R$) or 8318 J/(kg $\cong$ K). The specific heat ratio $C_P/C_v$ for air is 1.4.

(d) Oil trapped in the vapor return line such that $\Delta P$ across the constriction is 10 psi (68.9 kPa). Assume evaporator temp. $= 0\,^{\circ}C$.

10.6. Chopped onions are frozen in a continuous belt freezer at using $-50\,^{\circ}C$ air at high velocity. When onions with a moisture content of 86% are loaded on the belt with a thickness of 2 cm., it took 20 min for the temperature to drop from $10\,^{\circ}C$ to $-20\,^{\circ}C$. Onion juice is added and mixed with the chopped onions in a ratio 0.10 parts juice to 0.90 parts of the chopped onions. The freezing point of the chopped onions is $-0.5\,^{\circ}C$. The juice contains 1.5% solids (all soluble). The juice has a freezing point of $-0.16\,^{\circ}C$. Assuming that the rate of heat transfer is the same, calculate the time required to freeze the onions with the added juice.

10.7. A MAP pouch with a volume of 2 l empty is filled with chicken thigh meat at a temperature of $5\,^{\circ}C$. The pouch is positioned inside a vacuum packaging machine inside a processing room where the room temperature is the same as that of the chicken meat. A vacuum of 25 in. Hg is drawn, and the pouch is back flushed with a gas mixture consisting of 30% $CO_2$ and 70% $N_2$. By volume. If the vacuum drawn was 25 in. Hg, and the ratio of void volume to meat volume in the pouch is 1:2 by volume, calculate the resulting gas composition in the pouch after the pouch is sealed. Atmospheric pressure is 101 kPa, and air originally inside the pouch is 21% oxygen and 70% nitrogen by volume.

## Suggested Reading

ASHRAE. (1965). *ASHRAE guide and data book. Fundamentals and equipment for 1965 and 1966.* Atlanta: American Society for Heating, Refrigerating and Air Conditioning Engineers.

ASHRAE. (1966). *ASHRAE Guide and Data Book. Applications for 1966 and 1967.* Atlanta: American Society for Heating, Refrigerating and Air Conditioning Engineers.

ASHRAE. (2000). *ASHRAE Standard. Designation and safety classification of refrigerants.* Atlanta: American Society for Heating, Refrigerating and Air Conditioning Engineers.

Charm, S. E. (1971). *Fundamentals of food engineering* (2nd ed.). Westport: AVI Publishing Co..

Ciobanu, A., & Dincer, I. (1997). *Heat transfer in food cooling operations.* Washington, DC: Taylor and Francis.

Floros, J. D., & Matsos, K. R. (2005). Modified atmosphere packaging of fresh produce, meats, and ready to eat products. In J. H. Han (Ed.), *Innovation in food packaging.* San Diego: Academic Press e-Book.

Green, D. W., & Mahoney, J. D. (1997). *Perry's chemical engineers handbook* (7th ed.). New York: McGraw-Hill Book Co..

Johnson, A. T. (1999). *Biological process engineering.* New York: John Wiley & Sons.

Kader, A. A., Zagory, D., & Kerbel, E. L. (1989). Modified atmosphere packaging of fruits and vegetables. *Critical Reviews in Food Science and Nutrition., 28*(1), 1–30.

Lascu, G., Bercescu, V., & Niculescu, L. (1976). *Cooling technology in the food industry.* Tunbridge Wells: Abacus Press.

Singh, R. P., & Heldman, D. (2001). *Introduction to food engineering.* San Diego: Academic Press.

# Evaporation

11

The process of evaporation is employed in the food industry primarily as a means of bulk and weight reduction for fluids. The process is used extensively in the dairy industry to concentrate milk; in the fruit juice industry to produce fruit juice concentrates; in the manufacture of jams, jellies, and preserves to raise the solids content necessary for gelling; and in the sugar industry to concentrate sugar solutions for crystallization. Evaporation can also be used to raise the solids content of dilute solutions prior to spray drying, vacuum belt drying, or freeze-drying.

Evaporation is used to remove water from solutions with or without insoluble suspended solids. If the liquid contains only suspended solids, dewatering can be achieved by either centrifugation or filtration. The process of evaporation involves the application of heat to vaporize water at the boiling point. Its simplest form is atmospheric evaporation where the liquid in an open container is heated and the vapors driven off are dispersed into the atmosphere. Atmospheric evaporation is simple but it is slow and is not very efficient in the utilization of energy. Furthermore, because most food products are heat sensitive, prolonged exposure to high temperature in atmospheric evaporation causes off-flavors, color changes, or degradation of overall quality. In addition, because food also contains volatile compounds, vapors produced by evaporation could generate nuisance odors; therefore, they must be contained by condensation. Evaporators

used on food products remove water at low temperatures by heating the product in a vacuum. Efficient energy utilization can be designed into the system by using heat exchangers to extract heat from the vapors to preheat the feed or by using multiple effects where the vapors produced from one effect are used to provide heat in the succeeding effects.

Problems in evaporation involve primarily heat transfer and material and energy balances, the principles of which have been discussed earlier.

## 11.1 Single-Effect Evaporators

Figure 11.1 is a schematic diagram of a single-effect evaporator. The system consists of a vapor chamber where water vapor separates from the liquid, a heat exchanger to supply heat for vaporization, a condenser to draw out the vapors from the vapor chamber as rapidly as they are formed, and a steam jet ejector for removing noncondensible gases from the system. Each vapor chamber is considered an effect.

### 11.1.1 The Vapor Chamber

This is usually the largest and most visible part of the evaporator. Its main function is to allow separation of vapor from liquid and prevent carry-over

**Fig. 11.1** Schematic
diagram of a single-effect
evaporator

of solids by the vapor. It is also a reservoir for the product. The temperature inside an evaporator is determined by the absolute pressure in the vapor chamber. The vapor temperature is the temperature of saturated steam at the absolute pressure inside the chamber. When the liquid is a dilute solution, vapor and liquid temperatures will be the same. However, concentrated solutions exhibit a boiling point rise resulting in a higher boiling temperature than that of pure water. Thus, vapors leaving the liquid will be superheated steam at the same temperature as the boiling liquid. Depending on the extent of heat loss to the surroundings around the vapor chamber, the vapor may be saturated at the absolute pressure within the vapor chamber or superheated steam at the boiling temperature of the liquid.

In most food products, the soluble solids are primarily organic compounds, and the boiling point rise can be expressed as follows:

$$\Delta T_b = 0.51 \, m \qquad (11.1)$$

where $\Delta T_b$ in °C is the increase in the boiling point of a solution with molality m, above the boiling point of pure water at the given absolute pressure. In addition to the boiling point rise due to the presence of solute, the pressure at the bottom of a liquid pool is higher than the absolute pressure of the vapor, and this pressure difference can add to the temperature of the liquid pool. The pressure exerted by a column of liquid of height, h, and density, ρ, is:

$$P = \rho(h)\frac{g}{g_c} \qquad (11.2)$$

where $g_c = 1$ when using SI units.

**Example 11.1** Calculate the boiling temperature of liquid containing 30% soluble solids at a point 5 ft (1.524 m) below the surface inside an evaporator maintained at 20 in. Hg vacuum (33.8 kPa absolute). Assume the soluble solids are hexose sugars and the density of the liquid is 62 lb/ft$^3$ (933 kg/m$^3$). Atmospheric pressure is 30 in. Hg (101.5 kPa).

**Solution:**

The absolute pressure in lb$_f$/in.$^2$ corresponding to 20 in. Hg vacuum is (30–20) (0.491) = 4.91 psia. From the steam tables (Appendix A.3), the temperature corresponding to 4.91 psia is (by interpolation) 161.4 °F (71.9 °C). The molecular weight of a hexose sugar is 180. The molality of 30% soluble solids will be:

$$m \frac{\text{moles solute}}{1000 \text{ g solvent}} = \frac{0.3/180}{0.7/1000} = 2.38$$

Using Eq. (11.1):

$$\Delta T_b = 0.51(2.38) = 1.21°C \text{ or } 2.2°F$$

The absolute pressure at the level considered is the sum of the absolute pressure of the vapor and the pressure exerted by the column of liquid. Expressed in lb$_f$/in.$^2$, this pressure is:

$$P = 4.91 + \rho h \frac{g}{g_c} = 4.91 + 62 \frac{lb_m}{ft^3}(5ft) \cdot \frac{lb_f}{lb_m} \cdot \frac{1ft^2}{144 \text{ in}^2} = 7.06 \text{ psia}$$

From the steam tables, the boiling temperature corresponding to 7.06 psia is (by interpolation) 175.4 °F (79.7 °C).

The boiling temperature of the liquid will be 175.4 + 2.2 = 177.6 °F (80.9 °C).

The significance of the boiling point rise is that the liquid leaving the evaporator would be at the boiling point of the liquid rather than at the temperature of the vapor. The boiling temperature at a point submerged below a pool of liquid would have the effect of reducing the $\Delta T$ available for heat transfer in the heat exchanger if the heat exchange unit is submerged far below the fluid surface.

## 11.1.2 The Condenser

Two general types of condensers are used. A surface condenser is used when the vapors need to be recovered. This type of condenser is actually a heat exchanger cooled by refrigerant or by cooling water. The condensate is pumped out of the condenser. It has a high first cost and is expensive to operate. For this reason, it is seldom used if an alternative is available. Condensers used on essence recovery systems of fruit juice concentrators fall in this category.

The other type of condenser is one where cooling water mixes directly with the condensate. This condenser may be a barometric condenser where vapors enter a water spray chamber on top of a tall column. The column full of water is called a barometric leg, and the pressure of water in the column balances the atmospheric pressure to seal the system and maintain a vacuum. The temperature of the condensate-water mixture should be in the order of 5 °F (2.78 °C) below the temperature of the vapor in the vapor chamber to allow continuous vapor flow into the condenser. The height of the barometric leg must be sufficient to provide sufficient positive head at the base to allow the condensate and cooling water mixture to flow continuously out of the condenser at the same rate they enter. A jet condenser is one where part of the cooling water is sprayed in the upper part of the unit to condense the vapors, and the rest is introduced down the throat of a venturi at the base of the unit to draw the condensed vapor and cooling water out of the condenser. The jet condenser uses considerably more water than the barometric condenser, and the rate of water consumption cannot be easily controlled.

The condenser duty $q_c$ is the amount of heat that must be removed to condense the vapor:

$$q_c = V(h_g - h_{fc}) \tag{11.3}$$

where V is the quantity of vapor to be condensed, $h_g$ is the enthalpy of the vapor in the vapor chamber of the evaporator, and $h_{fc}$ is the enthalpy of the liquid condensate.

For direct contact condensers, the amount of cooling water required per unit amount of vapor condensed can be determined by a heat balance:

$$W(h_{fc} - h_{fw}) = V(h_g - h_{fc}) \qquad (11.4)$$

$$\frac{W}{V} = \frac{h_g - h_{fc}}{h_{fc} - h_{fw}} \qquad (11.5)$$

where W is quantity of cooling water required, and $h_{fw}$ is enthalpy of cooling water entering the condenser.

The enthalpy of the condensate-water mixture, $h_{fc}$, should be evaluated, in the case of barometric condensers, at a temperature 5 °F (2.7 °C) lower than the vapor temperature.

**Example 11.2** Calculate the ratio of cooling water to vapor for a direct contact barometric condenser for an evaporator operating at a vapor temperature of 150 °F (65.55 °C). What would be the minimum height of the water column in the barometric leg for the evaporator to operate at this temperature? Cooling water is at 70 °F (21.1 °C). Atmospheric pressure is 760 mm Hg.

**Solution:**
At a temperature of 150 °F, the absolute pressure of saturated steam is 3.7184 psia (25.6 kPa). $h_g = 1126.1$ BTU/lb or 2.619 MJ/kg. The condensate-cooling water mixture must be at 5 °F below 150 or 145°F (62.78 °C). $h_{fc} = 112.95$ BTU/lb or 0.262 MJ/kg. The enthalpy of the cooling water $h_{fw} = 38.052$ BTU/lb or 0.088 MJ/kg. Basis: V = one unit weight of vapor. Using Eq. (11.5):

$$\frac{W}{V} = \frac{1126.1 - 12.95}{112.95 - 38.052} = 13.52$$

The atmospheric pressure is 760 mm Hg or 101.3 kPa. From the steam tables, the density of water at 145 °F is $1/V = 61.28$ lb/ft$^3$ or 981.7 kg/m$^3$. The pressure that must be counteracted by the column of water in the barometric leg is the difference between barometric pressure and the absolute pressure in the system:

$$\Delta P = 101.3 - 25.6 = 75.7\,\text{kPa} = \rho gh$$

$$h = \frac{75,700\,\text{kg} \cdot m}{s^2 \cdot m^2} \frac{1}{981.7\ \text{kg} \cdot m^{-3}} \frac{1}{9.80\,\text{m} \cdot s^{-2}}$$

$$= 7.868\,\text{m or } 25.8\ \text{ft}$$

### 11.1.3   Removal of Noncondensible Gases

A steam jet ejector is often used. Figure 11.2 is a schematic diagram of a single-stage ejector. High-pressure steam is allowed to expand through a jet, which increases its velocity. The movement of steam through the converging-diverging section at high velocity generates a zone of low pressure in the suction chamber, and noncondensible gases can be drawn into the ejector. The noncondensible gases mix with the high velocity steam and are discharged into the atmosphere. Steam jet ejectors are more effective than vacuum pumps in that water vapor present in the noncondensible gases does not interfere with its operation. If suction absolute pressures are 4 in. Hg (13.54 kPa) or lower, multistage ejectors are used. The capacity of jet ejectors is dependent

**Fig. 11.2** Schematic diagram of a single-stage ejector

HIGH
PRESSURE
STEAM

DISCHARGE

SUCTION

**Fig. 11.3** Solubility of air
in water at atmospheric
pressure at various
temperatures (Source: Data
from Perry, et al. 1963.
Chemical Engineers
Handbook. McGraw-Hill,
New York)

on the design of the ejector, the pressure of the
high-pressure steam, and the pressure differential
between the suction and discharge.

Capacity charts for steam jet ejectors are usually provided by their manufacturers, and the
capacity is expressed as weight of air evacuated
per hour as a function of suction pressure and
steam pressure.

The amount of noncondensible gases to be
removed from a system depends upon the extent
of leakage of air into the system and the amount
of dissolved air in the feed and in the cooling
water. In addition to the noncondensible gases,
jet ejectors also have to remove the water vapor
that is present along with the noncondensible
gases in the condenser. Air leakage has been
estimated at 4 g air/h for every meter length of
joints. The solubility of air in water at atmospheric pressure at various temperatures can be

determined from Fig. 11.3. The amount of water
vapor with the noncondensible gases leaving the
condenser can be calculated as follows:

$$W_v = \frac{P_c(18)}{(P_v - P_c)29} \qquad (11.6)$$

where $W_v$ is kg water vapor/kg air, $P_c$ is vapor
pressure of water at temperature of condensate-
cooling water mixture, and $P_v$ is absolute pressure
inside the evaporator.

**Example 11.3** Calculate the ejector capacity
required for an evaporator that processes
100 kg/h of juice from 12% to 35% solids. The
evaporator is operated at 65.6 °C (150 °F).
Cooling water is at 21.1 °C (70 °F). Product
enters at 35 °C (89.6 °F). The condenser temperature is maintained at 2.78 °C (5 °F) below the
vapor temperature.

**Solution:**

From Example 11.2, the ratio of cooling water to vapors for an evaporator operated under the identical conditions specified in this problem was determined to be 13.52 kg of water at 21.1EC. A material balance gives:

$$\text{Wt concentrate} = \frac{100(0.12)}{0.35} = 34.28 \text{ kg/h}$$

$$\text{Wt vapors} = 100 - 34.28 = 65.71 \text{ kg.h}$$

The weight of cooling water $W_c$ is 13.52 (65.71) = 888 kg/h. From Fig. 11.3, the solubilities of air in water are $2.37 \times 10^{-4}$ and $1.96 \times 10^{-4}$ kg air/kg water at 21.2 °C and 35 °C, respectively. The amount of air, $M_a$, introduced with the condenser water and with the feed is:

$$M_a = 100\left(1.96 \times 10^{-4}\right) + 888\left(2.37 \times 10^{-4}\right)$$
$$= 0.23 \text{ kg/h}$$

Using Eq. (11.6), a vapor pressure at 62.82 °C of 22.63 kPa (3.2825 psia) and a pressure of 25.63 kPa (3.7184 psia) at the vapor temperature of 65.6 °C:

$$W_v = \frac{(22.63)\ 18}{(25.63 - 22.63)(29)} = 4.68 \frac{\text{kg water}}{\text{kg air}}$$

The total ejector load excluding air leakage is equivalent to 4.68 + 0.23 or 4.91 kg/h.

The majority of the ejector load is the water vapor that goes with the noncondensible gases.

### 11.1.4  The Heat Exchanger

The rate of evaporation in an evaporator is determined by the amount of heat transferred in the heat exchanger. Variations in evaporator design can be seen primarily in the manner in which heat is transferred to the product. Considerations like the stability of the product to heat, fouling of heat exchange surfaces, ease of cleaning, and whether the product could allow a rapid enough rate of heat transfer by natural convection dictate the design of the heat exchanger for use on a given product. The schematic diagram of an evaporator

shown in Fig. 11.1 shows a long tube, vertical, forced circulation evaporator. This type of design for the heat exchanger is usually used when a single effect concentrates a material that becomes very viscous at the high solids content. Because of the forced circulation, heat transfer coefficients are fairly high even at the high viscosity of the concentrate. Some evaporators would have the heat exchanger completely immersed in the fluid being heated inside the vessel that constitutes the fluid reservoir and vapor chamber. Heat is transferred by natural convection. This type of heat exchange is suitable when the product is not very viscous and is usually utilized in the first few effects of a multiple-effect evaporator.

Heat exchangers on evaporators for food products have the food flowing inside the tubes for ease of cleaning. In very low-temperature evaporation such as in fruit juice concentration, the operation has to be stopped regularly to prevent microbiological buildup and also to clean deposits of food product on the heat exchange surfaces.

The capacity of an evaporator is determined by the amount of heat transferred to the fluid by the heat exchanger. If $q$ = the amount of heat transferred, $P$ = mass of concentrated product, $C_c$ = specific heat of the concentrate, $V$ = mass of the vapor, $h_g$ = enthalpy of the vapor, $h_f$ = enthalpy of the water component of the feed that is converted to vapor, $T_1$ = feed inlet temperature, and $T_2$ = liquid temperature in the evaporator, a heat balance would give:

$$q = PC_c(T_2 - T_1) + V(h_g - h_f) \qquad (11.7)$$

The rate of heat transfer can be expressed as:

$$Q = U\ A\Delta T \qquad (11.8)$$

A material balance would give:

$$P = \frac{Fx_f}{x_p} \qquad (11.9)$$

and

$$V = F\left(1 - \frac{x_f}{x_p}\right) \qquad (11.10)$$

Equations (11.7) and (11.10) can be used to calculate the capacity of an evaporator in terms of a rate of feed F, knowing the initial solids content, $x_f$, the final solids content $x_p$, and the amount of heat transferred in the heat exchanger expressed in terms of the heat transfer coefficient U, the area available for heat transfer, A, and the temperature difference between the boiling liquid in the evaporator and the heating medium, $\Delta T$.

In evaporators where heat transfer to the product is by natural convection, products that have tendencies to form deposits on the heat exchange surface foul the heat exchange surface and reduce the overall heat transfer coefficient, U. When evaporation rate slows down considerably such that it seriously affects production, the operation is stopped and the evaporator is cleaned. In evaporators used on tomato juice, the temperatures are high enough that microbiological buildup is not a factor, and shutdown for cleanup is usually done after about 14 days of operation. In some models of evaporators used on orange juice, on the other hand, operating time between cleanups is much shorter (2–3 days) because of the problem of microbiological buildup at the lower temperatures used.

Fouling of heat exchange surfaces is minimized with reduced $\Delta T$ across the heat exchange surface and by allowing the product to flow rapidly over the heat exchange surfaces. Although forced recirculation through the heat exchanger results in rapid heat transfer, a disadvantage is the long residence time of the product inside the evaporator.

For products that are heat sensitive and where low-temperature differentials are allowable in the heat exchanger, the falling film heat exchanger is used. Figure 11.4 is a schematic diagram of a falling film-type heat exchanger used extensively in the concentration of fruit juices. The product flows in a thin film down heated tubes where heat is transferred and vapor is removed. The product passes through the heat exchange tube of one effect only once, and this short time of contact with a hot surface minimizes heat-induced flavor or color changes and nutrient degradation.

The coefficient of heat transfer, U, in evaporator heat exchangers is of the order

**Fig. 11.4** Schematic diagram of a falling film evaporator

200 BTU/(h · ft$^2$ · °F) or 1136 W/(m$^2$ · K) for natural convection and 400 BTU/h(ft$^2$) (°F) or 2272 W/(m$^2$ · K) for forced convection. The effect of increased viscosity on heat transfer can be estimated by using the relationship: heat transfer coefficient is proportional to the viscosity raised to the power − 0.44. Thus, $U_1/U_2 = (\Phi_1/\Phi_2)^{-0.44}$ where $U_1$ is the heat transfer coefficient corresponding to viscosity $\Phi_1$ and $U_2$ is the heat transfer coefficient corresponding to the viscosity $\Phi_2$. This relationship would be useful in estimating how much of a change in evaporator performance would be expected with a change in operating conditions that would result in variations in product viscosity.

**Example 11.4** A fruit juice is to be concentrated in a single-effect forced recirculation evaporator from 10% to 45% soluble solids. The feed rate is 5500 lb/h or 2497 kg/h. Steam condensing at 250 °F (121.1 °C) is used for heating. The vapor temperature in the evaporator should be at 130 °F

(54.4 °C). Assume the soluble solids are hexose sugars in calculating the boiling point rise. Use Siebel's formula for calculating the specific heat of the juice. The feed is at 125 °F (51.7 °C). The heat transfer coefficient, U, is 500 BTU/(h · ft$^2$ · °F) or 2839 W/(m$^2$ · K). Calculate the steam economy to be expected and the heating surface area required.

**Solution:**

Equations (11.7, 11.8, 11.9, and 11.10) will be used. The steam economy is defined as the ratio of vapor produced to steam consumed. $C_c$ will be calculated using Siebel's formula:

$$C_c = 0.8(.55) + 0.2$$
$$= 0.64 \text{ BTU/lb}(°F) \text{ or } 2679 \text{ J/(kg.K)}.h_g$$
$$= \text{enthalpy of vapor at } 130°F$$
$$= 1117.8 \text{ BTU/(lb)}$$
$$= 2.679 \text{ MJ/(kg)}h_f$$
$$= \text{enthalpy of the water component}$$
$$\text{of the feed at } 125°F.$$
$$= 92.96 \text{ BTU/lb or } 0.216 \text{ MJ/kg}.$$

The temperature of the concentrate leaving the evaporator, $T_2$, will be the sum of the vapor temperature and the boiling point rise, $\Delta T_b$. Using Eq. (11.1), m for 45% soluble solids is:

$$m = \frac{45/180}{55/1000}$$
$$= 4.545 \text{ moles sugar/1000 g water}$$
$$\Delta T_b = 0.51(4.545) = 2.32°C \text{ or } 4.2°F$$
$$T_2 = 130 + 4.2 = 134.2°F \text{ or } 56.72°C$$

Substituting in Eq. (11.10):

$$V = 5500\left[1 - \frac{0.10}{0.45}\right]$$
$$= 4278 \text{ lb/h or } 1942 \text{ kg/h}$$

Using Eq. (11.9):

$$P = 5500\left[\frac{0.10}{0.45}\right] = 1222 \text{ lb/h or } 555 \text{ kg/h}$$

Using Eq. (11.7):

$$q = 1222(0.64)(134.2 - 125)$$
$$+ 4278(1117.8 - 92.96)$$
$$= 4,391,500 \text{ BTU/h or } 1.2827 \text{ MW}$$

Equation (11.8) can be used to calculate the heat transfer surface area:

$$A = \frac{q}{U\Delta T} = \frac{4,391,500}{500(250 - 134.2)}$$
$$= 75.8 \text{ ft}^2 \text{or } 7.04 \text{ m}^2$$

Note that the liquid boiling temperature was used in determining the heat transfer $\Delta T$ rather than the vapor temperature. The enthalpy of vaporization of steam at 250 °F is 945.5 BTU/lb or 2.199 MJ/kg. Steam required $= q/h_{fg} = 4,391,500/945.5 = 4645$ lb/h or 2109 kg/h. Steam economy $= 4278/4645 = 0.92$.

## 11.2   Improving the Economy of Evaporators

Poor evaporator economy results from wasting heat present in the vapors. Some of the techniques used to reclaim heat from the vapors include use of multiple effects such that vapors from the first effect are used to heat the succeeding effects, use of vapors to preheat the feed, and vapor recompression.

### 11.2.1   Vapor Recompression

Adiabatic recompression of vapor results in an increase in temperature and pressure. Figure 11.5 is a Mollier diagram for steam in the region involved in vapor recompression for evaporators. Recompression involves increasing the pressure of the vapor to increase its condensing temperature above the boiling point of the liquid in the evaporator. Compression of saturated steam would result in superheated steam at high pressure. It would be necessary to convert this vapor to saturated steam by mixing with liquid water before introducing it into the heating element of the evaporator. Superheated steam in the heat exchanger could lower the overall heat transfer

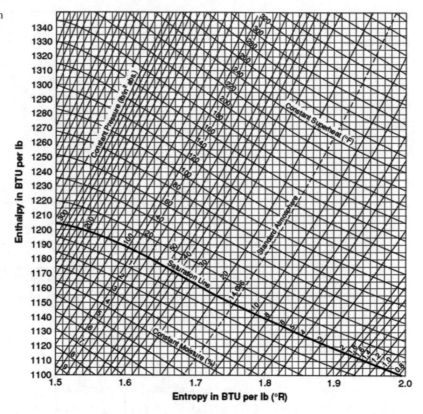

**Fig. 11.5** Mollier diagram for steam in the region involved in vapor recompression for evaporators (Courtesy of Combustion Engineering Inc.)

coefficient. In Chap. 10, the work involved in adiabatic compression was found to be the difference in the enthalpy of the low-pressure saturated vapor and the high-pressure superheated vapor. The ratio between the latent heat of the saturated steam produced from the hot vapors and the work of compression is the coefficient of performance of the recompression system.

**Example 11.5** In Example 11.4, determine the coefficient of performance for a vapor recompression system if used on the unit.

**Solution:**

From Fig. 11.5, the initial point for the compression is saturated vapor with an enthalpy of 1118 BTU/lb or (2.600 MJ/kg). Isentropic compression to the pressure of saturated steam at 250 °F (121.1 °C) would give a pressure of 29.84 psia (206 kPa) an enthalpy of 1338 BTU/lb (3.112 MJ/kg) and a temperature of 612 °F (322 °C).

Figure 11.6 is a schematic diagram of how these numbers were obtained from Fig. 11.5.

If condensate from the heat exchanger at 250 °F is used to mix with the superheated vapor after compression to produce saturated steam at 250 °F, the amount of saturated steam produced will be:

$$\text{Wt saturated steam} = 1 + \frac{h_{g1} - h_{g2}}{h_{fg}}$$

$$= 1 + \frac{1338 - 1164}{945.5}$$

$$= 1.184 \, \text{lb} \ (0.537 \, \text{kg})$$

$$\text{COP} = \frac{1.184(945.5)}{1(1338 - 1118)} = 5.09$$

The coefficient of performance in vapor recompression systems is high. COP will be higher if the $\Delta T$ is kept to a minimum. $\Delta T$ in vapor recompression systems are usually of the order 10 °F (5.6 °C). Although an increase in COP is achieved with the low $\Delta T$, increased area for heat

**Fig. 11.6** The vapor recompression process on a Mollier diagram

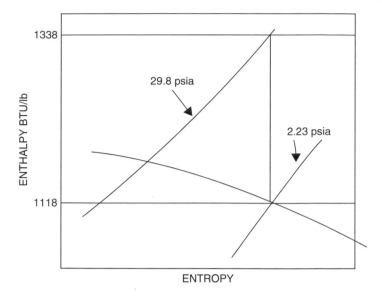

**Fig. 11.7** Schematic diagram of a triple-effect evaporator with forward feed

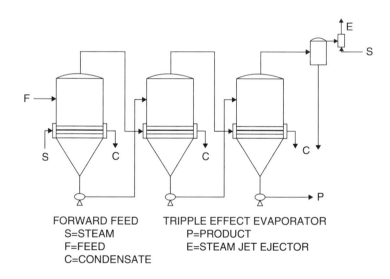

FORWARD FEED              TRIPPLE EFFECT EVAPORATOR
S=STEAM                        P=PRODUCT
F=FEED                         E=STEAM JET EJECTOR
C=CONDENSATE

transfer in the evaporator heating unit is also required.

## 11.2.2  Multiple-Effect Evaporators

Steam economy can also be improved by using multiple evaporation stages and using the vapors from one effect to heat the succeeding effects. Steam is introduced only in the first effect. Figure 11.7 shows a triple-effect evaporator with forward feed. This type of feeding is used when the feed is at a temperature close to the vapor temperature of the effect where it is introduced. If substantial amounts of heat are necessary to bring the feed temperature to the boiling temperature, other types of arrangements such as backward feed may be used. In a backward feed arrangement, the flow of the feed is countercurrent to the flow of vapor.

Multiple-effect evaporators are often constructed with the same heat transfer surface areas in each effect. The governing equation for evaporation rate in multiple-effect evaporators is the heat transfer equation (Eq. 11.8) as in single-effect evaporators. However, the $\Delta T$ in each effect of a multiple-effect evaporator is only a fraction of the total $\Delta T$; therefore, for the same rate of evaporation and the same total $\Delta T$, a multiple-effect evaporator with "n" effects would require approximately "n" times the heat exchange area for a single-effect evaporator. The savings in energy costs with the improvement in steam economy is achieved only with an increase in the required heat transfer surface area.

Only two parameters affecting heat transfer $\Delta T$ can be controlled in the operation of a multiple-effect evaporator. These are the temperature of the vapor in the last effect and the temperature of steam in the first effect. The vapor temperature and pressure in the first and in the intermediate effects will develop spontaneously according to the heat balance occurring within these various effects. The total $\Delta T$ in a multiple-effect evaporator would be the difference between the steam temperature and the vapor temperature in the last effect. A boiling point rise would decrease the available $\Delta T$:

$$\Delta T = T_s - T_{vn} - \Delta T_{b1} - \Delta T_{b2} - \cdots \Delta T_{bn}$$

$$(11.11)$$

where $\Delta T$ = the total available temperature drop for heat transfer, $T_s$ = steam temperature, $T_{vn}$ = vapor temperature in last effect, and $\Delta T_{b1}$, $\Delta T_{b2} \ldots \Delta T_{bn}$ = boiling point rise in effects, 1, 2, ... n.

Multiple-effect evaporator calculations are done using a trial-and-error method. A $\Delta T$ is assumed for each effect, and by making heat and material balances in each effect, the rate of heat transfer in each effect is compared with the heat input necessary to achieve the desired evaporation rate in each effect. Adjustments are then made on the assumed $\Delta T$s until the heat input and the heat requirement for each effect are in balance. Multiple-effect evaporator calculations are tedious and time-consuming. They are best done on a computer.

Reasonable approximations can be made on capacity for a given multiple-effect evaporator knowing the heat transfer surface areas and the heat transfer coefficients by assuming equal evaporation in each effect and making a heat balance as if the evaporator is a single effect with a heat transfer area equal to the sum of all the areas. The overall heat transfer coefficient can be calculated as follows:

$$\frac{1}{U} = \frac{1}{U_1} + \frac{1}{U_2} + \cdots \frac{1}{U_n} \qquad (11.12)$$

**Example 11.6** In Example 11.4, the evaporator is a triple-effect evaporator with forward feed. Assume U is the same in all effects. Calculate the approximate heat transfer surface areas in each effect and the steam economy.

**Solution:**

Given:

Feed rate (F) = 5500 lb/h or 2497 kg/h

Steam temperature $(T_s)$ = 250 °F (121.1 °C)

Vapor temperature in last effect $(T_{v3})$ = 130 °F (54.4 °C)

Feed temperature $(T_f)$ = 125 °F (51.7 °C)

$$U_1 = U_2 = U_3$$

$$= 500 \frac{BTU}{h \cdot ft^2 \cdot °F} \quad or \quad 2839 \frac{W}{m^2 \cdot K}$$

Solids content of feed $(x_f)$ = 10%

Solids content of product $(x_p)$ = 45%

Molecular weight of solids (M) = 180

First, determine the overall heat transfer, $\Delta T$, using Eq. (11.11). It will be necessary to determine the boiling point rise in each effect. Using Eq. (11.10):

$$\text{Total evaporatiion (V)} = 5500 \left[ 1 - \frac{0.10}{0.45} \right]$$
$$= 4278 \text{ lb/h or } 1942 \text{ kg/h}$$

Assuming equal evaporation, $V_1 = V_2 = V_3 = 1426$ lb/h or 647.3 kg/h. The solids content in each effect can be calculated by rearranging Eq. (11.10). The subscripts n on V and refer to the feed to and vapor from effect n:

$$x_n = \frac{x_f F_n}{F_n - V_n} \quad x_1 = \frac{0.10(5500)}{5500 - 1426} = 0.135$$

$$x_2 = \frac{0.135(5500 - 1426)}{5500 - 1426 - 1426} = \frac{0.135(4074)}{2648} = 0.208$$

$$x_3 = x_p = 0.45$$

The molalities are:

$$m_1 = \frac{0.135(1000)}{(1 - 0.135)(180)} = 0.067 \, \text{molal}$$

$$m_2 = \frac{0.208(1000)}{(1 - 0.208)(180)} = 1.459 \, \text{molal}$$

$$m_3 = \frac{0.45(1000)}{(1 - 0.45)(180)} = 4.545 \, \text{molal}$$

Using Eq. (11.1), the boiling point rises are:

$$\Delta T_{h1} = 0.51(0.067) = 0.0342 \,^\circ\text{C or } 0.06 \,^\circ\text{F}$$

$$\Delta T_{h2} = 0.51(1.459) = 0.744 \,^\circ\text{C or } 1.34 \,^\circ\text{F}$$

$$\Delta T_{h3} = 0.51(4.545) = 2.32 \,^\circ\text{C or } 4.17 \,^\circ\text{F}$$

The total $\Delta T$ for heat transfer is calculated using Eq. (11.11):

$$\Delta T = 250 - 130 - 4.17 - 1.34 - 0.06$$

$$= 114.4 \,^\circ\text{F For } 63.57 \,^\circ\text{C}$$

Overall U is calculated using Eq. (11.12):

$$\frac{1}{U} = \frac{1}{500} + \frac{1}{500} + \frac{1}{500} = 0.006$$

$$U = 166.67 \, \text{BTU} / \left( \text{h} \cdot \text{ft}^2 \cdot \text{EF} \right) \text{or} \, 946.3 \, \text{W} / (\text{m}^2 \cdot \text{K})$$

From section "The Heat Exchanger," the required heat transfer rate for this evaporator was determined to be 4,391,500 BTU/h or 1.2827 MW. Using Eq. (11.8):

$$A = \frac{4.391,500}{166(114.4)} = 230.3 \text{ft}^2 \text{ or } 21.4 \text{ m}^2$$

The heat transfer surface area for each effect will be 230.3/3 ft$^2$ = 76.77 ft$^2$ or 7.13 m$^2$.

Calculations of steam economy can only be done using the trial-and-error procedure necessary to establish the $\Delta T$ and the vapor temperature in each effect. The steam economy of multiple-effect evaporators is a number slightly less than the number of effects.

## 11.2.3 Entrainment

When the liquid to be evaporated contains suspended solids, the liquid has a tendency to foam. The level of the foam may rise much higher than the normal liquid level resulting in carry-over of solids with the vapors. Entrainment not only results in loss of valuable food solids, but the solids in the condensate may result in problems with condensate water treatment particularly if a direct contact condenser is used and the condensate is cooled in a cooling tower and recirculated.

## 11.2.4 Essence Recovery

A major problem in concentration of fruit juices is the loss of essence during the evaporation process. With condensers where the cooling water directly contacts the vapor, it is not possible to recover the flavor components that are vaporized from the liquid. In the past, the problem of essence loss was solved in the orange juice industry by concentrating the juice to a higher concentration than is desired and diluting the concentrate with fresh juice to the desired solid concentration. The essence in the fresh juice gives the necessary flavor to the concentrate.

One method for essence recovery is by flashing the juice into a packed or perforated plate column maintained at a very low absolute pressure. Flash evaporation is a process where hot liquid is introduced into a chamber that is at an absolute pressure where the boiling point of the liquid is below the liquid temperature. The liquid will boil immediately upon exposure to the low pressure, vapor is released, and the liquid temperature will drop to the boiling point of the liquid at the given absolute pressure.

The feed is preheated to 120 °F–150 °F (48.9 °C–65.6 °C) and is introduced into a column maintained at an absolute pressure of approximately 0.5 psia (3.45 kPa). There is no heat input in the column; therefore, evaporative cooling drops the temperature of the liquid. The vapors rise up the packed column continually getting richer in the volatile components as they proceed up the column. A surface condenser cooled by a refrigeration system traps the volatile components. The essence concentrate recovered is blended with the concentrated product.

In a multiple-effect evaporator, a backward feed arrangement is used, and the vapors from the last effect are condensed using a surface condenser. The condensate containing the essence is flashed into the essence recovery unit.

### 11.2.5 Temperature-Accelerated Short-Time Evaporator (TASTE)

The popularity of frozen juice concentrates and consumer demand for good tasting beverage juices resulted in the development of energy-efficient evaporators which subjected the juices to low temperature and short time of exposure to high temperatures during the concentration process. Another feature of TASTE is improved recovery of essence to add back to the concentrate making it possible for the reconstituted concentrate to have similar flavor as freshly squeezed juice. TASTE is widely used to produce frozen orange, lemon, grape, and almost any other juice derived from fruits grown in a large scale. The quantity of fresh juice available for processing is a limitation on the use of TASTE since the system is suitable only for large-scale operations.

A unique feature of TASTE is that the juice passes through a heat exchanger at each effect only once, and the heated juice is discharged through nozzles into a distribution chamber where flash evaporation occurs to produce an atomized, fog-like mixture of liquid and vapor. This mixture then travels down the inside of tubes positioned at the lower level of the distribution chamber. The outer surface of the tubes is heated by steam flashed out of the liquid when the liquid-vapor mixture exits the tubes. The vapors then exit the vacuum chamber to be used as the energy source for heating the liquid entering the next effect. The liquid pool at the bottom of an effect is pumped through a heat exchanger to enter the next effect. The volume of liquid in an effect is minimized thus minimizing the residence time in an effect. TASTE system effects are small-diameter, tall columns to allow rapid movement of liquid through the effects reducing the time of residence at high temperature in the system. Another feature of the system is the use of a hydrocyclone in each effect to separate liquid from vapor as this mixture leaves the tubes inside an effect. Essence recovery is done on vapor vented out of the heat exchanger used to heat the liquid entering second and third effects by leading the vapor leaving the vent into a condenser.

Another feature of the TASTE system is the use of one of the effects as a pasteurizer for the juice concentrate. Thus, pasteurization and preheating the liquid entering the pasteurizing effect can be done simultaneously. The concentrates produced are thoroughly pasteurized such that the product can be filled directly into barrels or totes, frozen and stored frozen until an order is placed for the product.

---

## Problems

11.1. A single-effect falling film-type evaporator is used to concentrate orange juice from 14% to 45% solids. The evaporator utilizes a mechanical refrigeration cycle using ammonia as refrigerant, for heating and for condensing the vapors. The refrigeration cycle is operated at a high pressure of 200 psia (1.379 MPa) and a low side pressure of 50 psia (344.7 kPa). The evaporator is operated at a vapor temperature of 90 °F (32.2 °C). Feed enters at 70 °F (21.1 °C). The ratio of insoluble to soluble solids in the juice is 0.09, and the soluble solids may be considered as glucose and sucrose in 70:30 ratio. Consider the $\Delta T$ as the log mean $\Delta T$ between the liquid refrigerant temperature and the feed temperature at one point and the hot refrigerant gas temperature and the concentrated liquid boiling temperature at the other point. The evaporator has a heat transfer surface area of 100 ft² (9.29 m²), and an overall heat transfer coefficient of 300 BTU/(h $ ft² $ °F) or 1703 W/(m² $K) may be expected. Calculate:

(a) The evaporator capacity in weight of feed per hour

(b) Tons of refrigeration capacity required for the refrigeration unit based on the heating requirement for the evaporator

(c) Additional cooling required for condensation of vapors if the refrigeration unit is designed to provide all of the heating requirements for evaporation

11.2. Condensate from the heating unit of one effect in a multiple-effect evaporator is flashed to the pressure of the heating unit in one of the succeeding effects. If the condensate is saturated liquid at 7.511 psia (51.79 kPa) and the heating unit contains condensing steam at 2.889 psia (19.92 kPa), calculate the total available latent heat that will be in the steam produced from a unit weight of the condensate.

11.3. A single-effect evaporator was operating at a feed rate of 10,000 kg/h concentrating tomato juice at 160 °F (71.1 °C) from 15% to 28% solids. The ratio of insoluble to soluble solids is 0.168, and the soluble solids may be assumed to be hexose sugars. Condensing steam at 29.840 psia (205.7 kPa) was used for heating, and the evaporator was at an absolute pressure of 5.993 psia (41.32 kPa). It is desired to change the operating conditions to enable the efficient use of a vapor recompression system. The steam pressure is to be lowered to 17.186 psia (118.37 kPa). Assume there is no change in the heat transfer coefficient because of the lowering of the heating medium temperature.
Calculate:

(a) The steam economy for the original operating conditions

(b) The capacity in weight of feed per hour under the new operating conditions

(c) The steam economy of the vapor recompression system

Express the steam economy as the ratio of the energy required for concentration of the juice to the energy required to compress the vapor assuming a mechanical efficiency of 50% for the compressor. Assume condensate from the heating element is added to the superheated steam to reduce temperature to saturation.

## Suggested Reading

Bennet, C. O., & Myers, J. E. (1962). *Momentum, heat, and mass transport.* New York: McGraw-Hill Book.

Charm, S. E. (1971). *Fundamentals of food engineering* (2nd ed.). Westport: AVI Publishing Co.

Cook Machinery Company. (2004). http://cookmachineryllc.com/juice-evaporators-principles-of-operation-and-troubleshooting. Accessed 15 Feb 2018.

Foust, A. S., Wenzel, E. A., Clump, C. W., Maus, L., & Andersen, L. B. (1960). *Principles of unit operations.* New York: John Wiley & Sons.

Green, D. W. (1997). *Perry's chemical engineers handbook* (7th ed.). New York: Mcgraw- Hill Book.

Heldman, D. R. (1973). *Food process engineering.* Westport: AVI Publishing Co.

McCabe, W. L., & Smith, J. C. (1967). *Unit operations of chemical engineering* (2nd ed.). New York: McGraw-Hill Book.

Perry, R. H., Chilton, C. H., & Kirkpatrick, S. D. (1963). *Chemical engineers handbook* (4th ed.). New York: McGraw-Hill Book.

Saravacos, G. D., & Kostaropoulos, A. E. (2002). *Handbook of food processing equipment.* New York: Kluwer Academic/Plenum Publishers.

Singh, R. P., & Heldman, D. R. (2001). *Introduction to food engineering* (3rd ed.). San Diego: Academic Press.

# Dehydration

Dehydration is a major process for food preservation. The reduced weight and bulk of dehydrated products and their dry shelf stability reduce product storage and distribution costs. As dehydration techniques that produce good-quality convenience foods are developed, more dehydrated products will be commercially produced. At present, instant beverage powders, dry soup mixes, spices, and ingredients used in further processing are the major food products dehydrated.

## 12.1 Water Activity

Dehydrated foods are preserved because water activity is at a level where no microbiological activity can occur and where deteriorative chemical and biochemical reaction rates are reduced to a minimum. Water activity ($a_w$) is measured as the equilibrium relative humidity (ERH), the percent relative humidity (RH) of an atmosphere in contact with a product at the equilibrium water content. $a_w$ is also the ratio of the partial pressure of water in the headspace of a product (P) to the vapor pressure of pure water ($P^0$) at the same temperature:

$$a_w = ERH = \frac{P}{P^0} \qquad (12.1)$$

The relationship between $a_w$ and the rate of deteriorative reactions in food is shown in Fig. 12.1. Reducing $a_w$ below 0.7 would prevent microbiological spoilage. However, although microbial spoilage would not occur at $a_w = 0.7$, prevention of other deteriorative reactions needed to successfully preserve a food product by dehydration requires $a_w$ to be reduced to $= 0.3$.

A food material may also be dehydrated only for weight or bulk reduction and finally preserved using other techniques. Readers should pay attention to this cautionary note. *Dehydration itself does not inactivate pathogens. Although viable pathogenic microorganisms can no longer grow in a properly dehydrated product, the possibility that they can grow to numbers adequate to cause human illness when the dehydrated product is subsequently used as an ingredient in formulated products, has been demonstrated to occur. An example is Salmonella in dried whey used as an ingredient in crackers, breakfast cereal, and other products has caused a massive recall of these products because of the potential health safety problem. To be safe, any food product that may be a potential pathogen carrier, must be adequately treated to eliminate pathogenic microorganisms prior to the dehydration process. Procedures that can be used to eliminate pathogens in a moist raw product prior to dehydration can be found in Chap. 8 – Thermal Process Calculations.*

### 12.1.1 Thermodynamic Basis for Water Activity

The first and second laws of thermodynamics are discussed in Chap. 3 Sect. 3.2—

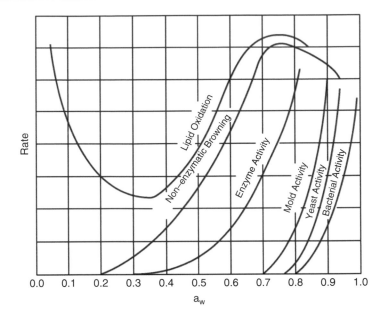

Thermodynamics. In addition to the thermodynamic variables of enthalpy (H), internal energy (E), work (W), and heat (Q) are discussed under Thermodynamics in Sect. 4.2. Other thermodynamic variables related to chemical changes will be discussed in this section.

F = free energy, another form of energy in a system that is different from PV work. This form of energy is responsible for chemical or electrical work and is responsible for driving chemical reactions:

$$F = H - TS \qquad (12.2)$$

$\Phi$ = chemical potential. It is directly related to the free energy. In any system undergoing change, $d\Phi = dF$:

Because : dF = dH − T dS − S dT

and : dH = dE + P dV + V dP

and : dE = T dS − P dV

Then : dF = V dP − S dT;

$d\Phi$ = V dP − S dT

At constant T, $d\Phi/dP = V$.

In a system in equilibrium, the chemical potential is equal in all phases. In the gas phase, the ideal gas equation for one mole of gas is V = RT/P. Thus:

$$d\Phi = RTd(\ln P) \qquad (12.3)$$

If the vapor is nonideal, the fugacity, f, may be substituted for P. At low pressures of near ambient or in a vacuum, f is almost the same as P, and the two may be interchanged in use with very little error.

Thus, the chemical potential can be measured directly by the vapor pressure of a component. The activity of a component is defined as:

$$a = \frac{f}{f^0}$$

where $f^0$ is the fugacity of the pure component at the same temperature. If the component is water:

$$a_w = \frac{f}{f^0} \equiv \frac{P}{P^0} \qquad (12.4)$$

The thermodynamic basis for the water activity is the chemical potential of water, its ability to participate in chemical reactions.

## 12.1.2  Osmotic Pressure

One consequence of reduced water activity is an increase in osmotic pressure, which interferes with food, water, and waste transport between a cell and its surroundings. The osmotic pressure is related to the activity coefficient as follows:

$$\pi = \frac{-RT}{V}\ln x\gamma \qquad (12.5)$$

where $\pi$ = osmotic pressure (atm), R = gas constant (82.06 mL · atm/gmole · K), $\gamma$ = activity coefficient, V = molar volume = 18 mL/gmole for water, x = mole fraction of water, and T = absolute temperature. The argument of the logarithmic term in Eq. (12.5) is the water activity. As the water activity approaches 1, the logarithmic term approaches zero and the osmotic pressure approaches zero.

### 12.1.3   Water Activity at High Moisture Contents

At high moisture contents, depression of the water vapor pressure by soluble solids is similar to vapor pressure depression by solutes in solution. The presence of insoluble solids is ignored, and the solution phase determines the water activity.

For ideal solutions, the water activity is equal to the mole fraction of water, $x_w$:

$$a_w = x_w \qquad (12.6)$$

For nonideal solutions, the activity coefficient, $\gamma$, corrects for deviation from nonideality:

$$a_w = \gamma\ x_w \qquad (12.7)$$

The mole fraction of water, $x_w$, expressed in terms of weight fraction of water ($x'_w$) and solute ($x'_s$) having molecular weights of 18 and $M_s$, respectively, is:

$$x_w = \frac{\text{moles water}}{\text{moles water} + \text{moles solute}} \qquad (12.8)$$

$$x_w = \frac{\frac{x'_w}{18}}{\frac{x'_w}{18} + \frac{x'_s}{M_s}}$$

Equation (12.6) is Raoult's law. In the form of Eq. (12.8), Raoult's law is often cited to govern the dependence of water activity on water and solute concentration in foods. Qualitatively, Raoult's law explains adequately the dependence

of water activity on water and solute content in foods. Solutes with low molecular weights also provide larger reduction of vapor pressure per unit weight of solute than those with high molecular weights. However, $a_w$ calculated using Eq. (12.8) will differ from actual $a_w$ because the solution phase in most foods is nonideal.

The deviation of $a_w$ from Raoult's law is related to the heat of mixing involved in the dissolution of solute. Let $\Delta H_{m1}$ = partial molal heat of mixing, defined as the change in enthalpy of mixing with the addition of or removal of 1 mole of component 1 at constant temperature and pressure. Equation (12.9) can be derived (Perry et al. 1963) by assuming that the excess entropy of mixing is zero:

$$\overline{\Delta H_{m1}} = \frac{d(\Delta H_m)}{dN_1} = RT\ \ln\ \gamma_1 \qquad (12.9)$$

Solutes in food systems exhibit negative heat of mixing with water (heat is released in mixing), and $\gamma$ in Eq. (12.9) is less than 1. Solutions with negative heat of mixing will have $a_w$ less than that calculated using Raoult's law.

The activity coefficient of a regular solution can be derived using Van Laar's approximations based on van der Waals equation of state for one mole of substance:

$$P = \frac{RT}{V - b} - \frac{a}{V^2} \qquad (12.10)$$

The constant a in van der Waals equation of state represents the magnitude of attractive forces between molecules, and the constant b represents the reduction in the volume as a result of these attractive forces. In a mixture containing $N_1$ and $N_2$ moles of components 1 and 2, respectively, a for the mixture will be determined by the constant $a_1$ for component 1 and $a_2$ for component 2. Because a is associated with attraction between molecules, the rules of permutation will give $N_1$ ways in which $N_1$ molecules can interact. If each of these interactions yields $a_1$, the total yield of $N^2_1$ interactions will be $a_1N^2_1$. The yield of all interactions between $N_2$ molecules of component 2 will be $a_2N^2_2$. Molecules of component 1 and 2 also interact with each other, and the yield of each of these interactions will be the geometric mean of $a_1$ and $a_2$. The number of ways in which $N_1$ molecules of component 1 and $N_2$ molecules of component 2 can interact with each other (excluding interactions between like molecules) will be $2N_1N_2$, and the yield of these interactions will be $(a_1a_2)^{1/2}(2N_1N_2)$. Thus, for a mixture, the van der Waals constant a can be expressed in terms of the constants of the components according to:

$$a = a_1N_1^2 + a_2N_2^2 + 2N_1N_2(a_1a_2)^{1/2} \quad (12.11)$$

Constant b is associated with the volume of each molecule; therefore, b for a mixture can be expressed in terms of $b_1$ and $b_2$ as:

$$b = b_1 N_1 + b_2N_2 \quad (12.12)$$

Van Laar's first approximation assumes the ratio a/b to be the internal energy of the mixture that approximates the heat of vaporization. Furthermore, in solutions, the volume changes on mixing are small such that internal energy, E, and enthalpy, H, are equivalent. The heat of mixing will be the difference between the sum of the internal energies of pure component 1 and pure component 2 and the internal energy of the mixture:

$$-\Delta H_m = E_1 + E_2 - E_{12} \quad (12.13)$$

$E_1$ is $a_1/b_1$, $E_2$ is $a_2/b_2$, and $E_{12}$ is a/b for the mixture determined using Eqs. (12.11) and (12.12). Equation (12.13) becomes:

$$-\Delta H_m = \frac{a_1}{b_1} + \frac{a_2}{b_2} - \frac{a_1N_1^2 + a_2N_2^2 + 2N_1N_2(a_1a_2)^{1/2}}{N_1b_1 + N_2b_2}$$

Simplifying:

$$-\Delta H_m = \frac{N_1N_2b_1b_2}{N_1b_1 + N_2b_2}\left[\frac{a_1^{1/2}}{b_1} - \frac{a_2^{1/2}}{b_2}\right]^2$$

If the value of b for the two components are equal:

$$-\Delta H_m = \frac{N_1N_2}{(N_1+N_2)b}\left(a_1^{1/2} - a_2^{1/2}\right)^2 \quad (12.14)$$

The partial molal heat of mixing can be determined by differentiating Eq. (12.14) with respect to $N_1$, keeping $N_2$ constant:

$$\overline{\Delta H_{m1}} = \frac{-d\Delta H_m}{dN_1}$$

$$= \frac{N_2(a_1^{1/2} - a_2^{1/2})^2}{b} \frac{d}{dN_1}\left[\frac{N_1}{N_1+N_2}\right]$$

$$= \frac{N_2(a_1^{1/2} - a_2^{1/2})}{b}\left[\frac{N_2}{(N_1+N_2)^2}\right]$$

Because $a_1, a_2$, and b are all constants, $k' = \frac{a_1^{1/2} - a_2^{1/2}}{b}$ and:

$$\begin{aligned}-\Delta H_{m1} &= k'\left[\frac{N_2}{N_1+N_2}\right]^2 \\ &= k' \cdot (x_2)^2\end{aligned} \quad (12.15)$$

Combining Eqs. (12.9) and (12.15):

$$RT \ln \gamma_1 = -k'x_2^2 \quad (12.16)$$

Because from Eq. (12.7), $\gamma_w = a_w/x_w$, and in a two-component system, $x_2 = 1 - x_w$. At constant temperature, Eq. (12.16) can be written as:

$$\log\frac{a_w}{x_w} = -k(1 - x_w)^2 \quad (12.17)$$

Equation (12.17) shows that a plot of $\log a_w/x_w$ against $(1 - x_w)^2$ will be linear with a negative slope. Equation (12.17) has been used successfully by Norrish to predict the water activity of sugar solutions. Values of the constant k in Eq. (12.17) for various solutes are shown in Table 12.1.

**Example 12.1** Calculate the water activity of a 50% sucrose solution.

**Solution**
From Table 12.1, the k value for sucrose is 2.7. Sucrose has a molecular weight of 342. The mole fraction of water is:

$$x_w = \frac{50/18}{50/18 + 50/342} = 0.95$$

Using Eq. (12.17):

$$\log a_w = \log x_w - 2.7(1 - x_w)^2$$
$$= \log 0.95 - 2.7(0.05)^2$$
$$a_w = 0.935$$

**Table 12.1** Values of the constant k for various solutes in Norrish's equation for water activity of solutions

| Sucrose | 2.7 |
|---|---|
| Glucose | 0.7 |
| Fructose | 0.7 |
| Invert sugars | 0.7 |
| Sorbitol | 0.85 |
| Glycerol | 0.38 |
| Propylene glycol | −0.12 |
| Nacl | 15.8($x_2 < 0.02$)<br>7.9($x_2 > 0.02$) |
| Citric acid | 6.17 |
| d-Tartaric acid | 4.68 |
| Malic acid | 1.82 |
| Lactic acid | −1.59 |

*Sources*: Norrish, R. S., *J. Food Technol. 1:25,* 1996; Toledo, R. T., Proc. *Meat. Int Res. Conf,* 1973; Chuang, L., M.S. thesis, University of Georgia, *1974;* Chirife, J., and Ferro-Fontain, C., *J. Food Sci. 45:802,* 1980

### 12.1.3.1   Gibbs-Duhem Equation

The chemical potential is a partial molal quantity that can be represented as the change in the free energy with a change in the number of moles of that component, all other conditions being maintained constant $\Phi_1 = (dF/dN_1)$:

$$dF = \left(\frac{dF}{dN_1}\right)dN_1 + \left(\frac{dF}{dN_2}\right)dN_2$$
$$= \mu_1 \; dN_1 + \mu_2 \; dN_2$$

But:

$$F = \Phi_1 N_1 + \Phi_2 N_2$$
$$dF = \Phi_1 \; dN_1 + N_1 \; d\Phi_1 + \Phi_2 \; dN_2 + N_2 \; d\Phi_2$$

and:

$$N_1 \; d\Phi_1 + N_2 \; d\Phi_2 = 0 \qquad (12.18)$$

Equation (12.18) and similar expressions in Eqs. (12.19) and (12.20) are different forms of the Gibbs-Duhem equation:

$$x_1 \left[\frac{dlna_1}{dx_1}\right] + x_2 \left[\frac{dlna_2}{dx_1}\right] = 0 \qquad (12.19)$$

or:

$$x_1 dlna_1 + x_2 \; dlna_2 = 0 \qquad (12.20)$$

Let $(a_w)^0$ represent the activity of water in a system with only one solute and water in the mixture. Integration of Eq. (12.20) gives:

$$\ln (a_w)^o = -\int_{x_1}^{x_2} dln(a_2)$$

In a multicomponent system involving two solutes:

$$dln(a_w) = -d\frac{x_2}{x_1} \; dln(a_2) - \frac{x_3}{x_1} \; dln(a_3)$$

$$\ln (a_w) = -\int_{x_1}^{x_2}\ln (a_2) - \int_{x_1}^{x_3} dln(a_3)$$
$$(12.21)$$

$$a_w = (a_{w2}{}^o)(a_{w3}{}^o)$$

The water activity of a mixture involving several components can be determined from the product of the water activity of each component separately if all the water present in the mixture is mixed with individual components. Eq. (12.21) was derived by Ross (Food Technol. 3:26, 1975).

**Example 12.2** Calculate the water activity of a fruit preserve containing 65% soluble solids, 2% insoluble solids, and the rest water. The soluble solids may be assumed to be 50% hexose sugars and 50% sucrose.

**Solution**

Basis: 100 g of fruit preserve:

> g hexose sugars = 65(0.5) = 32.5 g
> g sucrose = 65(0.5) = 32.5 g
> g water = 33 g

Consider sucrose dissolves in all of the water present:

$$x_s = \frac{32.5/342}{33/18 + 32.5/342} = 0.0492$$
$$x_w = 1 - 0.0492 = 0.9507$$

Using Eq. (12.17) and a k value of 2.7 for sucrose:

$$\log(a_{w1})^0 = \log 0.9507 - 2.7(0.0492)^2$$
$$(a_{w1})0 = 0.9365$$

Consider that hexose dissolves in all of the water present:

$$x_i = \frac{32.5/180}{32.5/180 + 33/18} = 0.0896$$
$$x_w = 1 - 0.0896 = 0.9103$$

Using Eq. (12.17) and a k value for hexose of 0.7:

$$\log(a_{w2})^0 = \log 0.9103 - 0.7(0.0896)^2$$
$$(a_{w2})^0 = 0.8986$$

Using Eq. (12.21), the $a_w$ of the mixture is:

$$a_w = (a_{w1})^0 (a_{w2})^0$$
$$= 0.9365\,(0.8986) = 0.841$$

### 12.1.3.2 Other Equations for Calculating Water Activity

#### 12.1.3.2.1 Bromley's Equation

For salts and other electrolytes, Bromley's equation (Bromley AIChE J. 19:313, 1073) accounts for ionic dissociation and nonideality:

$$a_w = \exp\left(-0.0183\,m_i\phi\right) \quad (12.22)$$

where $\phi$ = osmotic coefficient, $m_i$ = moles of ionic species i per kilogram of solvent, and $m_i = (I \cdot m)$ and where I = number of ionized species per mole and m = molality. The osmotic coefficient, $\phi$, is calculated using Eq. (12.23):

$$\phi = 1 + 2.303[F_1 + (0.06 + 0.6B)F_2 + 0.5BI] \quad (12.23)$$

The parameter $F_1$ in Eq. (12.23) is calculated using Eq. (12.24):

$$F_1 = F_{id}\left[-0.017z\ I^{0.5}\right] \quad (12.24)$$

The parameter $F_{id}$ in Eq. (12.24) is calculated using Eq. (12.25):

$$F_{id} = 3\ I^{-1.5}\left[1 + I^{0.5} - \frac{1}{1 + I^{0.5}} - 2\ln\left(1 + I^{0.5}\right)\right] \quad (12.25)$$

The parameter $F_2$ in Eq. (12.23) is calculated using Eq. (12.26):

$$F_2 = \frac{z}{aI}\left[\frac{1 + 2aI}{(1 + aI)^2} - \frac{\ln(1 + aI)}{aI}\right] \quad (12.26)$$

The parameter I is the ionic strength, evaluated as half the sum of the product of the molality of dissociated ions and the square of their charge:

$$I = 0.5\sum\left(n_i \cdot \xi^2\right)$$

where n = stoichiometric molar number of ions/mole and $\xi$ is the ionic charge.

For example, for $MgCl_2$, $n_i$ for Mg = 1, and $n_i$ for Cl = 2. $\xi$ for Mg is +2, and $\xi$ for Cl is $-1$. Thus, $I = 0.5\ m\ [(1)(2)^2 + 2(-1)^2] = 3$ m.

B and a are constants for each salt obtained from a regression of activity coefficient data against ionic strength. A comprehensive listing of a and B values for different salts is given by Bromley (1973). Chirife and Ferro-Fontan (J. Food Sci. 45:802, 1980) reported a value of B for sodium lactate of 0.050 kg/gmol. z is the charge number, the ratio of the sum of the product of the number of ions, and its charge to the stoichiometric number of ions, for example, for $MgCl_2$, $z = [1(2) + 2(1)]/3 = 4/3 = 1.333$. For NaCl, KCl, sodium acetate, and salts with monovalent cation and anion, $z = [1(1) + 1(1)]/2 = 1.0$

The values of the parameters z, I, B, and a for various salts are as follows:

For NaCl: z = 1, I = m, B = 0.0574, a = 1.5
For KCl: z = 1, I = m, B = 0.0240, a = 1.5
For $KNO_3$: z = 1, I = m, B = $-0.0862$, a = 1.5
For $MgCl_2$: z = 1.33, I = 3 m, B = 0.01129, a = 1.153

#### 12.1.3.2.2 Lang-Steinberg Equation

This equation (J. Food Sci. 46:670, 1981) is useful for mixtures of solids, where there is no solute-insoluble solid interaction. The moisture content of each component $X_i$ was considered to be linear when plotted against log $(1 - a_{wi})$, with a slope $b_i$ and intercept $a_i.a_{wi}$ as the water activity of component i at moisture content $X_i$. If X is the equilibrium moisture content of the mixture, g water/g dry matter, and S is the total mass of dry matter in the mixture, the $a_w$ of the mixture can be calculated from the mass of dry matter in each component, $S_i$, as follows:

$$\log(1 - a_w) = \frac{XS - \sum (a_i S_i)}{\sum (b_i X_i)} \quad (12.27)$$

### 12.1.4   Water Activity at Low Moisture Contents

If the water content of a food is plotted against the water activity at constant temperature, a sigmoid curve ally results. The curve is known as the sorption isotherm for that product. Figure 12.2 shows a sorption isotherm for raw beef at 25 °C.

The sorption isotherm can be subdivided into three zones, each zone representing a different mechanism for water sorption. In zone C, the influence of insoluble solids on the $a_w$ is negligible. $a_w$ is dependent on the solute and water content of the solution phase and can be calculated using Eqs. (12.17) and (12.21).

In zone B, the influence of insoluble solids on $a_w$ becomes significant. The sorption isotherm flattens out, and very small changes in moisture content are reflected by very large changes in the water activity. In this zone, water is held in the solid matrix by capillary condensation and multilayer adsorption. Some of the solutes may also be in the form of hydrates. Some of the water may still be in the liquid phase, but its mobility is considerably restricted because of attractive forces with the solid phase. The quantity of water present in the material that would not freeze at the normal freezing point usually is within this zone.

Zone A represents adsorption of water on the surface of solid particles. None of the water is in the liquid phase anymore. The heat of vaporization of water in this zone is higher than the heat of

**Fig. 12.2** Sorption isotherm of dried raw beef at 25 °C, showing the three segments of the BET plot

vaporization of pure water because both heat of vaporization and heat of adsorption must be supplied to remove the water molecules from the solid surface.

The relationship between water activity and moisture content in zone A is best described by the Brunauer-Emmett-Teller (BET) equation. If x is the moisture content in g $H_2O$/g dry matter:

$$\frac{a_w}{x(1-a_w)} = \frac{1}{x_m C} + \frac{C-1}{x_m C}(a_w) \qquad (12.28)$$

$x_m$ in Eq. (12.28) is the mass fraction of water in the material equivalent to a unimolecular layer of water covering the surface of each particle. C is a constant at constant temperature and is related to the heat of adsorption of water on the particles. C is temperature dependent.

A plot of $a_w/x(1-a_w)$ against $a_w$ is called the BET plot. The plot would be linear, and the slope and intercept of the line can be used to determine the constant $x_m$, the moisture content at which the water molecules coat the surface of the solid particles in a monomolecular layer.

The region of maximum stability for a food product is usually in zone B in Fig. 12.2. When the moisture content in the product drops to a level insufficient to cover the solid molecules in a monomolecular layer, the rate of lipid oxidation increases. Determining the moisture content for maximum shelf stability of a dehydrated material would involve determining the sorption isotherm and calculating the value of $x_m$ in Eq. (12.28) from a BET plot.

The dashed lines in Fig. 12.2 are BET plots of the sorption data for raw beef. The line in zone A represents monomolecular adsorption and will be used for calculating $x_m$. The slope and intercept of the line are 13.7 and 0.98, respectively. From Eq. (12.28), the intercept is $1/x_m C$. $x_m C = 1/0.98 = 1.02$. The slope is $(C-1)/x_m C$:

$$\frac{C-1}{1.02} = 13.7$$
$$C = 14.974$$
$$x_m = \frac{1.02}{14.974}$$
$$= 0.068 \ gH_2O/g \ dry \ matter$$

### 12.1.4.1 The GAB (Guggenheim-Anderson-de Boer) Equation

This equation gives a better fit than the BET over a wider range of moisture contents. Let X = moisture content on a dry solids basis and $X_m$ = the moisture content on a dry basis, equivalent to a monomolecular layer of water.

Then:

$$X = \frac{X_m C \ k \ a_w}{(1 - k \ a_w)(1 - k \ a_w + C \ k a_w)} \qquad (12.29)$$

The GAB equation is a three-parameter equation with k, C, and $X_m$ as constants. C and $X_m$ have similar significance as in the BET equation. k is a third parameter that corrects for the difference in properties of adsorbed water relative to liquid water and permits the GAB equation to hold over a wider range of moisture content than the BET. Both k and C are temperature dependent.

Evaluation of $X_m$ using Eq. (12.29) is more precise than that using the BET equation, because the fit of Eq. (12.29) to the data extends over a wide range of moisture contents. The narrow range of moisture contents to which Eq. (12.28) fits often presents some problems in the determination of $X_m$ using the BET equation, particularly when data points are not close enough at the lower moisture contents.

Rearranging Eq. (12.29) by taking the reciprocal:

$$\frac{1}{X} = \frac{1}{X_m}\left[\frac{1}{Ck \ a_w} - \frac{1}{C}\right][1 - k \ a_w + C \ k \ a_w]$$
$$\frac{a_w}{X} = \left[\frac{k}{X_m C}\right](1-C)a_w^2 + \left[\frac{C-2}{X_m C}\right]a_w + \frac{1}{X_m C \ k}$$

Thus, a polynomial nonlinear regression of $(a_w/X)$ against $(a_w)$ will give values for $\alpha$ the coefficient of the quadratic term, $\beta$ the coefficient of the linear term, and $\gamma$ the intercept:

$$\beta = \frac{C-2}{X_m C}; \quad \alpha = \left[\frac{k}{X_m C}\right][1-C]; \quad \gamma = \frac{1}{X_m Ck}$$

These three equations can be used to evaluate C, k, and $X_m$.

Multiple nonlinear regression techniques must be used to fit and evaluate the parameters for this equation from moisture-water activity data.

**Example 12.3 Fitting GAB Equation to Sorption Isotherm Data** The following data ($a_w$, X) was obtained for the sorption isotherm of potatoes at 25 °C. X is the moisture content in g water/g dry matter. Calculate the GAB parameters $X_m$, C, and k. Data: (0.112, 0.035), (0.201, 0.057), (0.327, 0.08), (0.438, 0.105), (0.529, 0.13), (0.577, 0.145), (0.708, 0.19), (0.753, 0.204), (0.843, 0.27), (0.903, 0.37).

**Solution**
Use Excel to calculate $a_w/X$ and tabulate against $a_w$. The following are the results:

| Independent variable ($a_w$) | Dependent variable ($a_w/X$) |
|---|---|
| 0.112 | 3.2 |
| 0.201 | 3.52 |
| .327 | 4.09 |
| .438 | 4.17 |
| .529 | 4.07 |
| .577 | 3.98 |
| .708 | 3.73 |
| .753 | 3.69 |
| .843 | 3.12 |
| .903 | 2.44 |

Perform a nonlinear regression (using a statistical software package, e.g., Systat) to fit the following model to the data:

$$a_w/X = \alpha(a_w)^2 + \beta(a_w) + \gamma$$

Regression results: $\alpha = -8.24$, $\beta = 7.69$, and $\gamma = 2.39$. The $r^2$ value of the regression = 0.969 indicating a good fit throughout the full data set. Values of k, C, and $X_m$ can then be calculated.

Solving for $X_m$ in the expressions for $\alpha$, $\beta$, and $\gamma$ in the GAB equation:

$$X_m = [k(1 - C)/C\alpha]$$
$$= [(C - 2)/C\beta] = 1/Ck\gamma$$

Solving the equalities will result in two equations:

$$k^2(1 - C) = \alpha/\gamma; C = 2 + \beta/\gamma k$$

Substituting C: $k^2(1 - 2 - \beta/\gamma k) = \alpha/\gamma$ and $k^2 + (\beta/\gamma)k + \alpha/\gamma = 0$. Using the quadratic equation and taking the positive root: k = 0.848, C = 2 + β/γk = 5.79, and $X_m = 1/Ck\gamma = 0.0852$.

### 12.1.4.2   Other Equations for Sorption Isotherms of Foods

Other equations that have been used to fit the sorption isotherms of foods are Eqs. (12.30), (12.31), (12.32), (12.33), (12.34), (12.35), (12.36), (12.37), (12.38), and (12.39). Iglesias and Chirife (1982) fitted sorption isotherms to these equations and reported the best-fitting equations with the corresponding value of the equation parameters. Halsey's (Eq. 12.34), Henderson's (Eq. 12.35), Oswin's (Eq. 12.38), and Smith's (Eq. 12.39) were reported to fit more water sorption/desorption data among these equations.

Caurie equation:

$$\ln(1 - X) = A - r \ln(a_w) \tag{12.30}$$

X = moisture content dry basis
A and r are constants obtained from the regression equation for $\ln(1 - X)$ versus $\ln(a_w)$.

Chen equation:

$$a_w = \exp[-K \exp(-bX)]$$
$$\ln[-\ln(a_w)] = \ln(K) - bX \tag{12.31}$$

K and b are constants obtained from a linear regression of $\ln[-\ln(a_w)]$ versus X.

Chong-Pfost equation:

$$\ln(a_w) = \frac{-A}{T}\exp(-BX) \tag{12.32}$$

A and B are constants.
Day and Nelson equation:

$$a_w = 1 - \exp[P_1 \ T^{P_2}X^{P_3}] \tag{12.33}$$

$P_1$, $P_2$, and $P_3$ are constants; T = absolute temperature.

Halsey equation:

$$a_w = \exp\left[\frac{-A}{T}\theta^b\right] \tag{12.34}$$

$\theta = X/X_m$ and a and b are constants.
Henderson equation:

$$1 - a_w = \exp[-aX^b] \tag{12.35}$$

a and b are constants.
Iglesias and Chirife equation:

$$X = B_1 \left[ \frac{a_w}{1 - a_w} \right] + B_2 \qquad (12.36)$$

$B_1$ and $B_2$ are constants.
Kuhn equation:

$$X = \frac{A}{\ln (a_w)} + B \qquad (12.37)$$

Oswin equation:

$$X = A \left[ \frac{a_w}{1 - a_w} \right] + B \qquad (12.38)$$

Smith equation:

$$X = A - B \, \ln (1 - a_w) \qquad (12.39)$$

These equations are empirical and are usually used when the fit to the GAB equation is not very good. The advantage of the GAB and BET equations is that they are based on physical adsorption phenomena; therefore the parameters have physical meaning. Expressing sorption isotherms in equation form avoids the need for graphs in representing water activity of foods as a function of moisture content. These equations will also facilitate the calculation of equilibrium moisture content and water activity in individual particulate component when mixing particulate solids with different water activities.

## 12.2   Mass Transfer

During dehydration, water is vaporized only from the surface. The transfer of water vapor from the wet surface to a stream of moving air is analogous to convection heat transfer; therefore a mass transfer coefficient is used. Moisture flux is proportional to the driving force, which is the difference in vapor pressure on the surface and the vapor pressure of water in air surrounding the surface. At the same time that water is removed from the surface, water diffuses from the interior of a solid toward the surface. The latter is a general form of diffusion that is analogous to conduction heat transfer. The differential equations for conduction also apply for diffusion,

but a mass diffusivity is used in place of the thermal diffusivity.

### 12.2.1   Mass Diffusion

For an infinite slab, Eq. (7.89) in the section "The Infinite Slab," Chap. 7, can be used for the dimensionless moisture change with time. Expressing these equations in terms of moisture content and mass diffusivity:

$$\theta = \frac{X - X_m}{X_0 - X_m}$$

where $X$ is moisture content at any time, dry matter basis (kg water/kg dry matter), $X_0$ is initial moisture content, and $X_m$ is equilibrium moisture content. The equation for moisture content at any point $y$ in the solid measured from the center, at any time $t$, when the moisture content $X$ was originally uniform at $X_0$ is:

$$\theta = 2 \sum_{n=0}^{\infty} \frac{(-1)^n}{(n + 0.5)\pi} \left[ \left( e^{-(n+.5)^2 \pi^2 D_m \frac{t}{L^2}} \right) \right.$$
$$\left. \left( \cos \left( \frac{(n + 0.5)\pi y}{L} \right) \right) \right]$$

$$(12.40)$$

Assumptions used in the derivation of this equation are constant diffusivity, and constant surface moisture at the moisture content in equilibrium with the drying air, during the process. The value of $X_m$ is determined from the sorption isotherm of the solid at $a_w$ = decimal equivalent of percent relative humidity of the drying air.

During the early stages of dehydration, moisture is transferred and moves from the center toward the surface by capillary action. This mechanism is more rapid than diffusion, and rate of surface evaporation controls the rate of drying. However, in the later stages of drying, diffusion controls the rate of moisture migration within the solid. Diffusivity may be constant if cells do not collapse and pack together. Firm solids such as grain may exhibit constant diffusivity, but high-moisture products such as fruits and vegetables may exhibit variable diffusivity with moisture content depending on the physical changes that occur as water is removed.

The mass diffusivity $D_m$ has the same units as thermal diffusivity and can be used directly to substitute for $\alpha$ (thermal diffusivity) in the heat transfer equations. Of interest in dehydration processes is the average moisture content in the slab at any time during the drying process. Let W and Z represent the width and length of the slab. The total moisture in the slab of half-thickness, L is:

$$\text{Total moisture} = \rho W Z \int_0^L X \, dy$$

The mean moisture $\bar{X}$ = total moisture/total mass:

$$\bar{X} = \frac{1}{L} \int_0^L X \, dy$$

The dimensionless moisture ratio $\bar{\theta}$ based on the average moisture content is:

$$\bar{\theta} = \frac{\bar{X} - X_m}{X_0 - X_m}$$

Substituting the expression for X obtained from Eq. (12.40) into the above integral and integrating:

$$\bar{\theta} = 2 \sum_{n=0}^{\infty} \left[ \frac{(-1)^n}{(n+0.5)^2 \pi^2} [e]^{-(n+0.5)^2 \pi^2 D_m \frac{t}{L^2}} \right. \\ \left. [\sin (n+0.5)\pi] \right]$$

(12.41)

The squared term in the denominator of Eq. (12.41) results from the integration of the cosine function of Eq. (12.40).

Table 12.2 shows the values of the dimensionless mean moisture ratio $\bar{\theta}$ as a function of $(D_m t)/L^2$. L is the half thickness of the slab. For three-dimensional diffusion, $\overline{\theta_{L1}}$ is obtained for $(D_m t/L^2{}_1)$, $\overline{\theta_{L2}}$ is obtained for $(D_m t/L^2_2)$, and $\overline{\theta_{L3}}$ is obtained for $(D_m t/L^2_3)$. The composite $\bar{\theta}$ is the product of $\overline{\theta_{L1}}$, $\overline{\theta_{L2}}$, and $\overline{\theta_{L3}}$.

When $(D_m t)/L^2 > 0.1$, the first three terms in the series in Eq. (12.41) are adequate for series convergence. Taking logarithms of both sides of Eq. (12.41):

**Table 12.2** Dimensionless mean moisture ratio as a function of $D_m t/L^2$

| $\frac{D_m t}{L^2}$ | $\bar{\theta}$ | $\frac{D_m t}{L^2}$ | $\bar{\theta}$ | $\frac{D_m t}{L^2}$ | $\bar{\theta}$ |
|---|---|---|---|---|---|
| 0 | 1.0000 | 0.10 | 0.6432 | 1.10 | 0.0537 |
| 0.01 | 0.8871 | 0.20 | 0.4959 | 1.20 | 0.0411 |
| 0.02 | 0.8404 | 0.30 | 0.3868 | 1.30 | 0.0328 |
| 0.03 | 0.8045 | 0.40 | 0.3021 | 1.40 | 0.0256 |
| 0.04 | 0.7743 | 0.50 | 0.2360 | 1.50 | 0.0200 |
| 0.05 | 0.7477 | 0.60 | 0.1844 | 1.60 | 0.0156 |
| 0.06 | 0.7236 | 0.70 | 0.1441 | 1.70 | 0.0122 |
| 0.07 | 0.7014 | 0.80 | 0.1126 | 1.80 | 0.0095 |
| 0.08 | 0.6808 | 0.90 | 0.0879 | 1.90 | 0.0074 |
| 0.09 | 0.6615 | 1.00 | 0.0687 | 2.00 | 0.0058 |

$$\log(\bar{\theta}) = \frac{\pi^2 D_m}{L^2} t \log(e) + B$$

$$B = \log(2) + \log A_1 (e)^{0.25} + \log A_2 (e)^{2.25}$$

$$+ \log A_3 (e)^{6.25}$$

$$A_1, A_2, A_3 = \frac{(-1)^n \sin[(n+0.5)\pi]}{(n+0.5)^2 \pi^2}$$

$$\text{for } n = 0, 1, 2$$

(12.42)

Equation (12.42) shows that a semi-log plot of the dimensionless mean moisture ratio against time of drying will be linear if $D_m$ is constant. The mass diffusivity, $D_m$, can be calculated from the slope.

**Example 12.4** The desorption isotherm of apples was reported by Iglesias and Chirife (1982) to best fit Henderson's equation with the constants a = 4.471 and b = 0.7131. Experimental drying data for apple slices showed that when the average moisture content was 1.5 kg water/kg dry matter, the drying rate, which may be assumed to be diffusion controlled, was $8.33 \times 10^{-4}$ kg water/[kg dry matter (s)]. The apple slices were 1.5 cm thick and 2.5 cm wide and were long enough to consider diffusion to occur from two dimensions. The drying air has a relative humidity of 5%.

(a) Calculate the mass diffusivity of water at this stage of drying.

(b) Calculate the drying rate and moisture content after 1 h of drying from when the moisture content was 1.5 kg water/kg dry matter.

**Solution**

(a)  Henderson's equation will be used to calculate the equilibrium moisture content, $X_m$. Solving for $X_m$ by substituting $X_m$ for X in Henderson's equation:

$$\ln(1 - a_w) = -aX^b; \quad X_m = \left[\frac{-\ln(1 - 0.05)}{4.471}\right]^{1/0.7131}$$

$$X_m = 0.0019$$

Equation (12.42) for diffusion from two dimensions is:

$$\log(\overline{\theta}) = \left[\frac{1}{(L_1)^2} + \frac{1}{(L_2)^2}\right]\pi^2 D_m \log(e)t + B$$

(12.43)

Differentiating Eq. (12.43) with respect to t:

$$\frac{d}{dt}\left[\log(\overline{\theta})\right] = \left[\frac{1}{(L_1)^2} + \frac{1}{(L_2)^2}\right]\pi^2 D_m \log(e)$$

(12.44)

Differentiating the expression for $\overline{\theta}$ with respect to t:

$$\log(\overline{\theta}) = \log\left(\frac{\overline{X} - X_m}{\overline{X}_o - X_m}\right)$$

$$\frac{d}{dt}\left[\log(\overline{\theta})\right] = \frac{1}{(\overline{X} - X_m)\left[\ln(10)\right]}\frac{d\overline{X}}{dt}$$

(12.45)

Combining Eqs. (12.44) and (12.45) and solving for $D_m$:

$$D_m = \frac{d\overline{X}/dt}{(\overline{X} - X_m)\left[\frac{1}{(L_1)^2} + \frac{1}{(L_1)^2}\right]\pi^2}$$

(12.46)

Substituting known quantities:

$$D_m$$
$$= \frac{8.33 \times 10^{-4}}{(1.5 - 0.0019)\left[(1/0.0075)^2 + (1/0.0125)^2\right]\pi^2}$$
$$= 2.3302 \times 10^{-9}\,\mathrm{m^2/s}$$

(b)  Solving for $D_m t/L^2$ for $t = 3600$ s:

$$L = 0.0075; \left(D_m t/L^2\right) = 0.149$$
$$L = 0.0125; \left(D_m t/L^2\right) = 0.054$$

From Table 12.2:
(When $D_m t/L^2$) = 0.149, by interpolation:

$$\overline{\theta_{L1}} = [0.6432 - (0.6432 - 0.4959)(0.049)]$$
$$/0.1 = 0.571$$

When $(D_m t/L^2) = 0.054$, by interpolation:

$$\overline{\theta_{L1}} = [0.7477 - (0.7477 - 0.7236)(0.044)]$$
$$/0.01 = 0.738$$

For diffusion from two directions:

$$\overline{\theta} = 0.571(0.738) = 0.4214$$
$$\overline{X} = 0.4214(1.5 - 0.019) + 0.019$$
$$= 0.633\,\mathrm{kg\ water/kg\ dry\ matter}$$

The drying rate is $d\overline{X}/dt$. Using Eq. (12.46):

$$\frac{d\overline{X}}{dt} = D_m(\overline{X} - X_m)\left[\frac{1}{(L_1)^2} + \frac{1}{(L_2)^2}\right]\pi^2$$

$$= \left[2.330 \times 10^{-9}(0.633 - 0.0019)(\pi^2)\right]$$

$$\left[\frac{1}{(0.0075)^2} + \frac{1}{(0.0125)^2}\right]$$

$$= 0.000351\,\mathrm{kg\ water/(s \cdot kg\ dry\ matter)}$$

## 12.2.2  Mass Transfer from Surfaces to Flowing Air

When air flows over a wet surface, water is transferred from the surface to air. The equations governing rate of mass transfer is similar to that for heat transfer. By analogy, the driving force for mass transfer is a concentration difference, and the proportionality constant between the mass flux and the driving force is the mass transfer coefficient:

$$\frac{dW_w}{A\,dt} = k_g M_w(a_{ws} - a_{wa})$$

(12.47)

where $W_w$ is the mass of water vapor transferred from the surface to the moving air, A is the surface area exposed to air, $M_w$ is the molecular weight of water, $a_{ws}$ is water activity on the surface, and $a_{wa}$ is the water activity of the drying air.

Using the dimensionless $a_w$ difference as the driving force results in the mass transfer coefficient, $k_g$, having units in a general form widely used in the literature, kg moles/(m$^2 \cdot$ s).

Determination of the mass transfer coefficient is analogous to that in heat transfer, which involves the use of dimensionless groups. The equivalent of the Nusselt number in mass transfer is the Sherwood number (Sh):

$$Sh = \frac{k_g D}{D_{wm}} \qquad (12.48)$$

where D is the diameter or characteristic length and $D_{wm}$ is the diffusivity expressed in kg mole/(m $\cdot$ s).

The equivalent of the Prandtl number in mass transfer is the Schmidt number expressed in either the mass diffusivity, $D_{mx}$ in m$^2$/s, or $D_{wm}$ in kg mole/(m $\cdot$ s). Physical property terms in the Schmidt number for dehydration processes are those for air. $M_a$ is the molecular weight of air, 29 kg/kg mole. $D_m$ is diffusivity of water in air $= 2.2 \times 10^{-5}$ m$^2$/s:

$$Sc = \frac{\mu}{\rho D_m} = \frac{\mu}{M_a D_{wm}} \qquad (12.49)$$

Correlation equations for the mass transfer coefficient are similar to those for heat transfer. Typical expressions are as follows.

Gilliland and Sherwood equation:

$$Sh = 0.023 \, Re^{0.81} \, Sc^{0.44} \qquad (12.50)$$

Equation (12.50) was derived for vaporization from a water film flowing down a vertical tube to air flowing upward through the tube. The Reynolds number of flowing air ranged from 2000 to 35,000, and pressure ranged from 0.1 to 3 atm.

Colburn j factor:

$$j = \frac{kg}{G} Sc^{0.666} = 0.023 \, Re^{-0.2} \qquad (12.51)$$

G is the molar flux of air, kg moles/(m$^2 \cdot$ s) $=$ PV/RT; P $=$ pressure in Pa; V $=$ velocity, m/s; R $= 8315$ N $\cdot$ m / (kg mole $\cdot$ K); and T $=$ absolute temperature. Eq. (12.51) may be used for mass transfer into air flowing through particles in a

packed bed. Eq. (12.50) would be suitable for air flowing parallel to the surface of a bed of particles.

Ranz and Marshall's equation:

$$Sh = 2 + 0.6 \, Re^{0.5} \, Sc^{0.33} \qquad (12.52)$$

Equation (12.52) is suitable for mass transfer from surfaces of individual particles such as in fluidized beds.

In dehydration, mass transfer is not the rate-limiting mechanism, particularly at high air velocities needed to maintain low humidity in the drying air.

**Example 12.5** Calculate the rate of dehydration expected from mass transfer when 1 cm carrot cubes having a density of 1020 kg/m $^3$ are dried in a fluidized bed with air at 2% relative humidity flowing at the rate of 12 m/s. Atmospheric pressure is 101 kPa, and air temperature is 80 °C. Express dehydration rate as kg/(s $\cdot$ kg dry matter) when moisture content is 5 kg water/kg dry matter.

**Solution**
At 80 °C, the viscosity of air from the *Handbook of Chemistry and Physics* is 0.0195 cP. The characteristic length of a cube may be calculated as the diameter of a sphere having the same surface area:

$$D = L(6/\pi)^{0.5} = 1.382 \, L$$

The mass flux is calculated using the ideal gas equation by substituting air velocity, $\bar{V}$, for volume:

$$G = \frac{101,000(12)(29)}{8315(353)} = 11.975 \frac{kg}{m^2 \cdot s}$$

$G = P\bar{V}M_a/RT \cdot M_a$ is the molecular weight of air, 29 kg/kg mole.

The Reynolds number is DG/$\Phi = 1.382$ LG/$\Phi$:

$$Re = \frac{1.382(0.01)(11.975)}{0.0195(0.001)} = 8487$$

The density, $\rho$, is obtained using the ideal gas equation:

$$\rho = \frac{PM_a}{RT} = \frac{101,000(29)}{8315(353)} = 0.998 \frac{kg}{m^3}$$

$$Sc = \frac{\mu}{\rho D_m} = \frac{0.0195(0.001)}{0.998(2.2 \times 10^{-5})} = 0.888$$

$$Sh = 2 + 0.6(8487)^{0.5}(0.888)^{0.33} = 55.15$$

Equation (12.49):

$$= 7.57 \times 10^{-7} \text{kgmole}/(m \cdot s)$$

Equation (12.48):

$$kg = Sh \left( \frac{D_{wm}}{D} \right)$$

$$= \frac{55.150(7.57 \times 10^{-7})}{0.01(1.382)} = 0.00302 \frac{\text{kgmole}}{m^2 \cdot s}$$

$$D_{wm} = \left( \frac{\rho D_m}{M_a} \right) = \frac{0.998(2.2 \times 10^{-5})}{29}$$

Express area as m²/kg dry matter (DM):

$$A = \frac{6(0.01)^2}{(0.01)^3(1020)(1/6)} = 3.529 \frac{m^2}{\text{kg DM}}$$

Equation (12.47):

$$\frac{dW}{dt} = [3.529][0.003021(18)(1 - 0.02)]$$

$$= 0.188 \frac{kg}{s \cdot \text{kg DM}}$$

The drying rate, based on surface mass transfer, is 0.188 kg water/(s · kg DM).

## 12.3   Psychrometry

### 12.3.1   Carrying Capacity of Gases for Vapors

The mass of any component of a gas mixture can be calculated from the partial pressure using the ideal gas equation. The ideal gas equation can be written for the whole mixture or for a single component as follows:

$$P_t V = n_t RT \quad \text{for the whole mixture}$$

or:

$$P_a V = n_a RT \quad \text{for component A,} \qquad (12.53)$$

where $P_a$ and $n_a$ are partial pressure and number of moles of component A in the mixture, respectively. If the mixture consists only of components A and B:

$$P_t = P_a + P_b \qquad (12.54)$$

For component B:

$$(P_t - P_a)(V) = n_b RT \qquad (12.55)$$

Dividing Eqs. (12.53) by (12.55):

$$\frac{P_a}{P_t - P_a} = \frac{n_a}{n_b}$$

If $M_a$ is the molecular weight of component A and $M_b$ is the molecular weight of component B, the mass ratio of the two components can be determined from the partial pressure as follows:

$$\frac{W_b}{W_a} = \frac{n_b M_b}{n_a M_a} = \frac{P_b}{(P_t - P_b)} \frac{M_b}{M_a} \qquad (12.56)$$

If component A is water and component B is air, the mass ratio is known as the "humidity" or "absolute humidity." If $P_a$ is equal to the vapor pressure of water at the given temperature of the air, the mass ratio of components A and B is the saturation humidity. If $P_a$ is less than the vapor pressure, the ratio $P_a/P_s$, where $P_s$ is the saturation partial pressure or the vapor pressure, is the saturation ratio. This ratio expressed as a percentage is also known as the "relative humidity."

**Example 12.6** Dry air is passed through a bed of solids at the rate of 1 m³/s at 30 °C and 1 atm. If the solids have an equilibrium relative humidity of 80%, and assuming that the bed is deep enough such that equilibrium is attained between the solids and the air before the air leaves the bed, determine the amount of water removed from the bed per hour. Use 29 for the average molecular weight of air. Atmospheric pressure is 101.325 kPa.

**Solution**

At equilibrium, the partial pressure of water in the air should be 80% of the vapor pressure of water at 30 °C. From Appendix Table A.4, the vapor pressure of water at 30 °C is 4.2415 kPa. The partial pressure of water in the air is 80% of 4.215 or 3.393 kPa. The molecular weight of water is 18.

The absolute humidity can be calculated using Eq. (12.56):

$$\frac{W_w}{W_a} = \frac{3.394}{(101.325 - 3.394)29} \cdot 18$$

$$= 0.0215 \frac{\text{kg water}}{\text{kg dry air}}$$

In order to calculate the total amount of water removed in 1 h, the total amount of dry air that passed through the bed per hour must be calculated.

For the dry air component:

$$PV = nRT = \frac{W_a}{M_a}RT$$

$$W_a = \frac{P_a M_a}{RT} V$$

Substituting values for $P_b$, $M_b$, R, and T and using R = 8315 N · m/(kg mole) (K):

$$W_a$$
$$= \frac{(101,325\,\text{N} \cdot \text{m}^{-2})(29)(\text{kg})(\text{kg mole})^{-1}}{8315\text{N} \cdot \text{m}(\text{kg mole})^{-1}(\text{K})^{-1}(30 + 273)\text{K}} \left[1\frac{\text{m}^3}{\text{s}}\right]$$
$$= 1.166\,\text{kg/s of dry air.}$$

The weight of water removed per hour is:

$$W_w = 1.166\frac{\text{kg dry air}}{\text{s}} 0.0215\frac{\text{kg water}}{\text{kg dry air}} \frac{3600\,\text{s}}{\text{h}}$$

$$= 90.24\,\text{kg/h of water}$$

## 12.3.2 The Psychrometric Chart

Psychrometry is a study of the behavior of mixtures of air and water. In the preceding section, the determination of saturation humidities from the vapor pressure and the determination of the humidity from the partial pressure of water in air were discussed. A graph of humidity as a function of temperature at varying degrees of saturation forms the main body of a psychrometric chart. For processes that involve loss or gain of moisture by air at room temperature, the psychrometric chart is very useful for determining changes in temperature and humidity.

Another main feature of a psychrometric chart is the wet bulb temperature. When a thermometer is fitted with a wet sock at the bulb and placed in a stream of air, evaporation of water from the sock cools the bulb to a temperature lower than what would register if the bulb is dry. The difference is known as the wet bulb depression and is a function of the relative humidity of the air. The more humid air allows less vaporization, thus resulting in a lower wet bulb depression.

The various quantities that can be determined from a psychrometric chart are as follows:

Humidity (absolute humidity) (H): the mass ratio of water to dry air in the mixture.

Relative humidity (% RH): the ratio of partial pressure of water in the air to the vapor pressure of water, expressed in a percentage.

Dry bulb temperature ($T_{db}$): actual air temperature measured using a dry temperature-sensing element.

Wet bulb temperature ($T_{wb}$): air temperature measured using a wet-sensing element that allows cooling by evaporation of water.

Dew point ($T_{dp}$): temperature to which a given air-water mixture needs to be cooled to start condensation of water. At the dew point, the air is saturated with water vapor. The dew point is also that temperature where the vapor pressure of water equals the partial pressure of water in the air.

A psychrometric chart has its axes temperature on the abscissa and humidity on the ordinate. Two parameters are necessary to establish a point on the chart that represents the condition of air. These parameters could be any two from relative humidity, dry bulb temperature, wet bulb

**Fig. 12.3** Diagram showing the use of a psychometric chart for determining properties of air-water mixtures (Courtesy of Proctor and Schwartz, Inc. Copyrighted by Proctor and Schwartz. Inc.)

temperature, and dew point or absolute humidity. Dew point and absolute humidity are not independent, knowing one establishes the other.

Psychrometric charts in metric and English units are in the Appendix (Figs. A5 and A7). Figure 12.3 is a diagrammatic representation of how quantities are read from a psychrometric chart. Figure 12.3a shows how dew point is determined from the humidity. Figure 12.3b shows how the wet bulb temperature is determined from the humidity and the dry bulb temperature. Figure 12.3c shows how the percentage of relative humidity is determined knowing the dry bulb and the humidity and how dry bulb temperature is determined from humidity and % relative humidity. Figure 12.3d shows how humidity and dew point are determined from the wet and dry bulb temperatures.

**Example 12.7** Air has a dew point of 40 °C and has a relative humidity of 50%. Determine (a) the absolute humidity, (b) the wet bulb temperature, and (c) the dry bulb temperature.

**Solution**

Using a psychrometric chart, draw a vertical line through T = 40°. At the intersection with the line representing 100% relative humidity (% RH), draw a horizontal line, and read the absolute humidity represented by this horizontal line at the abscissa (see Fig. 12.3a). (a) H = 0.049 kg water/kg dry air.

Extend the horizontal line representing the humidity until it intersects the diagonal line representing 50% RH. Draw a vertical line through this intersection, and read the dry bulb temperature representing this vertical line at the abscissa (see Fig. 12.3c). (b) Dry bulb temperature = 53 °C.

From the point represented by H = 0.049 and $T_{tb}$ = 53 °C, draw a line that parallels the wet bulb lines. Project this line to its intersection with the 100% RH line. Draw a vertical line at this intersection, project to the abscissa, and read the wet bulb temperature at the abscissa (see Fig. 12.3b). (c) $T_{tb}$ = 42 °C.

**Example 12.8** Air in a smokehouse has a wet bulb temperature of 54 °C and a dry bulb temperature of 71 °C. Determine (a) the humidity, (b) the dew point, and (c) the percentage of relative humidity.

**Solution**

Draw vertical lines through T = 54 °C and the line representing 100% RH; draw a line that parallels the wet bulb lines. From the intersection of this drawn wet bulb line with the vertical line representing T = 71 °C, draw a horizontal line and project to the abscissa. Read the humidity represented by this horizontal line (see Fig. 12.3d). (a) H = 0.1 kg water/kg dry air.

From the intersection of the horizontal line representing H = 0.1 with the 100% RH line, draw a vertical line and connect to the abscissa. (b) $T_{dp}$ = 53 °C.

From the intersection of the line representing T = 71 °C and that representing H = 0.1, interpolate between the diagonal lines representing 40% and 50% RH, and estimate the % RH (see Fig. 12.3b). (c) % RH = 43%.

### 12.3.3  Use of Psychrometric Chart to Follow Changes in the Properties of Air-Water Mixtures Through a Process

Figure 12.4 shows the path of a process on a psychrometric chart for heating, cooling, and adiabatic humidification.

When the temperature of air is increased at constant pressure and there is no water added or removed from the air, the process is a constant humidity process. $T_{db}$ increases, $T_{wb}$ also increases, but % RH decreases. The process is represented by A in Fig. 12.4. When the temperature of air is decreased above the dew point, the process is a constant humidity process represented by B in Fig. 12.4. Both $T_{db}$ and $T_{wb}$ decrease, and % RH increases. Cooling below the dew point results in condensation of water and humidity drops. The % RH remains at 100%, and the temperature and humidity drop following the

**Fig. 12.4** Diagram showing paths of cooling, heating, condensation, and adiabatic humidification on a psychometric chart line representing 100% RH. This is shown by C in Fig. 12.4.

Adiabatic humidification is a process where water is picked up by air and the heat required to vaporize the added water comes from the sensible heat loss that results from a reduction of the temperature of the air. The path traced by the temperature and humidity of the air is represented by D in Fig. 12.4, and this path parallels the wet bulb lines. The wet bulb temperature of the air remains constant, humidity and % RH increase, and $T_{db}$ decreases.

This is the process that occurs when air is passed over or through a bed of wet solids in drying or when air is passed through water sprays in a water cooling tower.

**Example 12.9** Ambient air at 25 °C and 50% RH is heated to 175 °C. Determine (a) the % RH and (b) the wet bulb temperature of the heated air.

**Solution**
Use the psychrometric chart, and draw a vertical line representing $T_{db} = 25$ °C. Project this line until it intersects the diagonal line representing 50% RH. Draw a horizontal line, and project to the ordinate to determine the absolute humidity. H = 0.0098 kg/water/kg dry air. Project the horizontal line representing H = 0.0098 until it intersects a vertical line representing $T_{db} = 175$ °C. At the intersection, the % RH can be interpolated between the diagonal lines representing 0.15% RH and 0.2% RH. (a) % RH = 0.18%.

From the intersection, draw a line parallel to the wet bulb line, and project to its intersection with the line representing 100% RH. Project the intersection to the abscissa, and read (b) $T_{wb} = 45$ °C.

## 12.4  Simultaneous Heat and Mass Transfer in Dehydration

In dehydration, moisture is removed by evaporation. Heat must be transferred to equal the heat of vaporization. If the rate of mass transfer exceeds heat transfer needed to supply the heat

of vaporization, the material will drop in temperature. This process is called evaporative cooling. The mass transfer equation (Eq. 12.47) can be written for a material undergoing dehydration in the form:

$$\frac{dW_w}{Adt} = k_{gw}(H_i - H) \qquad (12.57)$$

$dW_w/dt$ is mass of water transferred, $H_i$ is the humidity at the interface where water is vaporized, and H is the humidity of the drying air. $H_i$ is the equilibrium humidity at the interface at the temperature at the interface. When the temperature at the interface drops, $H_i$ will also drop and the rate of mass transfer will slow down according to Eq. (12.57).

Rate of heating needed to vaporize water at the rate $dW_w/dt$ will be:

$$q = \frac{dW_w}{dt} h_{fg}$$

where $h_{fg}$ is the enthalpy of vaporization at the temperature of the interface. Heat transfer needed for vaporization will be;

$$q = k_{fg}(A)(H_i - H)(h_{fg}) \qquad (12.58)$$

At equilibrium, the heat for vaporizing the mass transferred (Eq. 12.58) will equal the rate of heat transfer (Eq. 12.59). The temperature and humidity at the interface will be $T_s$ and $H_s$, respectively:

$$q = hA(T - T_s) \qquad (12.59)$$

Equating Eqs. (12.58) and (12.59):

$$(h_{fg})k_{gw}(A)(H_s - H) = hA(T - T_s)$$
$$H = H_s - \frac{hT_s}{k_{gw}(h_{fg})} + \frac{h}{k_{gw}(h_{fg})}T \qquad (12.60)$$

Equation (12.60) is the equation of a wet bulb line, the relationship between equilibrium temperature and humidity when a thermometer bulb is wrapped with a wet sock and exposed to a flowing stream of air. The line will go through the points $T_s$ and $H_s$, the wet bulb temperature and the saturation humidity at the wet bulb temperature, respectively.

Food products during drying will follow the relationship between heat and mass transfer as expressed in Eqs. (12.58), (12.59), and (12.60).

The difference between the dry bulb, T, and the wet bulb temperature, $T_s$, is known as the wet bulb depression, an index of the drying capacity of a stream of air.

If the temperature of a product during dehydration is higher than the wet bulb temperature, then the rate of heat transfer (Eq. 12.59) exceeds evaporative cooling by mass transfer (Eq. 12.58), and mass transfer controls the drying rate. This phenomenon is often observed in the later stages of drying when the interface for vaporization is removed from the surface requiring vapor to flow through the pores of the dried material close to the surface, before mixing with the drying air at the surface.

The air temperature drops as it undergoes adiabatic humidification. Vaporization of water to increase the humidity requires energy which comes from a drop in the sensible heat of air. The heat balance is as follows.

Heat of vaporization, $q_v$, to increase humidity from H to $H_s$ is:

$$q_v = (H_s - H)(h_{fg}) \qquad (12.61)$$

The loss in sensible heat $q_s$ for air with a specific heat $C_p$ is:

$$q_s = C_p(T - T_s) \qquad (12.62)$$

Equating Eqs. (12.61) and (12.62):

$$C_p(T - T_s) = (H_s - H)(h_{fg})$$
$$H = H_s + \frac{C_p}{h_{fg}}T_s - \frac{C_p}{h_{fg}}T \qquad (12.63)$$

Equation (12.63) is the adiabatic humidification line, the change in humidity of air as it drops in temperature during adiabatic humidification. Equation (12.63) will be exactly equal to Eq. (12.60) if the ratio of the heat transfer coefficient to the mass transfer coefficient, $h/k_{gw}$, equals the specific heat of air, $C_p$. Indeed, this has been shown to be true for water vaporizing into air at one atmosphere and at moderate temperatures. Thus, in air drying, wet bulb lines

on the psychrometric chart coincide exactly with the adiabatic humidification lines.

**Example 12.10**   Room air at 80 °F (26.7 °C) and 50%RH is heated to 392 °F (200 °C) and introduced into a spray drier where it leaves at a temperature of 203 °F (95 °C).

Determine the humidity and relative humidity of the air leaving the drier. Assume adiabatic humidification in the drier.

**Solution**

Using a psychrometric chart, locate the point that represents T = 80 °F and 50% RH. The humidity at this point is 0.011. When air is heated, the temperature increases at constant humidity. Air leaving the heater will have a humidity of 0.011 and a temperature of 392 °F (200 °C). Starting from T = 392 °F (200 °C) and H = 0.011, draw a curve that approximates the closest wet bulb curve until a temperature of 203 °F (95 °C) is reached. The humidity at this point is 0.055, and the relative humidity is 10%.

## 12.5   The Stages of Drying

Drying usually occurs in a number of stages, characterized by different dehydration rates in each of the stages. Figure 12.5 shows the desorption isotherm and the rate of drying of apple slices in a cabinet drier with air flowing across the trays containing the slices at 0.3 m/s, 150 °F (65.6 °C), and 20% RH. The drying rate curve shows a constant drying rate (line AB) from a starting moisture content of 8.7 g $H_2O$/g dry matter to a moisture content of 6.00 kg $H_2O$/kg dry matter. At this stage in the drying cycle, vaporization is occurring at the product surface, and free water with $a_w$ of 1.00 is always available at the surface to vaporize. This stage of drying is the constant rate stage. The rate of drying is limited by the rate at which heat is transferred into the material from air. The product temperature is usually at the wet bulb temperature of the drying air.

From a moisture content of 6.00 g $H_2O$/g dry matter, drying rate decreased linearly with the

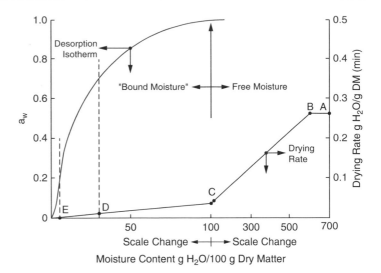

**Fig. 12.5** Desorption isotherm of raw apple slices and drying rates at different moisture contents

moisture content. Point B is the critical moisture content, and line BC represents the first falling rate period. The first falling rate period is characterized by a slight increase in the temperature of the product, although this temperature may not be very much higher than the wet bulb temperature. Free moisture is no longer available at the surface, and the rate of drying is controlled by moisture diffusion toward the surface. Most of the water in the material is still free water with $a_w$ of 1. However, diffusion toward the surface is necessary for vaporization to occur.

From a moisture content of 1.20 g $H_2O$/g dry matter, the slope of the drying rate against moisture content changes, and the rate of drying goes through a second falling rate period. The second falling rate period starts at point C, where the equilibrium relative humidity for the material begins to drop below 100%. The moisture content at point C is the bound water capacity of the material. During the second falling rate period, dehydration proceeds through the portion of the sorption isotherm where water in the material is held by multimolecular adsorption and capillary condensation. Heat of vaporization of water in this stage of the dehydration process is higher than the heat of vaporization of pure water because the heats of adsorption and vaporization must be provided. Vaporization at this stage

occurs at the interior rather than the surface, and water vapor has to diffuse to the surface before it mixes with the flowing stream of air.

At a moisture content of 0.30 g $H_2O$/g dry matter, the drying rate goes through another falling rate zone. This stage of drying corresponds to the region of the sorption isotherm where water is held in mono- or multimolecular layers. Dehydration should be terminated at any point along the line DE. Point E represents the equilibrium moisture content where dehydration stops because $a_w$ of the surface and air are equal.

## 12.6  Prediction of Drying Times from Drying Rate Data

### 12.6.1  Materials with One Falling Rate Stage Where the Rate of Drying Curve Goes Through the Origin

The drying rate curve for these materials will show a constant rate $R_c$ from the initial moisture content $X_o$ to the critical moisture content $X_c$. The drying rate then falls in a linear relationship with decreasing moisture content until it becomes 0 at $X = 0$. The drying times required to reach a moisture content X in either of the stages of drying are:

$$\text{Constant rate}: -\frac{dX}{dt} = R_c \qquad (12.64)$$

The total time, $t_c$, for the constant rate stage is:

$$t_c = \frac{X_0 - X_c}{R_c} \qquad (12.65)$$

At the falling rate period : $-\dfrac{dX}{dt} = \dfrac{R_c}{X_c}(X)$

$$(12.66)$$

$$\int_{t_c}^{t} dt = \frac{X_c}{R_c}\int_{X_c}^{X}\frac{dX}{X}$$

$$(12.67)$$

$$t - t_c = \frac{X_c}{R_c}\ln\frac{X_c}{X}$$

The total time from $X_0$ to $X$ in the falling rate stage can be calculated by substituting $t_c$ in Eq. (12.65) into Eq. (12.67):

$$t = \frac{X_0 - X_c}{R_c} + \frac{X_c}{R_c}\ln\frac{X_c}{X} \qquad (12.68)$$

Equation (12.68) shows that if a material exhibits only one falling rate stage of drying and the drying rate is 0 only at $X = 0$, the drying time required to reach a desired moisture content can be determined from the constant rate $R_c$ and the critical moisture content $X_c$. For these materials, $X_c$ is usually the moisture content when $a_w$ starts to drop below 1.0 in a desorption isotherm.

**Example 12.11** A material shows a constant drying rate of 0.15 kg $H_2O$/min (kg dry matter) and has an $a_w$ of 1.0 at moisture contents >1.10 kg $H_2O$/kg dry matter. How long will it take to dry this material from an initial moisture content of 75% (wet basis) to a final moisture content of 8% (wet basis)?

**Solution**
Converting the final and initial moisture contents from wet to a dry basis:

$$X_0 = \frac{0.75\,\text{kg water}}{0.25\,\text{kg dry matter}}$$
$$= 3.0\,\text{kg water/kg dry matter}$$
$$X = \frac{0.08\,\text{kg water}}{0.92\,\text{kg dry matter}}$$
$$= 0.0869\,\text{kg water/kg dry matter}$$
$$X_c = 1.10\,\text{kg water/kg dry matter}$$
$$R_c = 0.15\,\text{kg water/min (kg dry matter)}$$

Using Eq. (12.68):

$$t = \frac{3.0 - 1.10}{0.15} + \frac{1.10}{0.15}\ln\frac{1.10}{0.0869}$$
$$= 12.7 + 18.6 = 31.3\,\text{min}$$

## 12.6.2 Materials with More than One Falling Rate Stage

Most food solids would exhibit this drying behavior. The drying rate curve shown in Fig. 12.5 for apple slices is a typical example. The drying time in the constant rate follows Eq. (12.65). However, because the rate of drying against moisture content plot no longer goes to the origin from the point $(X_c, R_c)$, Eq. (12.66) cannot be used for the falling rate stage. If the rate versus moisture content line is extended to the abscissa, the moisture content where rate is 0 may be designated as the residual moisture content $X_r$ and for the first falling rate period:

$$\frac{d(X - X_{r1})}{dt} = \frac{R_c}{X_{c1} - X_{r1}}(X - X_{r1}) \qquad (12.69)$$

Integrating Eq. (12.69) and using Eq. (12.65) for drying time in the constant rate zone, drying to a moisture content $X$ in the first falling rate stage would take:

$$t = \frac{X_0 - X_{c1}}{R_c} + \frac{X_{c1} - X_{r1}}{R_c}$$
$$\times \ln\frac{X_{c1} - X_{r1}}{X - X_{r1}} \qquad (12.70)$$

where $X_{c1}$ and $X_{r1}$ represent the critical moisture content and the moisture content a the end of the first falling rate stage of drying. The time required to dry to moisture content X in the second falling rate stage would be:

$$
t = \frac{X_0 - X_{c1}}{R_c} + \frac{X_{c1} - X_{r1}}{R_c} \ln \left[ \frac{X_{c1} - X_{r1}}{X_{c2} - X_{r1}} \right]
$$
$$
+ \left[ \frac{X_{c1} - X_{r1}}{R_c} \right] \left[ \frac{X_{c2} - X_{r2}}{R_c} \right]
$$
$$
\ln \left[ \frac{X_{c2} - X_{r2}}{X - X_{r2}} \right]
$$

$$(12.71)$$

**Example 12.12** Figure 12.6 shows the drying curve for apple slices blanched in 10% sucrose solution and dried in a cabinet drier using air in parallel flow at a velocity of 3.65 m/s, at $T_{db} = 170$ °F (76.7 °C) and $T_{wb} = 100$ °F (37.8 °C) for the first 40 min and $T_{db} = 160$ °F (71.1 °C) and $T_{wb} = 110$ °F (43.3 °C) for the rest of the drying period. Calculate the drying time to reach a moisture content of 0.15 kg water/kg dry matter.

**Solution**
The drying rate curve was obtained by drawing tangents to the drying curve at the designated moisture contents and determining the slopes of the tangents. The drying time required to obtain a moisture content of 0.15 kg water/kg dry matter will be:

$$
t = \frac{5.3 - 2.5}{0.163} + \frac{2.5 - 0.35}{0.163} \ln \left[ \frac{2.5 + 0.35}{1.0 - 0.35} \right]
$$
$$
+ \left[ \frac{2.5 - 0.35}{0.163} \right] \left[ \frac{1 - 0.1}{0.163} \right] \ln \left[ \frac{1 - 0.1}{0.15 - 0.35} \right]
$$
$$
= 20.2 + 15.8 + 52.7 = 88.7 \text{min}
$$

The drying curve in Fig. 12.6 shows a drying time of 90 min at X = 0.15.

### 12.6.3  The Constant Drying Rate

The constant drying rate $R_c$ is heat transfer controlled and can be calculated using a heat balance. Let $\rho_s$ = the dry solid density, kg dry solids/m$^3$ of wet material. $\rho_s$ = wet material density x mass fraction dry solids in the wet material.

**Fig. 12.6** Drying curve and drying rate as a function of the moisture content of blanched apple slices, showing several breaks in the drying rate

If L is the depth of the innermost section of the material from the drying surface (if drying occurs from both sides, L will be half the total thickness of the solid), $M_s$ is the mass of dry solids, and A is the area of the top surface of the solid, the volume V of material will be:

$$V = \text{surface area (depth)} = A(L) = \frac{M_s}{\rho_s}$$

$$\frac{A}{M_s} = \frac{1}{L(\rho_s)}$$

(12.72)

Heat balance: latent heat of evaporation = heat transferred:

$$\frac{dX}{dt} M_s h_{fg} = h \, A(T_a - T_s)$$

(12.73)

where $h_{fg}$ = latent heat of vaporization at the surface temperature of the material, $T_s$, $T_a$ = the dry bulb temperature of the air, and h is the heat transfer coefficient. The surface temperature during the constant rate period is also the wet bulb temperature ($T_{wb}$) of the air. $T_s = T_{wb}$. $dX/dt = R_c$.

$$R_c = \frac{h(T_a - T_s)}{h_{fg}} \cdot \frac{A}{M_s}$$

(12.74)

Substituting Eq. (12.72) in Eq. (12.74):

$$R_c = \frac{h(T_a - T_s)}{h_{fg} L \rho_s}$$

(12.75)

Equation (12.75) can be used to calculate the constant rate of drying from the heat transfer coefficient and the wet and dry bulb temperatures of the drying air for a bed of particles with drying air flowing parallel to the surface.

Expressions similar to Eq. (12.75) may be derived using the same procedure as above for cubes with sides L evaporating water at all sides:

$$R_c = \frac{6h(T_a - T_s)}{h_{fg} L \rho_s}$$

(12.75a)

For a brick-shaped solid with sides "a" and "2a" and thickness L:

$$R_c = \frac{h(T_a - T_s)}{h_{fg} \, \rho_s} \left[ \frac{3}{a} + \frac{2}{L} \right]$$

(12.75b)

For a spherical solid: $R_c = 3 \, h(T_a - T_s)/(Rh_{fg})$.

The heat transfer coefficient can be calculated using the following correlation equations (Sherwood, Ind. Eng. Chem. 21:976, 1029):

If air flow is parallel to the surface:

$$h = 0.0128G^{0.8}$$

(12.76)

where h = the heat transfer coefficient in BTU/(h ·ft$^2$·°F) and G is the mass rate of flow of air, $lb_m$/(hft$^2$). In SI units, Eq. (12.76) is:

$$h = 14.305G^{0.8}$$

(12.77)

where h = W/(m$^2$· K) and G is in kg/(m$^2$· s). If flow is perpendicular to the surface:

$$h = 0.37G^{0.37}$$

(12.78)

where h is in BTU/(h · ft$^2$·°F) and G is in $lb_m$/(ft$^2$· h). In SI units, Eq. (12.78) is:

$$h = 413.5G^{0.37}$$

(12.79)

where h is in W/(m$^2$· K) and G is in kg/(m$^2$· s).

When air flows through the bed of solids, Ranz and Marshall's equation (Chem. Eng. Prog. 48 (3):141, 1956; Appendix Table A.12 for particles in a gas stream) may be used for determining the heat transfer coefficient.

**Example 12.13** Calculate the constant drying rate for blanched apple slices dried with air flowing parallel to the surface at 3.65 m/s. The initial moisture content was 85.4% (wet basis), and the slices were in a layer 0.5 in. (0.0127 m) thick. The wet blanched apples had a bulk density of approximately 35 lb/ft$^3$ (560 kg/m$^3$) at a moisture content of 87% (wet basis).

Dehydration proceeds from the top and bottom surfaces of the tray. Air is at 76.7 °C (170 °F) db and 37.8 °C (100 °F) wb:

$$\rho_s = \frac{560 \, kg}{m^3} \frac{0.13 kg \, DM}{kg}$$

$$= 72.8 \frac{kg \, DM}{m^3} \text{ or } 4.55 \frac{lb \, DM}{ft^3}$$

$$V = 3.65 \text{ m/s or } 12.0 \text{ ft/s}$$

**Solution**

Using the ideal gas equation, R = 8315 Nm/kgmole · K, T = 76.7 °C, P = 1 atm, = 101.3 k Pa, and M = 29 kg/kg mole:

$$\frac{\text{kg air}}{\text{m}^3} = \frac{P(M)}{R(T)} = \frac{(101,300)(29)}{8315(349.7)}$$

$$= 1.01\frac{\text{kg}}{\text{m}^3} \text{ or } 0.063\frac{\text{lb}}{\text{ft}^3}$$

$$G = \frac{\text{kg air}}{\text{m}^3} \times \text{velocity} = 1.01\frac{\text{kg}}{\text{m}^3}3.65\frac{\text{m}}{\text{s}}$$

$$= 3.687\text{kg}/(\text{m}^2 \cdot \text{s}) \text{ or } 2713 \text{ lb}/(\text{ft}^2 \cdot \text{h})$$

Using Eq. (12.77):

$$h = 14.305(3.687)^{0.8}$$

$$= 40.6\,\text{W}/(\text{m}^2 \cdot \text{K}) \text{ or } 7.15 \text{ BTU}/(\text{h} \cdot \text{ft}^2 \cdot {}^{\circ}\text{F})$$

$$T_a - T_s = 76.7 - 37.8 = 38.9\,{}^{\circ}C \text{ or } 70\,{}^{\circ}F$$

$$h_{fg} = \text{heat of vaporization at } 37.8\,{}^{\circ}\text{C}\,(100\,{}^{\circ}\text{F})$$

$$= 1037.1 \text{ BTU/lb or } 2.4123 \text{ MJ/kg}$$

Because drying occurs on top and bottom surfaces, L = 0.0127/2 = 0.00635.

Using Eq. (12.75):

$$R_c = \frac{40.6(38.9)}{2.4123 \times 10^6(0.00635)(72.8)}$$

$$= 0.00146\frac{\text{kg water}}{\text{s} \cdot \text{kg DM}}$$

$$= 5.098 \text{ kg water}/(\text{h} \cdot \text{kg DM})$$

## 12.7  Spray Drying

Spray drying is a process where a liquid droplet is rapidly dried as it comes in contact with a stream of hot air. Figure 12.7 is a schematic diagram of a spray drier where the atomized feed travels concurrent with the drying air. The small size of the liquid droplets allows very rapid drying and the residence time of the material inside the spray drier is in the order of seconds. The dried material is separated from air in a cyclone separator. The dried material is continuously withdrawn and cooled. Heat could damage the product if contact with the high temperature drying air is prolonged.

While the droplet is drying, the temperature remains at the wet bulb temperature of the drying air. For this reason, very high temperatures of the drying air can be tolerated in a drier with a minimum of damage to the heat-sensitive components. Furthermore, rate of degradative reactions in foods slows down at low moisture contents. Thus, the portion of the drying process where product temperatures go higher than the wet bulb temperature does not result in severe heat damage to the product.

**Fig. 12.7** Schematic diagram of a spray drier

A major requirement of successful spray drying is the reduction of the moisture content of a liquid droplet to a dryness level that would prevent the particle from sticking to a solid surface, as the particle impinges on that surface. The rate of drying of the particles must be such that from the time the particle leaves the atomizer to the time it impinges upon the walls of the spray drier, the particle is dry. The trajectory and velocity of the particles determine the available drying time. The rate of drying and the time required to dry are dependent upon the temperature of the drying air, the heat transfer coefficient, and the diameter of the droplet being dried.

A constant rate and a falling rate drying stage are also manifested in a spray-drying process. As the wet droplets leave the atomizer, their surfaces rapidly lose water. Solidified solute and suspended solids rapidly form a solid crust on the surface of each particle. The diameter of the particle usually decreases as drying proceeds. The formation of the solid crust constitutes the constant rate stage of drying. When the crust becomes sufficiently thick to offer considerable resistance to movement of water toward the surface, the drying rate drops, and the rate of drying is controlled by the rate of mass transfer. The temperature of the particle increases, and the liquid trapped in the interior of the particle vaporizes and generates pressure. Eventually, a portion of the crust breaks, and the vapor is released. Spray-dried particles consist of hollow spheres or fragments of spheres. This shape of the particles is responsible for the excellent rehydration properties of spray-dried powders.

### 12.7.1   Drying Times in Spray Drying

Drying rates in the constant rate period are generally heat transfer controlled, and the heat balance is given in Eq. (12.73). If $\rho_L$ is the density of the liquid being dried, r is the radius of the droplet, and $X_0$ is the initial moisture content (dry basis, kg water/kg dry matter):

$$M_s = \frac{4\pi r^3 (\rho_L)}{3(1 + X_0)}$$

and Eq. (12.73) becomes:

$$\frac{dX}{dt} \cdot \frac{4\pi r^3 (\rho_L)(h_{fg})}{3(1 + X_0)} = h(\pi r^2)(T_a - T_s)$$

The constant drying rate in Eq. (12.75) can be determined if the heat transfer coefficient h and the radius of the liquid droplets are known. $T_s$ during the constant drying stage would be the wet bulb temperature of the drying air. Integrating Eq. (12.80), the constant rate drying time $t_c$ is:

$$\frac{dX}{dt} = \frac{3(1 + X_0)(h)(T_a - T_s)}{4r\rho_L h_{fg}} \qquad (12.80)$$

$$t_c = \frac{4(X_0 - X_c)(r)(\rho_L)h_{fg}}{3(1 + X_0)(h)(T_a - T_s)} \qquad (12.81)$$

For water vaporizing from a very small spherical particle in slow-moving air, the following relationship has been derived for the limiting case of very small Reynolds number in Froessling's boundary later equations for a blunt-nosed solid of revolution:

$$\frac{hr}{k_f} = 1.0; \quad h = \frac{k_f}{r} \qquad (12.82)$$

$k_f$ in Eq. (12.82) is the thermal conductivity of the film enveloped around the particle. In spray drying, $k_f$ may be assumed to be the thermal conductivity of saturated air at the wet bulb temperature. Substituting Eq. (12.82) in Eq. (12.81):

$$t_c = \frac{4(X_0 - X_c)(r^2)(\rho_L)h_{fg}}{3k_f(1 + X_0)(T_a - T_s)} \qquad (12.83)$$

$t_c$ represents the critical time in spray drying that must be allowed in a particles' trajectory before it impinges on a solid surface in the drier. The drying time in the falling rate period derived by Ranz and Marshal is:

$$t_c = \frac{h_{fg}(\rho_L)(r_c^2)(X_c - X)}{3k_f(\Delta T)} \qquad (12.84)$$

where $r_c$ is the radius of the dried particle and $\Delta\overline{T}$ is the mean temperature between the drying air and the surface of the particle during the falling rate period. $\Delta\overline{T}$ may be considered as a log mean between the wet bulb depression and the difference between the exit air and product temperatures. In $\rho$ or liquids that contain a high concentration of suspended solids or crystallizable solutes, there is very little change in the droplet diameter during spray drying. Both r and $r_c$ in Eqs. (12.83) and (12.84), therefore, may be approximated to be the diameter of the liquid droplet leaving the atomizer.

For centrifugal atomizers, the diameter of the droplets as a function of peripheral speed of the atomizer is shown in Fig. 12.8. For pneumatic atomizers, a graph of drop diameter as a function of atomizing air pressure at different liquid flow rates is shown in Fig. 12.9.

**Example 12.14** Calculate the drying time for a liquid atomized in a centrifugal atomizer at a feed rate of 15 lb/min (6.8 kg/min) at a peripheral speed of 200 ft/s. Base the drying time for a particle size representing a diameter larger than that of 90% of the total droplets produced. Assume there is no change in droplet diameter with drying. The liquid originally has a density of 61 lb/ft$^3$ (993 kg/m$^3$) and a moisture content of 8% (wet basis) using air at 347 °F (175 °C) and a humidity of 0.001 H$_2$O/dry air. The critical moisture content is 2.00 g H$_2$O/g dry matter. The dried solids have a density of 0.3 g/cm$^3$. Exit air temperature is 220 °F (104.4 °C). Product exit temperature is 130 °F (54.4 °C).

**Fig. 12.8** Droplet size as a function of peripheral speed of a centrifugal atomizer (From Marshall, W. R. Jr., 1954. Chem. Eng. Prog. Monogr. Ser. 50(2):71. AIChE, New York. Used with permission)

**Fig. 12.9** Droplet size as a function of pressure in a pneumatic atomizer (From Marshall, W. R. Jr., 1954. Chem. Eng. Prog. Monogr. Ser. 50(20):79. AIChE, New York. Used with permission)

**Solution**
From Fig. 12.8, the correction factor for a 15 lb/ min feed rate is 0.9. At a peripheral speed of 200 ft/s, the diameter corresponding to 90% cumulative distribution is 310 microns. Using the correction factor, the diameter is 279 microns. Using Eqs. (12.48) and (12.49):

$$X_0 = \frac{89}{11} = 8.09 \, \text{kg H}_2\text{O/kg dry matter}$$

$$X_c = 2.00$$

$$X = \frac{0.08}{0.92} = 0.087 \, \text{kg H}_2\text{O/kg dry matter}$$

From a psychrometric chart, $T_s = T_{wb} = 109\,°F$ (43 °C). $h_{fg}$ at 109 °F is 1031.4 BTU/lb or 2.3999 MJ/kg. The thermal conductivity of the gas film enveloped around the particle can be calculated. It is the thermal conductivity of saturated air at 109 °F (43 °C). The humidity is 0.056 kg $H_2$O/kg dry matter. The thermal conductivity of dry air is 0.0318 W/m · K and that of water vapor is 0.0235 W/m · K:

$$k_f = \frac{.0318}{1.056} + \frac{(0.0235) \cdot (0.056)}{1.056} = 0.0314 \, \frac{\text{W}}{\text{m} \cdot \text{K}}$$

$$r = \frac{279}{2} \times 10^{-6}\text{m} = 193.5 \times 10^{-6}\text{m}$$

Equation (12.81) in SI units:

$$t_c = \frac{(4) \cdot (8.09 - 2) \cdot (1.35 \times 10^{-6}) \cdot (993) \cdot (2.399 \times 10^{-6})}{(3) \cdot (0.0314) \cdot (1 + 1.809) \cdot (175 - 43)}$$

$$t_c = 10 \, \text{s}$$

For Eq. (12.84), the $\Delta \overline{T}$ is:

$$\overline{\Delta T} = \frac{(175 - 43) \cdot (104.4 - 54.4)}{\ln\left(\dfrac{142}{50}\right)} = 84.5\,°F$$

$$t_f = \frac{(2.399 \times 10^5) \cdot [(0.3) \cdot (1000)] \cdot (193.5 \times 10^{-6})^2 \cdot (2.0 - 0.087)}{(3) \cdot (0.0314) \cdot (84.5)} = 6.5 \, \text{s}$$

## 12.8    Freeze Drying

Dehydration carried out at low absolute pressures will allow the vaporization of water from the solid phase. Figure 12.10 shows the vapor pressure of water over ice at various temperatures below the freezing point of water. To carry out freeze drying successfully, the absolute pressure in the drying chamber must be maintained at an absolute pressure of at least 620 Pa.

Figure 12.11 is a schematic diagram of a freeze drier. The absolute pressure inside the drying chamber is determined by the temperature at which the vapor trap is maintained. This pressure corresponds to the vapor pressure over ice at the vapor trap temperature. The vacuum pump is designed primarily to exhaust the vacuum chamber at the start of the operation and to remove noncondensing gases and whatever air leaked into the system. The volume of vaporized water at the low absolute pressures in freeze drying is very

**Fig. 12.10** Heat of sublimation ($\Delta H_s$) and vapor pressure of water above ice ($P_i$) (Based on data from Charm, S. E. 1971. Fundamentals of Food Engineering. 2nd ed. AVI Publishing Co. Westport, CT)

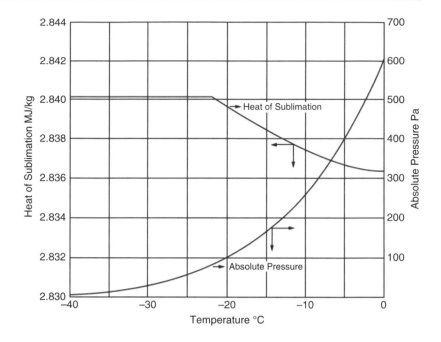

**Fig. 12.11** Schematic diagram of a freeze drier showing a vapor trap external to the vacuum chamber

large; therefore, removal of the vapor by the vacuum pump alone would require a very large pump. Condensing the vaporized water in the form of ice in the vapor trap is an efficient means of reducing the volume of gases to be removed from the system by the vacuum pump.

Heat must also be supplied to the material being dried to provide the energy of vaporization. This is accomplished by the use of hollow shelves through which a heated liquid is circulated. The temperature of the shelves can be regulated by regulating either the temperature or the amount of heat transfer medium supplied to the shelves. The material to be dried rests on the top of the heated shelves. Heat transfer occurs by conduction from the heated shelves, by convection from the air inside the drying chamber to the exposed surfaces, and by radiation.

## 12.8.1 Drying Times for Symmetrical Drying

Analysis of freeze drying is different from that in conventional drying in that drying proceeds from the exposed surfaces toward the interior. The outer layers are completely dry as the ice core recedes. Vaporization of water occurs at the surface of the ice core. Heat of sublimation is conducted to the surface of the ice core through the dried outer layer. Vaporized water diffuses through the pores of the dried outer layer before it leaves the solid and goes to the atmosphere in the drying chamber. Symmetrical drying occurs when the rate at which the ice core recedes is equal at both top and bottom of the material. To simplify the calculations, unidirectional heat transfer is assumed.

Let $W$ = kg water/m$^3$ of wet material. If $\rho$ is the density of the wet material and $X_0$ is the initial moisture content on a dry basis, kg water/kg dry matter, $W = (\rho) X_0/(1 + X_0)$.

If drying is symmetrical, the mass of water evaporated, $M_c$, expressed in terms of a dried layer $\Delta L$, is:

$$M_c = WA\,(\Delta L)\,(2)$$

Let $X'$ = fraction of water evaporated = $M_c$/ total water:

$$X' = \frac{W(A)(\Delta L)(2)}{W(A)(L)} = \frac{\Delta L(2)}{L}$$

$X'$ is also $\dfrac{X_0 - X}{X_0}$

where $L$ = the thickness of the solid and $X$ = moisture content, kg H$_2$O/kg dry matter:

$$\Delta L = \frac{L'X}{2}$$

The overall heat transfer coefficient calculated from the sum of the resistance to heat transfer of the dried layer and the heat transfer coefficient at the surface is:

$$\frac{1}{U} = \frac{1}{h} + \frac{\Delta L}{k} = \frac{1}{h} + \frac{LX'}{2k} \tag{12.85}$$

$$U = \frac{k}{k/h + LX'/2} \tag{12.86}$$

If drying is symmetrical, heat is transferred from both sides. Let $T_s$ = temperature of the shelf and $T_f$ is the temperature of the frozen core surface. The heat transferred to the ice core will be:

$$q = UA\Delta T$$
$$= \frac{k}{k/h + LX'/2}(2A)(T_a - T_f) \tag{12.87}$$

For simultaneous heat and mass transfer, heat transferred = heat of vaporization. $\Delta H_s$ is the heat of sublimation of ice at $T_f$:

$$\Delta H_s(W)(A)(L)\frac{d'X}{dt} = \frac{2A(T_a - T_f)(k)}{k/h + LX'/2} \tag{12.88}$$

Simplifying and integrating Eq. (12.88):

$$t = \frac{\Delta H_s(W)(L)}{2k(T_s - T_f)}\left[\frac{k'X}{h} + \frac{LX'^2}{4}\right] \tag{12.89}$$

$W$ in Eq. (12.89) can be expressed in terms of either the density of the wet material, $\rho$, or the density of the dried material $\rho_s$:

$$W = X_{0\rho_s} \quad \text{or} \quad W = (\rho)\frac{(X_0)}{1 + X_0}$$

$$t = \frac{\Delta H_s(X_0)(\rho_s)(L)}{2k(T_s - T_f)}\left[\frac{k'X}{h} + \frac{LX'^2}{4}\right] \tag{12.90}$$

or:

$$t = \frac{\Delta H_s(\rho)(X_0)(L)}{2k(1 + X_0)(T_s - T_f)}\left[\frac{k'X}{h} + \frac{LX'^2}{4}\right]$$

$$(12.91)$$

Either Eqs. (12.90) or (12.91) can be used to calculate the time of drying depending on whether $\rho$ or $\rho_s$ is known.

**Example 12.15** The density of a sample of beef is 60 lb/ft$^3$ (965 kg/m$^3$). How long will it take to dry a 1 in. (2.54 cm)-thick strip of this sample from an initial moisture of 75% to a final moisture content of 4% (wb). Freeze drying is carried out at an absolute pressure of 500 microns of mercury. Assume that the shelf temperature and the air in the drying chamber are equal at 80 °F (26.7 °C). Assume symmetrical drying. The thermal conductivity of the dried meat is 0.0692 W/m · K. Estimate the heat transfer coefficient by assuming that a 3 mm thick of vapor (k of water = 0.0235 W/m · K) envelopes the surfaces where drying occurs and that the heat transfer coefficient is of equivalent resistance to the resistance of this vapor film:

$$h = \frac{k}{x} = \frac{0.0235}{0.003} = 7.833 \text{ W/m}^2\text{K}$$

The absolute pressure in the chamber is:

$$P = 500 \times 10^{-6}\text{m Hg}\frac{133.3 \times 10^3\text{Pa}}{\text{m Hg}} = 66.65\,\text{Pa}$$

From Fig. 12.10, the temperature of ice in equilibrium with a pressure of 66.65 Pa

is −24.5 °C. The heat of sublimation $H_s = 2.8403$ MJ/kg. The final moisture content on a dry basis is $X = 0.04/0.96 = 0.0417$ kg water/kg dry matter. The initial moisture content on a dry basis is $X_0 = -0.75/0.25 = 3.00$. The fraction of water remaining at the completion of the drying period is $X' = X_0 - X/X_0 = 3–0.0417/3.0 = 0.986$.

Substituting in Eq. (12.91):

$$t = \left[\frac{2.8403 \times 10^6(965)(3.00)(0.0254)}{2(0.0692)(1 + 3)(26.7 + 24.5)}\right]$$

$$\left[\frac{0.0692(0.986)}{7.833} + \frac{0.0254(0.986)^2}{4}\right]$$

$$= 109,673\ \text{s} = 30.46\ \text{h}$$

## 12.9  Vacuum Belt Drier

*Principle of Operation* A vacuum belt drier is used to dry heat-sensitive pasty materials and concentrated solutions. The system (Fig. 12.12) consists of a long cylinder with a continuous conveyor belt traversing the diameter the whole length of the cylinder. Product is deposited on the belt at one end of the unit, and water is evaporated as the belt moves forward until the dried product falls off the belt into a receiver tank at the opposite end.

*Energy Transfer* Energy to vaporize water is applied to the product on the belt by conduction

**a – bell; b – top heater plate; c – swivel liquid feed port; d – dried product receiver**

**Fig. 12.12** Picture of a pilot plant size vacuum belt drier from the bottom and radiation from the top. A heated plate under the belt heats the product by conduction. Above the conveyor is another heated plate that transfers energy to the paste by radiation. The temperatures of the top and bottom plates are independently controlled and are set to the maximum temperature that would permit complete drying of the product when the belt makes a turn into the opposite direction. Temperatures must be optimized to avoid heat-induced degradation reactions.

*Vacuum* Initial evacuation of air from the unit must be done rapidly, and a low absolute pressure must be maintained so that water vapor removal from the product is heat transfer rather than mass transfer controlled. Water vapor and noncondensible gases entering the cylinder must be evacuated using a liquid ring seal mechanical vacuum pump or a steam jet ejector. Typically, an absolute pressure of not more than 50 mm Hg is maintained inside the cylinder. Inadequate vacuum pump capacity will be manifested by condensation of water vapor on the cylinder walls. To facilitate evacuation of water vapor, dry air from the outside is continuously bled into the unit at a slow rate.

*Continuous Feeding of Wet Product* The paste to be dried is pumped into the unit and is discharged on the surface of the conveyor belt through a nozzle at the tip of a wand that slowly oscillates from one side of the conveyor belt to the other. Typically, the rate of addition of the paste and the speed of side-to-side movement of the wand are such that a thin uniform layer of paste is deposited on the belt. As the paste on the belt receives energy from the heated plates, vapor is generated because of the elevated temperature and low absolute pressure. Product on the belt will foam. If the foam forms a thick layer, a porous layer of dried solid forms at the top and bottom hindering heat transfer into the wet middle layer. Ideally, the thickness of the foam should be such that drying will continue throughout the whole thickness all the way to the end of the belt. Spattering of wet material on the belt must also be avoided because the spatter will be deposited on the surface of the plate above the belt and interfere with radiant heat transfer. The combination, vacuum, upper and lower heated plate temperature, and thickness of the deposited layer of wet product on the belt must be carefully regulated to dry successfully in this system.

*Operating the Drier* Drying time can be controlled by slowing down the speed of the belt. However, to operate at maximum drying capacity, heated plate temperatures must be elevated, and too long a residence time of the wet product in the heated zone could result in scorching of the product. Conditions must be optimized for each product dried. The paste density, foaming tendency, thermophysical properties, and radiant energy absorption all play a role in the rate of dehydration.

## Problems

12.1. Pork has $a_w$ of 1.00 at a moisture content of 50% (wet basis) and higher. If pork is infused with sucrose and NaCl and dehydrated such that at the end of the dehydration process, the moisture content is 60% (wet basis) and the concentration of sugar and NaCl are 10% and 3%, respectively, calculate the $a_w$ of the cured product.

12.2. What concentration of NaCl in water would give the same water activity as a 20% solution of sucrose?

12.3. The following data were obtained on the dehydration of a food product: initial moisture content = 89.7% (wet basis).

| Drying time (min) | Net weight (kg) |
|---|---|
| 0 | 24.0 |
| 10 | 17.4 |
| 20 | 12.9 |
| 30 | 9.7 |
| 40 | 7.8 |
| 50 | 6.2 |
| 60 | 5.2 |
| 70 | 4.5 |
| 80 | 3.9 |
| 90 | 3.5 |

Draw the drying curve for this material, and construct a curve for the drying rate as a function of the moisture content.

(a)  What is the critical moisture content for each of the falling rate zones?

(b)  What is the constant drying rate?

(c)  Determine the residual moisture content for each of the falling rate stages.

(d)  The dehydration was conducted at an air flow rate of 50 m/s at a dry bulb temperature of 82 °C and a wet bulb temperature of 43 °C. The wet material has a density of 947 kg/m³ and was dried in a layer 2.5 cm thick. If the same conditions were used but the initial moisture content was 91% (wet basis) and a thicker layer of material (3.5 cm) was used on the drying trays, how long will it take to dry this material to a final moisture content of 12% (wb)?

12.4.  A continuous countercurrent drier is to be designed to dry 500 kg/h of food product from 60% (wet basis) moisture to 10% (wet basis) moisture. The equilibrium moisture content for the material is 5% (wet basis), and the critical moisture content is 30% (wet basis). The drying curve of the material in preliminary drying studies showed only one falling rate zone. Air at 66 °C dry bulb and 30 °C wet bulb will be used for drying. The exit air relative humidity is 40%. Assume adiabatic humidification of the air. The drying air is drawn from room temperature at 18 °C and 50% RH. The wet material has a density of 920 kg/m³. The drying tunnel should use trucks that hold a stack of 14 trays, each 122 cm wide, 76 cm deep along the length of the tunnel, and 5 cm thick. The distance between trays on the stack is 10 cm. The drying tunnel has a cross-sectional area of 2.93 m². The material in the trays will be loaded at a depth of 12.7 mm. Calculate:

(a)  The number of trays of product through the tunnel/h.

(b)  The rate of travel by the trucks through the tunnel. Assume distance between trucks is 30 cm.

(c)  The constant drying rate and the total time for drying.

(d)  The length of the tunnel.

(e)  If air recycling is used, the fraction of the inlet air to the drier that must come from recycled air.

(f)  The capacity of the heater required for the operation with recycling.

12.5.  A laboratory drier is operated with a wet bulb temperature of 115 °F and a dry bulb temperature of 160 °F. The air leaving the drier is at 145 °F dry bulb. Assume adiabatic operation. Part of the discharge air is recycled. Ambient air at 70 °F and 60% RH is heated and mixed with the recycled hot air. Calculate the proportion of fresh air and recycled hot air that must be mixed to achieve the desired inlet dry and wet bulb temperatures.

12.6.  If it takes 8 h to dry a material in a freeze drier from 80% $H_2O$ to 10% $H_2O$ (wet basis) at an absolute pressure of 100 $\Phi$m and a temperature of 110 °F (43.3 °C), how long will it take to dry this material from 80% to 40% water if the dehydration is carried out at 500 $\Phi$m and 80 °F (26.7 °C)? The material is 25 mm thick and has a density of 950 kg/m³, and the thermal conductivity of the dried material is 0.35 W/m · K. Thermal conductivity and heat transfer coefficients are independent of plate temperature and vacuum.

12.7.  Calculate the constant rate of drying in a countercurrent continuous belt dehydrator that processes 200 lb/h (90.8 kg/h) of wet material containing 80% water to 30% water. Air at 80° EF (26.7 °C) and 80% RH is heated to 180 °F (82.2 °C) in an electric heater, enters the drier, and leaves at 10% RH. The critical moisture content of the material is 28%. The drier is 4 ft (1.21 m) wide and the belt loaded to a depth of 2 in. (5.08 cm) of material, and the clearance from the top of the drier to the top of the material on the belt is 10 in. (25.4 cm). The density of the dry solids in the material is 12 lb/ft³ (193 kg/m³).

12.8.  In a spray-drying experiment, a sample containing 2.15% solids and 97.8% water was fed at the rate of 6.9 lb per hour (3.126 kg/h), and this sample was dried at

392 °F (200 °C) inlet air temperature. The exit air temperature was 200 °F (93.3 °C). The dried product was 94.5% solids, and the outside air was at 79 °F (26.1 °C) and 20% RH.

Calculate:

(a)  The weight water evaporated per hour.
(b)  The % RH of the exit air.
(c)  The mass flow rate of air through the drier in weight dry air/h.
(d)  In this same drier, if the inlet air temperature is changed to 440 °F (226.7 °C) and the % RH of the exit air was kept the same as in (b), weight of a sample containing 5% solids and 98% water can be dried to 2% water in 1 h. (Air flow rate is the same as before.) At what temperature would air exit the drier under the specified conditions of operation. Assume adiabatic drying.

12.9.  A dehydrator when operated in the winter where the outside air was 10 °F($-12.2$ °C) and 100% RH (H = 0.001) can dry 100 lb (45.5 kg) of fruit per hour from 90% water to 10% water. The inlet temperature of the air to the drier is 150 °F (65.6 °C) and leaves at 100 °F (37.8 °C). In the summer when the outside air is at 90 °F (32.2 °C) and 80% RH, determine the moisture content of the product leaving the drier if the operator maintains the same rate of 100 lb (45.4 kg) of wet fruit/h and the exit air from the drier has the same % RH as it was in the winter.

12.10.  The desorption isotherm of water in carrots at 70° EC is reported to fit Iglesias and Chirife's equation (Eq. 12.36) with the constants $B_1$ = 3.2841 and $B_2$ = 1.3923.

(a)  Determine the moisture contents where a shift in drying rate may be expected in the dehydration of carrots.
(b)  The following data represents the equilibrium water activity ($a_w$) for carrots at various moisture contents in kg water/kg dry matter (X): ($a_w$,

X); (0.02, 0.0045), (0.04, 0.009), (0.06, 0.0125), (0.08, 0.016), (0.10, 0.019), (0.12, 0.0225), (0.14, 0.025), (0.016, 0.028), (0.18, 0.031), and (0.20, 0.034). Fit this data to the BET isotherm, and determine the moisture content for a unimolecular layer, $X_m$.

(c)  Fit the data to the GAB equation, and determine the constants.

12.11.  The diffusivity of water in scalded potatoes at 69° EC and 80% moisture (wet basis) has been determined to be $0.22 \times 10^{-5}$ m$^2$/h. If 1 cm potato cubes are dried using air at 1.5 m/s velocity and 1% relative humidity, calculate the dry bulb temperature of the air that can be used such that the diffusion rate from the interior to the surface will be equal to the surface dehydration rate. Assume the air flows parallel to the cubes and that dehydration proceeds from all faces of each cube. The density of the potato cube is 1002 kg/m$^3$ at 80% moisture.

12.12.  Puffing can be induced during dehydration of diced carrots if the dehydration rate at the constant rate period is of the order 1 kg water/(min kg DM). In a fluidized bed drier where the air contacts individual particles at a velocity of 12 m/s, calculate the minimum dry bulb temperature of the drying air that would induce this rate of drying at the constant rate period. Assume drying air has a humidity of 0.001 kg water/kg dry air and surface temperature under these conditions is 5° EC higher than the wet bulb temperature.

Calculate the mass transfer rate under these conditions. Is dehydration rate heat or mass transfer controlled?

## Suggested Reading

Barbosa-Canovas, G. V., & Vega-Mercado, H. (1996). *Dehydration of foods*. New York: Chapman and Hall.

Charm, S. E. (1971). *Fundamentals of food engineering* (2nd ed.). Westport: AVI Publishing Co.

Foust, A. S., Wenzel, L. A., Clump, C. W., Maus, L., & Andersen, L. B. (1960). *Principles of unit operations.* New York: Wiley.

Goldblith, S. A., Rey, L., & Rothmayr, W. W. (1975). *Freeze drying and advanced food technology.* New York: Academic.

Green, D. W., & Mahoney, J. O. (Eds.). (1997). *Perry's chemical engineers handbook* (7th ed.). New York: McGraw-Hill Book Co.

Greensmith, M. (1998). *Practical dehydration.* Boca Raton: CRC Press.

Hartman, T. M. (1989). *Water and food quality.* New York: Elsevier.

Heldman, D. R. (1975). *Food process engineering.* Westport: AVI Publishing Co.

Hildebrand, J., & Scott, R. L. (1962). *Regular solutions.* Englewood Cliffs: Prentice-Hall.

Iglesias, H. A., & Chirife, J. (1982). *Handbook of food isotherms.* New York: Academic.

Leniger, H. A., & Beerloo, W. A. (1975). *Food process engineering.* Boston: D. Riedel Publishing Co.

Marshall, W. R., Jr. (1954). Atomization and spray drying. *American Institute of Chemical Engineers. Series: Chemical Engineering Progress Monograph Series, 50*(2), 50–56.

McCabe, W. L., & Smith, J. C. (1967). *Unit operations of chemical engineering* (2nd ed.). New York: McGraw-Hill Book Co.

McCabe, W. L., Smith, J. C., & Harriott, P. (1985). *Unit operations in chemical engineering* (4th ed.). New York: McGraw Hill.

Mujumdar, A. (Ed.). (2004). *Dehydration of products of biological origin.* Enfield: Science Publishers.

Norrish, R. S. (1966). An equation for the activity coefficient and equilibrium relative humidity of water in confectionery syrups. *Journal of Food Technology, 1,* 25–39.

Perry, R. H., Chilton, C. H., & Kirkpatrick, S. D. (1963). *Chemical engineers handbook* (4th ed.). New York: McGraw-Hill Book Co.

Rockland, L. B., & Stewart, G. F. (Eds.). (1981). *Water activity: Influences on food quality.* New York: Academic Press.

Ross, K. D. (1975). Estimation of $a_w$ in intermediate moisture foods. *Food Technology, 29*(3), 26–34.

Sandal, O. C., King, C. J., & Wilke, C. R. (1967). The relation-ship between transport properties and rates of freeze drying of poultry meat. *AICHE Journal, 13*(3), 428–438.

Watson, E. L., & Harper, J. C. (1989). *Elements of food engineering* (2nd ed.). New York: Van Nostrand Reinhold.

# Physical Separation Processes

Food technology has evolved from the practice of preserving products in very much the same form as they occur in nature to one where desirable components are separated and converted to other forms. Separation processes have been in use in the food industry for years, but sophistication in its use is a fairly recent occurrence. C urrent technology makes it possible to remove haze from wine and fruit juices or nectars, separate the proteins of cheese whey into fractions having different functional properties, separate foreign matter from whole or milled grains, and concentrate fruit juices without having to employ heat. Efficient separation processes have been instrumental in making economically viable the recovery of useful components from food processing wastes.

## 13.1 Filtration

Filtration is the process of passing a fluid containing suspended particles through a porous medium. The medium traps the suspended solids producing a clarified filtrate. Filtration is employed when the valuable component of the mixture is the filtrate. Examples are clarification of fruit juices and vegetable oil. If the suspended material is the valuable component (e.g., recovery of precipitated proteins from an extracting solution), and rapid removal of the suspending liquid cannot be carried out without the addition of a filter aid, other separation techniques must be used.

Surface filtration is a process where the filtrate passes across the thickness of a porous sheet, while the suspended solids are retained on the surface of the sheet. A sheet with large pores has low resistance to flow; therefore filtrate flow is rapid; however, small particles may pass through resulting in a cloudy filtrate. Surface filtration allows no cake accumulation. Flow stops when solids cover the pores. If the solids do not adhere to the filter surface, the filter may be regenerated by backwashing the surface. Filtration sterilization of beer using microporous filters is a form of surface filtration.

Depth filtration is a process where the filter medium is thick, and solids penetrate the depth of the filter. Eventually, solids block the pores and stop filtrate flow, or solids may break through the filter and contaminate the filtrate. Once filtrate flow stops or slows down considerably, the filter must be replaced. In depth filtration, particle retention may occur by electrostatic attraction in addition to the sieving effect. Thus, particles smaller than the pore size may be retained. Depth filters capable of electrostatic solids retention will be ideal for rapid filtration. Cartridge, fiber, and sand filters are forms of depth filtration. Depth filtration is not very effective when suspended solids concentration is very high.

Filter aid filtration involves the use of agents that form a porous cake with the suspended

R. T. Toledo et al., *Fundamentals of Food Process Engineering*,
https://doi.org/10.1007/978-3-319-90098-8_13

**Fig. 13.1** Diagram of filter system for filter aid filtration

solids. Filter aids used commercially are diatomaceous earth and perlite. Diatomaceous earth consists of skeletal remains of diatoms and is very porous. Perlite is milled and classified perlite rock, an expanded crystalline silicate. In filter aid filtration, the filter medium is a thin layer of cloth or wire screen that has little capacity for retention of the suspended solids and serves only to retain the filter aid. A layer of filter aid ($0.5-1$ kg/m$^2$ of filter surface) is precoated over the filter medium before the start of filtration. Filter aid, referred to as "body feed," is continuously added to the suspension during filtration. As filtration proceeds, the filter aid and suspended solids are deposited as a filter cake, which increases in thickness with increasing filtrate volume. Body feed filter aid concentration is usually in the range of one to two times the suspended solids concentration. A body feed concentration must be used that will produce a cake with adequate porosity for filtrate flow. Inadequate body feed concentration of filter aid will result in rapid decrease in filtrate flow and consequently shorten filtration cycles. Figure 13.1 is a diagram of a filtration system for filter aid filtration. The system is designed to make it easy to change from precoat operation to body feed filtration operation with the switch of a three-way valve. Some filters such as the vertical leaf filter can be backflushed to remove the filter cake. Some, like the plate and frame filter, are designed to be easily disassembled to remove the filter cake. The body feed may be a slurry of the filter aid that is

metered into the filter feed, or the dry filter aid may be added using a proportional solids feeder into the filter feed tank. In continuous vacuum filters, precoating is done only once at the start of the operation. Body feed is added directly to the vacuum filter pan. Figure 13.2 is a diagram of the most common filters used for filter aid filtration. Figure 13.2a is a vertical leaf filter, Fig. 13.2b is a plate and frame filter, and Fig. 13.2c is a continuous vacuum rotary drum filter.

### 13.1.1  Filtrate Flow Through Filter Cake

In filter aid filtration, filtrate flow through the pores in the cake is dependent on the pressure differential across the cake and the resistance to flow. The total resistance increases with increasing cake thickness; thus filtrate flow decreases with time of filtration. The resistance to filtrate flow across the filter cake is expressed as the specific cake resistance.

Figure 13.3 shows a section of a filter showing the filter medium, precoat, and filter aid. The total pressure drop across the filter is the sum of the pressure drop across the filter medium (the filter cloth and precoat), $\Delta P_m$, and that across the cake, $\Delta P_c$.

Let $v$ = velocity of filtrate flow, $\mu$ = filtrate viscosity, $A$ = filter area, and $m$ = mass of filter cake. The medium resistance, $R_m$, is:

**Fig. 13.2** Diagram of common filters for filter aid filtration. (**a**) Vertical leaf filter; (**b**) plate and frame filter; (**c**) rotary vacuum drum filter

**Fig. 13.3** Diagram of a filter section showing filter cloth, precoat, and filter cake

$$R_m = \frac{\Delta P_m}{\mu v} \quad (13.1)$$

The specific cake resistance, $\alpha$, is:

$$\alpha = \frac{\Delta P_c}{\mu v (m/A)} \quad (13.2)$$

The units of $R_m$ and $\alpha$ are $m^{-1}$ and m/kg, respectively, in SI.

Let $\Delta P$ = pressure differential across the filter:

$$\Delta P = \Delta P_m + \Delta P_c \quad (13.3)$$

Substituting Eqs. (13.1) and (13.2) in Eq. (13.3):

$$\Delta P = \alpha\mu v\frac{m}{A} + R_m\mu v \qquad (13.4)$$

Let V = volume of filtrate and c = concentration of cake solids in the suspension to be filtered. m = Vc. The filtrate velocity v = (1/A) dV/dt.

Substituting for m and v in Eq. (13.4):

$$\Delta P = \frac{\mu(dV/dt)}{A}\left[\frac{\alpha Vc}{A} + R_m\right] \qquad (13.5)$$

$$\frac{dt}{dV} = \frac{\mu}{A \cdot \Delta P}\left[\frac{\alpha Vc}{A} + R_m\right] \qquad (13.6)$$

Equation (13.6) is the Sperry equation, the most widely used model for filtrate flow through filter cakes. Equation (13.6) can be used to determine the specific cake resistance from filtration data. Filtration time, t, is plotted against filtrate volume, V, tangents to the curve are drawn at several values of V, and the slopes of the tangents, dt/dV, are determined. A plot of dt/dV versus V will have a slope equal to $\alpha c\mu/(A^2 \cdot \Delta P)$, and an intercept on the ordinate at V = 0 is equal to $\mu R_m/(A \cdot \Delta P)$. An easier method for determining $\alpha$ and $R_m$ will be shown in the next section.

## 13.1.2 Constant Pressure Filtration

When a centrifugal pump is used as the filter feed pump, the pressure differential across the filter, $\Delta P$, is constant, and Eq. (13.6) can be integrated to give:

$$t = \frac{\mu}{\Delta P}\left[\frac{\alpha c V^2}{2 A^2} + R_m\frac{V}{A}\right] \qquad (13.7)$$

Dividing Eq. (13.7) through by V:

$$\frac{t}{V} = \frac{\mu\,\alpha c}{2A^2\Delta P}V + \frac{\mu R_m}{A\Delta P} \qquad (13.8)$$

Equation (13.8) shows that a plot of t/V against V will be linear, and the values of $\alpha$ and $R_m$ can be determined from the slope and intercept. A common problem with the use of either Eqs. (13.6) or (13.8) is that negative values for $R_m$ may be obtained. This may occur when $R_m$ is much smaller than $\alpha$; if finely suspended material

is present that rapidly reduces medium porosity even with very small actual amount of cake solids deposited; or when $\alpha$ increases with time of filtration such that the least squares method of curve fitting weighs heavily the data during the later stages of filtration relative to those at the early stages. To avoid having negative values for $R_m$, Eq. (13.7) may be used in the analysis. If filtration is carried out using only filtrate containing no suspended solids, c = 0 and Eq. 13.7 becomes:

$$t = \frac{\mu R_m}{\Delta PA}V \qquad (13.9)$$

Because $R_m = \Delta P_m/(\mu \cdot V)$, and since during filtration with only the filter cloth and precoat, the pressure differential is $\Delta P_m$, and because A $\cdot$ V = the filtrate volumetric rate of flow, q, the coefficient of V in Eq. (13.9) is 1/q. Thus, Eq. (13.7) can be expressed as:

$$t = \frac{\mu\alpha c}{2\Delta PA^2}V^2 + \left(\frac{1}{q}\right)V \qquad (13.10)$$

q is evaluated separately as the volumetric rate of filtrate flow on the precoated filter at the pressure differential used in filtration with body feed. Because $\Delta P$ across the precoated filter is primarily due to the resistance of the deposited filter aid, q is primarily a function of the type of filter aid, the pressure applied, and the thickness of the cake. For the same filter aid and filtrate, q is proportional to the thickness and the applied pressure; therefore the dependence can be quickly established. If q is known, $\alpha$ can be easily determined from the slope determined by a regression of the function (t − V/q) against $V^2$ or from the intercept of a log-log plot of (t − V/q) against V.

The use of filtration model equations permits determination of the filtration constants $R_m$ and $\alpha$ on a small filter which can then be used to scale up to larger filtrations. An example of a laboratory filtration module is shown in Fig. 13.4. This is a batch filter with a tank volume of 7.5 L and a filter area of 20 cm². An opening at the bottom allows draining of the precoat suspension at the beginning of a filtration cycle and charging of the tank after precoating. To precoat, a suspension of filter aid is made at such a concentration that filtration

**Fig. 13.4** Diagram of a laboratory test filtration cell

**Table 13.1** Data for filtration clarification of apple juice

| Time t(s) | Filtrate volume V(m$^3$) $\times$ 10$^5$ | t/V $\times$ 10$^{-6}$ |
|---|---|---|
| 60 | 3.40 | 1.7647 |
| 120 | 5.80 | 2.0689 |
| 180 | 7.80 | 2.3077 |
| 240 | 9.60 | 2.5000 |
| 300 | 11.4 | 2.6316 |
| 360 | 13.1 | 2.7586 |
| 420 | 14.7 | 2.8669 |
| 480 | 16.1 | 2.9721 |
| 540 | 17.6 | 3.0769 |
| 600 | 19.0 | 3.1496 |
| 660 | 20.6 | 3.2117 |
| 720 | 22.0 | 3.2727 |
| 780 | 23.3 | 3.3476 |

of 1000 mL will give the precoating level of 1.0 kg/m$^2$. For example, 2.0 g dry filter aid in 1000 mL of water will be equivalent to a precoat of 1.0 kg/m$^2$ on a filter area of 20 cm$^2$ if 1000 mL of filtrate is allowed to pass through the filter from this suspension. Depending upon the fineness of the filter aid used, filter paper or cloth may be used as the filter medium.

**Example 13.1** Data in Table 13.1 were collected in filtration clarification of apple juice. The apple juice was squeezed from macerated apples, treated with pectinase, and after settling and siphoning of the clarified layer, the cloudy juice that remained in the tank was filter clarified using perlite filter aid. The cloudy juice contained 1.19 g solids/100 mL. Perlite filter aid was used which had a pure water permeability of 0.40 mL $\cdot$ s$^{-1}$ for 1 cm$^2$ filter area, 1 atm pressure drop, and 1 cm thick cake. Whatman No. 541 filter paper was used as the filter medium. The precoat was 0.1 g/cm$^2$ (1 kg/m$^2$). The filter shown in Fig. 13.4, which had a filtration area of 20 cm$^2$, was used. Flow of clear apple juice through the precoated filter was 0.6 mL/s at a pressure differential of

25 lb$_f$/in.$^2$ (172.369 kPa). The filtrate had a viscosity of 1.6 cP.

Calculate $\alpha$ and R$_m$. Calculate the average filtration rate in m$^3$/(min $\cdot$ m$^2$ filter area) if the filtration time per cycle is 30 min and the pressure differential across the filter is 30 lb$_f$/in.$^2$ (206.84 kPa). Assume the same solids and body feed concentration.

The cumulative filtrate volume with filtration time at a filter aid concentration of 2.5 g/100 mL of suspension is shown in Table 13.1.

**Solution**

Using Eq. (13.8), a regression of (t/V) against V was done using the transformed data in Table 13.1.

The regression equation was:

$$(t/V) = 7.61 \times 10^9 V + 1,684,000$$

The coefficient of V is $\alpha \cdot \mu \cdot c/(2\Delta P\ A^2)$. c = concentration of filter cake solids in the feed, which is the sum of the original suspended solids, 1.19 g/100 mL or 11.9 g/L, and the body feed filter aid concentration which is 2.5 g/100 mL or 25 g/L:

$$\alpha = \frac{7.61 \times 10^9 (2)(20 \times 10^{-4})^2 (172,369)}{1.6(0.001)925 + 11.9)}$$

$$= 1.78 \times 10^{11} m/kg$$

The intercept is R$_m\mu$/(A$\Delta$P):

$$R_m = \frac{1.684 \times 10^6 (20 \times 10^{-4})(172,369)}{0.001(1.6)}$$

$$= 3.63 \times 10^{11} m^{-1}$$

Similar values for $\alpha$ and $R_m$ were obtained using Eq. (13.10) by performing a regression on $(t - V/q)$ against $V^2$. q is $0.6 \times 10^{-6}$ m$^3$/s. A slightly better fit was obtained using Eq. (13.10) with $R^2 = 0.994$ compared with $R^2 = 0.964$ using Eq. (13.8). The regression equation using Eq. (13.10) was:

$$(t - V/q) = 7.29 \times 10^9 V^2$$

Values for $R_m$ and $\alpha$ were $3.59 \times 10^{11}$ and $1.7 \times 10^{11}$, respectively, in SI units. These values would be more accurate than those obtained using Eq. 13.8 because of the better $R^2$ value for the regression. However, the values are very close and would be considered the same for most practical purposes.

The average filtrate flow is calculated using Eq. (13.7). $t = 30$ min $= 1800$ s:

$$1800 = \frac{1.6(0.001)}{206,840}$$

$$\left( \frac{1.78 \times 10^{11} \times (36.9)}{2} \left[\frac{V}{A}\right]^2 + 3.63 \times 10^{11} \left[\frac{V}{A}\right] \right)$$

$$32.841 \times 10^{11} (V/A)^2 + 3.63 \times 10^{11} (V/A)$$
$$= 2.32698375 \times 10^{11}$$

Dividing through by $32.841 \times 10^{11}$ and solving for the positive root of the quadratic equation:

$$(V/A)^2 + 0.1105(V/A) - 0.070856 = 0$$

$$\frac{V}{A} = \frac{-0.1105 \pm \left([0.1105]^2 + 4(1)(0.070856)\right)^{0.5}}{2}$$

$$= 0.217 m^3/m^2 \text{filter area})$$

This is the volume of filtrate which passed through a unit area of the filter for a filtration cycle of 30 min. The average rate of filtration is:

$$(V/A)_{a\,vg}/t = 0.217/30$$

$$= 0.00722 m^3/(min \cdot m^2 \text{filter area})$$

The average filtration rate can be used to size a filter for a desired production rate if the same filtration cycle time is used.

Two approaches were used to evaluate $R_m$ and $\alpha$ in Example 13.1. In both approaches, the traditional graphical method for determining filtration rate by drawing tangents to the filtration curve was eliminated. In this example, the resistance of the filter medium was the same magnitude as the specific cake resistance; therefore, the Sperry equation worked well in determining $R_m$. There are instances, particularly when $R_m \ll \alpha$, when a negative intercept will be obtained using Eq. (13.8). Under these conditions, $R_m$ is best determined by filtration of solids-free filtrate to obtain q and using Eq. (13.10) to determine $\alpha$. In the above example, use of the values of $\alpha$ and $R_m$ calculated using Eq. (13.10) will give a filtrate volume per 30 min cycle of 0.221 m$^3$/m$^2$ of filter area or an average filtration rate of 0.00737 m$^3$/(min $\cdot$ m$^2$ filter area).

### 13.1.3   Filtration Rate Model Equations for Prolonged Filtration When Filter Cakes Exhibit Time-Dependent Specific Resistance

Equation (13.6) and variations on it such as Eq. (13.7) are used assuming that specific cake resistance is constant. These equations usually provide good fit with experimental filtration data for short filtration times as was shown in the previous example. However, when filtration time is extended, Eq. (13.8) usually overestimates the filtrate flow. Some filter cakes undergo compaction, or fine solids may migrate within the pores blocking flow and increase cake resistance as filtration proceeds.

### 13.1.4   Exponential Dependence of Rate on Filtrate Volume

The Sperry equation has been modified to account for changing resistance with increasing filtration time. de la Garza and Bouton (Am. J. Enol. Vitic. 35:189, 1984) assumed $\alpha$ to be constant and modified the Sperry equation such that the filtration rate is a power function of filtrate volume:

**Table 13.2** Filtration data for apple juice

| Time (s) | ln (t) | Filtrate volume $V (m^3) \times 10^5$ | $t/V \times 10^{-6}$ | $t - (V/q)$ |
|----------|--------|----------------------------------------|-----------------------|-------------|
| 84.3     | 4.353  | 9.74                                   | 8.659                 | 14.78       |
| 112.5    | 4.723  | 12.3                                   | 9.141                 | 24.59       |
| 140.6    | 4.946  | 14.4                                   | 9.793                 | 38.06       |
| 196.9    | 5.283  | 17.4                                   | 11.291                | 72.33       |
| 281.2    | 5.639  | 20.0                                   | 14.062                | 138.39      |
| 421.9    | 6.045  | 22.6                                   | 18.697                | 260.70      |
| 759.4    | 6.632  | 27.2                                   | 27.939                | 565.24      |
| 1068.8   | 6.974  | 29.7                                   | 35.932                | 856.29      |
| 1575     | 7.362  | 32.3                                   | 48.750                | 1344.2      |
| 2250     | 7.719  | 24.9                                   | 64.522                | 2000.9      |
| 3825     | 8.249  | 39.5                                   | 96.867                | 3542.9      |

$$\frac{dt}{dV} = \frac{\mu}{\Delta P \, A} \left( \alpha c \left[ \frac{V}{A} \right]^n + R_m \right) \qquad (13.11)$$

Integration of Eq. (13.11) gives:

$$t = \frac{\mu}{\Delta P \, A} \left( \frac{\alpha c}{n+1} \left[ \frac{V}{A} \right]^{n+1} + R_m V \right) \qquad (13.12)$$

Bayindirly et al. (J. Food Sci. 54:1003, 1989) tested Eq. (13.12) on apple juice and found that a common n in Eq. (13.11) could not be found to describe all data at different body feed concentrations. Consequently, an alternative equation was proposed which combines the specific cake resistance and solids concentration into a parameter k:

$$\frac{dt}{dV} = \frac{\mu}{A \Delta P} R_m [e]^{kV/A} \qquad (13.13)$$

Integration of Eq. (13.13) gives:

$$t = \frac{\mu}{k \Delta P} R_m [e]^{kV/A} \qquad (13.14)$$

Taking the natural logarithm of Eq. (13.14):

$$\ln (t) = \ln \left[ \frac{\mu R_m}{k \Delta P} \right] + \frac{k}{A} V \qquad (13.15)$$

A semi-log plot of t against (V/A) will be linear with slope k. $R_m$ is evaluated from the intercept.

**Example 13.2** Data in Table 13.2 was obtained from Bayindirly et al. on filtration of apple juice through a filter with a 30.2 cm$^2$ area using diatomaceous earth filter aid with an average particle size of 20 μm. The juice was said to have a viscosity just slightly greater than that of water (assume μ = 1.0 cP), and suspended solids in the juice was 0.3% (3.0 kg/m$^3$). Precoating was 0.25 g/cm$^2$ (2.5 kg/m$^2$) and body feed was 0.005 g/mL (5.0 kg/m$^3$). Pressure differential was 0.65 atm. Calculate the filtration parameters, k and $R_m$ and the average filtration rate in m$^3$/(min · m$^2$ of filter area), if a filtration cycle of 60 min is used in the filtration process.

**Solution**

Table 13.2 also shows the transformed data on which a linear regression was performed. Linear regression of (t/V) against V to fit Eq. (13.8) results in the following correlation equation:

$$t/V = 2.64 \times 10^{10} V - 3,107,386$$

The problem of a negative value for the term involving the medium resistance, $R_m$, is apparent in the analysis of this data. The correlation coefficient was 0.835 which may indicate reasonable fit; however, when filtration time against filtrate volume is plotted, the lack of fit is obvious. Thus, the Sperry equation could not be used on the date for this filtration.

A linear regression of ln(t) against V according to Eq. (13.15) results in the following regression equation:

$$\ln(t) = 13155.92V + 3.70293$$

The correlation coefficient was 0.998 indicating very good fit. A plot of filtration time against filtrate volume also shows very good agreement between values calculated using the correlation equation and the experimental data. The exponential dependence of filtration rate with filtrate volume appropriately described the filtration data.

The value of k is determined from the coefficient of V in the regression equation:

$$k/A = 13155.92; \quad k = 13155.92(30.2 \times 10^{-4})$$
$$k = 39.73 \text{ m}^{-1}$$

The constant 3.70293 in the regression equation is the value of $\ln(\mu R_m/k \cdot \Delta P)$:

$$R_m = \frac{e^{3.70293}(39.73)(0.65)(101,300)}{0.001(1)}$$
$$= 5.652 \times 10^{10} \text{m}^{-1}$$

The correlation equation will also be used to calculate the filtrate volume after a filtration time of 60 min:

$$V = \frac{\ln(3600) - 3.70293}{13155.92} = 0.000348 \text{m}^3$$

Average filtration rate $= (V/A)_{avg}/t$:

$$\frac{(V/A)_{avg}}{t} = \frac{0.000341}{(30.2 \times 10^{-4})(60)}$$
$$= 0.001882 \text{ m}^3/(\text{min} \cdot \text{m}^2 \text{ filter area})$$

### 13.1.5   Model Equation Based on Time-Dependent Specific Cake Resistance

Chang and Toledo (J. Food Proc. Pres. 12:253, 1989) modified Eq. (13.10), derived from the Sperry equation, by assuming a linear dependence of the specific cake resistance with time. The model fitted experimental data on body feed

filtration of poultry chiller water overflow for recycling, using perlite filter aid:

$$t = (k_0 + \beta t)V^2 + (1/q)V \qquad (13.16)$$

The ratio $(t - V/q)/V^2$ is $(k_0 + \beta t)$ at each filtration time t. The slope of the regression equation of $(t - V/q)/V^2$ against filtration time will have a slope of $\beta$ and an intercept of $k_0$. The filtrate volume at time t will be the positive root of Eq. (13.16):

$$V = \frac{-(1/q) + \left[(1/q)^2 + 4(k_0 + \beta t)t\right]^{0.5}}{2(k_0 + \beta t)} \qquad (13.17)$$

Filtration data may be analyzed using this procedure by using the raw volume vs. time data, and the filter area is not considered until the final analysis (i.e., calculated value of filtrate volume or filtration rate is converted to a per unit area basis). Another approach is to use filtrate volume per unit area of filter in the calculations, in which case the calculated value of filtrate volume will already be on a per unit area basis. The former approach is used in the following example to avoid having to manipulate very small numbers.

The units of $k_0$ and $\beta$ in Eq. (13.16) are $s \cdot m^{-6}$ and $m^{-6}$, respectively, because values of V used in the analysis have not been converted to volume/filter area.

**Example 13.3** Table 13.3 shows data on filtration of poultry chiller water overflow using perlite filter aid which has a rated pure water permeability of 0.4 mL/(s · m²) for a 1 cm cake and a 1 atm pressure differential. The filter has an area of 20 cm², and water at 2 °C was the filtrate. The precoat was 1.0 kg/m² of filter area. Whatman No. 541 filter paper was used as the filter medium. Body feed was 5 kg/m³ and suspended solids was 5 kg/m³. Pressure across the filter was 172 kPa. Filtrate flow across the precoated filter was 25.5 mL/s at 172 kPa pressure differential. Calculate the parameters for the time dependence of specific cake resistance, and determine the fit of Eq. (13.16) with experimental data. Calculate the average filtration rate if the cycle time is 20 min.

**Table 13.3** Data on filtration of poultry chiller water

| Time (s) | Volume of filtrate V(m³) | t − V/q | R = (t − V/q)/V² × 10⁸ |
|---|---|---|---|
| 60 | 0.000483 | 41.059 | 1.760 |
| 180 | 0.000823 | 147.7 | 2.181 |
| 300 | 0.001031 | 259.6 | 2.442 |
| 420 | 0.001173 | 374.0 | 2.718 |
| 540 | 0.001292 | 489.3 | 2.931 |
| 660 | 0.001385 | 605.7 | 3.156 |
| 780 | 0.001466 | 722.5 | 3.362 |
| 900 | 0.001538 | 839.7 | 3.550 |
| 1020 | 0.001601 | 957.1 | 3.734 |
| 1140 | 0.001658 | 1075.0 | 3.910 |
| 1200 | 0.001660 | 1134.9 | 4.118 |

$q = 25.5 \times 10^{-6}$ m³/s

**Table 13.4** Calculation of filtrate volume from cake resistance data

| Time (s) | $k_0 + \beta t \times 10^{-8}$ | 1/q | $C \times 10^{-10}$ | Filtrate volume (V) calculated from Eq. 13.17 (mL) |
|---|---|---|---|---|
| 60 | 1.94 | 39,215 | 4.82 | 464 |
| 180 | 2.17 | 39,215 | 15.8 | 824 |
| 300 | 2.40 | 39,215 | 29.0 | 1038 |
| 420 | 2.63 | 39,215 | 44.4 | 1191 |
| 540 | 2.86 | 39,215 | 62.0 | 1307 |
| 660 | 3.09 | 39,215 | 81.8 | 1399 |
| 780 | 3.32 | 39,215 | 104 | 1474 |
| 900 | 3.55 | 39,215 | 128 | 1537 |
| 1020 | 3.78 | 39,215 | 154 | 1591 |
| 1140 | 4.01 | 39,215 | 183 | 1637 |
| 1200 | 4.13 | 39,215 | 198 | 1659 |

$q = 2.55 \times 10^{-5}$; from Eq. 13.17: $C = (1/q)^2 + 4(k_o + \beta t)$; $V = [-(\text{column 2}) + (\text{column 3})^{0.5}]/[2 \text{ column 1}]$

## Solution

Table 13.3 also shows the values of $(t - V/q)/V^2$ on which a regression analysis was done against time to yield the following regression equation ($R^2 = 0.9883$):

$$k_0 + \beta t = 191,347t + 1.83 \times 10^8 \quad (13.17a)$$

Table 13.4 shows the calculation of filtrate volume against filtration time using the expression for $k_0 + \beta t$ in Eqs. (13.17a) and (13.17). A graph of calculated filtrate volume against filtration time shows good agreement between the model and experimental value.

For a filtration time of 20 min, the calculated value of V in Table 13.4 is 0.001659 m³. The average filtration rate is:

$$(V/A)_{avg}/t = 0.001659/[(20)(20 \times 10^{-4})]$$
$$(V/A)_{avg}/t = 0.0415 \text{m}^3/(\text{s} \cdot \text{m}^2 \text{filter area})$$

at 172 kPa pressure differential across the filter.

## 13.1.6 Optimization of Filtration Cycles

Filtration cycles are generally based on obtaining the fastest average filtrate flow. As filtrate flow drops with increasing cake thickness, a very long filtration time will result in the lowered average filtration rate. However, labor involved in disassembling the filter, removal of the cake, and precoating will be a major expense. Because precoating is needed before actual filtration, shortened filtration cycles will necessitate increased consumption of filter aid. Thus, optimum cycles based on maximum filtrate flow may not always be the ideal cycle time from the standpoint of economics. Filtrations with short cycle times for maximum filtrate flow per unit filter area may have to be done using a continuous rotary vacuum filter to prevent excessive filter aid use. The optimum cycle time for maximizing filtrate flow per cycle is derived as follows.

For filtrations that fit the Sperry equation (Eq. 13.7):

A cycle is the sum of filtration time, t, and the time to disassemble, assemble, and precoat the filter, $t_{DAP}$:

$$\frac{V}{Cycle} = \frac{V}{t + t_{DAP}}$$

Let $k_1 = \pi \alpha \, c/2\Delta P$ and $k_2 = R_m \pi/\Delta P$. Eq. 13.7 becomes:

$$t = k_1 (V/A)^2 + k_2 (V/A)$$

To obtain the optimum cycle time, the filtrate volume per cycle is maximized by taking the derivative and equating to zero as follows. Substitute equation for t as a function of V/A and take the first derivative

$$\frac{V}{t + t_{DAP}} = \frac{V}{k_1 \left(\dfrac{V}{A}\right)^2 + k_2 \left(\dfrac{V}{A}\right) + t_{DAP}}$$

Take the derivative $\left(\dfrac{d}{dV}\right)$ of the right side of the above equation. The derivative is equated to zero to obtain V/A for the maximum V/cycle:

$$\frac{d}{dV} = \frac{\left[k_1 \left(\dfrac{V}{A}\right)^2 + k_2 \left(\dfrac{V}{A}\right) + t_{DAP}\right] - V\left[2k_1 \left(\dfrac{V}{A}\right)^2 + \dfrac{k_2}{A}\right]}{\left[k_1 \left(\dfrac{V}{A}\right)^2 + k_2 \left(\dfrac{V}{A}\right) + t_{DAP}\right]^2}$$

For the derivative to equal zero, the numerator must equal zero. After combining like terms and simplifying, the root of the numerator in the equation becomes:

$$t_{DAP} - k_1 \left(\frac{V}{A}\right)^2 = 0$$

The term $\left(\dfrac{V}{A}\right)$ in the root of the derivative is the

maximum value $\left(\dfrac{V}{A}\right)_{max}$

$$\left(\frac{V}{A}\right)_{max} = \left[\frac{t_{DAP}}{k_1}\right]^{0.5}$$

The optimum cycle time, $t_{opt}$, is:

$$t_{opt} = k_1 \left[\frac{t_{DAP}}{k_1}\right] + k_2 \left[\frac{t_{DAP}}{k_1}\right]^{0.5}$$

$$t_{opt} = t_{DAP} + k_2 \left[\frac{t_{DAP}}{k_1}\right]^{0.5}$$

$$t_{opt} = t_{DAP} + R_m \left[\frac{2\mu t_{DAP}}{\alpha c \Delta P}\right]^{0.5} \qquad (13.18)$$

When filtration data fits Eq. (13.14), the (V/A) for maximum filtrate flow derived using the above procedure is the root of Eq. (13.19):

$$\mu R_m (e)^{k\left[\frac{V}{A}\right]} \left(\frac{k}{A^2} - \frac{V}{A}\right) - t_{DAP} = 0 \qquad (13.19)$$

When filtration data fits Eq. (13.16), t as a function of $k_0$, V, and q only is solved as follows:

$$t = \frac{k_0 V^2 + q^{-1} V}{1 - \beta V^2}$$

The optimum V for one filtration cycle is the root of Eq. (13.20):

$$V^4 (\beta k_0) + V^3 (2\beta q^{-1}) + V^2 (k_0 + 2\beta t_{DAP})$$
$$- t_{DAP} = 0$$
$$\qquad (13.20)$$

**Example 13.4** Calculate the optimum filtration cycle in Example 13.3 for poultry chiller water overflow filtration, to maximize filtrate flow, assuming 10 min. For disassembly, assembly, and precoating, use the values $k_0 = 1.83 \times 10^8$ s/m$^6$, $\beta = 191{,}347$ 1/m$^6$, and $q = 25.5 \times 10^{-6}$ m$^3$/s calculated in Example 13.3.

**Solution**
The values of $k_0$, $\beta$, q, and $t_{DAP} = 600$ s are substituted in Eq. (13.20), which is then solved

**Fig. 13.5** Visual BASIC program for calculating optimum filtration cycle times

for V. A Visual BASIC program will be used to solve for the optimum filtrate volume, V, and the filtration time. The first five lines in the program assign the known values of $\beta$, $k_0$, q, and $t_{DAP}$, followed by Eq. (13.20), and then the equation solves for t required for filtrate volume V to flow through the filter. Figure 13.5 shows the program and the output.

The value of $F = -1.046$ when $V = 0.001122$, and $F = 0.143$ when $V = 0.001123$. Thus, the optimum V for a cycle will be 0.001123 $m^3$ or 1123 mL. The optimum cycle time is 370 s.

The filtration behavior shown in this example results in very short cycle times because of very rapid loss of filter cake porosity. Thus, use of a batch filter in carrying out this filtration will result in excessive filter aid use. Use of a continuous rotary vacuum filter is indicated for this application.

### 13.1.7 Pressure-Driven Membrane Separation Processes

A form of filtration that employs permselective membranes as the filter medium is employed to separate solutes, macromolecules, and small suspended particles in liquids. A thin membrane with small pore size, which possesses selectivity for passing solute or solvent, is used. The solvent and small molecules pass through the membrane, and other solutes, macromolecules, or suspended solids are retained either by repulsive forces

**Fig. 13.6** Cross section of a membrane in ultrafiltration

acting on the membrane surface or by a sieving effect. Figure 13.6 shows a typical membrane filtration system. Cross-flow filtration is the most efficient configuration, as discussed later, and retentate recycling is needed to obtain both high cross-membrane fluid velocities and the desired final solids concentration in the product. The fluid crossing the membrane is the "permeate" and the fluid retained on the feed side of the membrane is the "retentate." The fluid entering the membrane is the "feed."

Pressure-driven membrane separation processes include:

Microfiltration (MF): particle size retained on the membrane is in the range of 0.02–10 μm. Sterilizing filtration is a MF process.

Ultrafiltration (UF): particle size retained on the membrane is in the range of 0.001–0.02 μm. Concentration of cheese whey and removal of lactose are a UF process.

Reverse osmosis (RO): solute molecules with molecular size less than 0.001 Φm (molecular weight < 1000 daltons) are retained on the membrane surface. In RO, solutes increase the osmotic pressure; therefore the transmembrane pressure is reduced by the osmotic pressure as the driving force for solvent flux across the membrane. Concentration of apple juice and desalination of brackish water are examples of RO processes.

Membranes are either isotropic or anisotropic. An isotropic membrane has uniform pores all the way across the membrane thickness, while an anisotropic membrane consists of a thin layer of permselective material on the surface and a porous backing. MF membranes may be isotropic or anisotropic, while permselective high-flux membranes used in UF and RO are primarily anisotropic. Membranes are formed by casting sheets using a polymer solution in a volatile solvent followed by evaporation of the solvent. Removal of the solvent leaves a porous structure in the membrane. Larger pore size membranes are produced by subjecting a nonporous polymeric membrane to high-energy radiation. The polymer is changed at specific points of entry of the radiation into the membrane. A secondary treatment to dissolve the altered polymer produces pores within the membrane. Dynamic casting may also be employed. A porous support is coated with the membrane material by pumping a solution containing the membrane material across its surface. As the solvent filters across the porous support, the membrane material is slowly deposited. Eventually, a layer with the desired permselectivity is produced.

Two major factors are important in the design and operation of pressure-driven membrane separation processes. These are transmembrane flux and solute rejection properties of the membrane. In general, transmembrane flux and rejection properties are dependent on properties of solute and suspended solids and membrane characteristics of (a) mean pore size, (b) range of pore size distribution, (c) tortuous path for fluid or particle flow across the membrane thickness or "tortuosity," (d) membrane thickness, and (e) configuration of pores. Operating conditions also affect flux and rejection of solutes, and this will be discussed in the succeeding sections.

## 13.1.8   Membrane System Configurations

The effectiveness of membrane systems may be measured by effectiveness in separating the component of interest at a high production rate. Regardless of whether the valuable component is in the retentate or in the permeate, transmembrane flux must be maximized. Because flux is a function of pressure and membrane area, and is constrained by fouling, the different membrane system configurations available are designed to maximize membrane area within a small footprint, operate at high pressure, minimize fouling, and facilitate cleaning. Figure 13.7 shows four different membrane system configurations. The spiral wound membrane (Fig. 13.7a) is the most commonly used configuration. The membrane module is tubular with channels between the spiral wound sheet for feed to flow across the membrane from one end of the tube to the opposite end. A spacer sheet between the membrane layer on the permeate side permits permeate to travel

around the spiral until it enters the permeate tube at the center of the spiral. The tubular spiral wound module is placed inside a tubular pressure vessel. This configuration permits the most effective packing of a large membrane area in a small space. The disadvantage is the tendency to trap solid particles in the spaces within the spiral thus making it difficult to clean. Feed to a spiral wound membrane system must be prefiltered to remove particles that could get trapped in the module. The use of spiral wound membranes has made it possible to economically produce potable water from seawater in large desalination plants.

Tubular membranes (Fig. 13.7b) made of ceramic material or inorganic solids deposited on a porous stainless steel support require a large footprint for a given membrane surface area. They have the advantage of being able to operate at high pressure and temperature, resist cleaning compounds, and permit backflushing to dislodge foulant that penetrated the pores. These membranes have high first cost, but membrane life is considerably longer than the polymeric membranes.

**Fig. 13.7** Membrane system configuration

Plate and frame membrane units (Fig. 13.7c) permit stacking of several membrane sheets to form a compact unit with large membrane surface area. Plates with the membrane material deposited on each surface are available, and when these plates are used, the unit is easily disassembled for cleaning. The manufacturer also supplies membrane support frames which permits a user to use flat sheet membranes on the unit.

A modular submerged module membrane unit is shown in Fig. 13.7d. The module consists of stacks of multiple envelope-like membrane elements arranged in a very compact configuration. Each element consists of two sheets of membrane material with a spacer between them. The edges of the membrane sheets are welded together to prevent feed from entering the element. An opening in the center of the element permits stacking them with gaskets in-between through a permeate removal tube. Because of the large membrane surface area, these units require very low transmembrane pressure to obtain the desired flux. Air scouring of the membrane surface prevents fouling.

Other configurations such as hollow fiber bundles, vibrating membrane units, and cartridge modules are also available and are used in industry.

### 13.1.9  Transmembrane Flux in Pressure-Driven Membrane Separation Processes (Polarization Concentration and Fouling)

Qualitatively, similarities exist in ordinary filtration and membrane separation processes. The net transmembrane pressure, which drives solvent flow across the membrane, is the sum of the pressure drop across the filter medium, $\Delta P_m$, and the pressure drop across the combined deposited solids on the surface and the boundary layer of concentrated suspension flowing across the membrane surface, $\Delta P_\delta$. The development of the equation for transmembrane flux will be similar to that for filtration rate in filter aid filtrations. The

pure water permeability of the membrane usually given by the membrane manufacturer is q in Eq. (13.10).

Transmembrane flux when a compressible deposit is on the membrane may be expressed by modifying the Sperry equation (Eq. 13.6) by letting $\alpha = \alpha_0 \Delta P^n$ as follows:

$$\frac{dV}{dt} = \frac{Aq\Delta P}{\mu \alpha_0 \Delta P^n cq(V/A) + A\Delta P} \quad (13.21)$$

n is a compressibility factor that is 0 for a completely incompressible and 1 for a completely compressible solid deposit. q is the pure water permeability at the transmembrane pressure used in the filtration, $\Delta P$.

If the second term in the denominator is much smaller than the first term, i.e., the resistance of the deposit and the fluid boundary layer at the membrane surface controls the flux, Eq. (13.21) becomes:

$$\frac{dV}{dt} = \frac{A^2 \Delta P^{1-n}}{\mu \alpha_0 cV} \quad (13.22)$$

Equation (13.22) shows that if n is 1 (i.e., the solid deposit is completely compressible), the transmembrane flux is independent of the transmembrane pressure. This phenomenon has been observed in UF and RO operations. For example, in UF of cheese whey, transmembrane flux is proportional to pressure at transmembrane pressure between 30 and 70 $lb_f/in.^2$ gauge (207–483 kPa) and is independent of pressure at higher pressures. The phenomenon of decreasing flux rates with increased permeate throughput V, shown in Eq. (13.22), has also been observed. Solids deposition on the membrane surface is referred to as "fouling."

Even in the absence of suspended solids, transmembrane flux with solutions will be different from the pure water permeability. This decrease in flux is referred to as "polarization concentration." In simplistic terms, Eq. (13.22) qualitatively expresses the phenomenon of polarization concentration in membrane filtrations. As solvent permeates through the membrane, the solids concentration, c, increases in the vicinity of the membrane. The transmembrane flux is inversely proportional to the solids concentration in Eq. (13.22).

The concentration on the membrane surface increases with increasing transmembrane flux; therefore, the phenomenon of flux decrease due to polarization concentration will be more apparent at high transmembrane flux. Flux decrease due to polarization concentration is minimized by increasing flow across the membrane surface by cross-flow filtration. In the absence of suspended solids which form fouling deposits, flux will be a steady state. In the presence of suspended solids which form compressible deposits, flux will continually decrease with increasing throughput of permeate, V.

Flux decrease due to polarization concentration is a function of the operating conditions, and if flow rate on the membrane surface, solids concentration in the feed, and transmembrane pressure are kept constant, flux should remain constant. Flux decline under constant operating conditions can be attributed to membrane fouling.

Porter (1979) presented theoretical equations for evaluation of polarization concentration in ultrafiltration and reverse osmosis. The theoretical basis for the equations is the mass balance for solute or solids transport occurring at the fluid boundary layer at the membrane surface.

In cross-flow filtrations, fluid flows over the membrane at a fast rate. A laminar boundary layer of thickness $\delta$ exists at the membrane surface. If C is the solids concentration in the boundary layer, and $C_4$ is the concentration in the liquid bulk, a mass balance of solids entering the boundary layer with the solvent and those leaving the boundary layer by diffusion will be:

$$\left(\frac{\partial V}{\partial t}\right)C = -D\left(\frac{\partial C}{\partial x}\right)$$

D is the mass diffusivity of solids. Integrating with respect to x, and designating the transmembrane flux, MV/Mt = J, and using the boundary conditions, $C = C_4$ at $x = \delta$:

$$\ln\left(\frac{C}{C_\infty}\right) = \frac{J}{D}(\delta - x) \qquad (13.23)$$

At the surface, $x = 0$ and $C = C_s$:

$$\ln\left(\frac{C_s}{C_\infty}\right) = \frac{J}{D}\delta \qquad (13.24)$$

Equation (13.24) shows that polarization concentration expressed as the ratio of surface to bulk solids concentration increases as the transmembrane flux and thickness of boundary layer increases. Equation (13.24) demonstrates that polarization concentration can be reduced by reducing transmembrane flux and decreasing the boundary layer thickness. Since flux must be maximized in any filtration, reduced polarization concentration can be achieved at maximum flux if high fluid velocities can be maintained on the membrane surface to decrease the boundary layer thickness, $\delta$.

Rearranging Eq. (13.24):

$$J = \frac{D}{\delta} \ln\left(\frac{C_s}{C_\infty}\right)$$

The ratio $D/\delta$ may be represented by a mass transfer coefficient for the solids, $k_s$. Thus:

$$J = k_s[\ln (C_s) - \ln (C_4)] \qquad (13.25)$$

Equation (13.25) shows that a semi-log plot of the bulk concentration $C_4$, against the transmembrane flux under conditions where $C_s$ is constant, will be linear with a negative slope of $1/k_s$. Thus, flux will be decreasing with increasing bulk solids concentration, and the rate of flux decrease will be inversely proportional to the mass transfer coefficient for solids transport between the surface and the fluid bulk.

Equations for estimation of mass transfer coefficients have been applied to determine the effects of cross-flow velocity on transmembrane flux, when polarization concentration alone controls flux. In Chap. 12, mass transfer coefficients were determined using the same equations as for heat transfer. The Sherwood number (Sh) corresponds to the Nusselt number; and the Schmidt number (Sc) corresponds to the Prandtl number.

The Dittus-Boelter equation can thus be used for estimation of mass transfer in turbulent flow:

$$Sh = 0.023 Re^{0.8} Sc^{0.33}$$

Because $Sh = k_s d_h/D$; $Re = d_h v \rho/\Phi$; and $Sc = \Phi/(D \cdot \rho)$:

$$
\begin{aligned}
k_s &= 0.023 \frac{D}{d_h} \frac{d_h^{0.8} v^{0.8} \rho^{0.8}}{\mu^{0.8}} \frac{\mu^{0.33}}{D^{0.33} \rho^{0.33}} \\
&= \frac{0.023 D^{0.67} v^{0.8} \rho^{0.47}}{d_h^{0.2} \mu^{0.47}}
\end{aligned}
\tag{13.26}
$$

Although the Dittus-Boelter equation has been derived for tube flow, an analogy with pressure drops through noncircular conduits will reveal that equations for tube flow can be applied to noncircular conduits if the hydraulic radius is substituted for the diameter of the tube. $D =$ mass diffusivity, $m^2/s$; $d_h =$ hydraulic radius $= 4$ (cross-sectional area)/wetted perimeter.

If fluid flows in a thin channel between parallel plates with width W and channel depth, 2b:

$$d_h = 4(2b)(W)/(4b + 4W) = 2bW/(b+w)$$

If $b << W$, $d_h = 2b =$ the channel depth.

In laminar flow, the Sieder-Tate equation for heat transfer can be used as an analog to mass transfer:

$$Sh = 1.86 [Re \cdot Pr \cdot (d/L)]^{0.33} \tag{13.27}$$

Substituting the expressions for Sh, Re, and Pr:

$$k_s = 1.86 \frac{v^{0.33} D^{0.66}}{d_h^{0.33} L^{0.33}} \tag{13.28}$$

L is the length of the channel.

Equations (13.25) and (13.28) can be used to predict transmembrane flux when polarization concentration occurs if the diffusivity D of the molecular species involved and the surface concentration are known. The form of these equations agrees with experimental data. However it is ineffective in predicting actual transmembrane flux by the inability to predict the concentration at the membrane surface, $C_s$. The value of these equations is in interpolating within experimentally observed values for flux and, in extrapolating within reasonable limits, fluxes

when data at one set of operating conditions are known. Porter (1979) suggests using the Einstein-Stokes equation (Eq. 13.29) to estimate D:

$$D = \frac{1.38 \times 10^{-23} T}{6 \pi \mu r} \tag{13.29}$$

where D is in $m^2/s$, $T = K$, $\Phi =$ medium viscosity in $Pa \cdot s$, and $r =$ molecular radius in meters.

**Example 13.5** The molecular diameter of β-lactoglobulin having a molecular weight of 37,000 daltons is $1.2 \times 10^{-9}$ m. When performing ultrafiltrations at 30 °C at a membrane surface velocity of 1.25 m/s, solids concentration in the feed of 12%, and transmembrane pressure of 414 kPa, the flux was 6.792 $L/(m^2 \cdot h)$. Calculate the concentration at the membrane surface and the flux under the same conditions but at a higher cross-membrane velocity of 2.19 m/s. The membrane system was a thin channel with a separation of 7.6 mm. Flow path was 30 cm long. The viscosity of the solution at 30 °C was 4.8 centipoise, and the density was 1002 $kg/m^3$.

**Solution**

The diffusivity using Eq. (13.29) is:

$$
\begin{aligned}
D &= \frac{1.38 \times 10^{-23}(30 + 273)}{6\pi(4.8)(0.001)\left(1.2 \times 10^{-9}\right)} \\
&= 3.851 \times 10^{-11} m^2/s
\end{aligned}
$$

The hydraulic radius, $d_h =$ channel depth $= 7.6 \times 10^{-3}$ m. The Reynolds number is:

$$
\begin{aligned}
Re &= 7.6 \times 10^{-3}(1.25)(1002)/[(4.8)(0.001)] \\
&= 1983
\end{aligned}
$$

For turbulent flow (Eq. 13.26):

$$
\begin{aligned}
k_s &= \frac{0.023\left(3.851 \times 10^{-11}\right)^{0.66}(1.25)^{0.33}(1002)^{0.47}}{\left(7.6 \times 10^{-3}\right)^{0.2}[(4.8)(0.001)]^{0.47}} \\
&= \frac{0.023\left(1.338 \times 10^{-7}\right)(1.076)(25.72)}{0.376(0.081317)} \\
&= 2.785 \times 10^{-6} m/s
\end{aligned}
$$

Given: $J = 6.792\ L/(m^2 \cdot h) = 1.887 \times 10^{-6}\ m^3/m^2 \cdot s)$

Using Eq. (13.25): $C_s = C_4[e]^{J/ks}$
$C_4$ is given as 0.12 g solids/g soln.

$C_s = 0.12[e]^{1.88 \times 10^{-6}/2.785 \times 10^{-6}} = 0.12\ (1.969)$
$= 0.236$ g solute/g soln.

For $v = 2.19$ m/s:

$$k_s = 2.785 \times 10^{-6} \left[\frac{2.19}{1.25}\right]^{0.8}$$

$$= 4.362 \times 10^{-6} \text{m/s}$$

Using Eq. (13.25):

$J = 4.362 \times 10^{-6}\ [\ln\ (0.236/0.12)]$
$= 2.95 \times 10^{-6} \text{m}^3/(\text{m}^2 \cdot \text{s})$ or $10.62$ L/$(\text{m}^2 \cdot \text{h})$

### 13.1.10  Solute Rejection

Solute rejection in membrane separations is dependent on the type of membrane and the operating conditions. A solute rejection factor, R, is used, defined as:

$$R = \frac{C_s - C_p}{C_s} \qquad (13.30)$$

where $C_s$ = concentration of solute on the membrane surface on the retentate side of the membrane and $C_p$ = concentration of solute in the permeate.

Equation (13.30) shows that polarization concentration, and increasing solute concentration in the feed, decreases the solute rejection by membranes.

Rejection properties of membranes are specified by the manufacturer in terms of the "molecular weight cutoff," an approximate molecular size that will be retained by the membrane with a rejection factor of 0.99 in very dilute solutions. The ideal membrane with a sharp molecular weight cutoff does not exist. On a particular membrane, the curve for rejection factor against molecular weight is often sigmoidal with increasing molecular weight of solute. Low molecular weight solutes will completely pass

through the membrane and R = 0. As the solute molecular weight increases, small increases in R will be observed until the molecular size reaches that which will not pass through a majority of the pores. When the molecular size exceeds the size of all the pores, the rejection factor will be 1.0. The pore size distribution, therefore, determines the distribution of molecular sizes which will be retained by the membrane. Some solutes will be retained because they are repelled by the membrane at the surface. This is the case with mineral salts on cellulose acetate membrane surfaces. This property of a membrane to repel a particular solute will be beneficial under conditions where separation from the solvent is desired (e.g., concentration) because high rejection factors can be achieved with larger membrane pore size therefore allowing solute rejection at high transmembrane flux.

Interactions between solutes also affect the rejection factor. For example, rejection factor for calcium in milk or whey is higher than aqueous calcium solutions. A protein to calcium complex will bind the calcium preventing its permeation through the membrane.

### 13.1.11  Sterilizing Filtrations

Sterilizing filtrations employ either depth cartridge filters or microporous membrane filters in a plate and frame or cartridge configuration. These filtrations are MF processes. Microporous membrane filters have pore sizes smaller than the smallest particle to be removed and perform a sieving process. They are preferred for sterilizing filtrations on liquid because the pressure drop across the membrane is much smaller than in a depth filter, and the possibility of microorganisms breaking through is minimal. Pore size of microporous membranes for sterilizing filtrations is less than 0.2 Φm.

The filtration rate in sterilizing filtrations has been found by Peleg and Brown (J. Food Sci. 41:805, 1976) to follow the following relationship:

$$\frac{dV}{dt} = k\frac{V^{-n}}{c} \qquad (13.31)$$

where c is load of microorganisms and suspended material that is removed by the filter, and k and n are constants which are characteristic of the fluid being filtered, the filter medium, the suspended solids, and fluid velocity across the membrane surface. k is also dependent on the transmembrane pressure. n is greater than 1; therefore, filtration rate decreases rapidly with increase in filtrate volume. No cake accumulation occurs in this type of filtration. The time for membrane replacement and pre-sterilization must be included in the analysis of the optimum cycle time and procedures for optimization will be similar to that in the section "Optimization of Filtration Cycles."

Cycle time can be increased when fluid is in cross-flow across the membrane surface at high velocities. This configuration minimizes solids deposition and membrane fouling. Fouled membrane surfaces may be rejuvenated by occasionally interrupting filtration through reduction of transmembrane pressure while maintaining the same velocity of fluid flow across the membrane. This procedure is called "flushing," as opposed to "backwashing" where the transmembrane pressure is reversed. Some membrane configurations are suitable for backwashing, while membrane fragility may restrict removal of surface deposits by flushing in other configurations. When rejuvenation is done by flushing or backflushing, the flushing fluid must be discarded or filtered through another coarser filter to remove the solids; otherwise, mixing with new incoming feed will result in rapid loss of filtration rate.

Pre-sterilization of filter assemblies may be done using high-pressure steam, or using chemical sterilants such as hydrogen peroxide, iodophors, or chlorine solutions. Care must be taken to test for filter integrity after pre-sterilization, particularly when heating of filter assemblies is used for pre-sterilization. One method to test for filter integrity in line is the "bubble point test," where the membrane is wetted, sterile air is introduced, and the pressure needed to dislodge the liquid from the membrane is noted. Intact membranes require specific pressures to dislodge the liquid from the surface, and any reduction in this bubble pressure is an indication of a break in the membrane.

Sterilizing filtrations are successfully employed in cold pasteurization of beer and wine, in the pharmaceutical industry for sterilization of injectable solutions, and in the biotechnological industry for sterilization of fermentation media and enzyme solutions.

**Example 13.6** Data on flux during membrane filtration to clarify apple juice (Mondor et al. Food Res. International 33:539–548, 2000) showed the flux to follow two different domains when plotted with time. The initial rapid flux decay was attributed to deposition of solids on the membrane pores, and the domain where flux appears to level off with time was attributed to be due to concentration polarization and formation of a gel layer by the deposited solids on the membrane surface. If $J_o$ is the flux rate at time $= 0$, $J_{41}$ is the flux corresponding to the point where flux levels off in the first part of the filtration curve, and $J_{42}$ is the flux corresponding to the time when flux leveled off in the second part of the filtration curve, and then the flux over the whole filtration cycle can be expressed as:

$$J = (J_o - J_{\infty 1})\exp(-\alpha t) \\ + (J_{\infty 1} - J_{\infty 2})\exp(-\beta t) + J_{\infty 2}$$

An experiment was conducted on cross-flow filtration of apple juice through a PVDF membrane system. The membrane had a pore size of 0.2 micrometers and a total surface area of $1.0 \text{ m}^2$. Cross-flow velocity was 3 m/s and transmembrane pressure was 1.5 Bars. The apple juice was obtained by pressing pectinase enzyme treated ground apples with rice hulls as filtration aid through multiple layers of cheese cloth in a bladder press. The data (time in hours, flux in liters/$\text{m}^2$ h) are as follows: (0.11, 32), (0.24, 28.6), (0.55, 22.7), (0.96, 18.1), (1.2, 15.1), (2.5, 9.8), (5.3, 5.9), (7. 4.7), (8, 4.7), and (9, 4.6).

(a) Assuming that a cleaned membrane will result in the same pattern of flux decay, how long would it take to recover 15 L of the clarified juice (permeate)?

**Solution**

The first term in the flux decay equation above is good for $t = 0$ to $t = t_{s1}$ where $t_{s1}$ corresponds to the end of the first stage of the filtration cycle. This time is determined by plotting $\ln(J - J_{42})$ against time to see where the linear portion deviates from the experimental data curve. A regression is then conducted for the range of time where the curve is linear. The second term in the above expression for J represents the flux decay from the start of the second stage in the filtration cycle. Again, the linear portion of the $\ln(J - J_{42})$ versus time curve is located, and a linear regression is conducted to obtain the slope. The time variable in the second term should be the difference between the actual time and the start of the linear plot close to the actual experimental curve.

$J_{42}$ from the above data = 4.6 L/m$^2$ h.

The transformed data is as follows:

| Time (h) | J (L/m$^2$h) | $\ln(J + J_{42})$ |
|----------|--------------|-------------------|
| 0.11 | 32 | 3.31 |
| 0.24 | 28.6 | 3.18 |
| 0.55 | 22.7 | 2.89 |
| 0.96 | 18.1 | 2.60 |
| 1.2 | 15.1 | 2.35 |
| 2.5 | 9.8 | 1.65 |
| 5.3 | 5.9 | 0.26 |
| 7 | 4.7 | −2.3 |
| 8 | 4.7 | −2.3 |

The first linear segment fitted $t = 0.11$ to $t = 2.5$. This segment has a slope of −0.69 and an intercept of 3.31. The value of $J_{41} = 4.7$ L/m$^2$ h. The second line segment fitted the curve from $t = 5.3$ to 8 and has a slope of −1.0. This line segment intersects the first line segment at $t = 3.5$. Thus the equation for J is as follows: $J = (32{-}4.7)$ $\exp(-0.69 \cdot t) + (4.7{-}4.6)\exp(-1.0(t - 3.5)) + 4.6$. $J = 27.3 \exp(-0.69 \cdot t) + 0.1 \exp(-(t - 3.5)) + 4.6$. $V =$ volume of permeate.

$$V = \int_0^t J\,dt = \int_0^t \{27.3[e]^{-0.69t} + 0.1[e]^{-(t-3.5)} + 4.6\}V$$

$$= \frac{-27.3}{0.69}[e]^{-0.69t} - 0.1[e]^{-(t-3.5)} + 4.6t$$

This equation will be solved using Excel.

Set up volume increments in 0.1 h. Suppose $t = 0.1$ h is in a3. The value of the integral will be $(-27.3/0.69)^* \exp(-0.79^*a3) - 0.1^* \exp(-(a3{-}3.5)) + 4.6^*a3 + (27.3/0.69) + 0.1$. The calculated values are as follows: $t = 0$, $V = 0$; $t = 0.1$, $V = 0.20$; $t = 0.3$, $V = 6.42$, $t = 0.4$, $V = 0.3$, $t = 0.5$, $V = 11.9$; $t = 0.6$, $V = 14.5$, $t = 0.7$, $V = 16.8$. Thus, it would take 0.62 h of filtration to obtain 15 L of filtrate.

### 13.1.12 Ultrafiltration

Ultrafiltration is widely employed in the increasingly important biotechnological industry for separation of fermentation products, particularly enzymes. Its largest commercial use is in the dairy industry for recovery of proteins from cheese whey and for pre-concentration of milk for cheese making. UF systems also have potential for use as biochemical reactors, particularly in enzyme or microbial conversion processes where the reaction products have an inhibitory effect on the reaction rate.

A large potential use for UF is in the extraction of nectar from fruits which cannot be pressed for extraction of the juice. The fruit mash is treated with pectinase enzymes, and the slurry is clarified by UF. UF applications in recycling of food processing wastewater and in recovery of valuable components of food processing wastes are currently in the developmental stages. More membranes for UF applications are available than for RO applications, indicating increasing commercial applications for UF in the food industry.

### 13.1.13 Reverse Osmosis

Reverse osmosis has potential for generating energy savings in the food processing industry, when used as an alternative to evaporation in the concentration of products containing low molecular weight solutes. The early work on RO has been devoted to desalination of brackish water, and a number of RO systems have now been developed which can purify wastewater for reuse after an RO treatment. In the food industry, RO has potential for juice concentration without the need for application of heat.

In RO, the transmembrane pressure is reduced as a driving force for solvent flux by the osmotic pressure. Since solutes separated by RO have low molecular weights, the influence of concentration on osmotic pressure is significant. Equation (12.5) in Chap. 12 can be used for determining the osmotic pressure. Because the product of the activity coefficient of water and the mole fraction of water is the water activity, this equation can be written as:

$$\pi = -\frac{RT}{V} \ln (a_w) \qquad (13.32)$$

where $\pi$ = osmotic pressure, R = gas constant, T = absolute temperature, and V = molar volume of water. $a_w$ = water activity.

The minimum pressure differential needed to force solvent across a membrane in RO is the osmotic pressure. Equations (12.17) and (12.21) in Chap. 12 can be used to determine a $_w$ of sugar solutions.

**Example 13.7** Calculate the minimum transmembrane pressure that must be used to concentrate sucrose solution to 50% sucrose at 25 °C.

**Solution**
From the Example 12.1 in Chap. 12, the water activity of a 50% sucrose solution was calculated to be 0.935.

R = 8315 N m /(kgmole · K); T = 303 K, and the density of water at 25 °C = 997.067 kg/m³ from the steam tables.

$$V = \frac{18kg}{kgmole} \frac{1}{997.067kg/m^3}$$

$$= 0.018053m^3/kgmole\pi$$

$$= \frac{-[8315Nm/(kgmole \cdot K)][303K]}{0.018053m^3/kgmole} \ln (0.935)$$

$$- 9379kPa$$

A minimum of 9379 kPa transmembrane pressure must be applied just to overcome the osmotic pressure of a 50% sucrose solution.

This example illustrates the difficulty in producing very high concentrations of solutes in reverse osmosis processes, and the large influence of concentration polarization in reducing transmembrane flux in RO systems.

Transmembrane flux in RO systems is proportional to the difference between the transmembrane pressure and the osmotic pressure Eq. (13.33).

$$J = \left(\frac{k_w}{x}\right)(P - \Delta\pi) \qquad (13.33)$$

where J = transmembrane flux, kg water/(s · m²); $k_w$ = mass transfer coefficient, s; x = thickness of membrane, m; P = transmembrane pressure, Pa; and $\pi$ = osmotic pressure, Pa

**Example 13.8** The pure water permeability of cellulose acetate membrane having a rejection factor of 0.93 for a mixture of 50% sucrose and 50% glucose is 34.08 kg/h of water at a transmembrane pressure of 10.2 atm (1034 kPa) for a membrane area of 0.26 m². This membrane system had a permeation rate of 23.86 kg/h when 1% glucose solution was passed through the system at a transmembrane pressure of 10.2 atm at 25 °C.

Orange juice is to be concentrated using this membrane system. In order to prevent fouling, the orange juice was first centrifuged to remove the insoluble solids, and only the serum which contained 12% soluble solids was passed through the RO system. After centrifugation, the serum was 70% of the total mass of the juice.

Assume that the soluble solids are 50% sucrose and 50% glucose. The pump that feeds the serum through the system operates at a flow rate of 0.0612 kg/s (60 mL/s of juice having a density of 1.02 g/mL). The transmembrane pressure was 2758 kPa (27.2 atm).

Calculate the concentration of soluble solids in the retentate after one pass of the serum through the membrane system.

**Solution**
This problem illustrates the effects of transmembrane pressure, polarization concentration and osmotic pressure on transmembrane flux, and

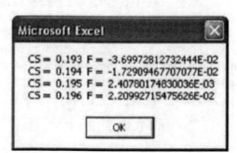

**Fig. 13.8** Visual BASIC program for problem involving reverse osmosis concentration of orange juice

solute rejection in reverse osmosis systems. The governing equations are Eq. (13.25) for polarization concentration and Eq. (13.33) for transmembrane pressure.

The reduction in the pure water permeability when glucose solution is passed through the membrane system can be attributed to polarization concentration. Equation (13.25):

$J = k_s \ln (C_s/C_4)$. Let $N_w$ = water permeation rate in kg/h.

$N_w = JA$ m$^3$/s (3600 s/h)(1000 kg/m$^3$):

$$J = \left(\frac{N_w}{A}\right)\left(\frac{1}{3.6}\right) \times 10^{-6}$$

The osmotic pressure at the membrane surface responsible for the reduction of transmembrane flux in the 1% glucose solution is calculated using Eq. (13.33).

$$N_w = \left(k_w \frac{A}{x}\right)(P - \pi)$$

The factor $(k_w A/x)$ can be calculated using the pure water permeability data, where $\pi = 0$:

$$\left(k_w \frac{A}{x}\right) = \frac{N_w}{P} = \frac{34.08}{10.2} = 3.34 \text{ kg}/(h \cdot atm)$$

For the 1% glucose solution, using Eq. (13.33):

$$N_w = \left(k_w \frac{A}{x}\right)(P - \pi)$$

$$23.86 = 3.34(10.2 - \pi)$$

$$\pi = 3.05 \text{ atm}.$$

A 1% glucose solution has a water activity of practically 1.0; therefore, the osmotic pressure must be exerted at the membrane surface. Thus, $C_s$ can be calculated as the glucose concentration that gives an osmotic pressure of 3.05 atm. Using Eq. (13.32), for $P = 3.05$ (101,325) = 309,041 Pa.:

$$309,041 = -(8315)(298/0.018)\ln(a_w)$$

$$a_w = e^{-0.00225} = 0.99775$$

For very dilute solutions, $a_w = x_w$. Let $C_s$ = solute concentration at the surface, g solute/g solution:

$$a_w = x_w = \frac{(1 - C_s)/18}{(1 - C_s)/18 - C_s/180}$$

$$a_w = \frac{(1 - C_s)(18)(180)}{18[180(1 - C_s) + 18C_s]} = 0.99775$$

$$321.271C_s + 3232.71 - 3232.71C_s = 3420 - 3420C_s$$

$$C_s = 7.29/330.561 = 0.022 \text{ g solute/g solution}$$

Solving for J from the permeation rate:

$$J = (23.86/0.26)(1/3.6) \times 10^{-6}$$

$$= 2.55 \times 10^{-5} \text{ m/s}$$

Using Eq. (13.25):

$$k_s = 2.55 \times 10^{-5}/\ln(0.022/0.01)$$

$$= 3.234 \times 10^{-5} \text{ m/s}$$

For the orange juice serum: $C_4 = 0.12$ g solute/g soln. Solving for $C_s$ using Eq. (13.25):

$$J = 3.234 \times 10^{-5} \ln(C_s/0.12)$$

Solving for $N_w$ from J:

$$N_w = J A (3.6)(10^6)$$

$$= [3.234 \times 10^{-5} \ln(C_s/0.12)](0.26)$$

$$\times (3.6 \times 10^6) = 30.27 \ln(C_s/0.12)$$

Solving for $N_w$ from the transmembrane pressure using Eq. (13.36):

$$N_w = (k_w A/x)(P - \pi) = 3.34(27.2 - \pi)$$

Equating the two equations for permeation rate:

$$30.27 \ln(C_s/0.12) = 3.34(27.2 - \pi)$$

$$\ln(C_s/0.12) = 0.1103(27.2 - \pi)$$

Solving for $\pi$ in atm using Eq. (13.32):

$$\pi = -(8315) \times (298/0.018) \ln(a_w)$$

$$[1 \text{ atm}/101,325 \text{ Pa}] = -1358.59 \ln(a_w), \text{atm}$$

Substituting in the equation for $C_s$:

$$\ln(C_s/0.12) = 0.1103[27.2 + 1358.59 \ln(a_w)]$$

$a_w$ can be solved in terms of $C_s$ as follows:
For sucrose, $x_{ws}$ = mole fraction of water in the sucrose fraction of the solution; $a_{ws}$ = water activity due to the sucrose fraction:

$$x_{ws} = \frac{(1 - 0.5C_s)/18}{(1 - 0.5C_s)/18 + 0.5C_s/342}$$

$$a_{ws} = x_{ws}[10]^{-2.7(1-x_{ws})^2}$$

For the glucose fraction of the mixture, let $x_{wg}$ = mole fraction of glucose; $a_{wg}$ = water activity due to the glucose fraction:

$$x_{wg} = \frac{(1 - 0.5C_s)/18}{(1 - 0.5C_s)/18 + 0.5C_s/180}$$

$$a_{wg} = x_{wg}[10]^{-0.7(1-x_{wg})^2}$$

$$a_w = a_{ws} \cdot a_{wg}$$

$C_s$ will be solved using a Visual BASIC program. In this program, a wide range of values of $C_s$ from 0.1 to 0.2 was first substituted with a larger step change of 0.01. The printout was then checked for the two values of $C_s$ between which the function F crossed zero. A narrow range of $C_s$ between 0.115 and 0.125 was later chosen with a narrower step change in order to obtain a value of $C_s$ to the nearest third decimal place which satisfied the value of the function F to be zero.

The value of $C_s$ will be 0.1949. The rejection factor will be used to calculate the permeate solute concentration:

$$R = (C_s - C_p)/C_s$$
$$C_p = C_s(1 - R) = 0.1949(1 - 0.93)$$
$$C_p = 0.0136 \text{ g solute/g permeate.}$$

Total mass balance:

$$F = P + R;$$
$$F = 0.0612 \text{ kg/s } (3600 \text{s/h}) = 220.3 \text{ kg/h}$$
$$P = N_w/(1 - C_p)$$
$$N_w = 30.27 \ln(0.1949/0.12) = 14.68 \text{ kg/h}$$
$$P = 14.68/(1 - 0.0134) = 14.87 \text{ kg/h}$$
$$R = 220.3 - 14.87 = 205.43 \text{ kg/h}$$

Solute balance, $FC_f = RC_r + PC_p$:

$$0.12(220.3) = 205.43(C_r) + 14.87(0.0134)$$
$$C_r = 0.128 \text{g solute/g retentate}$$

The effect of polarization concentration shown in this example problem demonstrates that increasing transmembrane pressure may not always result in a higher flux. A pressure will be reached beyond which polarization concentration will negate the effect of the applied pressure, and any further increase in transmembrane pressure will result in no further increase in flux.

Fouling remains a major problem in both RO and UF applications. Optimization of operating cycles using the principles discussed in the section "Optimization of Filtration Cycles" is an aspect of the operation to which an engineer can make contributions to successful operations.

Application of theories and equations on polarization concentration as discussed in the preceding sections will help in identifying optimum operating conditions, but absolute values of transmembrane flux are better obtained empirically on small systems and results used for scale-up than calculating theoretical values. Material balance calculations as discussed in Chap. 3 in the section "Multistage Processes," Example 3.21, will also help in analyzing UF systems, particularly in relation to recycling and multistage operations to obtain the desired composition in the retentate.

### 13.1.14 Temperature Dependence of Membrane Permeation Rates

Temperature affects liquid permeation rates through membranes in proportion to the change in viscosity of the fluid with temperature. If the solvent is water, and permeate flux at one temperature is given, the permeation rate at any other temperature can be calculated by:

$$J_{T1} = J_{T2}\frac{\mu_{T2}}{\mu_{T1}}$$

where $J_{T1}$ and $J_{T2}$ are fluxes at T1 and T2, respectively. The viscosity of water at temperature T in Celsius is:

$$\mu = \frac{1795 \cdot 10^{-6}}{1 + 0.036T + 0.000185T^2}$$

where $\mu$ = viscosity in Pa s and T is temperature in °C.

### 13.1.15 Other Membrane Separation Processes

Membranes are also used to remove inorganic salts from protein solutions or for recovery of valuable inorganic compounds from process wastes. Dialysis and electrodialysis are two commonly used processes.

Dialysis is a diffusion rather than pressure-driven process. A solute concentration gradient drives solute transport across the membrane. Dialysis is well known in the medical area, for removing body wastes from the blood of persons with diseased kidneys. In the food and biochemical industries, dialysis is used extensively for removal of mineral salts from a protein solution by immersing the protein solution contained in a dialysis bag or tube in flowing water. Solute flux in dialysis is very slow.

Electrodialysis is a process where solute migration across the membrane is accelerated by the application of an electromotive potential. Anion and cation selective membranes are used. The electromotive potential forces the migration of ionic species across the membrane toward the appropriate electrode. Electrodialysis is used to demineralize whey, but food industry applications of the process are still rather limited.

## 13.2   Sieving

Sieving is a mechanical size separation process. It is widely used in the food industry for separating fines from larger particles and also for removing large solid particles from liquid streams prior to further treatment or disposal. Sieving is a gravity-driven process. Usually a stack of sieves are used when fractions of various sizes are to be produced from a mixture of particle sizes.

To assist in the sifting of solids in a stack of sieves, sieve shakers are used. The shakers may be in the form of an eccentric drive that gives the screens a gyratory or oscillating motion, or it may take the form of a vibrator which gives the screens small amplitude high frequency up and down motion. When the sieves are inclined, the particles retained on a screen fall off at the lower end and are collected by a conveyor. Screening and particle size separation can thus be carried out automatically.

### 13.2.1   Standard Sieve Sizes

Sieves may be designated by the opening size, US-Sieve mesh or Tyler Sieve mesh. The Tyler mesh designations refer to the number of openings per inch, while the US-Sieve mesh designations is the metric equivalent. The latter has been adopted by the International Standards Organization. The two mesh designations have equivalent opening size although the sieve number designations are not exactly the same. Current sieve designations, unless specified, refer to the US-Sieve series. Sizes of particles are usually designated by the mesh size that retains particles that have passed through the next larger screen size. A clearer specification of particle size by mesh number would be to indicate by a plus sign before the mesh size that retains the particles and by a negative sign the mesh size that passed the particles. If a mixture of different sized particles is present, the designated particle size must be the weighted average of the particle sizes.

Table 13.5 shows US-Sieve mesh designations and the size of openings on the screen. A cumulative size distribution of powders can be made by sieving through a series of standard sieves and determining the mass fraction of particles retained on each screen. The particles are assumed to be spherical with a diameter equal to the mean of the sieve opening which passed the particles and that which retained the particles.

**Example 13.9**  Table 13.6 shows the mass fraction of a sample of milled corn retained on each of a series of sieves. Calculate a mean particle diameter which should be specified for this mixture.

**Solution**
The sieve opening is obtained from Table 13.6. The mean particle size retained on each screen is the mean opening size between the screens which retained the particular fraction and the one above it. The mass fraction is multiplied by the mean particle size on each screen, and the sum will be the weighted mean diameter of the particle. From Table 13.5, the mean particle diameter is 0.2741 mm.

This example clearly shows that the choice of sieves used in classifying the sample into the different size fractions will affect the mean particle diameter calculated. The use of a series of sieves of adjacent sizes in Table 13.5 is recommended.

**Table 13.5**  Standard US-Sieve sizes

| US-Sieve size (mesh) | Opening (mm) | US-Sieve size (mesh) | Opening (mm) |
|---|---|---|---|
| 2.5 | 8.00 | 35 | 0.500 |
| 3 | 6.73 | 40 | 0.420 |
| 3.5 | 5.66 | 45 | 0.354 |
| 4 | 4.76 | 50 | 0.297 |
| 5 | 4.00 | 60 | 0.250 |
| 6 | 3.36 | 70 | 0.210 |
| 7 | 2.83 | 80 | 0.177 |
| 8 | 2.38 | 100 | 0.149 |
| 10 | 2.00 | 120 | 0.125 |
| 12 | 1.68 | 140 | 0.105 |
| 14 | 1.41 | 170 | 0.088 |
| 16 | 1.19 | 200 | 0.074 |
| 18 | 1.00 | 230 | 0.063 |
| 20 | 0.841 | 270 | 0.053 |
| 25 | 0.707 | 325 | 0.044 |
| 30 | 0.595 | 400 | 0.037 |

Source: Perry, R. H., Chilton, C. H., and Kirkpatrick, S. D. 1963. *Chemical Engineers Handbook*, 4th ed. McGrawo-Hill Book Co., New York

**Table 13.6**  US-Sieve size distribution of power sample and calculation of mean particle diameter

| Mass fraction retained | US-Sieve screen | Sieve opening (mm) | Particle size | Fraction × size |
|---|---|---|---|---|
| 0 | 35 | 0.500 | | |
| 0.15 | 45 | 0.354 | 0.437 | 0.065 |
| 0.35 | 60 | 0.250 | 0.302 | 0.1057 |
| 0.45 | 80 | 0.177 | 0.213 | 0.095585 |
| 0.05 | 120 | 0.125 | 0.151 | 0.00755 |
| | | | | Sum = 0.2741 |

## 13.3  Gravity Separations

This type of separation depends on density differences between several solids in suspension or between the suspending medium and the suspended material. Generally, the force of gravity is the only driving force for the separation. However, the same principles will apply when the driving force is increased by application of centrifugal force.

### 13.3.1  Force Balance on Particles Suspended in a Fluid

When a solid is in suspension, an *external force*, $F_e$, acts on it to move in the direction of the force.

This external force may be gravitational force, in which case it will cause a particle to move downward, or it can be centrifugal force, which will cause movement toward the center of rotation. In addition, a *buoyant force*, $F_b$, exists that acts in a direction opposite the external force. Another force, the *drag force*, $F_d$, also exists, and this force is due to motion of the suspended solid and acts in the direction opposite the direction of flow of the solid. Let $F_t$ = the net force on the particle. The component of forces in the direction of solid motion will obey the following force balance equation:

$$F_t = F_e - F_d - F_b \qquad (13.34)$$

In applying Eq. (13.34), the sign on $F^t$ will determine its direction based on the designation of the direction of $F_e$ as the positive direction.

### 13.3.1.1  Buoyant Force

$F_b$ is the mass of fluid displaced by the solid multiplied by the acceleration provided by the external force. Let $a_e$ = acceleration due to the external force, m = mass of particle, $\rho_p$ = density of particle, and $\rho$ = density of the fluid.

The mass of displaced fluid = volume of the particle /density = $m/\rho_p$:

$$F_b = ma_e\left(\frac{\rho}{\rho_p}\right) \qquad (13.35)$$

The mass of displaced fluid = $m(\rho/\rho_p)$.

### 13.3.1.2  Drag Force

$F_d$ is fluid resistance to particle movement. Dimensional analysis will be used to derive an expression for $F_d$.

Assume that the $F_d/A_p$, where $A_p$ is the projected area perpendicular to the direction of flow; $F_d$ is a function of the particle diameter, d; $v_r$ is the relative velocity between fluid and particle; $\rho$ is the fluid density; and $\mu$ is the fluid viscosity.

The general expression used in dimensional analysis is that the functionality is a constant multiplied by the product of the variables each raised to a power:

$$\frac{F_d}{A_p} = C(v_r)^a\rho^b\mu^c d^e$$

Expressing in terms of the base units, mass, M; length, L; and time, t:

$$\frac{ML}{t^2}\frac{1}{L^2} = \frac{L^a}{t^a}\frac{M^b}{L^{3b}}\frac{M^c}{L^c\,t^c}L^e$$

Consider the coefficients of each base unit. To be dimensionally consistent, the exponents of each unit must the opposite side of the equation. Exponents of L:

$$-1 = a - 3b - c - e \qquad (13.i)$$

Exponents of t:

$$2 = a + c; \quad a = 2 - c \qquad (13.ii)$$

Exponents of M:

$$1 = b + c; \quad b = 1 - c \qquad (13.iii)$$

Combining Eqs. (13.ii) and (13.iii) in (13.i):

$$-1 = 2 - c - 3(1 - c) - c + e$$
$$e = -c$$

Thus, the expression for drag force becomes:

$$\frac{F_d}{A_p} = CV^{2-c}\rho^{1-c}\mu^c d^{-c}$$

Combining terms with the same exponent:

$$\frac{F_d}{A_p} = C\left[\frac{\mu}{dv_r\rho}\right]^n(v_r)^2\rho \qquad (13.36)$$

The first two terms on the right are replaced by a drag coefficient, $C_d$, defined as $C_d = C/Re^n$.

Equation (13.36) is often written in terms of $C_d$ as follows:

$$F_d = C_d\frac{(v_r)^2\rho}{2} \qquad (13.37)$$

Consider a spherical particle settling within a suspending fluid. A sphere has a projected area of a circle, $A_p = \pi d^2/4$. If particle motion through the fluid results in laminar flow at the fluid interface with the particle, $C_d = 24/Re$. The equation for drag force becomes:

$$F_d = \left(\frac{24}{Re}\right)\left(\frac{\pi d^2}{4}\right)(v_r)^2\rho = 3\pi\mu dv_r \quad (13.38)$$

## 13.3.2  Terminal Velocity

The terminal velocity of the particle is the velocity when $dv_r/dt = 0$, that is, settling will occur at a constant velocity. Substituting Eqs. (13.35) and (13.38) into Eq. (13.34):

$$F_t = ma_e - ma_e\left(\frac{\rho}{\rho_p}\right) - 3\pi\mu dv_r$$

$F_t = ma_p = m(dv_r/dt)$. The mass of the particle is the product of the volume and the density. For a sphere, $m = \pi d^3\rho/6$. Thus, at the terminal velocity, $v_t$:

$$ma_e\left(1 - \frac{\rho}{\rho_p}\right) = 3\pi\mu v_t d$$

Substituting for m:

$$v_t = \frac{a_e\left(\rho_p - \rho\right)d^2}{18\mu} \qquad (13.39)$$

Equation (13.39) is *Stokes' = law* for velocity of particle settling within the suspending fluid. For gravity settling, $a_e = g$. For centrifugation, $a_e = \omega^2 r_c$, where $r_c$ = radius of rotation and $\omega$ = the angular velocity of rotation.

### 13.3.3   The Drag Coefficient

$C_d$ used in the derivation of Eq. (13.39) is only good when $v_r$ is small and laminar conditions exists at the fluid-particle interface. In general, $C_d$ can be determined by calculating an index K as follows:

$$K = d\left[\frac{a_e\rho\left(\rho - \rho_p\right)}{\mu^2}\right]^{0.33} \qquad (13.40)$$

Data by Lapple and Shepherd (Ind. Eng. Chem. 32:605, 1940) shows that the value of $C_d$ can be determined for different values of K as follows:

If K > 0.33 or Re < 1.9; $C_d = 24/Re$

If 1.3 < K < 44 or 1.9 < Re < 500; $C_d = 18.5/Re^{0.6}$

If K > 44 or Re > 500; $C_d = 0.44$

If the Reynolds number at the particle to fluid interface exceeds 1.9, Eqs. (13.35) and (13.37) are combined in Eq. (13.34), and under conditions when the particle is at the terminal velocity:

$$ma_e\left[1 - \left(\frac{\rho}{\rho_p}\right)\right] = C_d A_p (v_r)^2 \frac{\rho}{2}$$

Substituting $A_p = \pi d^2/4$; and $m = \pi d^3 \rho_p/6$ and solving for $v_r$:

$$v_r = 1.155\left[\frac{a_e d\left(\rho_p - \rho\right)}{\rho C_d}\right]^{0.5} \qquad (13.41)$$

**Example 13.10**   In the processing of soybeans for oil extraction, hulls must be separated from the cotyledons in order that the soy meal eventually produced will have a high value as feed or as a material for further processing for food proteins. The whole soybeans are passed between a cracking roll which splits the cotyledons and produces a mixture of cotyledons and hulls. Figure 13.9 shows a system that can be used to separate hulls from the cotyledons by air classification. The projected diameters and the densities of cotyledons and hulls are 4.76 mm and 1003.2 kg/m³ and 6.35 and 550 kg/m³, respectively. The process is

**Fig. 13.9** Diagram of air classification system to separate soybean hulls from the cotyledons carried out at 20 °C. Calculate the terminal velocity of hulls and cotyledons in air. The appropriate velocity through the aperture will be between these two calculated terminal velocities.

**Solution**

The density of air at 20 °C is calculated using the ideal gas equation. The molecular weight of air is 29; atmospheric pressure = 101.325 kPa:

$$R = 8315 \, N \cdot M/(kgmole \cdot K)$$

$$\rho = \frac{p(29)}{RT} = \frac{101,325(29)}{8315(293)} = 1.206 kg/m^3$$

The viscosity of air at 20 °C is 0.0175 centipoise (from *Perry's Chemical Engineers Handbook*). Using Eq. (13.40):

$$a_e = \text{acceleration due to gravity} = 9.8 m/s^2.$$

For the cotyledons:

$$K = 0.00476 \left[ \frac{9.8(1.206)(1003.2 - 1.206)}{[(0.0175)(0.001)]^2} \right]^{0.33}$$

$$= 145$$

Because K > 44, $C_d = 0.44$.
Using Eq. (13.41), $a_e = 9.8$ m/s$^2$:

$$v_r = 1.155 \left[ \frac{9.8(0.00476)(1003.2 - 1.206)}{1.206(0.44)} \right]^{0.5}$$

$$= 10.84 m/s$$

For the hulls:

$$K = 0.00476 \left[ \frac{9.8(1.206)(550 - 1.206)}{[0.0175(0.001)]^2} \right]^{0.33}$$

$$= 118.9$$

K > 444; therefore, $C_d = 0.44$.
Using Eq. (13.41):

$$v_r = 1.155 \left[ \frac{9.8(0.00635)(550 - 1.206)}{1.206(0.44)} \right]^{0.5}$$

$$= 9.266 \, m/s$$

Thus air velocity between 9.266 and 10.84 m/s must be used on the system to properly separate the hulls from the cotyledons.

Equations (13.37), (13.38), (13.39), and (13.40) also apply in sedimentation, flotation, centrifugation, and fluidization of a bed of solids. In flotation separation where gravity is the only external force acting on the particles, Eq. (13.39) shows that solids having different densities can be separated if the flotation fluid density is between the densities of the two solids. The solid with lower density will float and that with the higher density will sink. Air classification of solids with different densities as shown in the example can be achieved by suspending the solids in air of the appropriate velocity, the denser solid will fall, and the lighter solid will be carried off by the air stream. Air classification is a commonly used method for removal of lighter foreign material from grains and in fractionation of milled wheat into flour with different protein and bran contents. Density gradient separations are employed in fractionating beans of different degrees of maturity and crab meat from scraps after manual removal of flesh.

In fluidization, the fluid velocity must equal the terminal velocity to maintain the particles in a stationary motion in the stream of air. Fluidized bed drying and freezing are commercially practiced in the food and pharmaceutical industries.

## Problems

13.1. Based on data for apple juice filtration in Table 13.2, calculate the filtration area of a rotary filter that must be used to filter 4000 L/h of juice having a suspended solids content of 50 kg/m$^3$. Assume k is proportional to solids content. The medium resistance increases with increasing precoat thickness. A 10-cm-thick precoat is usually used on rotary filters. Assume the medium resistance is proportional to the precoat thickness. The rotational speed of the filter is 2 rev/min, and the diameter is 2.5 m. A filtration cycle on a rotary filter is the time of immersion of the drum in the slurry, and in this particular system, 45% of the total circumference is immersed at any given time.

13.2. The following data were obtained in a laboratory filtration of apple juice using a mixture of rice hulls and perlite as the filter aid. The filtrate had a viscosity of 2.0 centipose. The suspended solid in the juice is 5 g/L, and rice hulls and perlite were added each at the same concentration as the suspended solids. The filter has an area of 34 cm³. Calculate:

(a) The specific cake resistance and the medium resistance

(b) The optimum filtration time per cycle and the average volume of filtrate per unit area per hour at the optimum cycle time

| Time(s) | Volume (mL) | Time (s) | Volume (mL) |
|---------|-------------|----------|-------------|
| 100     | 40          | 400      | 83          |
| 200     | 60          | 500      | 91          |
| 300     | 73          |          |             |

13.3. The following data were reported by Slack [Process Biochem. 17(4):7, 1982] on flux rates at different solids content during ultrafiltration of milk. Test if the flux vs. solids content follow that might be predicted by the theory on polarization concentration. What might be the reasons for the deviation? All flow rates are the same, and the same membrane was used in the series of tests.

| Mean solids content (%) | Flux (L/(m² ≅ h)) |
|-------------------------|--------------------|
| 12.3                    | 28.9               |
| 13.6                    | 25.5               |
| 14.9                    | 22.1               |
| 24.1                    | 23.8               |

13.4. The following analysis has been reported for retentate and permeate in ultrafiltration of skim milk. Retentate 0.15% fat, 16.7% protein, 4.3% lactose, and 22.9% total solids. Permeate 0% fat, 0% protein, 4.6% lactose, and 5.2% total solids.

(a) Calculate the rejection factor for lactose by the membrane used in this process.

(b) If the membrane flux in this process follows the data in Problem 3, calculate the lactose content of skim milk which was subjected to a diafiltration process where the original milk containing 8.8% total solids, 4.5% lactose, 3.3% protein, and 0.03% fat was concentrated to 17% total solids, diluted back to 10% total solids and reconcentrated by UF to 20% total solids.

13.5. Calculate the average particle diameter and total particle surface area assuming spherical particles, per kg solids, for a powder which has a bulk density of 870 kg/m³ and which has the following particle size distribution:

| +14, | −10 mesh | 10%   |
|------|----------|-------|
| +18, | −14 mesh | 14.6% |
| +25, | −18 mesh | 22.3% |
| +35, | −45 mesh | 30.2% |
| +60, | −45 mesh | 22.9% |

## Suggested Reading

Cheryan, M. (1986). *Ultrafiltration handbook*. Lancaster: Technomic Publishing Co.

Crewspo, J. G., & Boddeker, K. W. (1994). *Membrane processes in separation and purification*. Boston: Kluwer Academic Publishers.

Green, D. W., & Mahoney, J. O. (1997). *Perry's chemical engineers handbook*. (7th ed.). Chapter 18 and 22. McGraw Hill: New York.

Noble, R. D., & Stern, S. A. (1995). *Membrane separations technology: Principles and applications*. New York: Elsevier.

Poole, J. B., & Doyle, D. (1968). *Solid-liquid separation*. New York: Chemical Publishing Co.

Porter, M. C. (1979). Chapter 2: Membrane filtration. In P. A. Schweitzer (Ed.), *Handbook of separation techniques for chemical engineers* (pp. p2-1–p2-103). New York: McGraw-Hill.

Sourirajan, S. (1970). *Reverse osmosis*. New York: Logos Press.

# Extraction

One area in food processing that is receiving increasing attention is extraction. This separation process involves two phases. The solvent is the material added to form a phase different from that where the material to be separated originally was present. Separation is achieved when the compound to be separated dissolves in the solvent, while the rest of the components remain where they were originally. The two phases may be solid and liquid, immiscible liquid phases, or solid and gas. Solid-liquid extraction is also called leaching. In supercritical fluid extraction, gas at supercritical conditions contacts a solid or a liquid solution containing the solute. Extraction has been practiced in the vegetable oil industry for a long time. Oil from soybean, corn, and rice bran cannot be separated by mechanical pressing; therefore, solvent extraction is used for their recovery. In the production of olive oil, the product from the first pressing operation is the extra-virgin olive oil, the residue after first press may be repressed to obtain the virgin olive oil, and further recovery of oil from the cake is done by solvent extraction. Oil from peanuts is recovered by mechanical pressing and extraction of the pressed cake to completely remove the oil. One characteristic of solvent extracted oilseed meal is the high quality of the residual protein, suitable for further

processing into food-grade powders. They may also be texturized for use as food protein extenders.

Extraction of spice oils and natural flavor extracts has also been practiced in the flavor industry. Interest in functional food additives used to fortify formulated food products has led to the development of extraction systems to separate useful ingredients from food processing waste and medicinal plants.

Extraction is also used in the beet sugar industry to separate sugar from sugar beets. Sugar from sugarcane is separated by multistage mechanical expression with water added between stages. This process may also be considered a form of extraction. Roller mills used for mechanical expression of sugarcane juice are capital intensive, and when breakdowns occur, the downtime is usually very lengthy. It is also an energy-intensive process; therefore, modern cane sugar processing plants are installing diffusers, a water extraction process, instead of the multiple roller mills previously used.

In other areas of the food industry, water extraction is used to remove caffeine from coffee beans, and water extraction is used to prepare coffee and tea solubles for freeze or spray drying. Supercritical fluid extraction has been found to be effective for decaffeinating coffee and tea and for preparing unique flavor extracts from fruit and leaves of plants.

## 14.1  Types of Extraction Processes

Extraction processes may be classified as follows.

### 14.1.1  Single-Stage Batch Processing

In this process, the solid is contacted with solute-free solvent until equilibrium is reached. The solvent may be pumped through the bed of solids and recirculated, or the solids may be soaked in the solvent with or without agitation. After equilibrium, the solvent phase is drained out of the solids. Examples are brewing coffee or tea and water decaffeination of raw coffee beans.

### 14.1.2  Multistage Cross-Flow Extraction

In this process, the solid is contacted repeatedly, each time with solute-free solvent. A good example is soxhlet extraction of fat in food analysis. This procedure requires a lot of solvent, or in the case of a soxhlet, a lot of energy is used in vaporizing and condensing the solvent for recycling; therefore, it is not used in industrial separation process.

### 14.1.3  Multistage Countercurrent Extraction

This process utilizes a battery of extractors. Solute-free solvent enters the system at the opposite end from the point of entry of the unextracted solids. The solute-free solvent contacts the solids in the last extraction stage, resulting in the least concentration of solute in the solvent phase at equilibrium at this last extraction stage. Thus, the solute carried over by the solids after separation from the solvent phase at this stage is minimal. Solute-rich solvent, called the extract,

emerges from the system at the first extraction stage after contacting the solids that had just entered the system. Stage-to-stage flow of solvent moves in a direction countercurrent to that of the solids. The same solvent is used from stage to stage; therefore, solute concentration in the solvent phase increases as the solvent moves from one stage to the next, while the solute concentration in the solids decreases as the solids move in the opposite direction. A good example of a multistage countercurrent extraction process is oil extraction from soybeans using a carousel extractor. This system called the "rotocell" is now in the public domain and can be obtained from a number of foreign equipment manufacturers. A similar system produced by Extractionstechnik GmbH of Germany was described by Berk in a FAO publication. In this system (Fig. 14.1), two cylindrical tanks are positioned over each other. The top tank rotates, while the lower tank is stationary. Both top and bottom tanks are separated into wedges, such that the content of each wedge are not allowed to mix. Each wedge of the top tank is fitted with a swinging false bottom to retain the solids, while a pump is installed to draw out solvent from each of the wedges except one, in the lower tank. A screw conveyor is installed in one of the wedges in the

**Fig. 14.1** The rotating basket extractor (Source: Berk, Z. 1992. Technology for the production of edible flours and protein products from soybeans. FAO Agricultural Services Bulletin 97. Food and Agriculture Organization of the United Nations, Rome)

lower tank to remove the spent solids and convey them to a desolventization system. The false bottom swings out after the last extraction stage to drop the solids out of the top tank into the bottom wedge fitted with the screw conveyor. The movement of the wedges on the top tank is indexed such that with each index, each wedge will be positioned directly over a corresponding wedge in the lower tank. Thus, solvent draining through the bed of solids in a wedge in the top tank will all go into one wedge in the lower tank. Solvent taken from the wedge forward of the current wedge is pumped over the bed of solids, drains through the bed, and enters the receiving tank, from which another pump transfers this solvent to the top of the bed of solids in the preceding wedge. After the last extraction stage, the swinging false bottom drops down releasing the solids, the swinging false bottom is lifted in place, and the empty wedge receives fresh solids to start the process over again. A similar system although of a different design is employed in the beet sugar industry.

### 14.1.4   Continuous Countercurrent Extractors

In this system, the physical appearance of an extraction stage is not well defined. In its most simple form, an inclined screw conveyor may be pictured. The conveyor is initially filled with the solvent to the overflow level at the lower end, and solids are introduced at the lower end. The screw moves the solids upward through the solvent. Fresh solvent introduced at the highest end will move countercurrent to the flow of solids picking up solute from the solids as the solvent moves down. Eventually, the solute-rich solvent collects at the lowermost end of the conveyor and is withdrawn through the overflow. In this type of extraction system, the term "height of a transfer unit" (HTU) is used to represent the length of the conveyer where the solute transfer from the solids to the solvent is equivalent to one equilibrium stage in a multistage system.

Continuous conveyor type extractors are now commonly used in the oilseed industry. One type of extractor is a sliding cell basket extractor (Fig. 14.2a). The baskets affixed to a conveyor chain have false bottoms, which permit solvent sprayed at the top to percolate through and collect at a reservoir at the bottom of the unit. Pumps take the solvent from the reservoirs and take them to nozzles at the top of the baskets. The discharge point of the solvent at the top of the baskets is advanced such that the solvent weak in solute is fed to the baskets forward of the baskets from where the solvent had previously percolated.

Fig. 14.2 Continuous belt-type extractor. (a) An immersion-type multistage countercurrent extractor. (b) A percolation-type extractor

These units have been described in Berk's article and are produced by a German firm, Lurgi, GmbH. Another extractor suitable for not only oilseed extraction but also for extraction of health functional food ingredients from plant material is a perforated belt extractor. Figure 14.2b shows a perforated belt extractor produced in the United States by Crown Iron Works of Minneapolis, Minnesota. This unit is made to handle as small as 5 kg of solvent/h. A single continuous belt moves the solids forward, while solvent is sprayed over the solids. A series of solvent collection reservoirs underneath the conveyor evenly spaced along the length of the unit separates the solvent forming the different extraction stages. Each collection reservoir has a pump which takes out solvent from one stage, and this liquid is applied over the solids on the conveyor in such a manner that the liquid will drain through the bed of solids and collect in another collection reservoir of the preceding stage.

Most extractions in the food industry involve solid-liquid extraction; therefore, the discussion in this chapter will be limited to solid-liquid extraction.

## 14.2 General Principles

The following are the physical phenomena involved in extraction processes.

### 14.2.1 Diffusion

Diffusion is the transport of molecules of a compound through a continuum in one phase or through an interface between phases. In solid-liquid extraction (also known as leaching), the solvent must diffuse into the solid in order for the solute to dissolve in the solvent, and the solute must diffuse out of the solvent-saturated solid into the solvent phase. The rate of diffusion determines the length of time needed to achieve equilibrium between phases. In Chap. 12, section "Mass Diffusion," the solution to the differential equation for diffusion through a slab is presented in Eq. (12.41) in terms of the average

concentration of solute within a solid as a function of time. This equation may be used to calculate the time needed for equilibrium to be achieved when the quantity of solid is small relative to the solvent such that concentration of solute in the solvent phase remains practically constant. Qualitatively, it may be seen that the time required for diffusion to occur in order to reach equilibrium is inversely proportional to the square of the diffusion path. Thus, in solvent extraction, the smaller the particle size, the shorter the residence time for the solids to remain within an extraction stage.

Particle size, however, must be balanced by the need for the solvent to percolate through the bed of solids. Very small particle size will result in very slow movement of the solvent through the bed of solids and increases the probability that fines will go with the solvent phase interfering with subsequent solute and solvent recovery.

In soybean oil extraction, the soy is tempered to a certain moisture content in order that they can be passed through flaking rolls to produce thin flakes without disintegration into fine particles. The thin flakes have very short diffusion path for the oil, resulting in short equilibrium time in each extraction stage, and solvent introduced at the top of the bed of flakes percolates unhindered through the bed. The presence of small particle solids is not desirable in this system because the fine solids are not easily removed from the solvent going to the solvent/oil recovery system. The high temperature needed to drive off the solvent will result in a dark-colored oil if there is a large concentration of fine particles.

In cottonseed and peanut oil extraction, a pre-press is used to remove as much oil as can be mechanically expressed. The residue comes out of the expeller as small pellets which then goes into the extractor.

Some raw materials may contain lipoxygenase, which catalyzes the oxidation of the oil. Extraction of oil from rice bran involves the use of an extruder to heat the bran prior to extraction to inactivate lipoxygenase. The extruder produces small pellets which facilitate the extraction process by minimizing the amount of fines that goes with the solvent phase.

In cane sugar diffusers, hammer mills are used to disintegrate the cane such that the thickness of each particle is not more than twice the size of the juice cells. Thus, equilibrium is almost instantaneous upon contact of the particles with water. The cane may be pre-pressed through a roller mill to crush the cane and produce very finely shredded solids for the extraction battery.

## 14.2.2   Solubility

The highest possible solute concentration in the final extract leaving an extraction system is the saturation concentration. Thus, solvent to solid ratio must be high enough such that, when fresh solvent contacts fresh solids, the resulting solution on equilibrium will be below the saturation concentration of solute.

In systems where the solids are repeatedly extracted with recycled solvent (e.g., supercritical fluid extraction), a high solute solubility will reduce the number of solvent recycles needed to obtain the desired degree of solute removal.

## 14.2.3   Equilibrium

When the solvent to solid ratio is adequate to satisfy the solubility of the solute, equilibrium is a condition where the solute concentration in both the solid and the solvent phases is equal. Thus, the solution adhering to the solids will have the same solute concentration as the liquid or solvent phase. When the amount of solvent is inadequate to dissolve all the solute present, equilibrium is considered as a condition where no further changes in solute concentration in either phase will occur with prolonged contact time. In order for equilibrium to occur, enough contact time must be allowed for the solid and solvent phases.

The extent to which the equilibrium concentration of solute in the solvent phase is reached in an extraction stage is expressed as a stage efficiency. If equilibrium is reached in an extraction stage, the stage is 100% efficient and is designated an "ideal stage."

## 14.3   Solid-Liquid Extraction: Leaching

### 14.3.1   The Extraction Battery: Number of Extraction Stages

Figure 14.3 shows a schematic diagram of an extraction battery with n stages. The liquid phase is designated the overflow, the quantity of which is represented by V. The solid phase is designated the underflow, the quantity of which is represented by L. The extraction stage is numbered 1 after the mixing stage where the fresh solids first contact the solute laden extract from the other stages and n as the last stage where fresh solvent first entered the system and where the spent solids leave the system. The stage where extract from the first extraction stage contacts the fresh solids is called the mixing stage. It is different from the other stages in the extraction battery because at this stage the solids have to absorb a much larger amount of solvent than in the other stages.

**Fig. 14.3**  Schematic diagram of an extraction battery for multistage countercurrent extraction

Each of the stages is considered an ideal stage. Solute concentration in the underflow leaving stage n must be at the designated level considered for completeness of the extraction process. Residual solute in the solids fraction must be maintained at a low level. If it is a valuable solute, the efficiency of solute recovery is limited only by the cost of adding more extraction stages. On the other hand, if the spent solids is also valuable and the presence of solute in the spent solids affects its value, then the number of extraction stages must be adequate to reduce the solute level to a minimum desirable value. For example, in oilseed extraction, the residual oil in the meal must be very low; otherwise, the meal will rapidly become rancid.

Determination of the number of extraction stages may be done using a stage-by-stage material balance. Because the only conditions known are those at the entrance and outlet from the extraction system, the stage-by-stage material balance will involve setting up a system of equations that are solved simultaneously to determine the conditions of solute concentration in the underflow and overflow leaving each stage. The procedure will involve assumption of a number of ideal stages, solving the equations, and comparing if the calculated level of solute in the underflow from stage n matches the specified value. The process is very tedious. A graphical method is generally used.

### 14.3.2   Determination of the Number of Extraction Stages Using the Ponchon-Savarit Diagram

This graphical method for determining the number of extraction stages in a multistage extraction process involves the use of an X-Y diagram. The coordinates of this diagram are defined as follows:

$$Y = \frac{solid}{solute + solvent}$$

$$X = \frac{solute}{solute + solvent}$$

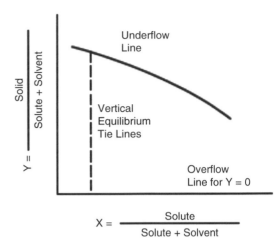

**Fig. 14.4** The underflow and overflow curves and equilibrium tie-lines on a Ponchon-Savarit diagram

where solid = concentration of insoluble solids. solute = concentration of solute, and solvent = concentration of solvent. Thus, the composition of any stream entering and leaving an extraction stage can be expressed in terms of the coordinates X,Y.

Figure 14.4 shows the X-Y diagram for a solid-liquid extraction process. The overflow line represents the composition of the solvent phase leaving each stage. If no solid entrainment occurs, the overflow line should be represented by $Y = 0$. The underflow line is dependent on how much of the solvent phase is retained by the solids in moving from one stage to the next. The underflow line will be linear if the solvent retained is constant and curved when the solvent retained by the solids varies with the concentration of solute. Variable solvent retention in the underflow occurs when the presence of solute increases significantly the viscosity of the solvent phase.

### 14.3.3   The Lever Rule in Plotting Position of a Mixture of Two Streams in an X-Y Diagram

Let two streams with mass, R and S, and with coordinates $X_r$ and $Y_r$, and $X_s$ and $Y_s$, respectively, be mixed together to form T with coordinates $X_t$ and $Y_t$. The diagram for the

**Fig. 14.5** Representation of a material balance on the X-Y diagram and the lever rule for plotting mass ratio of process streams as ratio of distances between points on the X-Y diagram

material balance is shown in Fig. 14.5(I). If R and S consist of only solute and solvent, solute balance results in:

$$RX_r + SX_s = (R + S)X_t \qquad (14.1)$$

A solid balance gives:

$$RY_r + SY_s = (R + S)Y_t \qquad (14.2)$$

Solving for R in Eq. (14.1) and S in Eq. (14.2) and dividing:

$$\frac{R}{S} = \frac{X_s - X_t}{X_t - X_r} = \frac{Y_s - Y_t}{Y_t - Y_r} \qquad (14.3)$$

Figure 14.5(II) represents the term involving Y in Eq. (14.3). The coordinate for the mixture should always be between those of its components. The ratio of the distance between the line $Y = Y_s$ and the line $Y = Y_t$, represented on the diagram by A, and the distance between the line $Y = Y_t$ and the line $Y = Y_r$, represented on the diagram by B, will equal the ratio of the mass of R and S.

Figure 14.5(III) represents the term involving X in Eq. (14.3). Again, the ratio of the masses of R and X equals the ratio of the distance C/distance D in the X-Y diagram. The composite of the material balance is drawn in Fig. 14.5(IV). The coordinates of points S and R when plotted in the X-Y diagram and joined together by a straight line will result in the point representing T to be in the line between S and R. The ratio of the distance on the line, between S and T, represented by E, and that between R and T, represented by F, is the ratio of the mass of R to S.

These principles show that a material balance can be represented in the X-Y diagram with each process stream represented as a point on the diagram, and any mixture of streams can be represented by a point on the line drawn between the coordinates of the components of the mixture. The exact positioning of the location of the point representing the mixture can be made using the lever rule on distances between the points as represented in Fig. 14.5(IV).

### 14.3.4 Mathematical and Graphical Representation of the Point J in the Ponchon-Savarit Diagram

Consider underflow, $L_b$, mixing with overflow, $V_a$, and the solid and solvent phases later separated to form the overflow stream $V_b$ and the underflow stream $L_a$. A total mass balance gives:

$$L_b + V_a = V_b + L_a = J \qquad (14.4)$$

The mixture of $L_b$ and $V_a$ forms the point J, shown in Fig. 14.6. A line drawn between the coordinates of $L_b$ and $V_a$ and another line drawn between the point $V_b$ and $L_a$ will intersect at point J. The point J may be used to help in plotting points representing the incoming and exiting streams in an extraction battery from the solvent to solid ratio.

### 14.3.5 Mathematical and Graphical Representation of the Point P

The point P is a mixture that results when the underflow leaving a stage mixes with the overflow entering that stage. It is also the mixture of overflow leaving a stage and underflow entering that stage. The mixture representing point P is shown schematically in Fig. 14.7. A total mass balance around the system represented by the dotted line in Fig. 14.6 gives:

$$-V_b + L_b = -V_a + L_a = -V_{n+1} + L_n$$
$$= P \qquad (14.5)$$

Thus, the point P will be an extrapolation of the line that joins $V_b$ and $L_b$, the line that joins $V_a$ and $L_a$, and the line that joins $V_{n+1}$ and $L_n$. All lines which join the underflow stream leaving a

**Fig. 14.6** Representation of the point J as the intersection of lines connecting four process streams entering and leaving a system

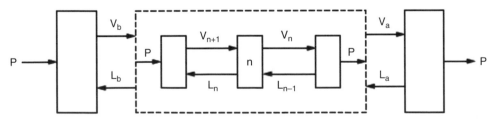

**Fig. 14.7** Representation of the point P for the common point through which all lines connecting the underflow leaving and the overflow entering an extraction stage must pass

stage and the overflow stream which enters that stage will all meet at a common point P. This is a basic principle used to draw the successive stages in a Ponchon-Savarit diagram for stage-by-stage analysis of an extraction process.

### 14.3.6  Equation of the Operating Line and Representation on the X-Y Diagram

Figure 14.8 represents an extraction battery with n cells. Each stage in the extractor may also be called an extraction cell. Cell $n + 1$ is the cell after cell n. The subscripts on the overflow stream, V, and the underflow stream, L, represent the cell from which the stream is leaving. Thus, the solvent phase entering cell n comes from cell $n + 1$ and is designated $V_{n+1}$, and the solid phase leaving cell n is designated $L_n$. Let V represent the mass of solute and solvent in an overflow stream and L represent the mass of solute and solvent in an underflow stream. A material balance around the battery of n cells is as follows:

Total mass balance:

$$V_{n+1} + L_a = L_n + V_a; \qquad V_{n+1}$$
$$= L_n + V_a - L_a \qquad (14.6)$$

Solute balance:

$$V_{n+1}X_{n+1} = V_aX_{Va} - L_aX_{La} + L_nX_n \quad (14.7)$$

Substituting Eq. (14.6) in Eq. (14.7) and solving for $X_{n+1}$:

$$X_{n+1}\frac{L_n}{L_n + V_a - L_a}X_n + \frac{V_aX_{Va} + L_aX_{La}}{L_n + V_a - L_a}$$
$$(14.8)$$

Equation (14.8) is the equation of an operating line. It shows that the point representing stage $n + 1$, which has the coordinate $X_{n+1}$, is on the same line drawn through the point representing stage n, which has the coordinate $X_n$.

It can be shown that the coordinates of point P represented by Eq. (14.5) can also satisfy Eq. (14.8); thus, a line drawn from a point representing stage n to point P will allow the determination of a point representing the coordinate of stage $n + 1$.

### 14.3.7  Construction of the Ponchon-Savarit Diagram for the Determination of the Number of Ideal Extraction Stages

Figure 14.9 shows the X-Y diagram for determination of the number of ideal stages. Known points representing the coordinates of solvent-saturated solids from the mixing stage, $L_a$, the extract, $V_a$, the spent solids, $L_b$, and the fresh solvent entering the system, $V_b$, are plotted first on the X-Y diagram. A total material balance must first be made to determine the coordinates of $L_b$ and $V_a$ from the solvent to solid ratio.

Point P is established from the intersection of lines going through $L_b$ and $V_b$ and $L_a$ and $V_a$.

The condition of equilibrium is represented by a vertical line. Equilibrium means that $X_{Ln}$ and $X_{Vn}$ are equal. The equilibrium line is also known as a "tie-line" and may not be vertical if equilibrium is not achieved. However, it is easier to assume equilibrium to occur, determine the number of ideal extraction stages, and incorporate the

**Fig. 14.8** Diagram representing the system used to make a material balance for deriving the equation of the operating line in a Ponchon-Savarit diagram

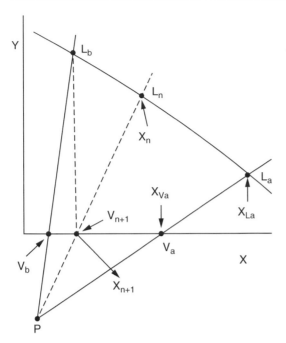

**Fig. 14.9** Plotting of point P, the operating line, and the equilibrium tie-line on a Ponchon-Savarit diagram

**Fig. 14.10** X-Y diagram for a single-stage extraction process

fact that equilibrium does not occur in terms of a stage efficiency.

The succeeding stages are established by drawing a line from the point representing stage n on the underflow line to point P. The intersection of this line with the overflow curve will determine $X_{n+1}$.

**Example 14.1** Draw a diagram for a single-stage extraction process involving beef (64% water, 20% fat, 16% nonextractable solids) and isopropyl alcohol in a 1:5 ratio. Isopropyl alcohol and water are totally miscible, and the mixture is considered to be the total solvent. Assume all fat dissolves in this solvent. Following equilibrium, the solids' fraction separated by filtration retained 10% by weight of the total solution of fat and solvent. Determine graphically the concentration of the fat in the extract, and calculate the fat content in the solvent-free solids.

**Solution:**
Basis: 100 kg beef. Because the X-Y diagram uses as a denominator for the X and Y coordinates only the mass of the solution phase rather than

total mass, the distances on the diagram should be scaled on the basis of the mass of solute and solvent present in each stream. Let $L_a$ represent the beef:

$$X_{La} = 20/(20 + 64) = 0.238$$
$$Y_{La} = 16/(20 + 64) = 0.190$$

The solvent, $V_b$, will have the coordinates (0,0) because it contains no solute nor solids. $L_a$ and $V_b$ are plotted in Fig. 14.10.

Point J, the point representing the mixture before filtration, is determined using the lever rule as follows: the ratio of the length of line segment $L_aJ$/length of line segment $V_bL_a$ equals the ratio of the mass of solvent and solute in $V_b$ to the total mass of solvent and solute in $V_b$ and $L_a$.

$$\text{Let}: V_bL_a = 100 \text{ units}$$

The ratio $V_b/(L_a + V_b) = 500/(20 + 64 + 500) = 0.856$. Therefore, the distance $V_bJ$ should be 85.6 units. For example, when plotting on a scale of 1 cm = 0.5 units for X and Y, length of $V_bL_a = 6.2$ cm. Thus, $L_aJ$ should be 6.2(0.856) = 5.31 cm long. These distances, when plotted, establish point J, which has coordinates 0.035, 1.1275. The vertical line X = 0.035 is the equilibrium line. Assuming no insoluble solids in the extract, $V_a$ has coordinates

**Table 14.1** Data for overflow composition during extraction of oil from soybeans

| $X = \frac{kg\,oil}{kg\,soln.}$ | Soln. retained kg/kg solids | $Y = \frac{Solids}{kg\,soln.}$ |
|---|---|---|
| 0 | 0.5 | 2.0 |
| 0.1 | 0.505 | 1.980 |
| 0.2 | 0.515 | 1.942 |
| 0.3 | 0.530 | 1.887 |
| 0.4 | 0.550 | 1.818 |
| 0.5 | 0.571 | 1.751 |
| 0.6 | 0.595 | 1.680 |
| 0.7 | 0.620 | 1.613 |

Solution represents the solvent phase, mass of solvent plus mass of dissolved solute

0.035, 0. Point $L_b$ is determined by plotting the distance $V_a L_b$ (5.5 cm), which is 10 times $V_a J$ (0.55 cm) because 10% of the solution is retained in the underflow. The ratio $X_{Lb}/Y_{Lb}$ = solute/ solids in $L_b$ = mass fraction solute in the solvent-free solids = 0.127.

**Example 14.2** Table 14.1 represents the amount of solution retained in soybean meal as a function of the oil concentration.
(a)  Draw the underflow curve.
(b)  If a solvent to soy ratio of 0.5:1 is used for extraction, and the original seed contains 18% oil, determine the number of extraction stages needed such that the meal after final desolventization will have no more than 0.01 kg oil/kg oil-free meal.

**Solution:**
(a)  The coordinates of points representing the underflow curve are given in Table 14.1 The parameter Y is the reciprocal of the solution retained/kg solids. The underflow curve is linear except for the last three points at high oil concentrations where there was a slight deviation from linearity. A regression analysis results in the equation of the best fit line as:

$$Y = -0.5775\ X + 2.036$$

Figure 14.11 shows the Ponchon-Savarit diagram for this problem. The underflow curve is plotted in Fig. 14.12.
(b)  The composition of the final extract and spent solids stream is calculated by performing a material balance around the whole system. Basis: 1 kg soybean; solvent = 0.5 kg. From the specified level of oil in the extracted meal:

$$kg\ oil\ in\ spent\ solids = \frac{0.01\,kg\,oil}{kg\,solids}0.82\ kg\ solids$$

$$= 0.0082$$

From the data in Table 14.1, at very low oil contents, the amount of solution retained by the solids is 0.5 kg/kg solids:

$$kg\ soln.in\ spent\ solids = \frac{0.5\,kg\,soln.}{kg\,solids}0.82\ kg\ solids$$

$$= 0.41$$

$$kg\ solvent = \frac{1\,kg\ solvent}{1.0082\,kg\,soln.}0.41\ kg\ soln.$$

$$= 0.4018\ kg\ solvent\ in\ spent\ solids$$

Mass of $L_n$, the spent solids $= 0.82 + 0.0082$

$$+0.4018 = 1.23\ kg$$

Coordinates of spent solids, $L_n$:

$$X = 0.0082/0.41 = 0.02\ kg\ solute/soln.$$

$$Y = 1\ kg\ solids/0.5\ kg\ soln.$$
$$= 2.0\ kg\ solids/kg\ soln.$$

Figure 14.13 shows the overall material balance with the mass and components of each stream entering and leaving the extraction battery:

Mass of extract $V_a = 1.5 - 1.23 = 0.27$ kg
Mass of oil $V_a = 0.82 - 0.0082 = 0.1718$ kg

The Ponchon-Savarit diagram can be constructed only for the cells following the mixing stage. Figure 14.14 shows the material balance for the mixing stage which is considered as cell 1 in this case and the stream entering and leaving stage 2. Material balance around cell 1:

**Fig. 14.11** The underflow curve and operating lines establishing the number of equilibrium stages needed to solve the example problem on continuous countercurrent oil extraction from soybeans

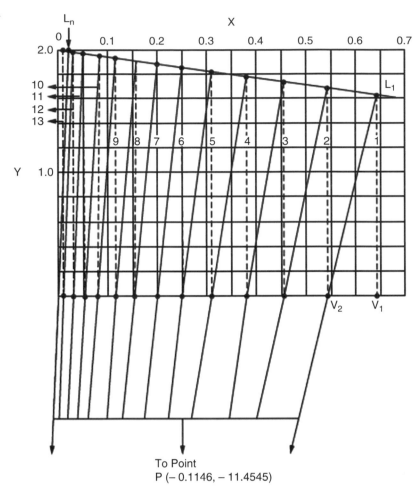

To Point
P (− 0.1146, − 11.4545)

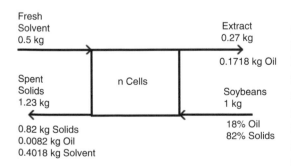

Fresh
Solvent
0.5 kg

Spent
Solids          n Cells
1.23 kg

0.82 kg Solids
0.0082 kg Oil
0.4018 kg Solvent

Extract
0.27 kg

0.1718 kg Oil

Soybeans
1 kg

18% Oil
82% Solids

**Fig. 14.12** Overall material balance to determine the extract mass and composition and spent solids' mass and composition for the example problem on soybean extraction

Final extract, $V_a$: X = 0.1718/0.27 = 0.636 kg oil/kg soln.

Because the last three data points deviated slightly from linearity, the amount of solution retained by the solids in stream $L_1$ will be calculated by interpolation from the tabulated data instead of the regression equation. Because final extract $V_a$ is in equilibrium with the solids in stream $L_1$, the X coordinate of stream $L_1$ must be the same as for the final extract. Thus, X for stream $L_1 = 0.636$ kg oil/kg soln. At the X value of $L_1 = 0.636$, the solution retained by the solids is obtained from Table 14.1 by interpolation.

$$= 0.595 + \frac{(0.620 - 0.595) \cdot (0.656 - 0.06)}{(0.7 - 0.6)}$$

$$= 0.609 \frac{kg\ soln.}{kg\ solids}$$

**Fig. 14.13** Diagram showing material balance around the mixing stage, cell 1, to establish the composition of overflow stream leaving and underflow stream entering the extraction battery after the mixing stage

**Fig. 14.14** Schematic diagram of a supercritical fluid extraction system

C = Control Valve
A = Heat Exchange Fluid
    for Heating or cooling

Total mass of $L_1 = 0.82$ kg solids $+ 0.609$ (0.82) $= 1.3194$ kg

Y of point $L_1$ in Fig. 14.11 has a value of $\frac{1}{0.609} = 1.642 \ \frac{kg \ solids}{kg \ sol.}$

Calculating mass of $V_2 : V_2 = L_1 + V_a - 1 = 1.3194 + 0.27 - 1 = 0.5894$ kg

$$Mass \ oil \ in \ L_1 = \left[ 0.609 \frac{kg \ soln}{kg \ solids} \cdot (0.82 \ kg \ solids) \right] \cdot \left[ 0.656 \frac{kg \ oil}{kg \ soln} \right]$$

Mass oil in $L_1 = 0.3276$ kg
Oil balance around cell 1:

Oil in $V_2 = 0.1718 + 0.3276 - 0.18$
$\qquad = 0.3194$ kg

X coordinate for $V_2$:

$\qquad X = 0.3194/0.5894 = 0.5419$

The streams to be considered to start plotting of the Ponchon-Savarit diagram are streams $V_2$ and $L_1$ and the solute-free solvent stream and the spent solids stream, $L_n$.

$V_2$ and $L_1$ are first plotted in Fig. 14.11. A line is drawn through the two points and extended beyond the graph. The equation of the line that connects the two points may also be determined. Points representing the solute-free solvent and the spent solids stream, $L_n$, are plotted next. The solute-free solvent is represented by the origin. A line is drawn through the two points and extended until it intersects the previous line drawn between $L_1$ and $V_2$. The equation for the line may also be determined, and the point of intersection, representing the point, P, can be calculated by solving the two equations simultaneously. The coordinate of point P is $-0.1146, -11.4545$.

An equilibrium line drawn from $V_2$ to the underflow curve will locate a point that represents $L_3$. Drawing a line from this point to connect with point P will locate an intersection on the overflow curve, for the point representing $V_3$. The process is repeated, until the last line drawn from the underflow curve to point P will intersect the overflow curve at a point that is equal to or less than the specified X for the spent solids, $L_n$. The number of extraction stages is the number of equilibrium tie-lines drawn in the diagram in Fig. 14.11. In this problem, 13 equilibrium stages are required.

## 14.4    Supercritical Fluid Extraction

Supercritical fluid extraction may be done on solids or liquids. The solvent is a dense gas at conditions of temperature and pressure where

further increase in pressure or a reduction in temperature will not result in a phase change from gas to liquid. The density of a supercritical fluid, however, is almost that of a liquid, but it is not a liquid. In addition, the solubility of solutes in a supercritical fluid approaches the solubility in a liquid. Thus, the principle of solute extraction from solids using a supercritical fluid is very similar to that for solid-liquid extractions discussed in Sect. 14.2.1 to 14.2.3.

### 14.4.1    Extraction Principles

Supercritical fluid extraction is done in a single-stage contractor with or without recycling of the solvent. When recycling is used, the process involves a reduction of pressure to allow the supercritical fluid to lose its ability to dissolve the solute, after which the solid is allowed to separate by gravity, and the gas at low pressure is compressed back to the supercritical conditions and recycled. Temperature reduction may also be used to drop the solute, and the solvent is reheated for recycling without the need for recompression.

Figure 14.14 shows a schematic diagram of a supercritical fluid extraction system. The basic components are an extractor tank and an expansion tank. Supercritical fluid conditions are maintained in the extractor. Temperature is

**Fig. 14.15**  Schematic diagram of a supercritical fluid extraction system using entrained ethanol in supercritical carbon dioxide

usually maintained under controlled conditions in both tanks. Charging and emptying the extractor is a batch operation. The pressure of the supercritical fluid is reduced by throttling through a needle valve or orifice after which it enters an expansion tank where the supercritical fluid becomes a gas. Because solute solubility in the gas is much less than in the supercritical fluid, solute separates from the gas in the expansion tank. The spent gas is then recompressed and recycled. Heat exchanges are needed to maintain temperatures and prevent excessive cooling at the throttling valve due to the Joule-Kelvin effect.

Two of the major problems of supercritical fluid extraction are channeling of solvent flow through the bed of solids and entrainment of the nonextractable component by the solvent. Time of solid-solvent contact is the quotient of extraction vessel volume divided by the solvent volumetric flow rate. The volume is calculated at the temperature and pressure inside the extraction vessel. Normally, volume of the solvent is measured at atmospheric pressure after the gas exited the expansion tank. From this measured volume, the number of moles of gas is calculated, and the volume of the supercritical fluid in the extraction vessel is then calculated using the equations of state for gases. The contact time should be adequate to permit solvent to penetrate solid particles and permit diffusion of solute from inside the solid particles into the solvent phase. To achieve equilibrium between the solution inside solid particles and the solvent phase, solvent flow must be adjusted to achieve the necessary contact time and to provide enough solvent such that concentration of dissolved solutes in the solvent phase will be below the solubility of solute in the solvent. A large quantity of solute to be extracted would require a larger rate of solvent flow to permit thorough solute extraction within a reasonable length of time. Supercritical fluid penetration into the interior of a solid is rapid, but solute diffusion from the solid into the supercritical fluid may be slow thus requiring prolonged contact time in the extraction vessel. Solvent flow rate, pressure, and temperature in the extraction vessel are the major supercritical fluid extraction process parameters.

## 14.4.2   Critical Points of Supercritical Fluids Used in Foods

Carbon dioxide is the most widely used supercritical fluid. The compound is nontoxic and nonflammable and is inexpensive and readily available. The critical point of carbon dioxide is 31.1 °C and 7.39 MPa (74 Bars). At temperature and pressure above the critical point, the fluid is supercritical. Ethanol may also be added in small amounts to supercritical carbon dioxide to change its polarity in some extraction operations. The critical point of ethanol is 243 °C and 6.38 MPa (64 Bars). Water may also be transferred from the solid to the solvent phase; therefore, the supercritical fluid in the extraction vessel may contain water. The critical point of water is 364 °C and 22.1 MPa (221 Bars).

## 14.4.3   Critical Point of Mixtures

The critical point of a mixture of compounds may be calculated as a mole fraction-weighted average of the individual components:

$$T_c = 3\left(x_1 T_{c1} + x_2 T_{c2} \ldots + x_n T_{cn}\right.$$
$$P_c = 3\left(x_1 P_{c1} + x_2 P_{c2} \ldots + x_n P_{cn}\right.$$

## 14.4.4   Properties of Supercritical Fluids Relative to Gases

The relative density of a supercritical fluid is in the range 0.1–1 s compared with a density of 1 for liquids and 0.001 for gases. The relative viscosity is 0.1–1 compared with 1 for liquids and 0.01 for gases. The relative diffusivity is 10–100 compared with 1 for liquids and $10^4$ for gases.

## 14.4.5   Supercritical Fluid Extraction Parameters

Separation of multiple solutes from a solid may be the objective of a supercritical fluid extraction process. For example, components of a fat or oil

**Table 14.2** Solubility of various compounds in supercritical $CO_2$ (in mg/g)

| Pressure (Bars)/temperature °C | 80/ 40 | 90/ 40 | 100/ 40 | 200/40 | 80/ 50 | 90/ 50 | 100/ 50 | 200/ 50 | 162/ 45 | 146/ 45 |
|---|---|---|---|---|---|---|---|---|---|---|
| Limonene | 5.8 | CM | | | | 17.1 | CM | | | |
| Lauric acid | | | 6.8 | 11.6 | | | | | | |
| Palmitic acid | | | 0.8 | 4.4 | | | | | | |
| Trilaurin | | | 0.1 | 26.8 | | | | | | |
| Tripalmitin | | | | 0.3 | | | | | | |
| Trimyristin | | | 0.5 | 5.6 | | | | | | |
| β-Carotene | | | | 0.0021 | | | | 0.0028 | | |
| Stearic acid[a] | | | | | | | | | 23.8 | 23.8 |
| Water | | | | | 1.6 | 1.9 | 2.8 | | | |

CM = completely miscible
[a]Solvent is $CO_2$ with acetic acid at 0.03 mole fraction as entrainer

high in polyunsaturated fatty acids may be removed from the mixture leaving triglycerides with saturated fatty acids with higher melting points and higher resistance to oxidation. Another example is the separation of phospholipids from the triglycerides in soybean oil to prevent gumming of the oil. Solubility of solutes in the supercritical solvent may be a function of pressure and/or temperature. Table 14.2 shows the solubility of some compounds in supercritical carbon dioxide. These data show that the right conditions may be selected to maximize solute solubility. Another method for changing solubility is the use of entrainers. For example, in Table 14.2, tristearin solubility in carbon dioxide with an acetic acid entrainer is very high at relatively low pressure. Data in Table 14.2 also shows that a pressure of 200 Bars and 40 °C should be good for extracting neutral lipids. Another example is the enhancement of the solubility of lecithin when ethyl alcohol is an entrainer in supercritical carbon dioxide. Other compounds with improved solubility in supercritical carbon dioxide-ethanol include limonoids from citrus seeds and phospholipids such as phosphatidylethanolamine and phosphatidylcholine from dried egg yolk.

Another technique used to separate solutes is by using a two-stage process. For example, one group of compounds may easily dissolve in the solvent under a given set of conditions while leaving the less soluble ones in the solid. A second extraction under conditions that favor the dissolution of the remaining solute in the solid

in the solvent will result in the isolation of these group of compounds from the other group removed in the first extraction. The following examples are optimized extraction conditions reported in the literature:

1. Preparation of ginger flavor extract from dried ginger. $CO_2$ at 15 g/min. First-stage extraction at 79 Bars and 30 °C, second stage at 246 Bars and 40 °C. (Yonei, Y, et al. J. Sup. Fl. 8:156–161, 1995)
2. Flavor extracts from coriander seed: 250 Bars and 40 °C. Celery powder and sage: 70 Bars and 20 °C (Catchpoole, O.J. et al. J. Sup. Fl. 9:273–279, 1996)
3. Egg phospholipids from dried egg yolk. First stage: $CO_2$ at 414 Bars and 45 °C. Second stage: $CO_2$ – 5% ethanol at 414 Bars and 45 °C (Shah A. et al. J. Sup. Fl. 30:303–313, 2004)

### 14.4.5.1 Supercritical Fluid Extraction of Spice Essential Oils

SFE extraction of spice oils is becoming commercially viable as the food industry continues to develop food formulations using only "natural" ingredients. For example, essential oil from the leaves of the plant rosemary (*Rosmarinus officinalis* L.) is widely used as a spice, flavoring agent, and antioxidant. The residue after SFE extraction for removal of essential oil can either be grounded into a powder and used as an antimicrobial agent in meat products, or a secondary

**Table 14.3** Supercritical $CO_2$ extraction parameters for various spices and herbs

| Material | Pressure, MPa | Temperature °C | $CO_2$ Flow Rate[a] Kg/h | Yield (kg extract/100 kg material) |
|---|---|---|---|---|
| Celery seed Dry | 14–35 | 40 | 1.5 | 6.0 |
| Ginger root 14% (w/w) $H_2O$ | 15–25 | 20–40 | 0.2016 | – |
| Oregano Dry leaves | 45 | 50 | 7.0 | 2.8 |
| Rosemary Essential oil | 11.5 | 40 | 0.3 | 0.676 |
| Rosemary Antioxidant | 35 | 100 | 0.3 | 1.33 |
| Sage Dry leaves | 9–12.8 | 25–50 | 0.003–0.021 | 4.8 |
| Turmeric 66.7% $H_2O$ | 30 | 30 | 0.147–0.169 | 8.0 |

Sources: Sovilj, M.N., Nikolovski, B.G., Spasojevic, M.D. 2011. Macedonian Journal of Chemistry and Chemical Engineering 30(2):197–220; Babovic, N., Djilas, S., Jadranin, M., Vajs, V., Ivanovic, J., Petrovic, S., Zizovic, I. 2010. Innovative Food Science and Emerging Technologies. 11:98–107
[a]$CO_2$ flow rate will depend on size of extraction vessel. Maximum yield was obtained when value of mass ratio $CO_2$/solid extracted $\cong 25$

SFE extraction of the phenolics components can be done to obtain an effective antimicrobial and antioxidant ingredient in meat products. The parameters for extraction of essential oil are different from when phenolic compounds are the extraction target. Use of cosolvents such as ethanol added to the $CO_2$ has been found to be effective when extracting compounds with low volatility. In addition, higher pressures and higher temperatures are used for extraction.

Table 14.3 shows a selection of SCFE extraction parameters for selected spices and herbs.

# Problems

14.1. Ground roasted coffee contains 5% soluble solids, 3% water, and 92% inert insoluble solids. In order to obtain an extract with high soluble solids content without having to concentrate it for spray drying, a countercurrent extraction process is to be used to prepare this extract. It is desired that the final extract contains 0.1 kg solubles/kg water and that the spent coffee grounds should have solubles not to exceed 0.005 kg/kg dry inert solids.

(a) Determine the water to coffee ratio to be used in the extraction.

(b) The coffee grounds carry 1 kg water/kg of soluble-free inert solids, and this quantity is constant with solute concentration in the extract. Calculate the number of extraction stages needed for this process.

14.2. A process for extracting sugar from sweet sorghum involves pressing the cane through a three-roll mill followed by shredding the fibrous residue (bagasse) and extracting the sugar out with water. The sorghum originally contained 20% fiber, 16% sugar, and 64% water. After milling, the moisture content of the bagasse is 55%. Because the fiber is used for fuel, after the extraction battery, the solids are squeezed to remove the absorbed solution, and the squeezed solution is added to the last stage of the extractor. The following are the constraints: The sugar recovery must be a minimum of 99%, and the concentration of sugar in the final extract must be 10%. The bagasse carries a constant amount of solution, 1.22 kg solution/kg fiber. Calculate:

(a) The water to solid ratio needed

(b) The number of ideal extraction stages

(c) The final sugar content if the extract is mixed with the juice first pressed out of the cane

14.3. In supercritical $CO_2$ extraction of resveratrol from grape seed and skin obtained as a by-product of the wine industry, 95% ethanol is mixed in with the $CO_2$ entering a SCFE extraction system by injecting it into the $CO_2$ line using a high-pressure syringe. The extraction is carried out at 48 MPa and 80 C. It is desired to use a SCFE extraction solvent containing 1 mole % ethanol in the ethanol-$CO_2$ mixture. Assume that the mass ratio $CO_2$/dry matter is set at 25. The grape-seed-skin material contains 60% (w/w) water. If the extraction vessel has a volume of 100 mL, and the density of the grape-seed-skin material is 900 kg/m$^3$, calculate:

(a) The mass of dry material in the wet grape-seed-skin material that can be extracted/batch

(b) The volume at 101 kPa and 25 °C per minute of $CO_2$ that will be used to set the desired mass ratio of $CO_2$ to dry grape-seed-skin if a $CO_2$ feed rate of 0.2 kg/h is used

(c) The rate at which the cosolvent, ethanol, must be injected into the $CO_2$ line to obtain the desired mole % of ethanol in the SCFE solvent.

(d) The required extraction time

## Suggested Reading

Berk, Z. (1992). *Technology for the production of edible flours and protein products from soybeans, FAO Agricultural Services Bulletin 97*. Rome: Food and Agriculture Organization of the United Nations.

Green, D. W., & Mahoney, J. O. (1997). *Perry's chemical engineers handbook* (7th ed.). New York: McGraw-Hill.

Johnston, K. P., & Penninger, J. M. L. (Eds.). (1989). *Supercritical fluid science and technology, ACS symposium series 406*. Washington, DC: American Chemical Society.

McCabe, W. L., Smith, J. C., & Harriott, P. (1985). *Unit operations in chemical engineering* (4th ed.). New York: McGraw-Hill.

Perry, R. H., & Chilton, C. H. (Eds.). (1973). *The chemical engineers handbook* (5th ed.). New York: McGraw-Hill.

Prabhudesai, R. K. (1979). Leaching. In P. A. Schweitzer (Ed.), *Handbook of separation techniques for chemical engineers*. New York: McGraw-Hill.

Worona, O., Rafinska, K., Mozenski, C., & Buszewski, B. (2017). Supercritical fluid extraction of bioactive compounds from plant materials. *Journal of AOAC International, 100*(6), 1624–1635.

# Emerging Food Processing Technologies 15

Heat treatment is commonly used to inactivate microorganisms and produce shelf-stable foods. These processes, when properly designed and operated, can ensure a high level of food safety. However, these processes often involve high-temperature long-time heating that causes deteriorations in food quality such as degradations in color, texture, and flavor and loss of vitamins and antioxidant capacities. Considerable effort has been made to develop new technologies that are able to effectively inactivate microbial while reducing the negative impact on quality attributes. A number of different processes have been studied extensively as alternatives to the conventional thermal processing. These methods can be categorized into thermal and nonthermal processing technologies, depending on their mechanism of the microbial inactivation. In this chapter, some of these emerging technologies are covered including thermal processing (microwave and radio-frequency heating) and nonthermal processing methods (high-pressure and pulsed electric field processing). The fundamentals of these technologies, some technical aspects, and research and application status are briefly introduced.

## 15.1 Microwave and Radio Frequency Heating

### 15.1.1 Fundamentals

Research on the use of microwave (MW) and radio frequency (RF) started during World War II as a by-product of the wide applications of radar technology. Nowadays, MW and RF heating is commonly used in the food industry for various applications. Dielectric and ionic heating are the major mechanisms involved in heating foods with MW or RF. A dielectric material is an electrical insulator that can be polarized by an applied electric field. When a dielectric is placed in the electric field, the weakly bonded molecules (such as free water) are polarized and reorient to align themselves with the polarity of the electric field via dipole rotation. As the electric field changes direction rapidly (e.g., $2.45 \times 10^9$ cycles per second in a 2450-MHz domestic oven), the quick rotation of the polar molecules creates frictional heat. As the dominant polar materials in food are water molecules, water content of the food is an important factor for the dielectric heating. Other food components such as salt, protein, and carbohydrates are also dipolar materials. As these components are volumetrically distributed within food material, dielectric heating results in volumetric heating, which is fast and more uniform when compared to conventional heating greatly based on heat conduction. At higher temperatures, the electric resistance heating from the dissolved ions also plays a role in the heating mechanisms.

In the electromagnetic spectrum, microwave and radio-frequency waves are at low-end frequency with long wavelength. The wavelength and the frequency have the following relation:

$$\lambda = \frac{c}{f} \qquad (15.1)$$

where $\lambda$ is the wavelength (m), f is the frequency (Hz), and c is the speed of light ($3 \times 10^8$ m/s). Specific frequencies are allocated for food applications, e.g., 2450 or 915 MHz is used for MW heating, corresponding to 12- or 34-cm wavelength. Domestic ovens are operated at 2450 MHz. RF heating uses frequencies of 13.56, 27.12, and 40.68 MHz, corresponding to wavelength 2211, 1105, and 737 cm, respectively, much greater than MW.

The dielectric properties of foods are the key parameters determining the coupling and distribution of electromagnetic energy, thus deciding the heating effectiveness of MW and RF. Dielectric properties are normally described by dielectric constant ($\varepsilon'$) and loss factor ($\varepsilon''$). $\varepsilon'$ describes the ability of a material to store energy in response to an applied electric field. $\varepsilon''$ describes the ability of the material to dissipate energy, i.e., the ability to generate heat. The relation between the two properties is given by:

$$\varepsilon'' = \varepsilon' \tan \delta \qquad (15.2)$$

where tan $\delta$ is the loss tangent, also called dissipation factor, indicating how much of the absorbed electric energy is dissipated as heat.

Dielectric properties increase with the moisture and salt content in food. Bulk density of powdered foods is also important as air has the lowest dielectric properties ($\varepsilon' = 1$, $\varepsilon'' = 0$); thus an increased air fraction in the mixture will lead to decreased dielectric constant and loss factor. Table 15.1 lists the dielectric properties of some common foods.

The dielectric properties also depend on temperature and frequency of applied electric field.

**Table 15.1**  Dielectric properties of common food products at room temperature (20–25 °C)[a]

| Foods | Dielectric constant | | | Loss factor | | |
|---|---|---|---|---|---|---|
| | 27.12 MHz | 915 MHz | 2450 MHz | 27.12 MHz | 915 MHz | 2450 MHz |
| Corn oil | | 2.6 | 2.5 | | 0.18 | 0.14 |
| Apple | 72.5 | 74.3 | | 120.4 | 8.5 | |
| Orange | 84 | 72.9 | | 223.3 | 16.5 | |
| Asparagus | | 74 | 71 | | 21 | 16 |
| Cherry | 91.2 | 73.7 | | 293.0 | 16.4 | |
| Carrot | | 73 | 72 | | 20 | 15 |
| Potatoes | | 65 | 54 | | 20 | 16 |
| White bread | 2.68 | 2.03 | | 3.90 | 0.59 | |
| Almond | 5.9 | 1.7 | | 1.2 | 5.7 | |
| Walnut | 4.9 | 2.2 | | 0.6 | 2.9 | |
| Salmon fillets | | 57 | 53 | | 23 | 16 |
| Cooked ham | | 61 | 60 | | 96 | 42 |
| Beef | | 62 | 61 | | 27 | 17 |
| Chicken breast | 75.0 | | 49.0 | 480.8 | | 16.1 |
| Sturgeon caviar salted | 129.8 | 29.8 | | 1349.4 | 40.5 | |
| Sturgeon caviar unsalted | 70.7 | 30.7 | | 470.8 | 18.7 | |

[a]Data are adapted from Tang J. (2015). J food sci. 80(8). Wang S, Tang J, Johnson JA, Mitcham E, Hansen JD, Hallman G, Drake SR, Wang Y. (2003) Biosystems Eng. 201–12. Al-Holy M, Wang Y, Tang J, Rasco B. (2005). J Food Eng. 564–570. Lyng J, Zhang L, Brunton N. (2005). Meat Sci. 589–602

**Fig. 15.1** Frequency-dependent dielectric constant (**a**) and loss factor (**b**) of broccoli powder with 14.9% moisture content at different temperatures (Adapted from Ozturk S, Kong F, Trabelsi S, Singh RK. (2016). J Food Eng. 169:91)

Generally, both $\varepsilon'$ and $\varepsilon''$ decrease with frequency and increase with temperature (Fig. 15.1). Therefore, regions of food with higher temperature will have higher dielectric properties, thus absorbing more supplied energy leading to faster temperature rise, a phenomenon called "runaway heating." A greater runaway heating can occur in RF than MW heating, due to the more rapid increase in loss factor with increasing temperature associated with lower frequencies.

### 15.1.2 Heat Generation by MW and RF in Food

The heat energy P (W/cm$^3$) generation in food when placed in an electric field can be calculated by using the equation:

$$P = \frac{q}{V} = 55.61 \times 10^{-14} fE^2 \varepsilon'' \qquad (15.3)$$

where f is applied frequency (Hz), E is the electric field strength (V/cm), and $\varepsilon''$ is the dielectric loss factor of the sample. f and E are set by the type of equipment used. Metal containers are opaque to microwave and radio waves; and they are reflected from the surface therefore none passes across to food contained inside. However, an electrically conductive metal with finite electrical resistance will heat up in the same manner as an electrical resistance wire will heat up when an electric current is passed through it. A continuous metal sheet with very low electrical resistance will not heat up in a microwave field. However, a discontinuous metal sheet such as metallized plastic contains many small areas of metal that presents a large resistance to electrical current flow; therefore intense heating occurs in these materials. Such materials called absorbers or intensifiers are used in microwavable packages of frozen breaded fried foods or pizza to ensure a crispy crust when heated in a microwave. Glass and plastic are practically transparent to microwave and radio-frequency waves, i.e., they transmit microwaves and very little energy is absorbed.

The frequency of microwave power generated by a microwave generator is declared on the name plate of the unit. The power output is also supplied by the manufacturer for each unit. The coupling efficiency of a microwave unit is expressed as the ratio of power actually supplied to the unit and the actual power absorbed by the material heated. When the quantity of material being heated is large, the power generated by the system limits the power absorbed, rather than the value predicted by Eq. (15.3). The time it takes to heat a large quantity of material can be used to determine the microwave power output of a unit.

If a maximum P is determined by varying the quantity of food heated until further reduction results in no further increase in P, this value will be the limiting power absorption and will be dependent on the dielectric loss properties of the material according to Eq. (15.3). If the dielectric loss factor of the material is known, it will be possible to determine the electromagnetic field strength which exists, and differences in heating rates of different components in the food mixture can be predicted using Eq. (15.3).

When the power output of a MW/RF unit limits the rate of energy absorption by the food, components having different dielectric properties will have different heating rates. Using subscripts 1 and 2 to represent component 1 and 2, Eq. (15.3) becomes:

$$q_1 = 55.61 \times 10^{-14} \frac{m_1}{\rho_1} fE^2 \varepsilon_1''$$

$$q_2 = 55.61 \times 10^{-14} \frac{m_2}{\rho_2} fE^2 \varepsilon_2''$$

Because $P = q_1 + q_2$, then:

$$P = 55.61 \times 10^{-14} fE^2 \left( \frac{m_1}{\rho_1} \varepsilon_1'' + \frac{m_2}{\rho_2} \varepsilon_2'' \right)$$

and

$$\frac{q_1}{P} = \frac{\rho_2 m_1 \varepsilon_1''}{\rho_2 m_1 \varepsilon_1'' + \rho_1 m_2 \varepsilon_2''} \quad (15.4)$$

Since $q_1 = m_1 C_{p1} \frac{dT_1}{dt}$, the equation can be rearranged to obtain the following form:

$$\frac{dT_1}{dt} = \frac{\rho_2 \varepsilon_1'' P}{C_{p1} (\rho_2 m_1 \varepsilon_1'' + \rho_1 m_2 \varepsilon_2'')}$$

Similarly

$$\frac{dT_2}{dt} = \frac{\rho_1 \varepsilon_2'' P}{C_{p2} (\rho_2 m_1 \varepsilon_1'' + \rho_1 m_2 \varepsilon_2'')}$$

The relative rate of heating is:

$$\frac{dT_1}{dT_2} = \frac{\rho_2 \varepsilon_1'' C_{p2}}{\rho_1 \varepsilon_2'' C_{p1}} \quad (15.5)$$

Similar expressions may be derived for more than two components.

**Example 15.1** The dielectric constant of beef at 23 °C and 2450 MHz is 28 and the loss tangent is 0.2. The density is 1004 kg/m$^3$ and the specific heat is 3250 J/(kg · K). Potato at 23 °C and 2450 MHz has a dielectric constant of 38 and a loss tangent of 0.3. The density is 1010 kg/m$^3$ and the specific heat is 3720 J/(kg · K).

(a) A microwave oven has a rated output of 600 W. When 0.25 kg of potatoes were placed inside the oven, the temperature rise after 1 min of heating was 38.5 °C. When 60 g of potato was heated in the oven, a temperature rise of 40 °C was observed after 20 s. Calculate the average power output of the oven and the mass of potatoes that must be present such that power output of the oven is limiting the rate of power absorption rather than the capacity of the material to absorb the microwave energy.

(b) If potatoes and beef are heated simultaneously, what would be the relative rate of heating?

**Solution**

(a) Assume that 0.25-kg mass of product is sufficient to make microwave power availability the rate-limiting factor for microwave absorption:

$$P = 0.25\,kg \left[\frac{3720\,J}{kg \cdot K}\right] \left[\frac{38.5\,K}{1\,min}\right] \left[\frac{1\,min}{60\,s}\right]$$

$$= 596.75\,W$$

When a small amount[of]material[is heated:]

$$P = 0.06\,kg \left[\frac{3720\,J}{kg \cdot K}\right] (40\,K) \left[\frac{1}{20\,s}\right]$$

$$= 446.4\,W$$

The amount absorbed with the small mass in the oven is much smaller than when a larger mass was present; therefore it may be assumed that power absorption by the material limits the rate of heating:

$$\frac{q}{V} = \left[\frac{446.4\,W}{0.06\,kg\dfrac{1\,m^3}{1010\,kg}}\right] \cdot \frac{1\,m^3}{10^{-6}\,cm^3} = 7.5144\,W/cm^3$$

Because, with a small load in the oven, the power absorption is only 7.5144 W/cm$^3$, this may be assumed to be $(q/V)_{lim}$, the maximum rate of power absorption by the material. If the power output of the oven as calculated above is 596.75 W, the mass present where the rate of power absorption equals the power output of the oven is:

$$\frac{q}{V}\bigg|_{lim} = \frac{P(\rho)10^{-6}}{m}$$

$$m = \frac{P(\rho)10^{-6}}{(q/V)_{lim}} = \frac{596.75(1010 \times 10^{-6})}{7.5144}$$

$$= 0.08\,kg$$

Thus, any mass greater than 0.08 kg will heat inside this microwave oven at the rate determined by the power output of the oven.

(b) Using Eq. (15.5): Let subscript 1 refer to beef and subscript 2 refer to the potato.

$$\frac{dT_1}{dT_2} = \frac{1010(28)(0.2)(3720)}{1004(38)(0.3)(3250)} = 0.565$$

Thus, the beef will be heating slower than the potatoes.

### 15.1.3 Penetration Depth of Microwave and Radio-Frequency Waves in Food

Attenuation factor ($\alpha'$) is a property of the material that determines the distribution of the energy within the material. A higher $\alpha'$ is related to a less uniformity of the energy distribution. It can be calculated with the equation:

$$\alpha' = \frac{2\pi}{\lambda} \left[\frac{\varepsilon'}{2}\left(\sqrt{1 + \tan^2\delta} - 1\right)\right]^{1/2} \quad (15.6)$$

Penetration depth (d) is another important property for dielectric heating. Penetration depth of electric field ($d_f$) is defined as the distance an incident electromagnetic wave can penetrate beneath the surface of a material as the electric field strength decreases to 1/e (e = 2.718) of its strength in the free space. The $d_f$ of the electric field is the inverse of the attenuation factor and can be calculated using the following equation:

$$d_f = \frac{\lambda}{2\pi} \left[\frac{2}{\varepsilon'\left(\sqrt{1 + \tan^2\delta} - 1\right)}\right]^{1/2} \quad (15.7)$$

The penetration depth of power ($d_p$) is defined as the distance at which the incident power is decayed to 1/e of the initial power. Lambert's expression is used to calculate power absorption:

$$P = P_0 e^{-2\alpha'd} \quad (15.8)$$

where $P_0$ is incident power entering the food and P is power at the depth d in food. Using this equation and $P/P_0 = 1/e$, the $d_p$ can be calculated as:

$$d_p = \frac{1}{2\alpha'} \quad (15.9)$$

$d_p$ is a few centimeters at 2450 MHz for a normal moist food and decreases at higher temperatures. The penetration depths at RF range are much greater than microwave

**Table 15.2** Penetration depth at different frequencies[a]

| Product | T (°C) | Penetration depth (cm) | | |
|---|---|---|---|---|
| | | 27 MHz | 915 MHz | 2450 MHz |
| Cheese sauce | 20 | 35.5 | 8.2 | 4.6 |
| | 80 | 22.8 | 4.5 | 2.6 |
| | 121.1 | 19.1 | 3.5 | 2.1 |
| Macaroni and cheese | 20 | 68.3 | 16.0 | 7.1 |
| | 80 | 39.8 | 9.1 | 5.3 |
| | 121.1 | 31.0 | 6.7 | 4.0 |

[a]Data adapted from Wang Y, Wig TD, Tang J, Hallberg LM. (2003). J Food Eng. 57: 257

frequencies (Table 15.2), which allow RF energy to penetrate food more deeply than MW energy. Therefore, MW heating is suitable for packages with relatively smaller thickness (e.g., 1–2 cm), while RF heating can be applied for larger-sized food packages.

**Example 15.2** Dielectric properties of the broccoli powder with 6.9% moisture content at 27.12 MHz are dielectric constant 3.72 and loss factor 0.05. The electric field strength was measured as 28.3 V/cm. Calculate the heat generated in the powder, the attenuation factor, and the penetration depth of electric field and power, respectively.

**Solution**
Using Eq. (10.3) to solve for the power dissipation:

$$P = 55.61 \times 10^{-14} f E^2 \varepsilon''$$

$$P = 55.61 \times 10^{-14} \times 27.12 \\ \times 10^6 \times (28.3 \text{ V/cm})^2 \\ \times 0.05 = 0.006 \text{ W/cm}^3$$

As $\tan \delta = \varepsilon''/\varepsilon' = 0.05/3.72 = 0.0134$

$$\alpha' = \frac{2\pi}{\lambda}\left[\frac{\varepsilon'}{2}\left(\sqrt{1 + \tan^2 \delta} - 1\right)\right]^{1/2} = 0.0073$$

$$d_f = 1/\alpha' = 1/0.0073 = 136 \text{ m}$$

$$d_p = 1/2\alpha' = 68 \text{ m}$$

**Fig. 15.2** Schematic diagram of a domestic oven

### 15.1.4   Technical Design

Microwave equipment mainly consists of microwave generator (magnetron), waveguides made of aluminum tubes, and a metal cavity where foods are heated. Figure 15.2 shows a diagram of 2450-MHz domestic microwave oven and the components. The magnetron is an oscillator converting the electric power into microwaves and emitting high-frequency radiant energy. Microwaves are directed to the cavity via a rectangular waveguide. The rotating stirrer scatters the transmitted energy throughout the oven cavity. The metallic wall of the cavity reflects the microwaves which are absorbed by the food and converted into heat. A turntable is used to enhance the uniformity of the microwaves in the cavity. The industry unit has the same components with longer waveguides. For a continuous operation, the cavity may be substituted with a tunnel fitted with a conveyor belt.

**Fig. 15.3** Schematic diagram of a pilot-scale 915-MHz microwave heating/drying system

**Fig. 15.4** Schematic diagram of a radio-frequency unit

Most of the MF heating are operated at 2450 MHz. 915-MHz system is also used due to its advantage in greater penetration depth. Figure 15.3 shows a 915-MHz pilot-scale microwave system built in Washington State University to study MW sterilization and pasteurization of foods. This system is comprised of a 5-kW power generator, a magnetron, a multimode stainless steel microwave cavity (1.07 m × 1.22 m × 1.47 m), rectangular waveguides, a stirrer (0.87-m diameter, 15 rpm), and a circulator to direct the reflected waves from the cavity to a matched water load. The generated microwave power can be adjusted between 0.2–5 kW, and power meters are used to monitor the microwave power input and reflected power from the cavity. This microwave unit incorporated an auxiliary hot air heating to improve heating uniformity and was used in disinfestation of insect pests in grains and fruits.

The RF source and applicator are usually built together (Fig. 15.4). An oscillating circuit, in which food itself is part of it, generates a rapidly alternating electric field inside a capacitor. The electric field strength is affected by the applied voltage, the distance between electrodes, and the frequency. The design of the electrodes is critical to create uniform electric field patterns and thus uniform heating patterns. The shape of electrodes and the distance between the two electrodes (electrode gap) are also important influencing the heating uniformity. The two commonly used electrodes are rod electrodes and plate electrodes. When electrode gap increases, the electric field strength decreases resulting in slower heat generation in food.

### 15.1.5   Research Status and Applications

As a volumetric heating process, MW and RF heating has the advantage of rapid heating, thus reducing heating time and the resultant quality changes in the treated foods compared to conventional heating. It is also possible to obtain greater heating uniformity in food than conventional heating. The process control is more convenient as the heating can be turned on and off instantaneously. MW and RF systems can be used to treat prepackaged food products because plastic packaging materials are transparent to the electromagnetic waves. MW and RF heating have found many applications in the food processing industry, including tempering of frozen foods, precooking of bacon, and finish drying of pasta, biscuits, and other cereal products.

MW and RF heating for food pasteurization and sterilization purposes have been widely explored. The major concern is heating nonuniformity that causes overheated or underheated areas in food. This nonuniform heating results from the discontinuous dielectric properties between foods and the surrounding medium (e.g., air or water), as well as from the variation in the dielectric properties of different food constituents. Composition and geometry of food and packaging materials also affect the heating uniformity. Studies have focused on developing predictive methods to quickly and reliably determine the location of cold spots and investigate methods to improve the heating uniformity. Digital imaging and computer simulations have been used to generate 3D heating patterns of food. Fiber-optic sensors are used to monitor the rapid temperature rise in foods during heating.

The commercial application of MW processing in food pasteurization has been successfully accomplished in the United States, Europe, and Japan, where MW heating is used to pasteurize and sterilize prepacked foods including cakes, breads, pasta, milk, and yoghurt. In the United States, FDA has set the rigorous food safety assurance requirements for low-acid foods in hermetically sealed containers. It was not until 2009 that FDA approved the first use of microwave heating to sterilize in-package food products, called "Microwave Assisted Thermal Sterilization (MATS™) system," developed by Washington State University. 915-MHz microwaves were selected instead of 2450 MHz due to its higher energy efficiency and the deeper penetration allowing heating of thicker packaged foods. To avoid edge heating, food packages is emerged in the water. The similarity between the dielectric properties of water and foods reduces reflection and refraction of microwaves at the interface between water and foods, thus reducing or eliminating edge heating of the foods. The system consists of four zones, i.e., for preheating, microwave heating, holding, and cooling; each has its own water circulation and external temperature control system. In operation, up to 48 single meal packages are moved on a microwave transparent conveyor from the preheating zone through the four cavities in a single file. The packages then move through the holding zone and, in the end, into the cooling zone. The heating time of food in the system is only 5–8 min, as compared with up to 1 h in conventional canning process.

RF heating has been used to control pathogens in products such as nut meats, flour, pasta, and fishmeal. It is also used to control insects in postharvest agricultural products. Due to the higher moisture content, insect has higher dielectric loss factor than host products, so that the product is heated more slowly than insect that can reduce quality loss. Various methods are developed to improve the RF heating uniformity including using hot air or water as surface heating, sample moving, and mixing. RF heating uniformity can be also improved by placing the samples in the middle of the two plate electrodes and using a similar dielectric material wrapping the samples.

Figure 15.5 shows a pilot-scale 27.12-MHz, 6-kW RF system used to pasteurize food powders. The electrode gap can be adjusted between 11 and 15 cm leading to different heating rate and uniformity. Food samples held in polyetherimide container are placed in between two parallel electrodes and heated to 80 °C, which is in the range of lethal temperature required to kill most insects and microorganisms. The change

**Fig. 15.5** Scheme view of a 6-kW 27.12 MHz RF oven with the fiber-optic sensor to monitor the central temperature in the sample and infrared camera system to capture temperature map. (Adapted from Ozturk S, Kong F, Singh RK, Kuzy JD, Li C. (2017). Innov. Food Sci. Emerg. Technol. 44:191)

in temperature of the sample during heating is recorded at the geometric center of the sample using a fiber-optic temperature sensor. The heating rates can be calculated from the temperature time history. The heating rate was found to be 10 °C/min for 3-kg corn flour to reach 80 °C in the RF oven (as a comparison, the heating rate at the geometric center of the same amount of material is 0.3 °C/min by hot air heating). An infrared camera is used to capture thermal images of the surface of the sample, and the heating uniformity is evaluated from the temperature map. The data of heating rate and uniformity are essential to design a RF heating protocol for food industry to inactivate microbials. Process validation using pathogenic bacteria or their surrogate is also required before the technology can be commercialized.

## 15.2 High-Pressure Processing

### 15.2.1 Fundamentals

High-pressure processing, also described as high hydrostatic pressure (HHP), or ultrahigh-pressure (UHP) processing is a process in which liquid and solid foods are subjected to very high pressures between 100 and 800 MPa to inactivate certain food microorganisms and enzymes. During the HHP process, the freshness of food can be largely kept due to minimal exposure of the food product to heat. HHP is thus categorized as a "nonthermal processing" to achieve high-quality microbiological safe products. As consumer's demand for fresh and nutritious food is increasing, the high-pressure processes have become one of the innovative technologies that are most accepted by consumers. High-pressure treatment to kill bacteria was observed in 1890 in milk. However the commercial application of HPP process was not started until 100 years later. In 1986, high pressure was used in Japan to treat acid food with refrigerated storage conditions.

High-pressure treatment alters morphology of microbial cells and disrupts cell membranes. Disruption of cell membrane causes leakage of intracellular constituent through the permeabilized cell membrane. The extent of microbial inactivation by HHP depends on a number of interacting factors, including magnitude and duration of high-pressure treatments, temperature, and type and number of microorganisms. Food composition, pH, and water activity can also influence the effect of microbial inactivation. HPP is an isostatic processing, in which high pressure acts

**Fig. 15.6** A batch HHP
system

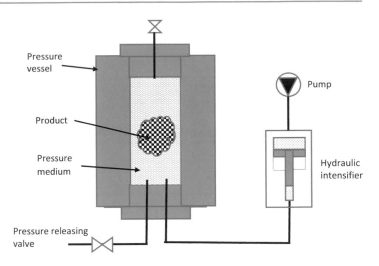

instantaneously and uniformly throughout mass
of food, so that food size, shape, and composition
are not factors for process determination. This is
another great advantage as compared to heat treatment which may require substantial time for heat
transfer to the slowest heating area in the food.

HHP processing has little effect in sensory,
physicochemical, and nutritional properties. It
affects only non-covalent bonds such as hydrogen, ionic, and hydrophobic bonds and has no
effect on small molecules such as amino acids,
vitamins, and flavor compounds. Thus there are
little changes in key food qualities such as flavor,
color, or nutritional content. However, high-
pressure processing can cause unfolding of protein chains and partial or complete protein dena-
turation which makes it an effective method
reducing the activity of food quality-related
enzymes such as oxidases.

High-pressure food preservation may be
applied as a pasteurization or a sterilization process. High-pressure pasteurization of high-acid
products may produce a product that is stable
during ambient storage. Use of the process on
low acid products may extend refrigerated shelf
life post-treatment. High-pressure sterilization
may be also achieved by combining with heat to
achieve an increased rate of inactivation of
microbes and enzymes.

The effect of pressure on a reaction at constant
temperature (i.e., 0 °C) can be written as follows:

$$\Delta V^* = \frac{2.3RT(\log k_{P1} - \log k_{P0})}{(P_1 - P_2)} \qquad (15.10)$$

where $k_{P1}$ and $k_{P0}$ are rates of the reaction at
the two pressures $P_1$ and $P_0$ (unit: atm). $\Delta V^*$ is
the volume change of activation between the
activated complex and the reactants. R is gas
constant $= 82$ (cm$^3$ atm/mole K), and T is the
temperature (K).

## 15.2.2   Technical Design

High-pressure food processing is mostly batch
or semicontinuous process. The HHP system
consists of a pressure vessel that holds the product
to be treated, a pressure intensifier that raises the
pressure of the pressurizing fluid to the target
operating pressure and pumps this high-pressure
fluid into the processing vessel, a hydraulic
pump to operate the intensifier, and appropriate
controls for pressurization, temperature control,
and depressurization (Fig. 15.6). Operation of a
batch HPP system is similar to the operation of a
batch thermal retort systems. The process
includes loading, compression, holding, decompression, and unloading. Product is loaded into
the process vessel, and the filled vessel is then
closed and compression starts to bring the vessel
to target pressure conditions. The vessel is held at
the pressure for required time length and then

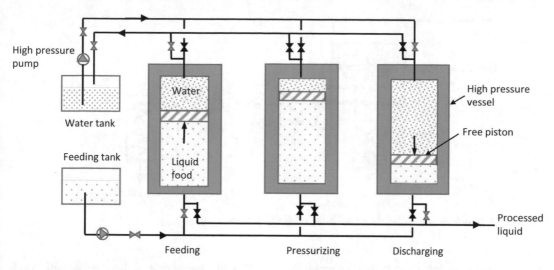

**Fig. 15.7** A semicontinuous HHP system

decompressed, and the processed products are unloaded. For most applications, products are held for 3–5 min at 300–600 MPa and approximately 5–6 cycles per hour.

HHP pressure vessels are equipped with seals that easily engage and disengage to shorten the time for loading and unloading the pressure vessel. For large systems, a product carrier is used to enable loading and unloading a full load of product rapidly. Product treated by HHP may be prepackaged and may contain large particulate material. Water is usually used as the pressurizing fluid, hence the term "hydrostatic."

The high-pressure intensifier consists of in-line dual cylinders where a drive piston is directly connected to the intensifier piston. The ratio of the area of the drive piston to that of the intensifier piston determines the pressure intensification. Standard hydraulic pumps that generate hydraulic fluid pressure of 20 MPa can be used to produce high-pressure fluid at 400 MPa if the drive and intensifier piston diameter ratio is 20:1.

Critical process factors to consider in the design of the HPP system include pressure, time at pressure, time to achieve treatment pressure, decompression time, and treatment temperature. Product properties include pH, composition, water activity, and packaging material integrity which are also important factors. Compression of foods may shift the pH of the food as a function of

imposed pressure. Lowering of pH in apple juice by 0.2 units per 100-MPa increase in pressure was observed. As pH is lowered, most microbes become more susceptible to HPP inactivation. The product volume, especially liquid product, can be significantly decreased at high pressure. A process pressure of 680 MPa can result in a 15% compression of the liquid treated.

HPP vessel can be seen as an adiabatic system. The compression and decompression can result in a transient temperature change in the product during treatment. The temperature of water increases about 3 °C for every 100 MPa of increased pressure at room temperature. Fats and oils have a heat of compression value of 8–9 °C/100 MPa, and proteins and carbohydrates have intermediate heat of compression values.

Semicontinuous process of HPP involves multiple vessels installed in parallel and operated in shift (Fig. 15.7). Only pumpable products can be treated with such system. As shown in Fig. 15.7, a low-pressure food pump is used to fill the pressure vessel. When filled, the inlet port is closed, and high-pressure process water is introduced behind the free piston to compress the liquid food. After an appropriate process hold time, the system is decompressed by releasing the pressure on the high-pressure process water. The treated liquid is discharged from the pressure vessel to a sterile hold tank through a sterile discharge port.

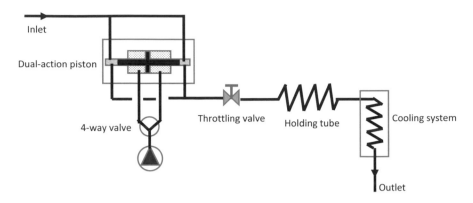

**Fig. 15.8** Continuous-flow high-pressure throttling (CFHPT) process

A low-pressure water pump is used to move the free piston toward the discharge port. The treated liquid food can be filled aseptically into presterilized containers.

A continuous high-pressure system was also developed that can be used only for homogeneous liquids (Fig. 15.8). The system consists of a feed pump, a pressure intensifier, a hydraulic system to operate the intensifier pistons, and a throttling device to reduce the pressure from high pressure to atmospheric pressure. The throttling device generates extremely high shear rates on the fluid leading to particle size reduction or homogenization during depressurization; the continuous high-pressure system may be considered a "high-pressure homogenization" system. In order to provide stable pressure, the continuous flow intensifiers incorporate two separate pistons operated with programmable logic controllers to time the cycling between the two pistons which have greatly reduced the amplitude of pressure oscillation. Since this process does not use hydrostatic pressure, it is different with the batch and semicontinuous HHP process. Pressure up to 45,000 psi (300 MPa) is available. Unlike the HHP processing system, the CFHPT system uses comparable lower pressure and does not need large volume vessels, and the initial and operating costs are lower.

Flexible packaging is best suited for batch in-container processing. Partially rigid packaging can also be used with at least one interface of the package being flexible enough to transmit the pressure. Continuous or semicontinuous systems are used for pumpable products, which are aseptically packaged after pressure treatment.

**Example 15.3** How long does it take to complete a 12D process for the inactivation of Z. *bailii* at 241-MPa pressure and 21 °C temperature? If the holding time needs to cut down to only 12 min during a 12D process, what pressure is required at 21 °C? The inactivation kinetics of Z. *bailii* at the pressure range from 241 to 345 MPa and 21 °C temperature can be described by:

$$\text{Log} \ (N/N_0) = -[0.0003 \exp(0.0266 \ P)] \ t$$

where N is the number of survivors after treatment, $N_0$ is the original number of bacteria, P is pressure in MPa, and t is the time of treatment in minutes.

**Solution**
As $D = t/\log(N_0/N) = 1/[0.0003 \exp.(0.0266 \ P)]$
    When $P = 241$, $D = 5.48$ min, $12D = 65.8$ min
    Therefore it would need 65.8 min to complete a 12D process.

To accomplish 12D in 12 min, D = 1 min, then use the above equation to solve P

P = 305 MPa.

### 15.2.3 Research Status and Applications

HHP pasteurization treatment for a short period of time at ambient temperature is able to inactivate vegetative pathogenic and spoilage microorganisms, yeasts, and molds. This process is used for preservation of acid foods (pH < 4.6). Data in Table 15.3 can be used to determine the time and temperature needed for successful HHP pasteurization of food products. Microbial inactivation at high pressure has been shown to be first order, and inactivation rate can be expressed as a D value as with thermal processes. Processes at 50 °C will take less time than those at ambient temperature. Minimum pressure for inactivating microorganisms by HHP is 300 MPa. Increasing pressure to 700 MPa rapidly decreases the D value, but a point of diminishing returns occurs at pressures greater than 700 MPa. Microbial inactivation should be at least 5D for pathogenic microorganisms and 8D for spoilage microorganisms. Using this criterion, HHP processing at 500 MPa and 50 °C for poultry products will require 15 min and for milk will require 12.5 min to obtain 5D reduction of *Staphylococcus aureus*. To obtain an 8-log reduction of

**Table 15.3** Resistance of microorganisms important in high-pressure pasteurization processes

| Microorganism | Substrate | P (MPa) | T (°C) | D value (min) | Reference |
|---|---|---|---|---|---|
| *S. typhimurium* | Pork | 414 | 25 | 1.48 | 2 |
| *S. typhimurium* | Milk | 300 | 25 | 9.21 | 12 |
| *L. monocytogenes* | Pork | 414 | 25 | 2.17 | 2 |
| *L. monocytogenes* | Pork | 414 | 50 | 0.37–0.63 | 3 |
| *L. monocytogenes* | Pork | 414 | 25 | 1.89–4.17 | 5 |
| *L. innocua* | Eggs | 450 | 20 | 3.0 | 8 |
| *S. aureus* | Poultry | 500 | 50 | 3 | 4 |
| *S. aureus* | Milk | 500 | 50 | 2.5 | 4 |
| *S. aureus* | Ham | 550 | 23 | 2.2 | 11 |
| *C. botulinum* type E (Beluga) | Crab meat | 758 | 35 | 3.38 | 6 |
| *C. botulinum* type E (Alaska) | Crab meat | 827 | 35 | 1.76 | 6 |
| *C. botulinum* type E (Alaska) | Buffer | 827 | 35 | 2.64 | 6 |
| *E. coli* | Milk | 400 | 50 | 1.0 | 7 |
| *E. coli* 0157:H7 | Milk | 400 | 50 | 3.0 | 4 |
| *E. coli* O157:H7 | Mango juice | 250 | 20–23 | 8.73 | 10 |
| *S. cerevisiae* | Orange juice | 400 | 37 | 0.97 | 9 |
| *S. cerevisiae* | Apple juice | 450 | 40 | 0.28 | 9 |
| *C. sporogenes* spores | Beef | 700 | 80 | 15.8 | 13 |
| *C. sporogenes* spores | Milk | 700 | 90 | 13.6 | 14 |
| *B. stearothermophilus* spores | Egg patty | 400 | 105 | 0.72 | 1 |
| *B. coagulans* spores | Tomato juice | 600 | 75 | 2.1 | 15 |

*References:* 1. Rajan S, Pandrangi S, Balasubramaniam VM, Yousef AE. (2006). LWT-Food Science and Technology. 39:844. 2. Ananth V, Dickson JS, Olson DG, and Murano EA (1998) J. Food Prot. 61:1649; 3. Kalchayanand N, Sikes A, Dunne CP, and Ray B. (1998) J. Food Prot. 61:425; 4. Patterson MF and Kilpatrick DJ. (1998) J. Food Prot. 61:432; 5. Murano EA, Murano PS, Brennan RE, Shenoy K and Moriera R. (1999). J. Food Prot. 62:480; 6. Reddy NR, Solomon HM, Fingerhut G, Balasubramanian VM, and Rodelhamel EJ. (1999) NCFST, Sumit-Argo, IL. 7. Gervilla R, Capellas M, Farragut V and Guamis B. (1997) J. Food Prot. 60:33; 8. Ponce E, Pla R, Mor-Mur M, Gervilla R and Guamis B. (1998) J. Food prot. 61:119; 9. Zook CD, Parish ME, Braddock RJ and Balaban MO. (1999) J. Food Sci. 64:533; 10. Hiremath N, Ramaswamy HS. (2012). J Food Processing Preservation 36: 113; 11. Tassou CC, Galiatsatou P, Samaras FJ, Mallidis CG. (2007) 8:478; 12. Erkmen O. (2009). Food and Bioproducts Processing. 87: 68. 13. Ramaswamy HS, Zaman SU, Smith JP. (2008). J Food Eng. 87:99. 14. Shao Y, Ramaswamy HS. (2011). Food and Bioprocess Technology. 4: 458. 15. Daryaei H, Balasubramaniam VM. (2013). Food Control. 30: 168

*Saccharomyces cerevisiae*, the HHP treatment will require 7.8 min at 400 MPa and 37 °C for orange juice, while 2.3 min at 450 MPa and 40 °C for apple juice.

Microbial inactivation kinetics in continuous flow pasteurization does not follow the same trend as HHP. Both high pressure and the shear force in the throttling valve contribute to the microbial inactivation. Hold time at high pressure prior to pressure reduction is only in the order of 0.5 s, yet substantial microbial inactivation results. Pressurizing to 242 MPa and releasing the pressure to 1 atm permit very rapid flow of fluid across the pressure-reducing valve resulting in a breakdown of microbial cell walls killing the microorganisms. Cells of *Saccharomyces cerevisiae* in orange juice were reduced by 8 log at 242 MPa. Cells of *Listeria innocua* are reduced 5 log. Cells of *Lactobacillus sake* were reduced 5 log by this treatment. Orange juice processed by continuous flow pasteurization at 242 MPa retained the fresh-squeezed orange juice flavor and was stable both microbiologically and biochemically when stored for 90 days at 4 °C. In addition, temperature increases instantaneously on reduction of pressure due to the conversion of potential energy at the high pressure into heat. Temperature rise is about 24 °C/100-MPa pressure. To minimize exposure to high temperature, feed temperature should be kept at the lowest temperature above the freezing point, and liquid must be cooled rapidly through a heat exchanger after exiting the throttling valve. This process is mild enough to permit the inactivation of 6 log of *Listeria innocua* in liquid whole eggs without coagulating the protein.

High pressure alone is inadequate to inactivate spores. It can combine high temperature above 121.1 °C for sterilization. For HHP, the temperature rise that occurs with pressurization can be utilized to advantage in sterilization processes. Most fluid foods that are high in moisture content such as fruit juices have similar temperature rise as water 3 °C/100 MPa. Therefore, a food product will exhibit a temperature rise of 21 °C going from one atmosphere to 700 MPa. If the product initial temperature prior to pressurization is 100 °C, temperature after pressurization will be 121 °C, a temperature that is lethal to spore-forming microorganisms. Of interest is inactivation of heat-tolerant spores of *Clostridium botulinum*. Type 62A in pH 7 buffer at 75 °C and 689 MPa has a D value of 10.59 min. Because sterilization will require at least 12D of *C. botulinum*, a 132-minute exposure time will be required.

The HPP is currently used in pasteurization of ready meals, meat and vegetables, guacamole, fruit juices, and smoothies and processing of various seafood including oysters, crabs, and shrimps. High-pressure pasteurized products are distributed under refrigerated conditions. Currently, high-pressure sterilization of low-acid food products is not commercialized in the United States due to limited knowledge on inactivation mechanisms of bacterial spores, process development, and validation.

## 15.3  Pulse Electric Field Processing

### 15.3.1  Fundamentals

Pulsed electric field (PEF) is another nonthermal processing extensively studied. In PEF processing, fluid foods (e.g., milk and juices) are placed between two electrodes and subjected to pulses of high voltage (typically 15–60 kV/cm) for very short time periods (1–100 μs) to kill microorganisms. Temperature rise is minimal, minimizing chemical or physical changes in the food material and reducing detrimental effect in quality of food like color, flavor, taste, and nutrient retention. It has been used as an alternative to traditional thermal pasteurization and is thus named cold pasteurization. PEF treatment has only limited effect on enzyme activity. It cannot inactivate bacterial spores. Therefore it has to be combined with other "hurdles" to enable sufficient commercial shelf life, such as heat, high pressure, and ultrasonication.

The pasteurizing effects of electric fields in foods were first observed in the early 1900s in electrically treated milk. In 1920s, a process called *electropure* was introduced in the United States to pasteurize milk with electricity, and the pasteurization was mostly due to thermal effects. The use of intensive electric field pulses on inactivation of microorganisms in foods was commercialized in the mid-1980s. Pilot-scale PEF equipment is available in processing apple juice, cider, and orange juice with extended shelf life and fresh-like quality.

Food containing ions is a good conductor of electricity. When a large flux of electrical current flows through the food for a very short period of time (on the order of microseconds), a high-voltage pulsed electric field is generated within the food. Capacitor is used to generate pulses. The capacitor slowly charges and then quickly discharges the stored electrical energy. The use of high electric field and short time pulses can effectively minimize heat generation due to Joule heating effect. When the microbial cell is placed in an electric field with a short pulse, the trans-membrane potential increases and pores are created in cell membranes, causing cells to swell and rupture, a phenomenon called electroporation. Electroporation results in a significant increase in the membrane permeability which is termed as electropermeabilization. Electropermeabilization can be reversible. When the applied electrical field strength exceeds a certain value, the cell wall ruptures releasing intracellular contents leading to death of the microbes. In addition, reactive free radicals or electrolysis products are generated during PEF processing that also contribute to the microbial inactivation.

The capability of PEF to inactivate microbes is affected by process parameters, nature of product, and properties of microbial cells. The strength of electric field, pulse length and shape, number of pulses, and temperature are critical factors influencing the microbial inactivation effect. Physical and chemical properties of the food also affect the rate of microbial decontamination. pH has significant influence on the inactivation kinetics of microbes. PEF is more effective in high-acid foods such as citrus juices. On the contrary, foods having high electrical conductivity show less inactivation of microorganisms after PEF treatment. Water activity has also a direct relation with microbial reduction by PEF

treatment. Microbial characteristics including type of microorganisms, species, and strains are also important. Generally, Gram-negative bacteria are less resistant to PEF than Gram-positive bacteria. *Salmonella* and *E. coli* are more susceptible to PEF as compared to *Listeria* and *Bacillus* species. Microbial growth conditions like concentration of nutrients of the treated medium also influence PEF efficiency.

In PEF processing, the value of the electric field E (in V/m) in the treatment chamber is defined by:

$$E = \frac{U}{d} \qquad (15.11)$$

where $d$ is the electrode gap (m) and $U$ is the voltage (V) across the electrodes. It is defined by:

$$U = R \cdot I$$

where $R$ ($\Omega$) is the sum of the resistances of the electric circuit and $I$ (A) the current intensity through the electrodes. The resistance of the food sample ($R_s$) is usually much greater than that of the electric components, so that the equation can be written as:

$$U = R_s \cdot I$$

With $R_s = \rho_s \cdot d \cdot S^{-1} = d \cdot (S \cdot \sigma_s)^{-1}$, $\rho_s$ is the electric resistivity ($\Omega \cdot m$), $S$ is the area of the electrode surface in contact with the food sample (m$^2$), and $\sigma_s$ is the electric conductivity of the material (S·m$^{-1}$).

The electrical energy $W$ (J) stored in the capacitor(s) with a capacity $C$ (F) is given by:

$$W = C \cdot U^2 / 2 \qquad (15.12)$$

The energy density W' (J·m$^{-3}$) dissipated in the sample per pulse is calculated by:

$$W' = \frac{\int u(t) \cdot i(t) \cdot dt}{V} = \frac{\int \left( u^2(t) \cdot R_s^{-1} \right) \cdot dt}{V} \qquad (15.13)$$

where u(t) and i(t) are instantaneous voltage across the electrodes and the instantaneous current crossing the sample, respectively. $V$ is the volume of the treatment chamber (m$^3$). The total cumulated treatment time $t_t$ (s) is defined by:

$$t_t = n \cdot \tau$$

where $\tau$ is the time width of a single pulse(s). The total energy density dissipated in the sample $W_t$ (J.m$^{-3}$) are:

$$W_t = n \cdot W'$$

The temperature increase $\Delta T$ (°C) after one pulse can be calculated from $W'$:

$$\Delta T = \frac{W' \cdot V}{m \cdot C_p} \qquad (15.14)$$

where m is the mass of the sample (Kg) and $Cp$ the specific heat capacity (J/Kg °C).

**Example 15.4** Apple juice is treated by PEF processing. Apple juice has electric conductivity 0.2200 S m$^{-1}$. The treatment chamber has a surface area of 27 cm$^2$. The gap of electrodes is 5 mm. The current intensity through the electrodes is 1000 A. Calculate the electric field E in the treatment chamber.

**Solution**

$$R_s = d \cdot (S \cdot \sigma_s)^{-1}$$
$$= 0.005\text{m} \times \left(0.0027\text{m}^2 \times 0.2200\,\text{Sm}^{-1}\right)^{-1}$$
$$= 8.42\ \Omega$$

$$U = R_s \cdot I = 8.42 \times 1000 = 8420\ \text{V}$$

$$E = \frac{U}{d} = \frac{8420\ V}{0.005\ m} = 1,684,000\ \text{V/m}$$
$$= 16.84\ \text{kV/cm}$$

## 15.3.2  Technical Design

The main components of a PEF system include a high-voltage pulse generator consisting of high-voltage power supply, capacitor, and switch to supply the high-voltage pulses with the required shape, duration, and intensity, a treatment chamber in which high-voltage pulses are applied to a pair of electrodes generating high-intensity electric fields between the electrodes where product is placed, a fluid handling system including pumps, water tanks, pipes and valves, and control and monitoring devices. A typical diagram of the PEF system is shown below (Fig. 15.9). To generate high-voltage pulses, energy from high-voltage power is stored in the capacitor, which is then discharged controlled by high-voltage switches to generate electrical field and release the high voltage in the forms of a pulse. The switches can be different types such as gas spark gap, thyratron, magnetic switch, or mechanical rotary switch.

Different forms of the pulses can be generated (Fig. 15.10). Exponential pulse is generated by a rapidly rising voltage which then decays slowly to zero. This pulse form involves relatively simple device, but excess heat can be generated by the long tail section of the pulse without microbial inactivation effect. Square pulse is able to maintain peak voltage for a long time. It requires more complicated networks in system, but is more power efficient. It generates less heat so that cooling is less needed as compared to exponential pulse and thus more commonly used. In addition, bipolar square wave pulse provides superior inactivation when compared to the monopolar pulse.

**Fig. 15.9** Pulsed electric field (PEF) system

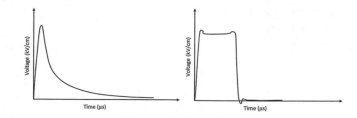

**Fig. 15.10** Exponential pulse (left) and square pulse (right)

**Fig. 15.11** Typical configuration of treatment chambers used in PEF processing: (**a**) parallel plate chamber; (**b**) coaxial chamber; (**c**) co-field chamber

The treatment chambers usually consist of two electrodes surrounded by insulating material. They are designed to attain high-intensity, spatially uniform electric fields in the treatment region for maximum microbial inactivation, with high electric resistance to reduce product heating and energy consumption. It is aimed to reduce nonuniformity to be less than 10%. The electric resistance decreases with the surface area of the electrodes and the electric conductivity of treated food. The electric resistance increases proportionally to the distance between the electrodes; however, increasing the distance also decreases the intensity of the electric field. It is thus important to test different configurations for optimal design of the treatment chamber.

Different types of treatment chamber are designed and can be categorized into parallel plate electrodes, coaxial cylinders, and co-field arrangements (Fig. 15.11). A parallel plate electrode treatment chamber can be used for batch process, which can give a quite uniform electric field distribution. The coaxial chamber has a radial geometry with two concentric cylindrical electrodes, and product can flow in between allowing a continuous process. The electric field strength, however, is not uniform and decreases from the inner to outer electrode. The uniformity can be improved by reducing the thickness of the gap or increase the radius of the chamber. The co-field chamber has two hollow cylindrical electrodes interconnected by an insulating tube in which the product continually flow through it. The co-field chamber has a relatively lower effective electrode area presenting a higher load resistance and a lower current. Figure 15.12 shows flow chart of a PEF food processing system with basic components.

A problem with PEF is that undesirable arcing or dielectric breakdown may occur in the food material. Arcing occurs when the applied field strength becomes equal to the dielectric strength

**Fig. 15.12** Flow chart of a
PEF food processing
system with basic
component

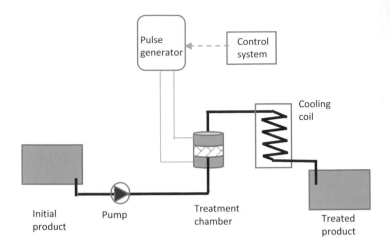

of the material. Vapor bubbles in a liquid food
subjected to PEF cause arcing as the gases or
vapors have a much lower dielectric strength
than do pure liquids. Dielectric breakdown can
be also caused by rough surface of the electrode.
Methods to avoid arcing include careful design
of the treatment chamber to improve uniform
electric field strength, using smooth electrode,
degassing of the material, and pressurizing the
liquid in the treatment chamber to prevent bubble
formation. The electrodes should be designed to
minimize local field enhancements as these
increase the probability of dielectric breakdown.

### 15.3.3   Research Status and Applications

PEF technology is mainly used for pasteurization
of various liquid food products like juices, milk
and dairy products, soup, and liquid eggs. Some
illustrative examples of foods treated with PEF
are shown in Table 15.4. PEF treatment of fresh
orange juice inactivated 99.9% of microbial flora,
with the square waves being most effective. Compared with heat pasteurization, the PEF-treated
orange juice retained more vitamin C and flavor.
Apple juice from concentrate treated with PEF at
50 kV/cm, 10 pulses with pulse width of 2 µs, and
maximum processing temperature of 45 °C had a
shelf life of 28 days. No significant difference was
observed in physical and chemical properties

(ascorbic acid, sugar) between untreated and electric field-treated juices.

High energetic demand is required for PEF
process to achieve satisfactory bacterial inactivation. This dissipated energy could significantly
increase the fluid temperature. Researchers have
found that the combination of thermal and PEF
treatment is more effective to kill bacteria. For
example, electric field strengths of 18–28 kV cm
$^{-1}$ for 17–235 µs at 4 °C did not reduce bacterial
numbers, but similar PEF treatment at 50–55 °C
gave a 5–6 log reduction for different pathogenic
microorganisms.

Low/medium electric field strength PEF can
disintegrate the biological tissue which is used as
a pretreatment method to improve efficiency of
food extraction, pressing, or drying. When plant
membranes are exposed to high enough PEF
(0.3–1 kV/cm) for a short time (0.025–0.1 s), an
irreversible rupture can be achieved, causing an
increase in mass transfer coefficients that result in
a faster transport of water to product surface.
Therefore, use of PEF pretreatment may lead to
significant saving of energy and better utilization
of production capacities in food dehydration and
separation processes.

### Problems

15.1.  Meat balls in tomato sauce are heated in a
       microwave. Dielectric loss factor, density,

**Table 15.4** Inactivation of microorganisms in fluid foods by pulsed electric field (PEF) treatment ($n$ is the number of pulses, $\tau$ is a pulse width, and $t$ is the total treatment time)

| Microorganism | Fluid food | Process parameters | Pulse type | Log CFU reduction | Ref. |
|---|---|---|---|---|---|
| *Leuconostoc mesenteroides* | Orange juice | E = 30 kV/cm. n = 6. $\tau$ = 2 $\mu$s. T = 54 °C | −/exponential | 5 | 1 |
| *Bacillus cereus* | Skim milk | E = 31 kV/cm. n = 20. t = 6 $\mu$s. T = 25 °C | −/exponential | 0.7 | 2 |
| *Enterobacter sakazakii* | Infant formula milk | E = 40 kV/cm. n = 144. $\tau$ = 2.5 $\mu$s. t = 360 $\mu$s. T = 40 °C | − | 1.2 | 3 |
| *Escherichia coli* | Whole milk | E = 30 kV/cm. n = 300. $\tau$ = 2.0 $\mu$s. t = 600 $\mu$s. T = 35 °C | −/square | 5 | 4 |
| *Escherichia coli* k12 | Apple juice | E = 31 kV/cm. n = 50.5. $\tau$ = 4.0 $\mu$s. t = 202 $\mu$s. T = 31 °C | Bipolar/square | 2.63 | 5 |
| *Escherichia coli* O157:H7 | Strawberry juice | E = 35 kV/cm. n = 425. $\tau$ = 4.0 $\mu$s. t = 1700 $\mu$s. T = 40 °C | Bipolar/square | 5.14 | 6 |
| *Lactobacillus brevis* | Orange juice | E = 25 kV/cm. n = 37.5. $\tau$ = 4.0 $\mu$s. t = 150 $\mu$s. T = 32 °C | Bipolar/square | 1.4 | 7 |
| *Listeria innocua* | Skim milk | E = 50 kV/cm. n = 30. $\tau$ = 2.0 $\mu$s. t = 60 $\mu$s. T = 28 °C | −/exponential | 2.7 | 8 |
| *Listeria monocytogenes* | Whole milk | E = 30 kV/cm. n = 400. $\tau$ = 1.5 $\mu$s. t = 600 $\mu$s. T = 25 °C | Bipolar/square | 3.0 | 9 |
| *Saccharomyces cerevisiae* | Beer (4% alc/vol) | E = 45 kV/cm. n = 46.3. $\tau$ = 1.5 $\mu$s. t = 70 $\mu$s. T = 53 °C | Bipolar/square | 2.1 | 10 |
| *Salmonella enteritidis* | Liquid egg yolk | E = 30 kV/cm. n = 105. $\tau$ = 2.0 $\mu$s. t = 210 $\mu$s. T = 40 °C | −/square | 5 | 11 |

*References*: 1. McDonald CJ, Lloyd SW, Vitale Ma, Petersson K, Innings F. (2000). J Food Sci. 65: 984. 2. Sobrino A, Rosell JR, Bendichio S, Sanchis V, Martín O. (2001). In E. Science, & N. F. Reserach (Eds.), NIZO Dairy Conference on Food Microbes (pp. 121). The Netherlands. 3. Pina-Pérez MC, Aliaga DR, Bernat CF, Enguidanos MR, López M. (2007). Int Dairy J. 17: 1441. 4. Zhao W, Yang R, Shen X, Zhang S, Chen X. (2013). Food Control. 32: 6. 5. Evrendilek GA, Zhang QH. (2005). J Food Eng. 68: 271. 6. Mosqueda-Melgar J, Raybaudi-Massilia RM, Martín-Belloso O. (2008). Innov Food Sci Emerg Technol, 9: 328. 7. Elez-Martínez P, Escolà-Hernández J, Soliva-Fortuny RC, Martín-Belloso O. (2005). Food Micro 22: 311. 8. Fernández-Molina JJ. (2001). In Barbosa-Cánovas, & Zhang (Eds.), Pulsed electric fields in food processing. Fundamental aspects and applications (pp. 149–166). Technomic Publishing Company Inc.: Lancaster. 9. Reina LD, Jin ZT, Zhang QH, Yousef AE. (1998). J Food Prot 61: 1203. 10. Milani EA, Alkhafaji S, Silva FVM. (2015). Food Control. 50: 223. 11. Amiali M, Ngadi MO, Smith JP, Raghavan GSV. (2007). J Food Eng. 79: 689

and specific heat capacity of meat ball and tomato sauce are 16.75 and 48.2, 1108 and 1022 kg/m³, and 3.35 and 3.68 kJ/kg K, respectively. When tomato sauce reaches 60 °C from 20 °C, what will be the temperature of the meatballs? Are there any ways you can use to increase heating uniformity?

15.2. A slice of smoked bacon with 100-g weight is being heated in a 2450-MHz microwave with a rated output of 800 W. Dielectric constant and loss factor are 31.1 and 10.3, respectively. Density is 1004 kg/m³ and specific heat 2.7 kJ/kg K. The temperature rises to 60 °C from 4 °C in 30 min. Assume it is heated at the rate determined by the power output of the oven; calculate the electric field strength inside the oven.

15.3. A company has developed a new microwave-ready product in 24-oz tray with mashed potato as the major component. The dielectric constant and loss factor are measured to be 55.5 and 16.1, respectively. What is the maximum thickness of the tray you recommend?

15.4. Inactivation kinetics for *L. innocua* at 330 atm, 20 °C follows the linear equation. Log n = 7.171−0.155 t
Calculate the D value at this condition.

15.5. HHP is used to inactivate orange pectin methylesterase (PME). At 18 °C, inactivation rate constants of PME are 0.024,

0.107, and 0.478 min$^{-1}$, at 400, 500, and 600 MPa, respectively, taking atmospheric pressure as reference. Assuming the inactivation follows a first-order model, what is the activation volume for PME in the experiment condition?

15.6. Z value model is used to describe the sensitivity of microbial to pressure inactivation. During HHP treatment of orange juice at constant temperature 25 °C, S. *cerevisiae ascospores* showed D values at different pressure, which are 0.18, 0.97, and 5.23 min for 500, 400, and 300 MPa, respectively. What is $z_p$ value of S. *cerevisiae* at 25 °C?

15.7. Orange juice is treated by PEF processing. A voltage of 40 kV is applied to the treatment chamber with a gap of 15 mm. Calculate the electric field in the treatment chamber.

15.8. During PEF treatment of apple juice, a voltage of 50 kV is applied in square wave to a treatment chamber with volume of 0.95 L, gap of 10 mm, and surface area of 25 cm$^2$. If resistance of the treatment chamber, pulse duration, and the number of pulses are 50 Ω, 20 μs, and 6, respectively, the electrical conductivity of the apple juice is 0.2200 S m$^{-1}$ at the processing temperature. Calculate the temperature increase in the product.

## Suggested Reading

Tang, J. (2015, August 1). Unlocking potentials of microwaves for food safety and quality. *Journal of food science, 80*(8).

Sosa-Morales, M. E., Valerio-Junco, L., López-Malo, A., & García, H. S. (2010, Oct 31). Dielectric properties of foods: Reported data in the 21st Century and their potential applications. *LWT-Food Science and Technology, 43*(8), 1169–1179.

Wang, Y., Wig, T. D., Tang, J., & Hallberg, L. M. (2003a, May 31). Dielectric properties of foods relevant to RF and microwave pasteurization and sterilization. *Journal of Food Engineering, 57*(3), 257–268.

Tang, J, & Chan, T. V. (2007, Jul 1). Microwave and radio frequency in sterilization and pasteurization applications. In S. Yanniotis, & B. Sunden, *Heat transfer in food processing* (pp. 101–142).

Hou, L., Johnson, J. A., & Wang, S. (2016, Mar 31). Radio frequency heating for postharvest control of pests in agricultural products: A review. *Postharvest Biology and Technology, 113*, 106–118.

Sharma, P., Bremer, P., Oey, I., & Everett, D. W. (2014, March 31). Bacterial inactivation in whole milk using pulsed electric field processing. *International Dairy Journal, 35*(1), 49–56.

Wang, S., Tang, J., Johnson, J. A., Mitcham, E., Hansen, J. D., Hallman, G., Drake, S. R., & Wang, Y. (2003b, June 30). Dielectric properties of fruits and insect pests as related to radio frequency and microwave treatments. *Biosystems Engineering, 85*(2), 201–212.

Rastogi, N. K., Raghavarao, K. S., Balasubramaniam, V. M., Niranjan, K., & Knorr, D. (2007, January 1). Opportunities and challenges in high pressure processing of foods. *Critical reviews in food science and nutrition, 47*(1), 69–112.

Huang, H. W., Wu, S. J., Lu, J. K., Shyu, Y. T., & Wang, C. Y. (2017, February 1). Current status and future trends of high-pressure processing in food industry. *Food Control, 72*, 1–8.

Barba, F. J., Parniakov, O., Pereira, S. A., Wiktor, A., Grimi, N., Boussetta, N., Saraiva, J. A., Raso, J., Martin-Belloso, O., Witrowa-Rajchert, D., & Lebovka, N. (2015, November 1). Current applications and new opportunities for the use of pulsed electric fields in food science and industry. *Food Research International., 77*, 773–798.

Barbosa-Canovas, G. V., & Altunakar, B. (2006). Pulsed electric fields processing of foods: An overview. In J. Raso & V. Heinz (Eds.), *Pulsed electric field technology for food industry: Fundamentals and applications* (pp. 153–194). New York: Springer.

Zeuthen, P., & Bøgh-Sørensen, L (Eds.). (2003, October 30). *Food preservation techniques*. Elsevier.

Stadler, R. H., & Lineback, D. R., (Eds). (2008, December 9). *Process-induced food toxicants: occurrence, formation, mitigation, and health risks*. Wiley.

Sahu, J. K., (Ed.). (2014, March 24). *Introduction to advanced food process engineering*. CRC Press.

# Appendices

## Appendix A.1: Conversion Factors Expressed As a Ratio

| Denominator | Numerators | | Denominator | Numerators | |
|---|---|---|---|---|---|
| I | II | III | I | II | III |
| Acre | 0.4046873 | Hectare | cPoise | 0.001 | Pa · s |
| Acre | 43560.0 | $(\text{ft})^2$ | $(\text{cm})^3$ | $3.531^*$ E–5 | $(\text{ft})^3$ |
| Atm (std) | 101.325.0 | Pascal | $(\text{cm})^3$ | 0.061023 | $(\text{in})^3$ |
| Atm (std) | 14.696 | $\text{lb}_f/\text{in}^2$ | $(\text{cm})^3$ | $2.642^*$ E–4 | Gal |
| Atm (std) | 29.921 | in Hg | $(\text{ft})^3$ | 7.48052 | Gal |
| Atm (std) | 76.0 | cm Hg @ 0°C | $(\text{ft})^3$ | 28.316 | Liter |
| Atm (std) | 33.899 | ft $H_2O$ @ 32.2 °F | Cup | $2.36588^*$ E–4 | $(\text{m})^3$ |
| | | | Dyne | $1.000^*$ E–5 | Newton |
| | | | Dyne·cm | $1.000^*$ E–7 | Newton·m |
| BTU | $1.0550^*$ E + 10 | Erg | Dyne | $2.248^*$ E–6 | $\text{lg}_f$ |
| BTU | 778.3 | ft·lb$_f$ | | | |
| BTU | $3.9292^*$ E–4 | Hp·h | Erg | $9.486^*$ E–11 | BTU |
| BTU | 1054.8 | Joule | Erg | $2.389^*$ E–8 | Cal |
| BTU | 252 | Cal | Erg | $1.000^*$ E–7 | Joule |
| BTU | $2.928^*$ E–4 | kW·h | | | |
| BTU/min | 0.023575 | Hp | Foot | 0.3048 | Meter |
| BTU/min | 0.01757 | kW | $(\text{ft})^3$ | $2.83168^*$ E–2 | $(\text{m})^3$ |
| BTU/min | 17.5725 | Watt | $(\text{ft})^2$ | $9.29030^*$ E–2 | $(\text{m})^2$ |
| BTU/s | 1054.35 | Watt | ft·lb$_f$ | 1.355818 | Joule |
| BTU/h | 0.29299 | Watt | | | |
| $\frac{\text{BTU}}{\text{h}(\text{ft}^2)(°\text{F})}$ | 5.678263 | $\text{W/m}^2 \cdot \text{K}$ | Gal (U.S.) | 0.13368 | $(\text{ft})^3$ |
| | | | Gal (U.S.) | 3.78541 | Liter |
| | | | Gal (U.S.) | $3.78541^*$ E–3 | $(\text{m})^3$ |
| $\frac{\text{BTU}}{\text{h}(\text{ft})(°\text{F})}$ | 1.730735 | $\text{W/m} \cdot \text{K}$ | Gram | $2.2046^*$ E–3 | Pound |
| BTU/lb | 2326.0 | J/kg | | | |
| | | | Hectare | 2.471 | Acre |
| $\frac{\text{BTU}}{\text{lb}(°\text{F})}$ | 4186.8 | $\text{J/kg} \cdot \text{K}$ | Hp | 42.44 | BTU/min |
| | | | Hp | 33,000 | ft · lb$_f$/min |
| Bushel | 1.2445 | ft$^3$ | Hp | 0.7457 | kW |

(continued)

© Springer International Publishing AG, part of Springer Nature 2018
R. T. Toledo et al., *Fundamentals of Food Process Engineering*,
https://doi.org/10.1007/978-3-319-90098-8

| Denominator | Numerators | | Denominator | Numerators | |
|---|---|---|---|---|---|
| I | II | III | I | II | III |
| Bushel | 0.035239 | m$^3$ | Hp(boiler) | 33,480 | BTU/h |
| Cal | 4.1868 | Joule | Inch | 2.5400$^*$ E−2 | Meter |
| Cal | 3.9684$^*$ E − 3 | BTU | in Hg @ 0 °C | 3.38638$^*$ E + 3 | Pascal |
| Cal | 4.1868$^*$ E + 7 | Erg | | | |
| centimeter | 0.3937 | Inch | in Hg@ 0 °C | 0.4912 | lb$_f$/in$^2$ |
| cm Hg@ 0 °C | 1333.33 | Pascal | | | |
| cm H$_2$O@ 4 °C | 98.0638 | Pascal | Joule | 9.48$^*$ E−4 | BTU |
| | | | Joule | 0.23889 | Cal |
| cPoise | 0.01 | g/cm·s | | | |
| cPoise | 3.60 | kg/m·h | | | |
| cPoise | 6.72$^*$ E−4 | lb/ft·s | | | |
| Joule | 10.000$^*$ E + 7 | Erg | Pound | 453.5924 | Gram |
| Joule | 0.73756 | ft·lb$_f$ | Pound | 0.45359 | kg |
| Joule | 2.77$^*$ E−4 | W·h | lb$_f$ | 4.44823 | Newton |
| | | | lb$_f$/in.$^2$ | 0.068046 | Atm (std) |
| kg | 2.2046 | Pound | lb$_f$/in.$^2$ | 68947 | dynes/cm$^2$ |
| km | 3281 | Foot | lb$_f$/in.$^2$ | 2.3066 | ft H$_2$O@39.2 °F |
| km | 0.6214 | Mile | lb$_f$/in.$^2$ | 2.035 | in Hg@0 °C |
| kW | 3413 | BTU/h | lb$_f$/ft$^2$ | 47.88026 | Pascal |
| kW·h | 3.6$^*$ E + 6 | Joule | lb$_f$/in.$^2$ | 6894.757 | Pascal |
| Liter | 0.03532 | (ft)$^3$ | Qt (U.S.) | 9.4635$^*$ E + 4 | (m)$^3$ |
| Liter | 0.2642 | Gal (U.S.) | Qt (U.S.) | 946.358 | (cm)$^3$ |
| Liter | 2.113 | Pint | Qt (U.S.) | 57.75 | (in.)$^3$ |
| | | | Qt (U.S.) | 0.9463 | liter |
| Meter | 3.281 | Foot | Qt (U.S.) | 0.25 | Gallon |
| Meter | 39.37 | Inch | | | |
| | | | Ton (metric) | 1000 | kg |
| Newton | 1.000$^*$ E + 5 | Dyne | Ton (metric) | 2204.6 | Pound |
| | | | Ton (short) | 2000 | Pound |
| Oz (liq) | 29.57373 | (cm)$^3$ | Ton (refri) | 12,000 | BTU/h |
| Oz (liq) | 1.803 | (in)$^3$ | Torr (mmHg @0 °C) | | |
| Oz (av) | 28.3495 | Gram | | 133.322 | Pascal |
| Oz (av) | 0.0625 | Pound | | | |
| | | | Watt | 3.413 | BTU/h |
| Pascal | 1.4504$^*$ E−4 | lb$_f$/in.$^2$ | Watt | 44.27 | ft·lb$_f$/min |
| Pascal | 1.0197$^*$ E−5 | kg$_f$/cm$^2$ | Watt | 1.341$^*$ E − 3 | Hp |
| Pint | 28.87 | (in.)$^3$ | Watt·h | 3.413 | BTU |
| Poise | 0.1 | Pa·s | Watt·h | 860.01 | Cal |
| lb$_f$ | 444823 | Dyne | Watt·h | 3600 | Joule |

To use: multiply quantities having units under Column 1 with the factors under Column II to obtain quantities having the units under Column III. Also use as a ratio in a dimensional equation Example: 10 acres = 10 × 0.4.046.873 ha. The dimensional ratio is (0.4046873) hectare/acre. The symbol $^*$E represents exponents of $10.9.486^*E-11 = 9.486 \times 10^{-11}$

## Appendix A.2: Properties of Superheated Steam

| Temp. °F | Absolute pressure lb $_f$/in$^2$ (psi) | | | | | |
|---|---|---|---|---|---|---|
| | 1 psi | | 5 psi | | 10 psi | |
| | $T_s = 101.74$ °F | | $T_s = 162.24$ °F | | $T_s = 193.21$ °F | |
| | v | h | v | h | v | h |
| 200 | 392.5 | 1150.2 | 78.14 | 1148.6 | 38.84 | 1146.6 |
| 250 | 422.4 | 1172.9 | 84.21 | 1171.7 | 41.93 | 1170.2 |
| 300 | 452.3 | 1195.7 | 90.24 | 1194.8 | 44.98 | 1193.7 |
| 350 | 482.1 | 1218.7 | 96.25 | 1218.0 | 48.02 | 1217.1 |
| 400 | 511.9 | 1241.8 | 102.24 | 1241.3 | 51.03 | 1240.6 |
| 450 | 541.7 | 1265.1 | 108.23 | 1264.7 | 54.04 | 1264.1 |
| 500 | 571.5 | 1288.6 | 114.21 | 1288.2 | 57.04 | 1287.8 |
| 600 | 631.1 | 1336.1 | 126.15 | 1335.9 | 63.03 | 1335.5 |
| Temp. °F | Absolute pressure lb $_f$/in$^2$ (psi) | | | | | |
| | 14.696 psi | | 15 psi | | 20 psi | |
| | $T_s = 212.00$ °F | | $T_s = 213.03$ °F | | $T_s = 227.96$ °F | |
| | v | h | v | h | v | h |
| 250 | 28.42 | 1168.8 | 27.837 | 1168.7 | 20.788 | 1167.1 |
| 300 | 30.52 | 1192.6 | 29.889 | 1192.5 | 22.356 | 1191.4 |
| 350 | 32.60 | 1216.3 | 31.939 | 1216.2 | 23.900 | 1215.4 |
| 400 | 34.67 | 1239.9 | 33.963 | 1239.9 | 25.428 | 1239.2 |
| 450 | 36.72 | 1263.6 | 35.977 | 1263.6 | 26.946 | 1263.0 |
| 500 | 38.77 | 1287.4 | 37.985 | 1287.3 | 28.457 | 1286.9 |
| 600 | 42.86 | 1335.2 | 41.986 | 1335.2 | 31.466 | 1334.9 |
| Temp. °F | Absolute pressure lb $_f$/in$^2$ (psi) | | | | | |
| | 25 psi | | 30 psi | | 35 psi | |
| | $T_s = 240.07$ °F | | $T_s = 250.34$ °F | | $T_s = 259.29$ °F | |
| | v | h | v | h | v | h |
| 250 | 16.558 | 1165.6 | | | | |
| 300 | 17.829 | 1190.2 | 14.810 | 1189.0 | 12.654 | 1187.8 |
| 350 | 19.076 | 1214.5 | 15.589 | 1213.6 | 12.562 | 1212.7 |
| 400 | 20.307 | 1238.5 | 16.892 | 1237.8 | 14.453 | 1237.1 |
| 450 | 21.527 | 1262.5 | 17.914 | 1261.9 | 15.334 | 1261.3 |
| 500 | 22.740 | 1286.4 | 18.929 | 1286.0 | 16.207 | 1285.5 |
| 600 | 25.153 | 1334.6 | 20.945 | 1334.2 | 17.939 | 1333.9 |

*Source*: Abridged from ASME. 1967. *Steam Tables. Properties of Saturated and Superheated Steam*—from 0.08865 to 15,500 lb per sq in. absolute pressure. American Society of Mechanical Engineers. NY. Used with permission

$v$ = specific volume in ft$^3$/lb; $h$ = enthalpy in BTU/lb

$T_s$ = saturation temperature at the designated pressure

## Appendix A.3: Saturated Steam
## Tables: English Units

| Temp. °F | Abs. pressure lb/in$^2$ | Specific Volume (ft$^3$/lb) | | | Enthalpy (BTU/lb) | | |
| | | Sat. liquid $v_f$ | Evap $v_{fx}$ | Sat. vapor $v_x$ | Sat. liquid $h_f$ | Evap. $h_{fg}$ | Sat. vapor $h_g$ |
|---|---|---|---|---|---|---|---|
| 32 | 0.08859 | 0.016022 | 3304.7 | 3304.7 | −0.0179 | 1075.5 | 1075.5 |
| 35 | 0.09998 | 0.016020 | 2950.5 | 2950.5 | 3.002 | 1073.8 | 1076.8 |
| 40 | 0.12163 | 0.016019 | 2445.8 | 2445.8 | 8.027 | 1071.0 | 1079.0 |
| 45 | 0.14753 | 0.016020 | 2039.3 | 2039.3 | 13.044 | 1068.2 | 1081.2 |
| 50 | 0.17796 | 0.016023 | 1704.8 | 1704.8 | 18.054 | 1065.3 | 1083.4 |
| 55 | 0.21404 | 0.016027 | 1384.2 | 1384.2 | 23.059 | 1062.5 | 1085.6 |
| 60 | 0.25611 | 0.016033 | 1207.6 | 1207.6 | 28.060 | 1059.7 | 1087.7 |
| 65 | 0.30562 | 0.016041 | 1022.8 | 1022.8 | 33.057 | 1056.9 | 1089.9 |
| 70 | 0.36292 | 0.016050 | 868.3 | 868.4 | 38.052 | 1054.0 | 1092.1 |
| 75 | 0.42985 | 0.016061 | 740.8 | 740.8 | 43.045 | 1051.3 | 1094.3 |
| 80 | 0.50683 | 0.016072 | 633.3 | 633.3 | 48.037 | 1048.4 | 1096.4 |
| 85 | 0.59610 | 0.016085 | 543.9 | 543.9 | 53.028 | 1045.6 | 1098.6 |
| 90 | 0.69813 | 0.016099 | 468.1 | 468.1 | 58.018 | 1042.7 | 1100.0 |
| 95 | 0.81567 | 0.016114 | 404.6 | 404.6 | 63.008 | 1039.9 | 1102.9 |
| 100 | 0.94924 | 0.016130 | 350.4 | 350.4 | 67.999 | 1037.1 | 1105.1 |
| 105 | 1.10218 | 0.016148 | 304.6 | 304.6 | 72.991 | 1034.3 | 1107.2 |
| 110 | 1.2750 | 0.016165 | 265.4 | 265.4 | 77.98 | 1031.4 | 1109.3 |
| 115 | 1.4716 | 0.016184 | 232.03 | 232.0 | 82.97 | 1028.5 | 1111.5 |
| 120 | 1.6927 | 0.016204 | 203.25 | 203.26 | 87.97 | 1025.6 | 1113.6 |
| 125 | 1.9435 | 0.016225 | 178.66 | 178.67 | 92.96 | 1022.8 | 1115.7 |
| 130 | 2.2230 | 0.016247 | 157.32 | 157.33 | 97.96 | 1019.8 | 1117.8 |
| 135 | 2.5382 | 0.016270 | 138.98 | 138.99 | 102.95 | 1016.9 | 1119.9 |
| 140 | 2.8892 | 0.016293 | 122.98 | 123.00 | 107.95 | 1014.0 | 1122.0 |
| 145 | 3.2825 | 0.016317 | 109.16 | 109.18 | 112.95 | 1011.1 | 1124.1 |
| 150 | 3.7184 | 0.016343 | 97.05 | 97.07 | 117.95 | 1008.2 | 1126.1 |
| 155 | 4.2047 | 0.016369 | 86.53 | 86.55 | 122.95 | 1005.2 | 1128.2 |
| 160 | 4.7414 | 0.016395 | 77.27 | 77.29 | 127.96 | 1002.2 | 1130.2 |
| 165 | 5.3374 | 0.016423 | 69.19 | 69.20 | 132.97 | 999.2 | 1132.2 |
| 170 | 5.9926 | 0.016451 | 62.04 | 62.06 | 137.97 | 996.2 | 1134.2 |
| 175 | 6.7173 | 0.016480 | 55.77 | 55.79 | 142.99 | 993.2 | 1136.2 |
| 180 | 7.5110 | 0.016510 | 50.21 | 50.22 | 148.00 | 990.2 | 1138.2 |
| 185 | 8.3855 | 0.016543 | 45.31 | 45.33 | 153.02 | 987.2 | 1140.2 |
| 190 | 9.340 | 0.016572 | 40.941 | 40.957 | 158.04 | 984.1 | 1142.1 |
| 195 | 10.386 | 0.016605 | 37.078 | 37.094 | 163.06 | 981.0 | 1144.1 |
| 200 | 11.526 | 0.016637 | 33.622 | 33.639 | 168.09 | 977.9 | 1146.0 |
| 205 | 12.776 | 0.016707 | 30.567 | 30.583 | 173.12 | 974.8 | 1147.8 |
| 210 | 14.132 | 0.016705 | 27.822 | 27.839 | 178.16 | 971.6 | 1149.8 |
| 212 | 14.696 | 0.016719 | 26.782 | 26.799 | 180.17 | 970.3 | 1150.5 |
| 220 | 17.186 | 0.016775 | 23.131 | 23.148 | 188.23 | 965.2 | 1153.4 |
| 225 | 18.921 | 0.016812 | 21.161 | 21.177 | 193.28 | 961.9 | 1155.2 |
| 230 | 20.791 | 0.016849 | 19.379 | 19.396 | 198.33 | 958.7 | 1157.1 |
| 235 | 22.804 | 0.016887 | 17.766 | 17.783 | 203.39 | 956.5 | 1158.9 |

(continued)

| Temp. °F | Abs. pressure lb/in $^2$ | Specific Volume (ft$^3$/lb) | | | Enthalpy (BTU/lb) | | |
|---|---|---|---|---|---|---|---|
| | | Sat. liquid $v_f$ | Evap $v_{fx}$ | Sat. vapor $v_x$ | Sat. liquid $h_f$ | Evap. $h_{fg}$ | Sat. vapor $h_g$ |
| 240 | 24.968 | 0.016926 | 16.304 | 16.321 | 208.45 | 952.1 | 1160.6 |
| 245 | 27.319 | 0.016966 | 14.998 | 15.015 | 213.52 | 948.8 | 1162.4 |
| 250 | 29.840 | 0.017006 | 13.811 | 13.828 | 218.59 | 945.5 | 1164.1 |
| 255 | 32.539 | 0.017047 | 12.729 | 12.747 | 223.67 | 942.1 | 1165.8 |
| 260 | 35.427 | 0.017089 | 11.745 | 11.762 | 228.76 | 938.6 | 1167.4 |
| 265 | 38.546 | 0.017132 | 10.858 | 10.875 | 233.85 | 935.2 | 1169.0 |
| 270 | 41.875 | 0.017175 | 10.048 | 10.065 | 238.95 | 931.7 | 1170.7 |
| 275 | 45.423 | 0.017219 | 9.306 | 9.324 | 244.06 | 928.2 | 1172.2 |
| 280 | 49.200 | 0.017264 | 8.627 | 8.644 | 249.17 | 924.6 | 1173.8 |
| 285 | 53.259 | 0.017310 | 8.0118 | 8.0291 | 254.32 | 920.9 | 1175.3 |
| 290 | 57.752 | 0.017360 | 7.4468 | 7.4641 | 259.45 | 917.3 | 1176.8 |

*Source*: Abridged from: ASME 1967. *Steam Tables. Properties of Saturated and Superheated Steam.* American Society of Mechanical Engineers, NY. Used with permission

## Appendix A.4: Saturated Steam Tables: Metric Units

| Temperature °C | Absolute pressure kPa | Enthalpy (MJ/kg) | | |
|---|---|---|---|---|
| | | Saturated liquid $h_f$ | (MJ/kg) Evaporation $h_{fg}$ | Saturated vapor $h_g$ |
| 0 | 0.6108 | −0.00004 | 2.5016 | 2.5016 |
| 2.5 | 0.7314 | 0.01049 | 2.4956 | 2.5061 |
| 5 | 0.8724 | 0.02100 | 2.4897 | 2.5108 |
| 7.5 | 1.0365 | 0.03151 | 2.4839 | 2.5153 |
| 10 | 1.2270 | 0.04204 | 2.4779 | 2.5200 |
| 12.5 | 1.4489 | 0.05253 | 2.4720 | 2.5245 |
| 15 | 1.7049 | 0.06292 | 2.4661 | 2.5291 |
| 17.5 | 2.0326 | 0.07453 | 2.4595 | 2.5342 |
| 20 | 2.3366 | 0.08386 | 2.4544 | 2.5381 |
| 22.5 | 2.7248 | 0.09780 | 2.4484 | 2.5428 |
| 25 | 3.1599 | 0.10477 | 2.4425 | 2.5473 |
| 27.5 | 3.6708 | 0.11522 | 2.4367 | 2.5518 |
| 30 | 4.2415 | 0.12566 | 2.4307 | 2.5563 |
| 32.5 | 4.8913 | 0.13611 | 2.4246 | 2.5609 |
| 35 | 5.6238 | 0.14656 | 2.4188 | 2.5653 |
| 37.5 | 6.4488 | 0.15701 | 2.4129 | 2.5699 |
| 40 | 7.3749 | 0.16745 | 2.4069 | 2.5744 |
| 42.5 | 8.4185 | 0.17789 | 2.4009 | 2.5788 |
| 45 | 9.5851 | 0.18834 | 2.3949 | 2.5832 |
| 47.5 | 10.8868 | 0.19880 | 2.3889 | 2.5877 |
| 50 | 12.3354 | 0.20925 | 2.3829 | 2.5921 |
| 52.5 | 13.9524 | 0.21971 | 2.3769 | 2.5966 |
| 55 | 15.7459 | 0.23017 | 2.3705 | 2.6000 |
| 57.5 | 17.7295 | 0.24062 | 2.3648 | 2.6054 |
| 60 | 19.9203 | 0.25109 | 2.3586 | 2.6098 |

(continued)

| Temperature °C | Absolute pressure kPa | Enthalpy (MJ/kg) | | |
|---|---|---|---|---|
| | | Saturated liquid $h_f$ | (MJ/kg) Evaporation $h_{fg}$ | Saturated vapor $h_g$ |
| 62.5 | 22.3466 | 0.26155 | 2.3525 | 2.6140 |
| 65 | 25.0159 | 0.27202 | 2.3464 | 2.6184 |
| 67.5 | 27.9479 | 0.28249 | 2.3402 | 2.6226 |
| 70 | 31.1622 | 0.29298 | 2.3339 | 2.6270 |
| 72.5 | 34.6961 | 0.30345 | 2.3276 | 2.6312 |
| 75 | 38.5575 | 0.31394 | 2.3214 | 2.6354 |
| 77.5 | 42.7706 | 0.32442 | 2.3151 | 2.6395 |
| 80 | 47.3601 | 0.33492 | 2.30879 | 2.64373 |
| 82.5 | 52.5777 | 0.34542 | 2.30251 | 2.64792 |
| 85 | 57.8159 | 0.34659 | 2.29611 | 2.65199 |
| 87.5 | 63.7196 | 0.36643 | 2.28971 | 2.65606 |
| 90 | 70.1059 | 0.37693 | 2.28320 | 2.66025 |
| 92.5 | 77.0489 | 0.38747 | 2.27669 | 2.66420 |
| 95 | 84.5676 | 0.39799 | 2.27023 | 2.66821 |
| 97.5 | 92.6379 | 0.40853 | 2.26349 | 2.67214 |
| 100 | 101.3250 | 0.41908 | 2.25692 | 2.67606 |
| 102.5 | 110.7410 | 0.42962 | 2.25035 | 2.67996 |
| 105 | 120.8548 | 0.44017 | 2.24354 | 2.68368 |
| 107.5 | 131.7114 | 0.45074 | 2.23674 | 2.68752 |
| 110 | 143.3489 | 0.46132 | 2.22994 | 2.69129 |
| 112.5 | 155.8051 | 0.47190 | 2.22313 | 2.69508 |
| 115 | 169.1284 | 0.48249 | 2.21615 | 2.69874 |
| 117.5 | 183.3574 | 0.49309 | 2.20929 | 2.70241 |
| 120 | 198.5414 | 0.50372 | 2.20225 | 2.70607 |
| 122.5 | 214.8337 | 0.51434 | 2.19519 | 2.70949 |
| 125 | 232.1809 | 0.52499 | 2.18807 | 2.71311 |
| 127.5 | 250.6391 | 0.53565 | 2.18083 | 2.71651 |
| 130 | 270.2538 | 0.54631 | 2.17365 | 2.71991 |
| 132.5 | 291.0837 | 0.55698 | 2.11632 | 2.72331 |
| 135 | 313.1771 | 0.56768 | 2.15899 | 2.72654 |

*Source*: Calculated from ASME 1967. *Steam Tables. Properties of Saturated and Superheated Steam*—from 0.08865 to 15.500 lb per sq in. absolute pressure. American Society of Mechanical Engineers, NY. Used with permission

## Appendix A.5: Flow Properties of Food Fluids

| Product | % Solids | Temperature °C | Flow constants | |
|---|---|---|---|---|
| | | | n (dimensionless) | k (dyne $\cdot$ s$^n$ /cm$^2$) |
| Applesauce | 11 | 30 | 0.34 | 116 |
| | | 82 | 0.34 | 90 |
| Apple juice | 50–65.5 Brix | 30 | 0.65 | – |
| | 10.5–40 Brix | 30 | 1.0 | – |
| Apricot puree | 16 | 30 | 0.30 | 68 |
| | | 82 | 0.27 | 56 |
| Apricot concentrate | 26 | 4.5 | 0.26 | 860 |
| | | 25 | 0.30 | 670 |
| | | 60 | 0.32 | 400 |
| Banana puree | – | 24 | 0.458 | 65 |
| Grape juice | 64 Brix | 30 | 0.9 | – |
| | 15–50 Brix | 30 | 1.0 | – |
| Orange juice concentrate | – | 15 | 0.584 | 11.9 |
| | | 0 | 0.542 | 18.0 |
| Orange juice concentrate | 30 Brix | 30 | 0.85 | – |
| | 60 Brix | 30 | 0.55 | 15.5 |
| | 65 Brix | 30 | 0.91 | 2.6 |
| Pear puree | 18.3 | 32 | 0.486 | 22.5 |
| | | 82 | 0.484 | 14.5 |
| | 26 | 32 | 0.450 | 62 |
| | | 82 | 0.455 | 36 |
| | 31 | 32 | 0.450 | 109 |
| | | 82 | 0.459 | 56 |
| | 37 | 32 | 0.479 | 355 |
| | | 82 | 0.481 | 160 |
| Peach puree | 12 | 30 | 0.28 | 72 |
| | | 82 | 0.27 | 58 |
| Plum juice | 14 | 30 | 0.34 | 22 |
| | | 82 | 0.34 | 20 |
| Tomato juice | 12.8 | 32 | 0.43 | 20 |
| | | 82 | 0.345 | 31.2 |
| | 25 | 32 | 0.405 | 129 |
| | | 82 | 0.43 | 61 |
| | 30 | 32 | 0.40 | 187 |
| | | 82 | 0.445 | 79 |
| Tomato catsup (0.15 g tomato solids/g catsup) | 36 | 30 | 0.441 | 81 |

*Source*: Holdsworth, S. D. Applicability of theological models to the interpretation of flow and processing behavior of fluid food products. J. Tex. Studies 2(4): 393–418, 1971

## Appendix A.6: Psychrometric Chart:
## English Units

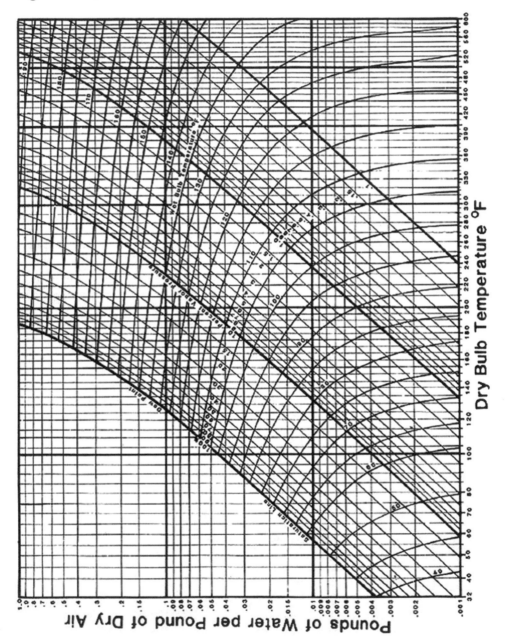

## Appendix A.7: Psychrometric Chart:
## Metric Units

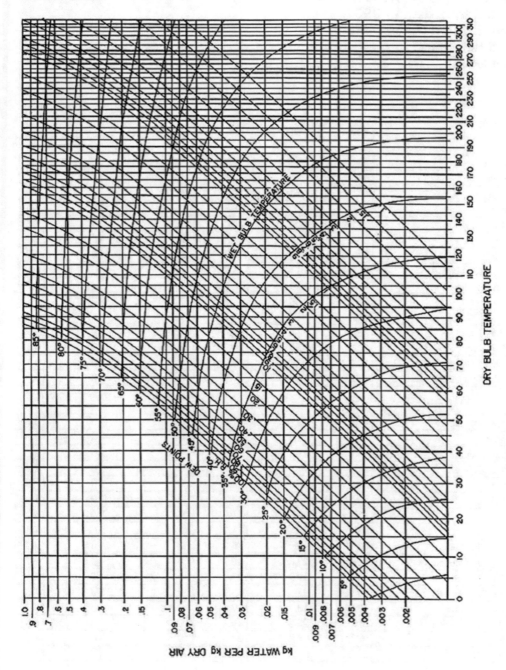

## Appendix A.8: Average Composition of Selected Foods in g/(100 g food). (Calculated from data in USDA National Nutrient Database for Standard Reference; formerly USDA Handbook 8)

| Product | Water | Protein | Fat | Carbohydrate | Fiber | Ash |
|---|---|---|---|---|---|---|
| Sausage | | | | | | |
| Beef Bologna | 55.3 | 12.2 | 28.5 | 0.8 | 0 | 3.2 |
| B/P Bologna | 54.3 | 11.7 | 28.3 | 2.8 | 0 | 3 |
| Pork Bologna | 60.6 | 15.3 | 19.9 | 0.73 | 0 | 3.5 |
| Beef Franks | 54.7 | 12 | 28.5 | 1.8 | 0 | 2.9 |
| B/P Franks | 53.9 | 11.3 | 29.2 | 2.6 | 0 | 3.2 |
| Chick. Franks | 57.5 | 12.9 | 19.5 | 6.8 | 0 | 3.3 |
| Turkey Franks | 63 | 14.3 | 17.7 | 1.5 | 0 | 3.5 |
| Canned Chopped Ham | 60.8 | 16.1 | 18.8 | 0.3 | 0 | 4.1 |
| Sliced Ham (Ex. Lean) | 70.5 | 19.4 | 5 | 1 | 0 | 4.2 |
| Sliced Ham (Reg.) | 64.6 | 17.6 | 10.6 | 3.1 | 0 | 4.1 |
| Salami, Beef (Cooked) | 59.3 | 14.7 | 20.1 | 2.5 | 0 | 4.4 |
| Salami, Pork, Dry | 36.2 | 22.6 | 33.7 | 1.6 | 0 | 5.9 |
| Salami, B/F, Dry | 34.7 | 22.9 | 34.4 | 2.6 | 0 | 5.5 |
| Smoked Link Sausage | 52.2 | 13.4 | 30.3 | 1.4 | 0 | 2.7 |
| Turkey Roll (light meat) | 71.6 | 18.7 | 7.2 | 0.5 | 0 | 2 |
| | | | | | | |
| Beef | | | | | | |
| Brisket (cooked) | 44.8 | 23.5 | 31.6 | 0 | 0 | 0.85 |
| Brisket (raw) | 55.1 | 16.9 | 26.5 | 0 | 0 | 0.8 |
| Lean, Prime, Raw | 68.6 | 21.2 | 9.7 | 0 | 0 | 1 |
| Lean, Prime, Cooked | 58.1 | 29 | 13 | 0 | 0 | 1.2 |
| Chuck, Raw | 61.8 | 18.5 | 18.4 | 0 | 0 | 0.9 |
| Chuck, Braised | 52.5 | 29.7 | 17 | 0 | 0 | 1 |
| Round , Raw | 65.9 | 20.4 | 11.6 | 0 | 0 | 1 |
| Round , Cooked | 58.1 | 27.4 | 11.7 | 0 | 0 | 1.3 |
| Top Sirloin, Ch. Raw | 62.7 | 19 | 16.2 | 0 | 0 | 0.9 |
| Top Sirloin, Ch. Cooked | 55.6 | 27.6 | 16.7 | 0 | 0 | 1.2 |
| Top Sirloin, Sel. Raw | 64.2 | 19.3 | 13.8 | 0 | 0 | 1 |
| Top Sirloin, Sel. Cooked | 57.6 | 28 | 13.9 | 0 | 0 | 1.3 |
| Gound, Ex-lean, Raw | 63.2 | 18.7 | 17.1 | 0 | 0 | 0.9 |
| Ground, Ex-lean, Cooked | 58.6 | 24.5 | 16.1 | 0 | 0 | 0.8 |
| Ground, Reg. Raw | 48.9 | 28.8 | 21.5 | 0 | 0 | 1.2 |
| Ground, Reg. Cooked | 54.2 | 24.1 | 20.7 | 0 | 0 | 1 |
| | | | | | | |
| Pork | | | | | | |
| Fresh Ham, Raw | 64.7 | 18.7 | 15.7 | 0 | 0 | 0.9 |
| Fresh Ham, Cooked | 56.8 | 28.9 | 14.3 | 0 | 0 | 1.1 |
| Loin, Raw | 72.2 | 21.4 | 5.7 | 0 | 0 | 1.1 |
| Loin, Cooked | 61.4 | 28.6 | 9.1 | 0 | 0 | 1.4 |

(continued)

| Product | Water | Protein | Fat | Carbohydrate | Fiber | Ash |
|---|---|---|---|---|---|---|
| Poultry | | | | | | |
| Chicken Dark, Raw, no Skin | 76 | 20.1 | 4.3 | 0 | 0 | 0.9 |
| Chicken Dark, Cooked, no Skin | 63.1 | 27.4 | 9.7 | 0 | 0 | 1 |
| Chicken LT, Raw, no Skin | 74.9 | 23.2 | 1.7 | 0 | 0 | 1 |
| Chicken, LT, Cooked, no Skin | 64.8 | 30.9 | 4.5 | 0 | 0 | 1 |
| Turkey, Dark, Raw, no Skin | 74.5 | 20.1 | 4.4 | 0 | 0 | 0 |
| Turkey, Dark, Cooked, no Skin | 63.1 | 28.6 | 7.2 | 0 | 0 | 1 |
| Turkey, LT, Raw, no Skin | 73.8 | 23.6 | 1.6 | 0 | 0 | 1 |
| Turkey, LT, Cooked, no Skin | 66.3 | 29.9 | 3.2 | 0 | 0 | 1.1 |
| Dairy/Eggs | | | | | | |
| Butter | 15.9 | 0.9 | 81.1 | 0.1 | 0 | 2.1 |
| Cheese, Cheddar | 36.8 | 24.9 | 33.1 | 1.3 | 0 | 3.9 |
| Cream, Half&Half | 80.6 | 3 | 11.5 | 4.3 | 0 | 0.7 |
| Cream, Whipping, Lt | 63.5 | 2.2 | 30.9 | 3 | 0 | 0.5 |
| Cream Whipping, Heavy | 57.7 | 2.1 | 37 | 2.8 | 0 | 0.5 |
| Egg, Whole | 75.3 | 12.5 | 10 | 1.2 | 0 | 0.9 |
| Egg, White | 87.8 | 10.5 | 0 | 1 | 0 | 0.6 |
| Egg, Yolk | 48.8 | 16.8 | 30.9 | 1.8 | 0 | 1.8 |
| Milk, Whole | 88 | 3.3 | 3.4 | 4.7 | 0 | 0.7 |
| Milk, Skim | 90.8 | 3.4 | 0.2 | 4.9 | 0 | 0.8 |
| Whey, acid | 93.4 | 0.8 | 0.1 | 5.1 | 0 | 0.6 |
| Whey, sweet | 93.1 | 0.9 | 0.4 | 5.1 | 0 | 0.5 |
| Fish/Shellfish | | | | | | |
| Catfish Raw | 75.4 | 15.6 | 7.6 | 0 | 0 | 1 |
| Catfish Cooked | 71.6 | 18.7 | 8 | 0 | 0 | 1.2 |
| Cod, Raw | 81.2 | 17.8 | 0.7 | 0 | 0 | 1.2 |
| Cod, Cooked | 75.9 | 22.8 | 0.9 | 0 | 0 | 1.5 |
| Halibut, Raw | 77.9 | 20.8 | 2.3 | 0 | 0 | 1.4 |
| Halibut, Cooked | 71.7 | 26.7 | 2.9 | 0 | 0 | 1.7 |
| Mackerel, Raw | 63.6 | 18.6 | 13.9 | 0 | 0 | 1.4 |
| Mackerel, Cooked | 53.3 | 23.9 | 17.8 | 0 | 0 | 1.5 |
| Salmon, Farmed, Raw | 68.9 | 19.9 | 10.9 | 0 | 0 | 1.1 |
| Salmon, Farmed, Cooked | 64.8 | 22.1 | 12.4 | 0 | 0 | 1.2 |
| Shrimp, Raw | 75.9 | 20.3 | 1.7 | 0.9 | 0 | 1.2 |
| Shrimp, Steamed | 77.3 | 20.9 | 1.1 | 0 | 0 | 1.6 |
| Oyster, Raw | 85.2 | 7.1 | 2.5 | 3.9 | 0 | 1.4 |
| Oyster, Steamed | 70.3 | 14.1 | 4.9 | 7.8 | 0 | 2.8 |
| Vegetables/Fruits | | | | | | |
| Beans, Lima, Raw | 70.2 | 6.8 | 0.9 | 20.2 | 1.9 | 1.9 |
| Beans, Lima, Boiled | 67.2 | 6.8 | 0.3 | 23.6 | 2.1 | 2.1 |
| Beans, Snap, Raw | 90.3 | 1.8 | 0.1 | 7.1 | 1.1 | 0.7 |
| Beans, Snap, Boiled | 89.2 | 1.9 | 0.3 | 7.9 | 1.4 | 0.7 |
| Beets, Raw | 87.6 | 1.6 | 0.2 | 0.6 | 0.8 | 1.1 |
| Beets, Boiled | 87.1 | 1.7 | 0.2 | 10 | 0.8 | 1.1 |
| Carrots, Raw | 87.8 | 1 | 0.2 | 10.1 | 1 | 0.9 |
| Carrots, Boiled | 87.4 | 1.1 | 0.2 | 10.5 | 1.5 | 0.9 |
| Potatoes, Raw (Flesh) | 79 | 2.1 | 0.1 | 18 | 0.4 | 0.9 |

(continued)

| Product | Water | Protein | Fat | Carbohydrate | Fiber | Ash |
|---|---|---|---|---|---|---|
| Potatoes, Baked (Flesh) | 75.4 | 2 | 0.1 | 21.6 | 0.4 | 1 |
| Fruits/Juices | | | | | | |
| Apples | 83.9 | 0.2 | 0.4 | 15.3 | 0.8 | 0.3 |
| Apple Juice, Bottled | 87.9 | 0.1 | 0.1 | 11.7 | 0.2 | 0.2 |
| Apricots | 86.4 | 1.4 | 0.4 | 11.1 | 0.6 | 0.8 |
| Avocados | 72.6 | 2.1 | 17.3 | 6.9 | 2.1 | 1.1 |
| Bananas | 74.3 | 1 | 0.5 | 23.4 | 0.5 | 0.8 |
| Cherries, Sour | 86.1 | 1 | 0.3 | 12.2 | 0.2 | 0.4 |
| Cherries, Sweet | 80.8 | 1.2 | 1 | 16.6 | 0.4 | 0.5 |
| Grapefruit, white | 90.5 | 0.7 | 0.1 | 8.4 | 0.2 | 0.3 |
| Grapefruit juice | 90 | 0.5 | 0.1 | 9.2 | 0 | 0.2 |
| Grape | 81.3 | 0.6 | 0.4 | 17.2 | 0.8 | 0.6 |
| Grape Juice | 84.1 | 0.6 | 0.1 | 15 | 0 | 0.3 |
| Peach | 87.7 | 0.7 | 0.1 | 11.1 | 0.6 | 0.5 |
| Pears | 83.8 | 0.4 | 0.4 | 15.1 | 1.4 | 0.3 |
| Pineapple | 86.5 | 0.4 | 0.4 | 12.4 | 0.5 | 0.3 |
| Strawberries | 91.6 | 0.6 | 0.4 | 7 | 0.5 | 0.4 |

# Appendix A.9: Thermal Conductivity of Construction and Insulating Materials

| | $\frac{BTU}{h(ft)(^{\circ}F)}$ | $\frac{W}{m(K)}$ |
|---|---|---|
| Building materials | | |
| Asbestos cement boards | 0.43 | 0.74 |
| Building brick | 0.40 | 0.69 |
| Building plaster | 0.25 | 0.43 |
| Concrete | 0.54 | 0.93 |
| Concrete blocks | | |
| Two oval core, 8 in. thick | 0.60 | 1.04 |
| Two rectangular core, 8 in. thick | 0.64 | 1.11 |
| Corkboard | 0.025 | 0.043 |
| Felt (wool) | 0.03 | 0.052 |
| Glass | 0.3–0.61 | 0.52–1.06 |
| Gypsum or plasterboard | 0.33 | 0.57 |
| Wood (laminated board) | 0.045 | 0.078 |
| Wood (across grain, dry) | | |
| Maple | 0.11 | 0.19 |
| Oak | 0.12 | 0.21 |
| Pine | 0.087 | 0.15 |
| Wood (plywood) | 0.067 | 0.12 |
| Rubber (hard) | 0.087 | 0.15 |
| Insulating materials | | |

(continued)

| | $\frac{BTU}{h(ft)(°F)}$ | $\frac{W}{m(K)}$ |
|---|---|---|
| Air | | |
| 32 °F (0 °C) | 0.014 | 0.024 |
| 212 °F (100 °C) | 0.0183 | 0.032 |
| 392 °F (200 °C) | 0.0226 | 0.039 |
| Fiberglass (9 lb/ft density) | 0.02 | 0.035 |
| Polystyrene | | |
| 2.4 lb/ft density | 0.019 | 0.032 |
| 2.9 lb/ft density | 0.015 | 0.026 |
| 1.6 lb/ft density | 0.023 | 0.040 |
| Polyurethane (5–8.5 lb/ft density) | 0.019 | 0.033 |
| Hog hair with asphalt binder (8.5 lb/ft density) | 0.028 | 0.048 |
| Mineral wool with binder | 0.025 | 0.043 |
| Metals | | |
| Aluminum | | |
| 32 °F (0 °C) | 117 | 202 |
| 212 °F (100 °C) | 119 | 205 |
| 572 °F (300 °C) | 133 | 230 |
| Cast iron | | |
| 32 °F (0 °C) | 32 | 55 |
| 212 °F (100 °C) | 30 | 52 |
| 572 °F (300 °C) | 26 | 45 |
| Copper | | |
| 32 °F (0 °C) | 294 | 509 |
| 212 °F (100 °C) | 218 | 377 |
| 572 °F(300 °C) | 212 | 367 |
| Steel (carbon) | | |
| 212 °F (100 °C) | 26 | 45 |
| 572 °F (300 °C) | 25 | 43 |
| Steel, stainless type 304 or 302 | 10 | 17 |
| Steel, stainless type 316 | 9 | 15 |

# Appendix A.10: Thermal Conductivity of Foods

| Food | Temp. °C | Thermal conductivity W/(m · K) | Food | Temp. °C | Thermal conductivity W/(m · K) |
|---|---|---|---|---|---|
| Apple juice | 80 | 0.6317 | Lemon | – | 1.817 |
| Applesauce | 29 | 0.5846 | Limes | | |
| Avocado | – | 0.4292 | Peeled | – | 0.4900 |
| Banana | – | 0.4811 | Margarine | – | 0.2340 |
| Beef | 5 | 0.5106 | Milk | | |
| | 10 | 0.5227 | 3% fat | – | 0.5296 |
| Beets | 28 | 0.6006 | 2.5% fat | 20 | 0.05054 |
| Broccoli | −6.6 | 0.3808 | Oatmeal, dry | – | 0.6404 |
| Butter | – | 0.1972 | Olive oil | 5.6 | 0.1887 |
| Butterfat | −10.6 | | | 100 | 0.1627 |

(continued)

| Food | Temp. °C | Thermal conductivity W/(m · K) | Food | Temp. °C | Thermal conductivity W/(m · K) |
|---|---|---|---|---|---|
| | to 10 | 0.1679 | Onions | 8.6 | 0.5746 |
| Cantaloupe | – | 0.5711 | Oranges | | |
| Carrots | | | Peeled | 28 | 0.5800 |
| Fresh | – | 0.6058 | Orange juice | –18 | 2.3880 |
| Puree | – | 1.263 | Peaches | 28 | 0.5815 |
| Corn | | | Peanut oil | 3.9 | 0.1679 |
| Yellow | – | 0.1405 | Pear | 8.7 | 0.5954 |
| Dent | – | 0.577 | Pear juice | 20 | 0.4760 |
| Egg white | – | 0.338 | | 80 | 0.5365 |
| Egg yolk | 2.8 | 0.5435 | Pea (Cowpea) | 2.8 to 16.7 | 0.3115 |
| Fish | –10 | 1.497 | | | |
| Cod | 3.9 | 0.5019 | Pineaple | – | 0.5486 |
| Salmon | –2.5 | 1.2980 | Plums | – | 0.5504 |
| | | | Pork | 6 | 0.4881 |
| Gooseberries | – | 0.2769 | | 59.3 | 0.5400 |
| Dry | – | 0.3288 | Potato, raw | – | 0.554 |
| Wet | – | 0.0277 | Poultry | – | 0.4119 |
| Frozen | | | Broiler | | |
| Grapefruit | – | 1.3500 | Sesame oil | – | 0.1755 |
| Mashed | | | Strawberries | 13.3 | 0.6750 |
| Honey | | 0.5019 | | –12.2 | 1.0970 |
| 80% water | 2 | 0.5019 | Tomato | – | 0.5279 |
| 80% water | 69 | 0.6230 | Turkey | 2.8 | 0.5019 |
| 14.8% water | 69 | 2.4230 | | –10 | 1.461 |
| Ice | –25 | 0.4500 | Turnips | – | 0.5625 |
| Lamb | 5.5 | 0.4777 | | | |
| | 61.1 | | | | |

*Source*: Excerpted from *Food Technol*. 34(11): 76–94, 1980

## Appendix A.11: Spreadsheet Program for Calculating Thermophysical Properties of Foods from Their Composition

## Appendix A.12: Correlation Equations for Heat Transfer Coefficients

Correlation equations for heat transfer coefficients:

*Very viscous fluids flowing inside horizontal tubes*

$$Nu = 1.62 \left[ prRe_L^d \right]^{0.33} \left[ 1 + 0.015(Gr)^{0.33} \right] \times \left[ \frac{\mu_f}{\mu_n} \right]^{0.33}$$

*Very viscous fluids flowing inside vertical tubes*

$$Nu = 0.255 Gr^{0.25} Re^{0.07} Pr^{0.37}$$

*Fluids in laminar flow inside bent tubes*

$$h = h_r \left[\frac{1 + 21}{Re^{0.14}}\right]\left[\frac{d}{D}\right]$$

$h_r = h$ in straight tube, $d$ = tube diameter, $D$ = diameter of curvature of bend.

*Evaporation from heat exchange surfaces*

$$\frac{q}{A} = 15.6 P^{1.156}(T_w - T_s)^{2.30/p^{0.24}}$$

$q/A$ = heat flux, $T_w$ = wall temperature, $T_s$ = saturated temperature of vapor at pressure $P$.

*Condensing vapors outside vertical tubes*

$$Nu_L = 0.925 \left[\frac{L^3 \rho^2 g \lambda}{\mu k \Delta T}\right]^{0.25}$$

*Condensation outside horizontal tubes*

$$Nu_{D0} = 0.73 \left[\frac{Do^3 \rho^2 g \lambda}{k \mu \Delta T}\right]^{0.25}$$

*Condensation inside horizontal tubes*

$$Nu_{Di} = 0.612 \left[\frac{Di^3 \rho_1 (\rho_1 - \rho_v) g}{k \mu \Delta T}\right]^{0.25}$$

*Condensation inside horizontal tubes*

$$Nu_{Di} = 0.024 (Re)^{0.8} (Pr)^{0.43} \rho_{corr}$$

$Re$ is based on the total mass of steam entering the pipe. $\rho_{corr} = [0.5/\rho_i](\rho_1 - \rho_v)(x_t - x_0)$. $x$ = steam quality, subscripts $i$ and $o$ refer to inlet and exit, and 1 and $v$ refer to condensate and vapor.

*Fluids in cross-flow to a bank of tubes*
$Re$ is based on fluid velocity at the entrance to the tube bank.

Tubes in line: $a$ = diameter to diameter distance between tubes.

$$Nu = \left[1.517 + 205 Re^{0.38}\right]^2 \left|\frac{4a}{(4a - \pi)}\right]$$

Tubes staggered, hexagonal centers; $a$ = distance between tube rows.

$$Nu = \left[1.878 + 0.256 Re^{0.36}\right]\left[\frac{4a}{(4a - \pi)}\right]$$

*Laminar flow in annuli*

$$Nu = 1.02 Re^{0.45} Pr^{0.5} \left(\frac{De}{L}\right)^{0.4} \left(\frac{D_2}{D_1}\right)^{0.8} Gr^{0.05} \left(\frac{\mu^{0.14}}{\mu_1}\right)$$

$De$ = hydraulic diameter; subscripts 1 and 2 refer to outside diameter of inner cylinder and inside diameter of outer cylinder, respectively.

*Turbulent flow in annuli*

$$Nu = 0.02 Re^{0.8} Pr^{0.33} \left(\frac{D_2}{D_1}\right)^{0.53}$$

*Finned tubes*
$Nu_d$, $Re_d$, and $A_0$ use the outside diameter of the bare tube. $A$ = total area of tube wall and fin.

Tubes in line: $Nu_d = 0.3 Re_d^{0.625} \left(\frac{A_0}{A}\right)^{0.375} Pr^{0.33}$

Staggered tubes: $Hu_d = 0.45 Re_d^{0.625} \left(\frac{A_0}{A}\right)^{0.375} Pr^{0.33}$
*Swept surface heat exchangers*
$N$ = rotational speed of blades, $D$ = inside diameter of heat exchanger, $V$ = average fluid velocity, $L$ = swept surface length.

$$Nu = 4.9 Re^{9.57} Pr^{0.47} \left(D\frac{N}{V}\right)^{0.17} \left(\frac{D}{L}\right)^{0.37}$$

*Individual particles*
*The Nusselt number and the Reynolds number are based on the particle characteristic diameter and the fluid velocity over the particle.*

Particles in a packed bed: $Nu = 0.015 Re^{1.6} Pr^{0.67}$
Particles in a gas stream: $Nu = 2 + 0.6 Re^{0.5} Pr^{0.33}$

Sources: Perry and Chilton, *Chemical Engineers Handbook*, 5th ed., McGraw-Hill Book Co., New York; Rohsenow and Hartnett, *Handbook of Heat Transfer*, McGraw-Hill Book Co., New York; Hausen, *Heat Transfer in Counterflow, Parallel Flow and Cross Flow*, McGraw-

Hill Book Co., New York; Schmidt, *Kaltechn.* 15:98, 1963 and 15:370, 1963; Ranz and Marshall, *Chem. Eng. Prog.* 48(3):141 1952.

## Appendix A.13: Visual BASIC Program for Evaluating Temperature Response of a Brick-Shaped Solid

```
Option Explicit
Dim X
Dim i As Integer
Dim BI(3) As Single
Dim Delta1(6) As Single
Dim Delta2(6) As Single
Dim Delta3(6) As Single
Dim YS, YC, TS, TC
Const TM = 177
Const T0 = 4
Const L1 = 0.0245
Const L2 = 0.0256
Const L3 = 0.0254
Const H1 = 6.5
Const H2 = 6.5
Const H3 = 6.5
Const K = 0.455
Const Rho = 1085
Const Cp = 4100
Dim TIMX
Dim d1, d2, d3
Dim YS1, YC1, YS2, YC2, YS3, YC3
Dim YXn, YCN, YCXn
Dim test
Dim Numx(6), DeNumx(6)
Dim YX(6), YCX(6)
Dim j As Integer
Dim fh
Dim LPHA As Variant
Sub TEMPSLAB()
LPHA = K / Rho / Cp
Debug.Print "alpha=" & LPHA
Worksheets("sheet1").Cells.Clear
BI(1) = H1 * L1 / K
Debug.Print "BI(1)" & BI(1)
BI(2) = H2 * L2 / K
Debug.Print "bi = " & BI(2)
BI(3) = H3 * L3 / K
Debug.Print "bi =" & BI(3)
Call toexc
```

```
j = 1
For TIMX = 0 To 4800 Step 600
Call DELN(BI(1), L1)
YS1 = YXn
YC1 = YCXn
d1 = Delta1(1)
Call DELN(BI(2), L2)
YS2 = YXn
YC2 = YCXn
d2 = Delta1(1)
Call DELN(BI(3), L3)
YS3 = YXn
YC3 = YCXn
d3 = Delta1(1)
YS = YS1 * YS2 * YS3
YC = YC1 * YC2 * YC3
TS = TM - YS * (TM - T0)
TC = TM - YC * (TM - T0)
If TIMX <> 0 Then Debug.Print "TIME ="
& TIMX, "TS =" & TS, "TC =" & TC
Call toexcel(j, TIMX / 60, Format(TS,
"0.000"), Format(TC, "0.000"))
j = j + 1
Next TIMX
fh = 1 / (LPHA * 0.4343 * ((d1^ 2 / L1 ^
2) + (d2 ^ 2 / L2 ^ 2) + (d3 ^ 2 / L3 ^
2)))
Debug.Print "FH=", fh
Call toexc1(fh, TM, T0, L1, L2, L3)
End Sub
Sub DELN(BI, L)  'DETERMINE ROOTS OF
DELTA1
Delta1(1) = DntanDn(0, 1.57, BI)
Delta1(2) = DntanDn(3.14, 4.71, BI)
Delta1(3) = DntanDn(6.28, 7.85, BI)
Delta1(4) = DntanDn(9.42, 11, BI)
Delta1(5) = DntanDn(12.5, 14.1, BI)
Delta1(6) = DntanDn(15.7, 17.3, BI)
For i = 1 To 6
Numx(i) = 2 * Sin(Delta1(i)) * Cos
(Delta1(i))
DeNumx(i) = Delta1(i) + Sin(Delta1(i))
* Cos(Delta1(i))
YX(i) = (Numx(i) / DeNumx(i)) * Exp
(-LPHA * TIMX * Delta1(i) ^ 2 / L ^ 2)
YCX(i) = YX(i) / Cos(Delta1(i))
Next i
YXn = YX(1) + YX(2) + YX(3) + YX(4) + YX
(5) + YX(6)
YCXn = YCX(1) + YCX(2) + YCX(3) + YCX
```

```
(4) + YCX(5) + YCX(6)
If YXn > 1 Then YXn = 1
If YCXn > 1 Then YCXn = 1
End Sub
Function DntanDn(Lo, Hi, BI)
Do
X = 0.5 * (Lo + Hi)
test = X * Tan(X) - BI
If test > 0 Then Hi = X Else Lo = X
If Abs(Hi - Lo) < 0.00001 Then GoTo
LABEL
Loop While Abs(test) > 0.001
LABEL:
DntanDn = X
End Function
Sub toexc()
With Sheets("sheet1")
.Cells(1, 1).Value = "alpha= "
.Cells(1, 2).Value = LPHA
.Cells(2, 1).Value = "bi(1)"
.Cells(2, 2).Value = BI(1)
.Cells(3, 1).Value = "bi(2)"
.Cells(3, 2).Value = BI(2)
.Cells(4, 1).Value = "(bi(3)"
.Cells(4, 2).Value = BI(3)
.Cells(5, 1).Value = "time": .Cells
(5, 2).Value = "tsurf": .Cells(5, 3).
Value = "tcent"
End With
End Sub
Sub toexcel(j, tx, surf, cent)
With Sheets("sheet1")
.Cells(j + 5, 1).Value = tx
.Cells(j + 5, 2).Value = surf
.Cells(j + 5, 3).Value = cent
End With
End Sub
Sub toexc1(fh, TM, T0, L1, L2, L3)
With Sheets("sheet1")
.Cells(j + 5, 1).Value = "fh = ": .
Cells(j + 5, 2).Value = fh
.Cells(j + 6, 1).Value = "Tm = ": .
Cells(j + 6, 2).Value = TM
.Cells(j + 7, 1).Value = "T0 = ": .
Cells(j + 7, 2).Value = T0
.Cells(j + 8, 1).Value = "L1 =": .Cells
(j + 8, 2) = L1
.Cells(j + 9, 1).Value = "L2 = ": .
Cells(j + 9, 2).Value = L2
.Cells(j + 10, 1).Value = "L3 = ": .
```

```
Cells(j + 10, 2).Value = L3
End With
End Sub
```

## Appendix A.14: Visual BASIC Program for Evaluating Local Heat Transfer

## Coefficient from Temperature Response of a Brick-Shaped Solid

```
Sub localh()
'calculate h from Fh
Fh = 33596
L1 = 0.0254
L2 = 0.0508
L3 = 0.1016
k = 0.455
RHO = 10855
Cp = 4100
ALPHA = k / (RHO * Cp)
h = 375
Do
Bi1 = h * L1 / k
Bi2 = h * L2 / k
Bi3 = h * L3 / k
Delta1 = getDelta(Bi1)
Delta2 = getDelta(Bi2)
Delta3 = getDelta(Bi3)
testfh2 = testfh
testfh = 1 - (Fh * ALPHA * 0.4343 *
((Delta1 ^ 2 / L1 ^ 2) + (Delta2 ^ 2 /
L2 ^ 2) + (Delta3 2 ^ L3 ^ 2)))
If testfh > 0 Then dh = 0.001 Else dh =-
0.001
h = h + dh
Debug.Print h, testfh
If testfh2 <> Empty Then If testfh /
testfh2 < 0 Then Exit Do
Loop While Abs(testfh) > 0.01
MsgBox " Heat transfer coef, H Test" &
Chr(13) & h - dh & " " &
testfh2 & Chr(13) &h & " " & testfh
End Sub
Function getDelta(Bi)
Lo = 0
Hi = 2
Do
```

```
x = 0.5 * (Lo + Hi)
Test = x * Tan(x) - Bi
If Test > 0 Then Hi = x Else Lo = x
If Abs(Hi - Lo) < 0.00001 Then Exit Do
Loop
getDelta = x
End Function
```

## Appendix A.15: Thermal Conductivity of Water As a Function of Temperature

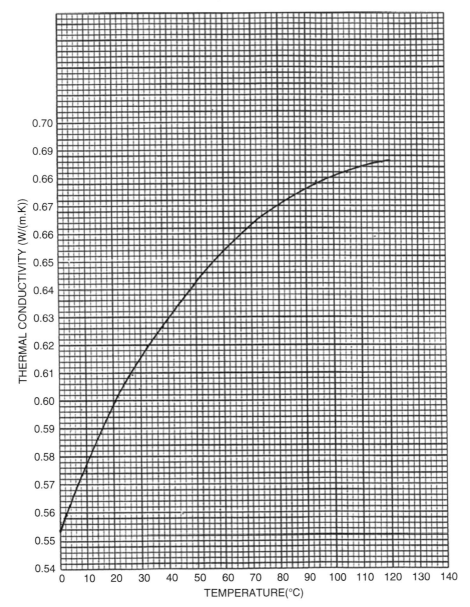

## Appendix A.16: Density of Water As a Function of Temperature

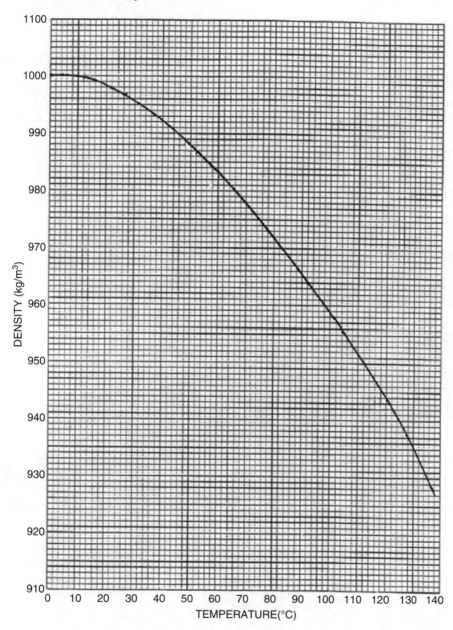

## Appendix A.17: Viscosity of Water as a Function of Temperature

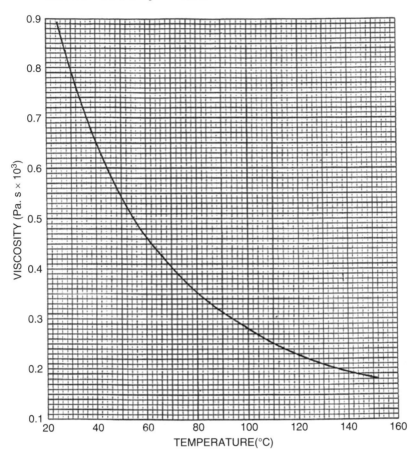

# Index

## A
Absolute pressure, 42, 43, 45, 48, 69, 70, 121, 122, 149, 196, 248, 278, 308–311, 318, 320, 347, 350–352
Absolute temperature, 42–43, 49, 94, 144, 149, 167, 189, 191, 323, 329, 333, 374
Adiabatic processes, gases, 50
Air classification, 381, 382
Arrhenius equation, 10, 93, 94, 189–190, 211
Ascorbic acid half-life, 189

## B
Back extrusion, fluids, 94, 95, 98
Bernoulli equation, 41, 120–123
BET equation, sorption, water activity, 328, 330
Black body radiation, 143, 147
Blending of ingredients, material balance formulation, 16, 28
Brick-shaped solids, 171, 173–174, 343
Broken heating curve, thermal process, 231–233, 237
Bromley's equation, water activity, 326
Buoyant forces, 163, 379, 380

## C
Channeling, 259, 261–263, 399
Clausius-Clapeyron equation, 53
Combined conduction and convection, 156–158
Commercial sterility, 195, 205, 245, 251, 253, 254, 271, 273
Component mass balance, equation formulation, 14
Compressor, refrigeration and power requirement, 290–292
Concentration changes material balance, 16–22
Conduction heat transfer, 135, 141, 239
Conservation of mass, law, 11, 13, 111
Consistency coefficient, 258, 259, 275
Constant rate air drying, heat and mass transfer, 349
Constant rate drying, 343, 345, 352
Constraints in material balance problems, 25–28
Continuity principle, 112–113
Continuous *vs.* batch process, 18
Continuous-flow high-pressure throttling (CFHPT) process, 414
Controlled atmosphere storage, 48, 296–301
Convection heat transfer, 142, 143, 149, 155, 156, 251, 330

Conversion factors, table, 5, 6, 42, 58
Correlation coefficient, 361, 362
Critical conditions, gases, 47
Crystallization, material balance, 11, 13, 15, 33
Crystallization process flow diagram, 12
Crystallizer, mass balance equation formulation, 15, 33

## D
D value, microbial inactivation, 189, 192, 209, 210, 216, 415
Dead space, 259, 261, 263
Deaerators, 245, 252
Determinants in Microsoft Excel, 191
Dialysis, 38, 377
Dielectric heating, 149–152, 403, 407
Dielectric properties, dielectric constant, loss factor, 404
Differential equations, 85
   finite difference, 49, 171
Differentiation formulas, 20, 85, 106, 108, 109, 173, 174, 186, 188, 299, 324, 330, 332, 388
Diffusion, in extraction, 388–389, 399
Dimensionless quantities, 162–164
Drag coefficient, 380–383
Drag force, 379, 380
Drying stage, 345
Drying times prediction from drying data, 340–344

## E
Ejector, steam jet, evaporator vacuum, 307, 351
Electric field strength, 405, 407–409, 419, 420
Electroporation, electropermeabilization, 417
Emerging food processing technologies, 403–422
Energy balance equation, 57, 58, 130
Energy exchange thermal radiation, 145–149
Enthalpy, 50–52, 57, 58, 63–72, 74, 75, 79, 177, 178, 281–288, 310, 312, 314, 315, 322–324, 338
Enthalpy change, freezing, 63–68, 177, 178, 295
Entrainment, evaporators, 318
Entropy, 49, 52, 56, 282, 285, 323
Enzyme reactions, 186–187
Equilibrium
   in extraction, 30
   thermodynamic, 49–52, 281
E(t) curve, 262–269, 275

© Springer International Publishing AG, part of Springer Nature 2018
R. T. Toledo et al., *Fundamentals of Food Process Engineering*,
https://doi.org/10.1007/978-3-319-90098-8

Printed in the United States
By Bookmasters